99 Springer Series in Solid-State Sciences

Edited by Hans-Joachim Queisser

Springer Series in Solid-State Sciences

Editors: M. Cardona P. Fulde K. von Klitzing H.-J. Queisser

Managing Editor: H. K. V. Lotsch

Volumes 1–89 are listed at the end of the book

90 **Earlier and Recent Aspects of Superconductivity**
Editors: J. G. Bednorz and K. A. Müller

91 **Electronic Properties of Conjugated Polymers III**
Basic Models and Applications
Editors: H. Kuzmany, M. Mehring, and S. Roth

92 **Physics and Engineering Applications of Magnetism**
Editors: Y. Ishikawa and N. Miura

93 **Quasicrystals**
Editors: T. Fujiwara and T. Ogawa

94 **Electronic Conduction in Oxides**
By N. Tsuda, K. Nasu, A. Yanase, and K. Siratori

95 **Electronic Materials**
A New Era in Materials Science
Editors: J. R. Chelikowsky and A. Franciosi

96 **Electron Liquids**
By A. Isihara

97 **Localization and Confinement of Electrons in Semiconductors**
Editors: F. Kuchar, H. Heinrich, and G. Bauer

98 **Magnetism and the Electronic Structure of Crystals**
By V. A. Gubanov, A. I. Liechtenstein, and A.V. Postnikov

99 **Electronic Properties of High-T_c Superconductors and Related Compounds**
Editors: H. Kuzmany, M. Mehring, and J. Fink

100 **Electron Correlations in Molecules and Solids**
By P. Fulde

H. Kuzmany M. Mehring
J. Fink (Eds.)

Electronic Properties of High-T_c Superconductors and Related Compounds

Proceedings of the International Winter School,
Kirchberg, Tyrol, March 3–10, 1990

With 194 Figures

Springer-Verlag
Berlin Heidelberg New York London
Paris Tokyo Hong Kong Barcelona

Professor Dr. Hans Kuzmany
Institut für Festkörperphysik, Universität Wien, Strudlhofgasse 4,
A-1090 Wien, Austria

Professor Dr. Michael Mehring
2. Physikalisches Institut, Universität Stuttgart, Pfaffenwaldring 57,
D-7000 Stuttgart, Federal Republic of Germany

Dr. Jörg Fink
Kernforschungszentrum Karlsruhe, Institut für Nukleare Festkörperphysik, Postfach 3640,
D-7500 Karlsruhe, Federal Republic of Germany

ISBN-13: 978-3-642-84347-1 e-ISBN-13: 978-3-642-84345-7
DOI: 10.1007/978-3-642-84345-7

Softcover reprint of the hardcover 1st edition 1990

2154/3140-543210 – Printed on acid-free paper

Preface

This volume contains the proceedings of the *International Winter School on Electronic Properties of High Temperature Superconductors*, which was held March 3–10, 1990, in Kirchberg (Tyrol), Austria. This course was a sequel to three meetings on the subject of conducting polymers held there one, three and five years previously. The new topic was taken up since many of the scientists working on conducting polymers are now also actively interested in high-T_c superconductors, a research area that continues to develop at a remarkable pace. There have been numerous specialist conferences on high-T_c superconductors since their discovery by Bednorz and Müller in 1986. The Kirchberg meetings are intended as a school where experienced scientists from universities, research laboratories, and industry have the opportunity to discuss their most recent results, and where young scientists and students can learn about the present status of research in this field. The 1990 winter school was again organized in cooperation with the "Bundesministerium für Wissenschaft und Forschung" in Austria and with the "Bundesministerium für Forschung und Technologie" in the Federal Republic of Germany.

The scientific interest was focused on the electronic structure of high-T_c superconductors and related compounds. At the beginning of each session, tutorial contributions were presented as an introduction for non-specialists. These were followed by research contributions elucidating the current state of our knowledge about high-T_c superconductors. Special attention was paid to chemical bonding, the doping mechanism, electronic structure, electron correlation effects, magnetic interactions, the electron–phonon interaction, and the theory of mechanisms of high-T_c superconductivity in cuprates. Contributions dealing with experimental methods concentrated on transport properties, microwave conductivity, optical spectroscopy, high-energy spectroscopies, and nuclear magnetic resonance. The discussion session was devoted to the problem of the existence of a Fermi liquid in cuprate superconductors, which is at present one of the most controversial subjects. Its solution is probably a prerequisite for the understanding of the mechanism of high-T_c superconductivity.

We thank all the authors for their contributions and all discussion speakers for their stimulating remarks, which played an essential role in making the winter school an exciting and informative event.

We are particularly grateful to the "Bundesministerium für Wissenschaft und Forschung" and the "Kulturamt des Landes Tirol" in Austria and to the

"Bundesministerium für Forschung und Technologie" in the Federal Republic of Germany, as well as to our sponsors from industry, for their financial support. This support was not merely a great help; it was, in fact, indispensable for the success of the winter school.

Finally, we thank the manager of the Hotel Sonnalp, Herr J.R. Jurgeit, and his staff for their continuous support and for their patience with the many special arrangements required during the meeting.

Vienna, Stuttgart, Karlsruhe
June 1990

H. Kuzmany
M. Mehring
J. Fink

Contents

Part I General Properties, Preparation, Doping, Structure

Part II Transport, Tunneling, Thermal Properties

Part IV Hyperfine Interactions, Magnetic Properties

Part V Optical Spectroscopy, Raman Scattering

Part I

General Properties, Preparation, Doping, Structure

Experimental Results on High-T_c Superconductors: A Critical Selection

B. Batlogg

AT&T Bell Laboratories, Murray Hill, NJ 07974, USA

Abstract Experimental progress in the study of superconductivity in cuprates has led to a clear picture of the unique properties of these compounds. As the sample quality improves, the results by different research groups converge and agree quantitatively in many cases. A summary of selected experiments is presented with a focus on charge and spin dynamics of the normal state and various probes of the superconducting state, such as anisotropic gap parameter, penetration depth and coherence length.

Significant progress has been made in the preparation, measurements and understanding of high T_c cuprate superconductors since the pioneering discovery by J. G. Bednorz and K. A. Müller.[1] By now, the number of publications has exceeded 10,000 and keeps growing with an impressive pace. More important than quantity is the high quality of experimental data, and the availability of single crystals is the basis of reproducible and definite studies of these highly anisotropic materials. Experimental progress in the more recent past is closely linked to an improved understanding of crystal-chemical properties and thus, to better materials. This contribution is based on an overview which appeared in print recently.[2]

1. Normal State

The cuprate superconductors crystallize in structures with pronounced layered character: layers of Cu-O squares, pyramids or octahedra are stacked to form "electronically active" building blocks comprising one, two or three (or more) closely connected Cu-O layers. Other building blocks separate these layers and provide for the charge transfer necessary to induce superconductivity. The electronic transport properties in the normal state are highly anisotropic: metallic-like in the planes and non-metallic perpendicular to the planes. The in-plane resistivities are remarkable in that the temperature dependence can be well described (in general) by $\rho(T)=\rho_0+\alpha T$. The slope α is surprisingly similar for several compounds, and no correlation between T_c and $\alpha(\equiv d\rho/dT)$ is observed. One would have expected such a correlation if conventional electron-phonon interaction were responsible for superconducting pairing. In Fig. 1, α is shown for superconductors with T_c's ranging from <10K to ~120K.

Studies of the frequency dependent conductivity in the infrared and far-infrared spectral range reveal that the T-dependence of $\rho_{DC}(T)\sim\alpha T$ is due to a

Springer Series in Solid-State Sciences, Vol. 99
Electronic Properties of High-T_c Superconductors and Related Compounds
Editors: H. Kuzmany · M. Mehring · J. Fink

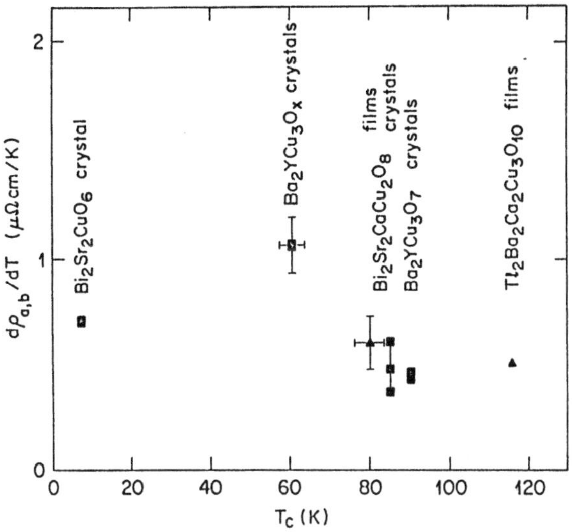

Fig. 1 Temperature coefficient of the in-plane resistivity of various cuprate superconductors. The similarity of $d\rho/dT$ rules out conventional electron-phonon interaction as the dominant pairing mechanism.

scattering rate effect, i.e. $1/\tau \sim T$. When a Drude-type relationship between σ_{DC}, τ and ω_p (plasma frequency) is assumed, then the scattering rate $\hbar/\tau$ is given by c.kT, with $c \simeq 1.4$. Actually, the measured $\sigma(\omega)$ is only crudely approximated by a Drude form ($\sim\omega^{-2}$), and detailed studies of the frequency dependence at various temperatures give insight into the charge dynamics specific to the layered cuprates. (Remark: due to a smaller plasma frequency $d\rho/dT$ in 60K T_c is higher than in the 90K T_c $Ba_2YCu_3O_x$. The scattering rate is the same within experimental uncertainties).

The conductivity perpendicular to the Cu-O layers is of non-metallic character and is up to 5 orders of magnitude smaller than parallel to the layers. The anisotropy at 100K ranges from $\sim 10^2$ in "123" to $>10^5$ in "Bi 2212".

Inelastic light scattering in the normal state reveals, in addition to phonon lines, a broad and featureless background of electronic origin. The electronic and phononic processes interfere and result in Fano-type resonances, from which one can extract information about electron-phonon interaction (at $q \cong 0$). This broad electronic background in the Raman experiments and the unusual temperature and frequency dependence of the conductivity $\sigma(\omega,T)$ are believed to be closely related and to be associated with a dynamic process particular to the highly correlated 2-d electron system in the cuprates.

The electronic states near the Fermi level are probed with high resolution angle resolved photo-electron spectroscopy. So far one can draw several basic conclusions. The data are consistent with the existence of a Fermi surface as calculated from band structure theory. The states near E_F are of mixed O 2p and Cu 3d parentage, and their cut-off at E_F is as sharp in energy as, e.g., in Pt metal. The shape of the spectral features and the broad background are of particular interest as the current discussion centers around the question of whether or not a Fermi surface of the traditional kind exists in the cuprates, and to what degree the Fermi liquid picture has to be modified.

2. Superconducting State

From several experiments, both of a macroscopic and microscopic nature, it became clear that superconductivity involves the pairing of electrons, i.e. $e^*=2e$. (ac-Josephson effect, flux jumps, flux lattice decoration, Little-Parks oscillations, Andreev-reflection). No agreement has been reached, however, on what pairing mechanism leads to T_c's above 100K in the cuprates.

$Ba_2YCu_3O_7$ is the most studied compound since crystals of acceptable quality have become available. From numerous measurements a more and more detailed description of the superconducting state emerges. The anisotropy of the normal state properties is reflected in the superconducting state. (Compared to "Bi 2212" and others, "123" is the least anisotropic one). The magnetic penetration depth λ is ~1500Å parallel to the Cu-O planes, and 5-10 times larger perpendicular to them. The coherence length is 12-15Å parallel and 1-3Å perpendicular to the Cu-O planes. Such a small coherence volume, which is a direct consequence of a high T_c (and a large superconducting gap), leads to easily observable fluctuation effects and to novel flux-dynamics in the Abrikosov state.

Spectroscopic studies of the superconducting gap tend to agree on a large ratio of $2\Delta/kT_c$, ranging from ~5 to 7 in the Cu-O planes and ~3–5 perpendicular to them. The spectroscopic techniques include tunneling, optical reflectivity, Andreev-reflection, photo-emission, and inelastic light scattering. The non-uniformity of $\Delta(\vec{k},T)$ is observed microscopically by Knight-shift and nuclear spin relaxation measurements on the two inequivalent Cu sites in "123". More detailed spectroscopic studies of $\Delta(k)$ are needed. The low-temperature dependence of $\lambda(T)$ and the spin susceptibility $\chi_s(T)$ suggest that the gap function $\Delta(\vec{k},T)$ has no nodes. An example of the consistency between different sets of data is shown in Fig. 2. Plotted are $H_{c1}(T)$ and $\lambda(T)$ for "123" on a

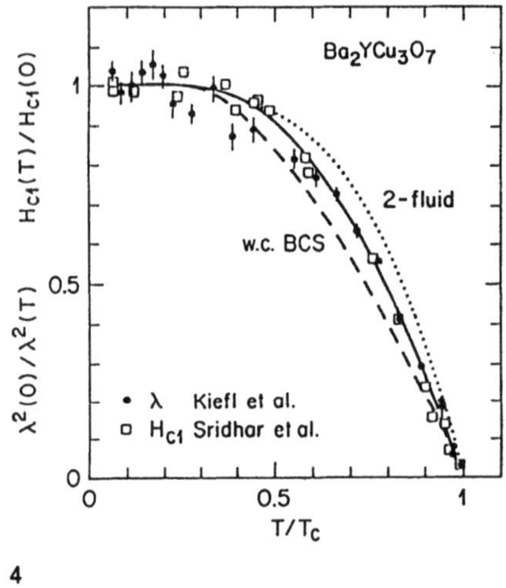

Fig. 2 Temperature dependence of the lower critical field H_{c1}, and the magnetic penetration depth λ (from μSR). $H_{c1}(T)$ and $\lambda(T)$ are normalized according to $H_{c1} \propto 1/\lambda^2$. The broken line represents the "weak-coupling BCS " and the 2-fluid model is an approximation to a "strong-coupling" temperature dependence.

relative scale with the ordinate chosen such that $H_{c1} \propto 1/\lambda^2$. The agreement is satisfactory, particularly for $T/T_c>0.5$, and the detailed temperature dependence is intermediate between the weak-coupling BCS and the two-fluid limits. Strong coupling effects are thus clearly discernible.

Although much progress has been made, many more questions remain to be answered, and the focus will be on the appropriate picture for the normal state and on the identification of the excitations which lead to superconducting pairing.

Acknowledgement

I am grateful to many colleagues for useful discussions, and particularly to R. J. Cava for stimulating collaborations and to L. W. Rupp, Jr. for technical support.

[1] J. G. Bednorz and K. A. Müller, *Z. Phys.* **64,** 189 (1986).

[2] For a collection of references and an extensive review of selected experiments see: B. Batlogg, p. 37-82, "High Temperature Superconductivity - The Los Alamos Symposium 1989", eds. K. Bedell, D. Coffey, D. Meltzer, D. Pines and J. R. Schrieffer, Addison-Wesley Publishing Company, 1990.

Double Chains in $YBa_2Cu_4O_8$ ($T_c = 80\,K$) and $Y_2Ba_4Cu_7O_{15\pm x}$ ($14 \leq T_c \leq 70\,K$) and Their Influence on the Physical Properties

E. Kaldis and J. Karpinski

Laboratorium für Festkörperphysik ETH, CH-8093 Zürich, Switzerland

Abstract. A brief overview is given of the changes of properties of $YBa_2Cu_3O_7$(123), when the CuO_4 double chains leading to the formation of $YBa_2Cu_4O_8$ (124) and $Y_2Ba_4Cu_7O_{15}$ (247) are introduced in the structure. Increase of the oxygen stability, lack of twinning, high Tc–pressure dependence, anomalous thermal expansion and compressibility, as well as changes of T_c are briefly discussed.

1. Synthesis and Structures

In bulk form 247 and 124 were observed in the fall of 1987 as the products of the decomposition of 123 under high oxygen pressure (2900 bar), during phase diagram studies [1,2]. Whereas the 247 had not been found before, the 124 had been observed as an unexpected phase during the deposition of 123 [3] two–phase thin films. Very soon after, electron diffraction on bulk ceramic samples showed an elongated c–axis [1] indicating the existence of a superstructure. The success of crystal growth (flux) under high oxygen pressure [1] enabled the structure determination of 247 [4]. Figure 1b shows this very unusual structure with alternating blocks of 123 and 124. Structures with larger periods have been found as intergrowths [5] but not yet as thermodynamically stable phases.

Soon after, the first bulk synthesis of polycrystalline 124 [6] and single crystals [7] succeeded, again as a result of the phase diagram studies. The x–ray structure proposed by [8] was refined by neutron diffraction [9] and its pressure/temperature dependence was measured [10]. Untwinned, nearly perfect single crystals of 124 were the first HTcS crystals to be subjected to a Fourier analysis [11]. Figure 1a shows the 124 structure and the b/2 shift along the b–axis which makes the doubling of the c–axis necessary. This is also true for the 247 where due to the existence of two structural blocks the c–axis becomes very large. A more detailed discussion of the properties of these compounds was given in a recent review [11a].The compounds 124 and 247 can be synthesized in polycrystalline form also under normal pressure, using powders with very fine particles [11b].

2. Change of Properties due to the Double Chains

The introduction of the double chains in the 123 structure influences strongly the physical properties from both the fundamental and applied points of view.

Springer Series in Solid-State Sciences, Vol. 99
Electronic Properties of High-T_c Superconductors and Related Compounds
Editors: H. Kuzmany · M. Mehring · J. Fink

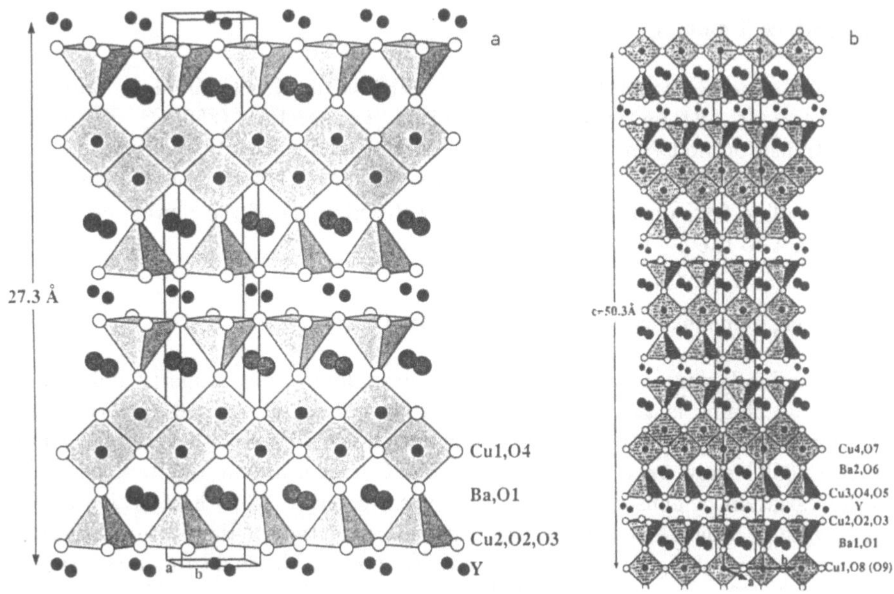

Figure 1. Cross sections of the b—c plane of the structures of a) 124 [10] and b) 247 [12], after A.W.Hewat. Space groups Ammm.

2.1 Oxygen Stability

Figure 2 shows the strong increase of the oxygen stability in 124. Decomposition starts only at T>850° C whereas 123 starts losing oxygen already at T>350°C. This tremendous increase of the oxygen stability takes place due to the increase of the coordination number of oxygen with copper, as can be seen from the structure of Fig. 1b. The increase of the oxygen stability is very important for applications and makes 124 a stoichiometric compound with a very small nonstoichiometric range. The effect also exists for the 247 but is of course not as strong as in 124 [13].

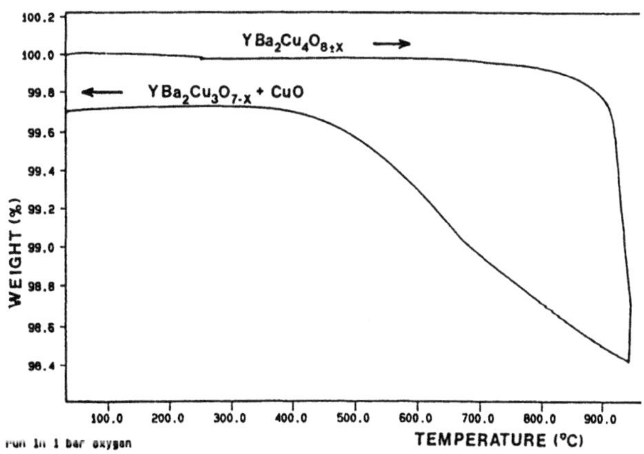

Figure 2. Thermal decomposition of 124 under P=1bar, to 123+CuO [6].

2.2 Lack of Twinning

Due to the oxygen stability of the double chains no structural phase transition takes place in 124 up to the decomposition temperature (850°C). Without a phase transition no twinning appears. The double chain fixes the b–axis.
An interesting anomaly, however, appears in single crystals and stoichiometric polycrystalline 247: up to 33% (statistically distributed) of the double chains are parallel not to the b– but to the a–axis [4,12].

2.3 Strong Dependence of T_c on Hydrostatic Pressure

From the fundamental point of view, the most interesting property induced by the double chains is the unexpectedly high value of $dT_c/dP=0.57K/kbar$ [14]. This result contradicted the previous assumption that HTcS with very high values of T_c (>60K) show very small dT_c/dP [15]. Our assumption was that, similarly to the variation of nonstoichiometry in oxygen deficient 123, where an increase of the oxygen occupancy of the chains leads to contraction of the Cu2–O1 distance and an increase of T_c [16], in 124 hydrostatic pressure leads also to a contraction of the Cu1–O2 and an increase of T_c. In other words, we assumed that oxygen deficiency has an analogous result to lattice compression.
Neutron diffraction under hydrostatic pressure of 10kbar showed exactly the expected effect [10], the apical oxygen shifting from the chains to the plains. Due to technical problems, these results showed the correct tendency but the absolute values were not accurate. In the meantime, however, x–ray high pressure single crystal work [17] has shown that the high pressure neutron data had the correct tendency, although the bond changes were not so large.
The above mentioned dT_c/dP measurements [14], were performed up to 10kbar. Very recently [18], the pressure range was increased up to 120kbar, increasing the T_c up to 108 K . In the range 50–70 kbar, a strong saturation of the effect takes place, probably due to exhaustion of the holes and/or due to possible structural changes of the material.
In most of the above work, it has been proposed that the increase of Tc associated with the contraction of the apical oxygen bond (Cu1–O2) is due to charge transfer from the chains to the plains [10,16]. This supports the early idea of K.A. Müller et al. [19,20] that this charge transfer leads to a pairing mechanism of the holes in the Cu2 planes.
Until recently only the valence bond sum had given a qualitative indication in this direction [16]. Then, however, the existence of the charge transfer between the chains and the planes was directly shown with inelastic neutron scattering both under normal and under increased pressure [21].

2.4 Anomalous Thermal Expansion

The investigation of the temperature dependence of the bond lengths in 124 [10] and 247 [22] showed, in contrast to 123 [23], strong effects near T_c. At temperatures lower than 100 K, the double chains stiffen and cannot contract further. Therefore, in the superconducting state of 124 the b–axis is essentially fixed. Similar effects appear in 247 [22].

A possible crystal–chemical explanation for this effect has been given recently [11]. Structure refinement of $SrCuO_2$ [24], which also contains double chains, showed that the Cu1–O1 distance is much larger in this compound (1.91 Å) than in the 124 (1.841). This may reflect the different kind of bonding in the two compounds and it gives a phenomenological explanation for the anomalies of thermal expansion and compressibilities of the double chain compounds 124 and 247.

2.5 Anomalous Compressibility

Recent x–ray high pressure work allowed the measurement of compressibilities and bulk modulus of the 124 [25]. The linear compressibilities along the three main crystallographic axes are strongly anisotropic, making the 124 (and possibly the 247) an exception among the HTc materials measured so far. Similarly to the thermal expansion, the b–axis, which contains the double chains, shows an anomalously small compressibility (1.0x $10^{-3}GPa^{-1}$). The a–axis compressibility is 2.8 times larger and the c–axis 4.5 times larger. The bulk modulus at 301 K is with 112 GPa quite low as compared to 123 with approx.150 GPa [26], and increases at 30 K to 122 GPa.

2.6 Variations of T_c

Due to the stability of the oxygen in the double chains, the 124 phase is nearly stoichiometric (7.86 $<x<$8.1) [7], therefore, variations of T_c due to nonstoichiometry have not been observed. Doping with Ca, however, increases the T_c of 124 to 90 K [27], contrary to its effect in 123 where T_c decreases [28]. The mechanism of this effect is not yet clear. The authors [27] suggest that Ca substitutes Y. The corresponding shift of charges should be, however, detectable with NQR, which is not the case [29]. Substitution of Ba would also be possible, however, the influence of Tc should then come from internal (chemical) pressure due to the much larger ionic radius of Ba. One would expect that such effects should be detected by appreciable contraction of the lattice constants, which is not the case. Our original assumption that the increase of T_c is associated with a contraction of the Cu1–O2 bond was able to be confirmed with neutron diffraction: a small decrease was detected [30].
Contrary to 124, the 247 has 50% single chains which present the unique phenomenon of localised nonstoichiometry. We have used that to selectively desorb oxygen from the single chains in order to see the influence on Tc. One interesting question is if the compound remains superconducting even when the single chains are depleted. In this case the 123 blocks of the 247 structure (Fig.1) are expected to become semiconducting. Our investigation showed that this is not the case and that coupling is possible over 11.5 Å between the 124 blocks, with T_c=14 K [31]. Figure 3 shows clearly that in the presence of double chains the T_c dependence on nonstoichiometry is quite different from that of 123.

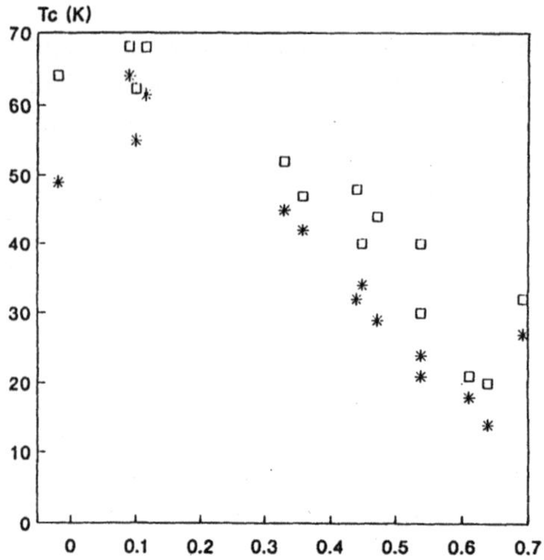

Figure 3. Dependence of T_c of 124 on the nonstoichiometry [31].

3. Conclusions

The interesting properties introduced by the double chains in 124 and 247 make these compounds model substances for the study of the importance of the chains in HTc superconductivity. In view of the fact that Ca–doped 124 has Tc=90 K, it is expected that very soon this compound, with its much higher oxygen stability, will be recognized as much more interesting than 123 from both the fundamental and applied points of view.

References

1.J.Karpinski,C.Beeli,E.Kaldis,A.Wisard,and E.Jilek, Physica C 153–155(1988)830
2.J.Karpinski,E.Kaldis, Nature 331(1988)242
3.A.Kapitulnik, Physica C 153 (1988)520
4.P.Bordet,C.Chaillout,J.Chenevas,J.L.Hodeau,M.Marezio,J.Karpinski,and E.Kaldis, Nature 334(1988)596
5.C.Chaillout,P.Bordet,J.Chenevas,J.L.Hodeau,M.Marezio,J.Karpinski,E.Kaldis, and S.Rusiecki, Solid State Commun.70(1989)275
6.J.Karpinski,E.Kaldis,E.Jilek,S.Rusiecki and B.Bucher, Nature 336(1988)660
7.J.Karpinski,E.Kaldis,E.Jilek,S.Rusiecki,E.Jilek,P.Fischer,P.Bordet,C.Chaillout,J. Chenevas,J.L.Hodeau,and M.Marezio, J.Less–Common Metals 150(1989)129
8.P.Marsh,R.M.Fleming,M.L.Mandich,A.M.De Santolo,J.Kwo,M.Hong, and L.J.Martinez–Miranda, Nature 334(1988)141
9.P.Fischer,J.Karpinski,E.Kaldis,E.Jilek,and S.Rusiecki, Sol.State Commun. 69(1989)531
10.E.Kaldis,P.Fischer,A.Hewat,E.Hewat,J.Karpinski, and S.Rusiecki, Physica C 159(1989)668
11.P.Bordet,J.L.Hodeau,R.Argoud,J.Muller,M.Marezio,J.C.Martinez,J.J.Prejean,J. Karpinski,E.Kaldis,S.Rusiecki,and B.Bucher, Physica C ,162–4 (1989)525
11a.E.Kaldis,J.Karpinski, Eur.J.Solid State Inorg.Chem.27(1990)143–190
11b.R.J.Cava,J.J.Krajewski,W.F.Peck,B.Batlogg,L.W.Lupp,R.M.Flemming,A.C.W .P.James, and P.Marsch, Nature 338(1989)328

12.A.W.Hewat,P.Fischer,E.Kaldis,J.Karpinski,S.Rusiecki,and E.Jilek, Physica C 167(1990)579
13.J.Karpinski,S.Rusiecki,E.Kaldis,B.Bucher, and E.Jilek, Physica C 160(1989)449
14.B.Bucher,J.Karpinski,E.Kaldis,and P.Wachter, Physica C 157(1989)478
15.C.W.Chu,Z.J.Huang,R.L.Meng,L.Gao,and P.H.Hor, Phys.Rev. B37(1988)9730
16.R.J.Cava,A.W.Hewat,E.A.Hewat,B.Batlogg,M.Marezio,K.M.Rabe,J.J.Krajewski ,W.F.Peck,and L.W.Rupp, Physica C 165(1990)419
17.J.S.Loveday,R.J.Nelmes,D.R.Allan,E.Kaldis,J.Karpinski,S.Rusiecki,and E.Jilek (to be published)
18.E.N. van Eenige,R.Griessen, R.J.Wijngaarden,J.Karpinski,E.Kaldis,S.Rusiecki, and E.Jilek, Physica C (to be published)
19.A.R.Bishop,R.L.Martin,K.A.Müller,and Z.Tesanovic', Z.Phys. B—Condensed Matter 76(1989)17
20.K.A.Müller, Z.Phys.B—Condensed Matter 50(1990)193
21.P.Allenspach,J.Mesot,U.Staub,A.Furrer,H.Blank,H.Mutka,C.Vettier,E.Kaldis,J. Karpinski,and S.Rusiecki, Europ.J.Solid State Chemistry (to be published)
22.A.W.Hewat,P.Fischer,E.Kaldis,J.Karpinski,S.Rusiecki,and E.Jilek, Physica C,(in print)
23.M.Francois,A.Junod,K.Yvon,A.W.Hewat,J.J.Capponi,P.Strobel,M.Marezio,and P.Fischer, Solid State Comm. 66(1988)1117
24.P.Bordet (private communication)
25.H.A.Ludwig,W.H.Fietz,M.R.Dietrich,H.Wühl,J.Karpinski,E.Kaldis,and S.Rusiecki, Physica C 167(1990)335
26.R.J.Wijngaarden and R.Griessen, in " Studies of High Tc Superconductors " A.V.Narlikar Ed.,Nova Science Publishers Inc.,New York 1989
27.T.Miyatake,S.Gotoh,N.Koshizuka, and S.Tanaka, Nature 341(1989)41
28.H.Kuzmany,T.Pichler,M.Matus,A.Hanninger,P.Rogl,and S.Pekker, this volume
29.I.Mangelshots,H.Zimmermann,M.Mali,J.Roos,D.Brinkmann,J.Karpinski, S.Rusiecki,and E.Kaldis, J.Less—Common Metals (1990) to be published
30.A.W.Hewat,S.Rusiecki,J.Karpinski,E.Kaldis,E.Jilek,to be published
31.J.Karpinski,S.Rusiecki,B.Bucher,E.Kaldis,and E.Jilek, Physica C 161 (1989)618

Constitution of the System Bi-Sr-Ca-Cu-O and Preparation of the Three Superconducting Compounds

B. Hettich, M. Schwarz, and H. Sixl

Hoechst AG, Postfach 800320, D-6230 Frankfurt 80, Fed. Rep. of Germany

Abstract. We report on phase equilibria, phase widths, compatibility regimes, stoichiometric competition and preparation of the three superconducting compounds in single phases. The knowledge of these phase equilibria, in which the superconducting compounds are involved, is the basis for the design of optimal ceramic procedures. The constitution of the multiphase system has been analyzed at different temperatures. A cycling temperature treatment has been optimized to obtain the triple layer compound as a single phase. Chemical and physical aspects of the solid state reactions are discussed.

1. Introduction

The most important structural features of the Bi-Sr-Ca-Cu-O superconducting phases are single **1** double **2** or triple **3** CuO_2-layers between BiO_2-double layers. As we reported previously [1-4] the transition temperatures are about 75 K **1**, 85 K **2** and 110 K **3**, respectively as shown in Figure 1. In order to obtain single phase systems and to avoid additional components like Pb as reported earlier [5, 6] it is necessary to know the exact ceramic preparation conditions and the competing phases. Therefore we made a complete constitution analysis of the quaternary oxide system and designed a cycling procedure for the preparation of the triple layer system. Analysis of the relevant data was performed using X-ray powder diffraction (XRD), SQUID magnetometer measurements (susceptibilities) and resistivity curves (R(T)) as shown in Figure 1 for the single layer **1**, double layer **2** and triple layer **3** compound.

2. Constitution analysis

On the surface of the $BiO_{1.5}$, CaO, CuO and SrO tetrahedron shown in Figure 2 a series of 13 binary oxide phases along the six edges, 7 ternary oxide phases at the four faces have been identified [7 - 13]. The three superconducting quaternary oxide phases within the tetrahedron are shown in Figure 2 together with one binary edge phase. These compounds with fixed Sr-Ca ratio can be represented by a triangular plane $Sr_2Bi_2O_5$-CaO-CuO. The phase equilibria within this plane are shown at three different temperatures. At 750°C the single layer **1** compound is the only stable phase. At 800 °C the double layer system **2** is added and finally at 850°C all three superconducting phases exist . In order to obtain for example the triple layer system the correct weight-in quantities and the correct temperatures have to be chosen and

Springer Series in Solid-State Sciences, Vol. 99
Electronic Properties of High-T_c Superconductors and Related Compounds
Editors: H. Kuzmany · M. Mehring · J. Fink

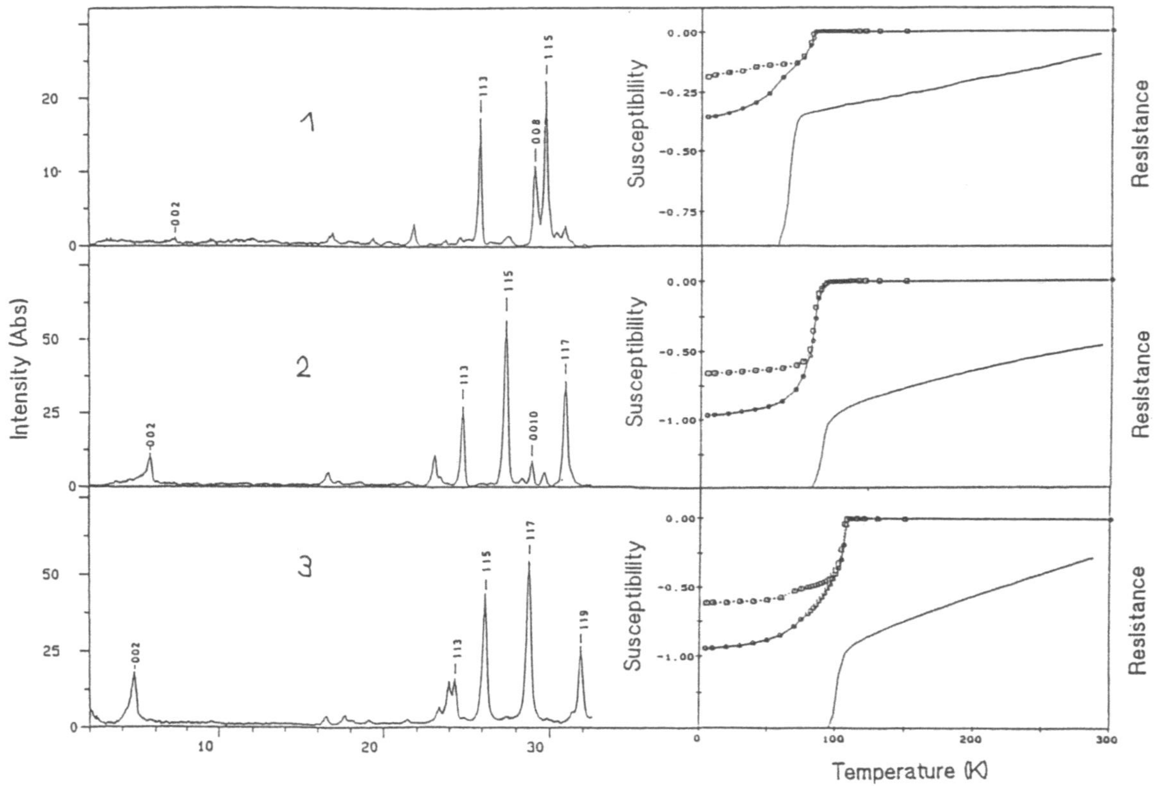

Figure 1 X-ray diffraction intensities, magnetic susceptibilities and electrical resistance measurements of the single layer 1, double layer 2 and triple layer 3 compound.

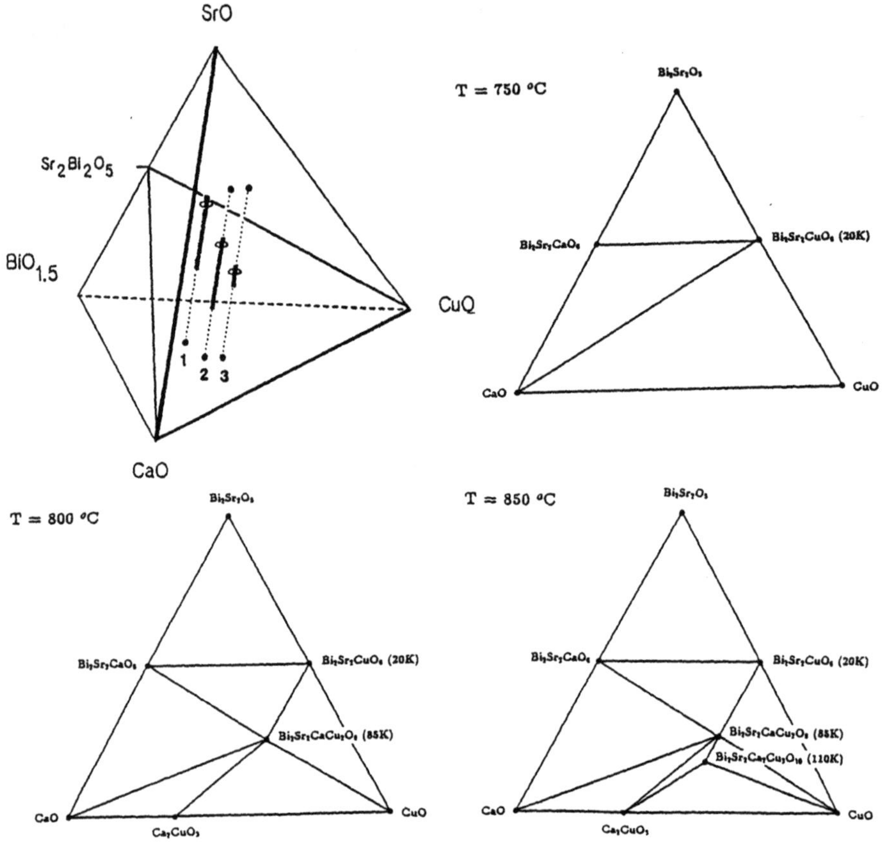

Figure 2 Quasi quaternary tetrahedron, showing the existence regimes of the three superconducting phases (full lines) and compatibility triangles at 750, 800 and 850 °C

finally it needs some time, of the order of days, to establish the phase equilibria. It is evident that the long lasting formation of **3**, which is due to kinetic effects, passes many phases especially those around **3** (**1**, **2**, CuO and Ca_2CuO_3). So these compounds act as intermediates until **3** is formed from the starting mixture of oxides.

The lines within the tetrahedron indicate the existence regimes of **1**, **2** and **3** when varying the Sr-Ca-ratio. The dotted lines consequently represent the non-existence regimes of the layer systems, where the originally superconducting single phases decay into two or more phases. The XRD and susceptibility analysis of Figure 3 clearly shows the phase equilibria when changing the Sr-Ca ratio. As expected the superconducting volume fraction as measured by XRD and by a SQUID magnetometer behave almost identically especially in the double **2** and in the triple **3** layer system. The phase widths of the single layer system **1** is given by $0 \leq x \leq 0.33$, in the double layer system **2** by $0.3 \leq x \leq 0.66$ and in the triple layer system **3** very sharply around $x \approx 0.5$.

The superconductor transition temperatures **1**, **2** and **3** are strongly dependent on the oxygen content with optimum at $1.6 \leq$

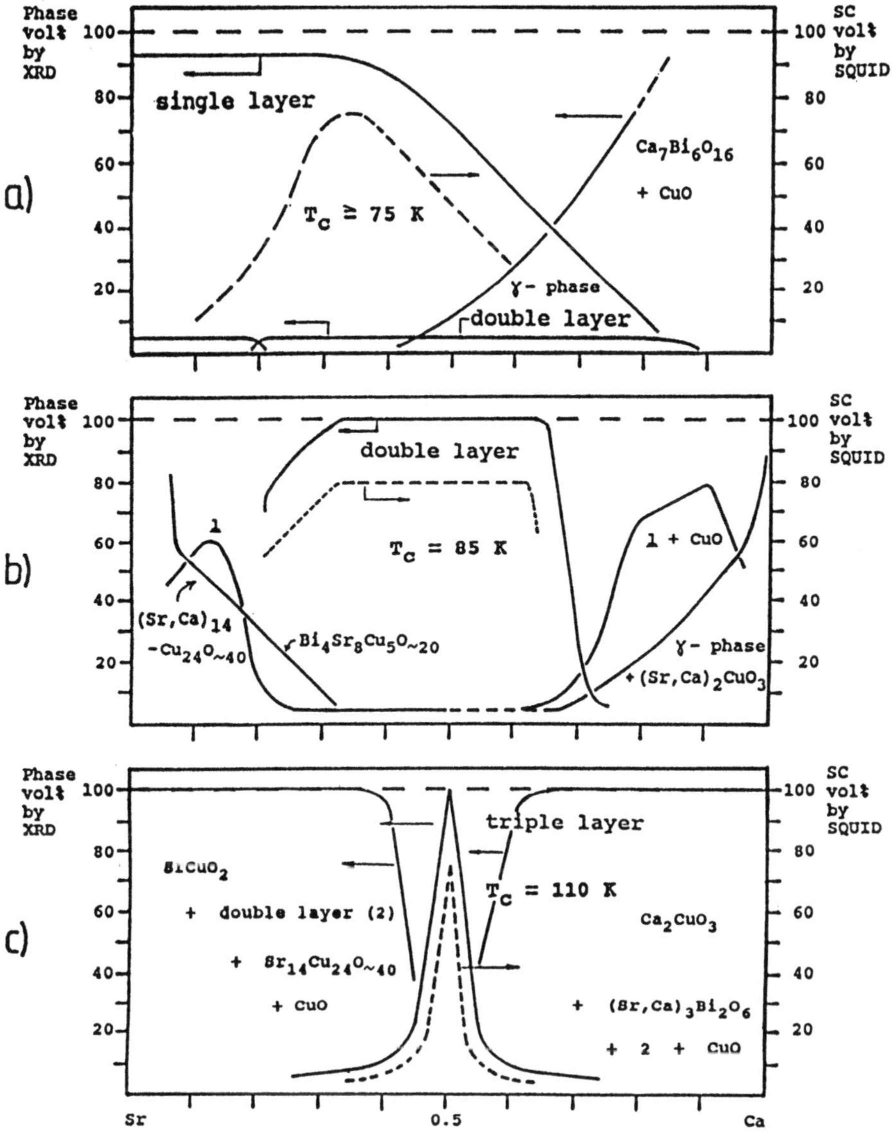

Figure 3 Volume ratio of the superconducting phases **1**, **2** and **3** as measured by X-ray diffraction (XRD) and by SQUID magnetic shielding effect. The competitive phases have not been measured in connection with the triple layer compound.

$\delta \leq 1.8$. Maximum T_c is obtained at ambient atmosphere with reduced oxygen pressure.

3. Preparation of the 2223 triple layer single phase

In a series of experiments the optimal reaction procedure for the formation of the triple layer system **3** has been analyzed using

Preparation of $Bi_2Sr_2Ca_2Cu_3O_{12-x}$ (T_c=110K) (3)

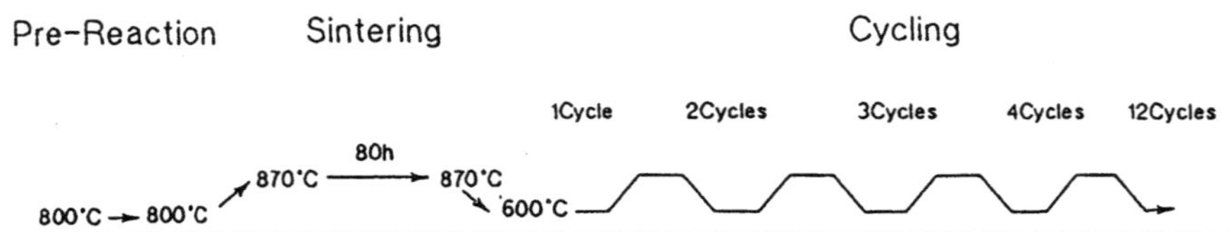

	Pre-Reaction	Sintering	1Cycle	2Cycles	3Cycles	4Cycles	12Cycles
X-Ray	1(30%) 2(70%)	1 (20%) 2 (80%)	2	2(85%) 3(15%)	1(5%) 2(10%) 3 (85%)	1(5%) 2(10%) 3(85%)	1(10%) 2(20%) 3(70%)
T_c From Resistivity Measurement (Surface)	85K (2)	85K (2)	85K(2)	110K(3)	110K(3)	110K(3)	85K(2)
T_c From DC-Susceptibility Measurement (Bulk)	75K (1;30%) 85K (2;70%)	75K (1;20%) 85K (2;80%)	85K(2)	85K(2;85%) 110K(3;15%)	75K(1;5%) 110K(3;95%)	75K(1;5%) 110K(3;80%)	85K(5%) 110K(75%)
Grain							

Figure 4 Preparation procedure of the triple layer compound including results of XRD, resistance and SQUID magnetometer measurements.

a nonstoichiometric mixture with excess of CuO to avoid decomposition reactions [5]. The comparison of X-ray diffraction, resistivity- and SQUID-measurements at different stages of the reaction is shown in Figure 4. After pre-reacting the four metal oxides at 800°C the material which at that stage consists of a mixture of **1** and **2**, is sintered sharply below its melting point at 870°C for 80 h. After sintering the composition remains almost unchanged whereas the first temperature cycle leads to the pure double layer system **2**. After the second cycle the X-ray diffraction diagram shows apparently still a pure double layer sample **2** but the resistivity measurement exhibits an onset at T_c = 110 K (best samples). This means that the sample contains a small amount of the triple layer **3** system, nucleated at the surface of the grain and thus forming a short-circuited surface pathway for the electrical current. After three cycles the X-ray diffraction shows the triple layer **3** system as the majority phase. Surprisingly, further cycling causes decomposition of the triple layer **3** system into **2**, **1** and CuO especially at the surface of the grain. In addition it is remarkable that the final cooling process after cycling of the samples has a significant influence on the superconducting properties of the grain surfaces of the material. Rapid cooling leads to phase segregation and semiconductor type behavior. Only slow cooling to room temperature leads to optimal electrical properties.

According to the results during the preparation procedure shown in Fig. 4, we clearly see that the individual superconducting phases are formed successively in the sequence **1**, **2**, **3**. The formation kinetics consequently is comparable to the sequence of the thermal stability ranges of **1**, **2** and **3** as shown in Figure 2 by the different compatibility triangles. According to STEM investigations it is obvious that, in the formation process of the triple layer compound, CuO_2-layers step by step are inserted into first the single layer compound **1** then into the double layer compound **2** to finally obtain the triple layer structure.

Acknowledgements

Work has been done within the Hoechst AG project team. We would like to thank Dr. Elschner for electrical measurements, Dr. Paulus for providing structural data and Dr. Schulze and Dr. Majewski for active work on phase diagrams.

References

[1] H.G. von Schnering, L. Walz, M. Schwarz, W. Becker, M. Hartweg, T. Popp, B. Hettich, P. Müller and G. Kämpf. Angew. Chemie Int. Ed. 27, 4 (1988) 574.

[2] M. Peuckert, W. Becker, J. Bock, B. Hettich, H.-W. Neumüller, M. Schwarz. Physica C 162 - 164 (1988) 893.

[3] M. Schwarz, B. Hettich and W. Becker. Proc. of the TMS Annual Meeting "Processing of HT-Superconductors", Anaheim, CA, (1990).

[4] J. Bock and E. Preisler. Solid State Comm. 72 (199) 453.

[5] H. Maeda, Y. Tanaka, m. Fukutomi and T. Asano. Japan. J. Appl. Phys. 27 (1988) L209.

[6] R.J. Cava, B. Batlogg, S.A. Sunshine et al.
Physica C 153 - 155 (1988) 560.

[7] P. Conflant, J.C. Boivin, G. Tridot.
C.R. Acad. Sc. Paris, t. 279 (1974) C-457.

[8] J.-C. Boivin, D. thomas, G. Tridot.
ibid., t. 276 (1973) C-1105.

[9] R. Guillermo, P. Conflant, J.-C. Biovin, D. Thomas.
Revue Chim. Min., t. 15 (1978) 153.

[10] C. Teske, H. Müller-Buschbaum in Z.a.a. Chem. 370 (1969) 134; ibid. 371 (1969) 325, ibid 379 (1970) 234.

[11] R.S. Roth, C.J. Rawn, J.J. Ritter, B.P. Burton.
J. Am. Ceram. Soc. 72 (8), (1988) 1545.

[12] H.-G. von Schnering.
personal communication, 7. Sept. 1988.

[13] T. Siegrist, L.F. Schneemeyer, S.A. Sunshine, J.V. Waszczak.
Mat. Res. Bull. 23 (1988) 1429.

Optimization of the Laser Ablation Process for Epitaxial YBCO Thin Film Growth

T. Freltoft and P. Vase

NKT A/S, Department of Corporate R&D, Sognevej 13,
DK-2605 Brøndby, Denmark

Abstract. High quality epitaxial thin films of $YBa_2Cu_3O_{7-x}$ on single crystal MgO substrates have been fabricated using the laser ablation process. The films have critical temperatures between 87 and 90 K and the critical current density measured at a 10 μm strip is above $10^6 A/cm^2$. Characterization of the films also includes Scanning Electron Microscopy (SEM), Energy Dispersive X-ray analysis (EDX), and X-ray diffraction. Extensive pole figure analysis has been performed for the X-ray data showing that only one crystal orientation is present in the films. The mosaic spread is about 1.5°. A systematic optimization of the ablation process has been performed varying deposition parameters such as laser power, substrate temperature, and oxygen pressure.

Introduction

During the last two years an extensive effort has been devoted to fabrication of high quality thin films of the copper-oxide superconductors. The aim has been to produce epitaxial films without grain boundaries in order either to study the new materials in more ideal forms than provided by bulk samples or to develop applications in the form of electronic devices, i.e. SQUIDs, sensors, amplifiers, mixers, micro-wave components etc. In this paper we describe the optimization of the laser ablation deposition method including structural and electrical characterization of the deposited thin films.

Experimental setup

Laser ablation is a relatively new process – one of the first reports came from Strinivasan and Mayne-Banton [1] in 1982. Since then, it has been extensively studied in relation to applications in laser surgery and polymer etching [2], but only after the break-through of ceramic superconductors was laser ablation really developed as a *deposition* process [3]. The advantage of laser ablation over other more traditional deposition processes like sputtering or thermal evaporation techniques is the ability to maintain a relatively complicated metal stoichiometry of the target material in the deposited film.

The laser ablation facility at NKT is shown schematically in Fig. 1. The laser is a Q-switched Nd-YAG (λ=1064 nm) which is frequency tripled to λ=365 nm. The pulse energy at 365 nm is around 130 mJ and with a pulse length of 5 ns this gives an average peak power of 26 MW. The laser beam is focused on a rotating bulk target of $YBa_2Cu_3O_7$ placed in a vacuum chamber with a controlled oxygen pressure. The high power laser pulse creates a plume of excited species from the target of primarily single metal atoms/ions or simple metal oxides. The kinetic energies involved are of the order of 50 eV, corresponding to particle velocities around 10^4 m/s. The light emitted from the plume has been analyzed [4] giving this information about the composition and kinetic energy of the excited particles.

A heated single crystal substrate is placed some distance across from the target. The substrate is attached to a stainless steel disk with silver glue, and the disk is

Springer Series in Solid-State Sciences, Vol. 99
Electronic Properties of High-T_c Superconductors and Related Compounds
Editors: H. Kuzmany · M. Mehring · J. Fink

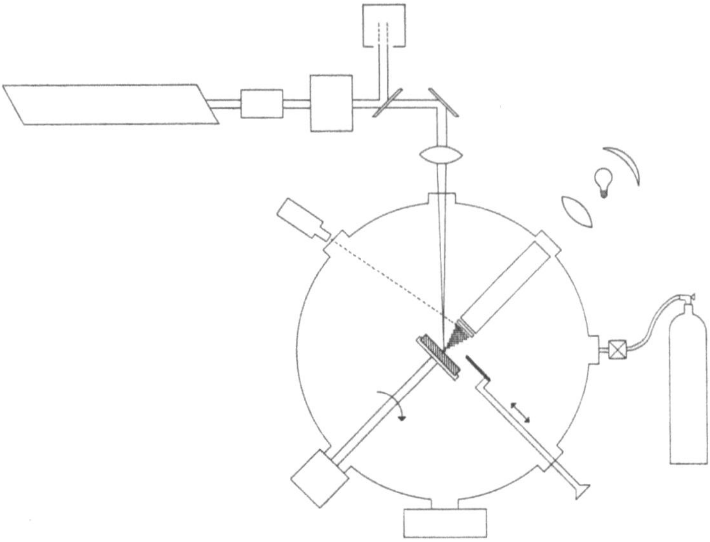

Fig. 1 The laser ablation facility at NKT shown schematically.

mounted on the tip of a rigid thermo-couple also using silver glue to ensure effective heat transfer between substrate, disk and thermo-couple. The heat source is a 1000 W halogen lamp (with reflector) placed *outside* the vacuum chamber. The light from the lamp is focused by a pair of aspherical condenser lenses at one end of a quartz rod inside the chamber. The rod transmits the light to the back side of the stainless steel disk enabling temperatures up to 950 °C to be reached at the substrate. In this way the heat source can be effectively cooled and is separated from the pure oxygen atmosphere inside the chamber.

The surface emissivity of the substrate/film can be monitored during deposition by a pyrometer placed outside the chamber. By a turbo molecular pump system, a base pressure of 10^{-6} mbar may be reached. Both oxygen pressure and substrate temperature are controlled automatically by feedback loops.

Film optimization

Having chosen a suitable single crystal substrate for the thin film deposition, the most important parameters for laser ablation of ceramic superconductors are the laser power density, the oxygen pressure, and the substrate temperature during deposition. Secondary parameters include the distance between target and substrate, and the laser beam spot size.

A typical deposition procedure was as follows: The 30 mm diameter $YBa_2Cu_3O_{7-x}$ target was sanded down to obtain a clean surface before each run. A single crystal substrate of MgO (10x10x0.5 mm^3 or 5x10x0.5 mm^3) polished one side and cut with the (100) axis perpendicular to the surface was mounted in the substrate holder and heated to 900 °C in ≈ 1 mbar oxygen for a few minutes in order to clean the surface of carbon contaminants. Then the substrate temperature, oxygen pressure, and laser power were adjusted to the relevant values, and pre-ablation was initiated to clean the target surface with a shutter in front of the substrate. After pre-ablation the shutter was removed and deposition initiated. The deposition rate varied significantly with deposition parameters, but for optimal conditions a typical value was 100 nm/min. The

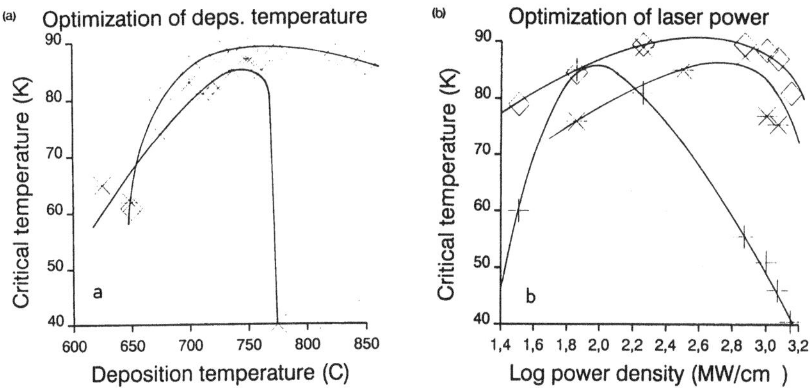

Fig. 2 Optimization of the laser deposition parameters. (a) Critical temperature versus deposition temperature for two different oxygen pressures, 0.10 mbar (stars) and 0.75 mbar (diamonds). (b) Critical temperature versus the average peak power (note the $\log_{10}$ scale) for 0.1 mbar (crosses), 1.0 mbar (diamonds), and 3.0 mbar (stars) oxygen pressure. The deposition temperature was here T = 750 °C for all films. All lines are guides to the eye.

oxygen pressure was increased to 1 atm after deposition before the temperature was ramped down, first quickly (-100 °C/min.) to 600 °C and then slowly (-20 °C/min.) to 150 °C where the heater was turned off.

Films were deposited with a systematic variation of laser power, oxygen pressure, and deposition temperature. Each film was characterized electrically by measuring the resistance versus temperature curve, and the metal composition was determined by EDX (Energy Dispersive X-ray analysis). Fig. 2 shows some of the data obtained in this optimization process.

Two film series for optimization of the deposition temperature are shown in Fig. 2a. The critical temperature (midpoint of transition) is plotted versus deposition temperature for two oxygen pressures, 0.10 and 0.75 mbar. The laser power density was 100 and 250 MW/cm^2 for the two series, respectively. In general, the laser power density was adjusted by moving the focusing lens, thereby changing the beam spot area, but keeping the pulse energy constant. For both pressures a deposition temperature around 750 °C is optimal, but the optimum is seen to be much broader for the higher pressure series. The sudden drop above 765 °C at 0.1 mbar oxygen happens because the perovskite structure becomes unstable. This is also known from the phase diagram [5] measured from bulk samples, and it shows that our temperature determination of the film/substrate is correct. In general it was found that independent of oxygen pressure and laser power density the optimal deposition temperature was around 750 °C.

Fig. 2b shows the critical temperature versus laser power density for a number of oxygen pressures. It is observed that also here we have a window in which high critical temperatures are achieved. At 0.1 mbar the optimum is obtained in a relatively narrow window around 100 MW/cm^2, while at higher oxygen pressures, the optimum moves up in power density and the window broadens. If the oxygen pressure becomes very high (>3.0 mbar), the window seems to become more narrow again. Another problem occurring if the oxygen pressure is too high is that the films become very inhomogeneous in thickness. In general it was found that an oxygen pressure between 0.5 and 1.0 mbar was optimal, corresponding to the laser power density 250 - 400 MW/cm^2.

Most authors give the laser power as a pulse energy, but since the ablation process is also crucially dependent on the very short pulse length we prefer to use the average peak power, i.e. pulse energy divided by pulse length (5 ns for our laser). Hence, 250 MW/cm^2 corresponds to 1.3 J/cm^2.

Film characteristics

The optimized films were characterized in terms of critical temperature, critical current density, structure (epitaxy), and microwave absorption at 35 GHz. Some of these characteristics are shown in Figs. 3 and 4.

Fig. 3 shows resistance versus temperature for a 10 by 500 μm^2 strip of thickness 300 nm. The transition takes place over ≈1 Kelvin and $T_{c,0}$ is ≈ 90 Kelvin. The critical current density for this strip at T = 77 K was 1.8 $10^6 A/cm^2$. The insert is a blow-up of the temperature range between 88 and 93 K.

Microwave surface resistance R_s was measured [6] at 35 GHz as a function of temperature. The measurements were performed by replacing one wall in a Cu cavity by a superconducting thin film and observing the change of the quality factor Q of the cavity. R_s for the film shows a sharp drop at T_c to a value which is around ten times lower than for Cu at T = 77 K. This is a result showing the potential applicability of the films for passive microwave devices.

The crystal structure was checked by x-ray diffraction in two ways. One is a standard $\theta - 2\theta$ scan with the momentum transfer perpendicular to the film surface. In this scan only (00l) reflections appear, showing that the films have perfect *c*-axis alignment. To investigate the grain orientation of the $a - b$ axes in the film plane, an extensive pole figure analysis was performed on a triple axis x-ray diffractometer [7].

A pole figure analysis is done by selecting a specific Bragg reflection from the crystal structure and setting the x-ray spectrometer angles to monitor only this reflection. Then the sample is turned in all possible directions while monitoring the count-rate of the Bragg peak as a function of sample orientation. In this way the distribution of crystal orientations in the film (texture) is determined. In Fig. 4 the result for the (113)

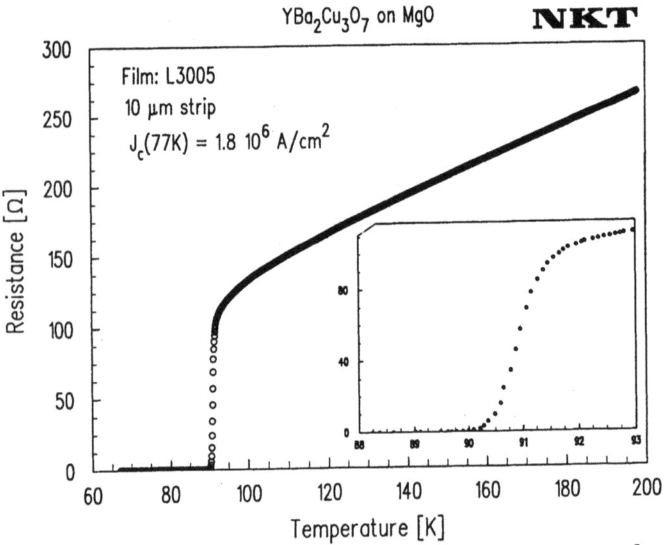

Fig. 3 Resistance versus temperature for a 10x500x0.3 μm^3 microstrip. The specific resistivity at T = 100 K is 76 $\mu\Omega$cm. The insert is a blow-up of the temperature range between T = 88 and 93 K showing a transition width of ≈ 1 K.

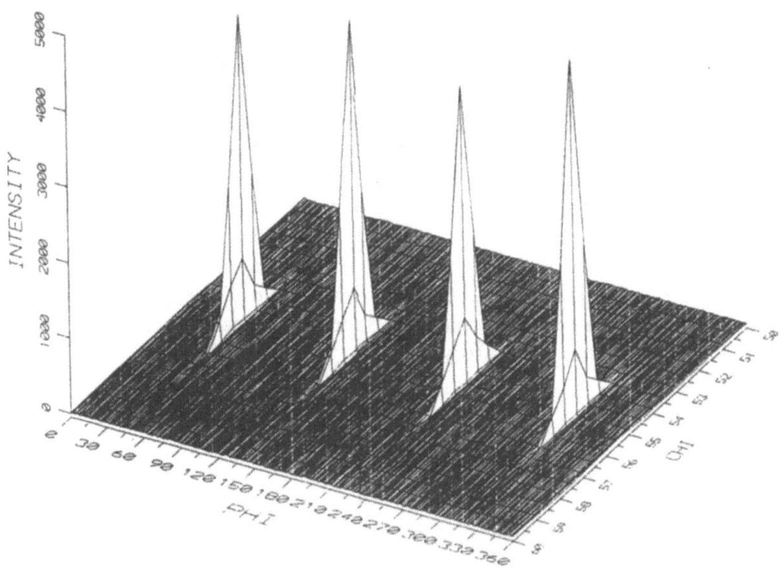

Fig. 4 Pole figure X-ray analysis of the (113) Bragg reflection from a YBCO thin film on single crystal MgO substrate. The data show that the film is epitaxially grown with the *a* and *b* axes along the MgO (010) and (001) directions, or vice versa due to twinning.

reflection is plotted. The presence of only four peaks positioned 90° apart shows that only two crystal orientations are present. The two orientations correspond to the two twin domains with the *a* and *b* axes interchanged. From the data the mosaic spread of the film has been calculated to be 1.3°.

The orientation relative to the underlying substrate is simply that the *a* and *b* axes lie along the MgO (010) and (001) directions, respectively, or vice versa depending on the twin domain. Hence, the films are epitaxially grown despite the fact that the lattice mismatch between MgO and YBCO is around 7 %. However, a lattice mismatch of this order of magnitude is traditionally prohibitive for epitaxial growth, and it is not yet understood why it still works in this case.

Summary

Optimization of the laser ablation process for depositing YBCO thin films on single crystal MgO substrates has been presented. It is found that on choosing the right combination of deposition parameters the optimum in critical temperature and film quality is fairly broad, so that a very accurate control of the parameters is not needed. The optimal deposition parameters found were: deposition temperature 750 °C, oxygen pressure 0.5-1.0 mbar, and average laser power density 250-400 MW/cm^2. The best target-substrate distance was 30 mm.

Films produced with the optimal deposition parameters have state-of-the-art characteristics. The transition temperature is 87-90 K and they are epitaxially grown with a mosaic spread of 1.3 degrees. The critical current density is $1.8 \times 10^6 A/cm^2$, and microwave losses at 35 GHz and 77 K are 10 times better than for Cu.

References

[1] R. Strinivasan and V. Mayne-Banton, Appl.Phys.Lett.**41**, 576 (1982).
[2] R. Strinivasan and B. Braren, Appl.Phys.A **45**, 289 (1988).
[3] D. Dijkkamp, T. Venkatesan, X.D. Wu, S.A. Shaheen, N. Jisrawi, Y.H. Min-Lee, W.L. McLean, and M. Croft, Appl.Phys.Lett. **51**, 619 (1987).
[4] J.P. Zheng, Z.Q. Huang, D.T. Shaw, and H.S . Kwok, Appl.Phys.Lett. **54**, 280, (1989).
[5] R. Bormann and J. Nölting, Appl.Phys.Lett. **54**, 2148 (1989).
[6] Microwave losses were measured by Claus Schelde Jacobsen and Niels Ebbe Dam, Physics Laboratory III, Technical University of Denmark.
[7] X-ray pole figure analysis was performed by Rasmus Kroman, Physics Department, RISØ National Laboratory.

New Dielectric Substrates for Microwave and Far-Infrared Applications of High-T_c Thin Films*

M. Berkowski[1], *A. Pajaczkowska*[1], *P. Gierłowski*[1], *W. Kula*[1], *R. Sobolewski*[1,**], *S.J. Lewandowski*[1], *B.P. Gorshunov*[2], *D.B. Lyudmirsky*[2], *and O.I. Sirotinsky*[2]

[1]Instytut Fizyki, Polska Akademia Nauk, PL-02669 Warszawa, Poland
[2]Institute of General Physics, USSR Academy of Sciences, SU-117924 Moscow, USSR
** Also at Department of Electrical Engineering, University of Rochester, Rochester, NY 14627, USA

Abstract. Preparation and far-infrared dielectric properties of $CaNdAlO_4$, $BaLaGa_3O_7$, and $SrLaGa_3O_7$ single crystals are described. We demonstrate that among these materials, $CaNdAlO_4$ is the best substrate for the epitaxial growth of high T_c thin films suitable for high-frequency applications.

1. Introduction

High frequency applications of high T_c superconducting films require high-quality dielectric substrates, which combine low microwave losses at low temperatures with a reasonable lattice match encouraging epitaxial growth of the films. Strontium titanate, lanthanum gallate, and lantanum aluminate are currently recognized as the best materials for epitaxial growth of high T_c films [1]. For microwave purposes $LaAlO_3$ is the best choice. Unfortunately, fabrication of large $LaAlO_3$ single crystals by the Czochralski method is difficult and expensive. $LaGaO_3$ wafers have serious twinning problems and, as we will show later, the material itself exhibits relatively high dielectric permittivity and losses. Ferroelectric properties of $SrTiO_3$ (as well as $KTaO_3$) limit its applicability to the low frequency regime.

Recently, some of us have proposed [2] and tested [3, 4] two new families of tetragonal single crystals suitable for the epitaxial growth of high-T_c thin films, namely ABC_3O_7 (gehlenites) and $ABCO_4$ (K_2NiF_4-type perovskites), where A stands for Ba, Sr, Ca; B = rare earth elements, and C = Ga, Al; any mixture of the elements within A, B, C, respectively, is also possible. In this paper we describe the crystal growth of these materials and fabrication of the substrates. We focus our attention on $CaNdAlO_4$, $BaLaGa_3O_7$, and $SrLaGa_3O_7$ and their dielectric properties at submillimeter-wave frequencies. Superconducting properties of the high-T_c films deposited on these substrates are also discussed.

2. Crystal Growth

Single crystals of $CaNdAlO_4$ were grown by the Czochralski method, using 36-mm-diameter iridium crucibles. The starting components were Nd_2O_3, Al_2O_3,

* The work in Warsaw was supported by the Polish Academy of Sciences under the grant RPBP 01.9.

and $CaCO_3$ powders (5N purity), mixed in a stoichiometric 1:1:1 molar ratio. Since crystal seeds were not available, the first boule was grown using an iridium wire for the nucleation. Then, the crystal was oriented and the seed crystals (2-3 cm long and 5 mm diameter cylinders) of the orientation (001), were core drilled. These cylindrical seeds were used for subsequent runs. The melting temperature was about 1820 °C. The pulling rate, 4 mm/hr, and the rotation rate, 40 - 50 r.p.m., were chosen to obtain a flat crystal-melt interface. The 17-mm-diameter single-crystal rods had neither iridium inclusions nor gas bubbles, and were cut and polished into 0.5-mm-thick (001) wafers. The lattice constant a = 0.368 nm of the $CaNdAlO_4$ tetragonal cell (c = 1.215 nm), is close to the a and b dimensions of the Y-Ba-Cu-O and Bi-Sr-Ca-Cu-O unit cells [4].

Gehlenite ($BaLaGa_3O_7$ and $SrLaGa_3O_7$) crystals were also grown by the Czochralski method. They exhibited a slight deviation from the ABC_3O_7 stoichiometry, having a $A_{1+x}B_{1+y}C_{3-x-y}O_{7-0.5x}$ chemical composition (x, y ~ 0.05). Thus, they could be considered as a solid solution of $(1-y)ABC_3O_7$ and $(x+y)AC_2O_4$. $BaLaGa_3O_7$ and $SrLaGa_3O_7$ single crystals of 17.5 mm diameter and up to 7.5 cm length were pulled from the iridium crucible. As previously, the pulling rate, 4 mm/hour, and the rotation rate were adjusted to obtain a flat crystal-melt interface. The melting points were 1580 °C and 1760 °C, for $BaLaGa_3O_7$ and $SrLaGa_3O_7$, respectively. Disk-shaped wafers (0.5-mm-thick) were cut from the crystals pulled in the ‹001› direction, and optically polished. The lattice constants of the $BaLaGa_3O_7$ and $SrLaGa_3O_7$ tetragonal unit cells are a = 0.8158 nm, c = 0.5399 nm, and a = 0.806 nm, c = 0.534 nm, respectively.

3. Dielectric Properties

Submillimeter-wave complex dielectric permittivity spectra (ε' and $\tan\delta = \varepsilon''/\varepsilon'$) of the $LaGaO_3$, $CaNdAlO_4$, $BaLaGa_3O_7$ and $SrLaGa_3O_7$ crystals are presented in Fig. 1. They were calculated from direct measurements of the complex transmission coefficients of the plane-parallel samples. The experiments were per-

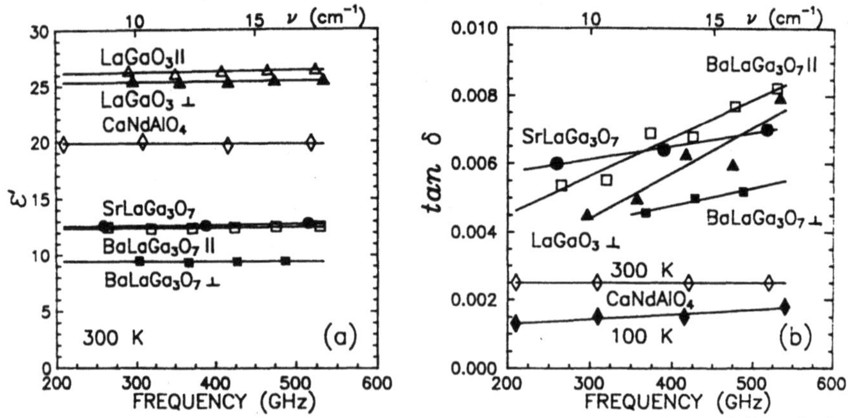

Fig. 1. Frequency dependence of the dielectric permittivity (a) and the loss tangent (b) for several dielectric substrates at 300 K. For comparison, the $LaGaO_3$ spectra and tan δ for $CaNdAlO_4$ at 100 K are also presented.

formed using the "Epsilon" spectrometer [5] in the frequency region 200 - 600 GHz, and in the temperature range 5 - 300 K. We have found that the ε' values are almost constant throughout the entire frequency range [see Fig. 1(a)], and are temperature independent. The dielectric losses at 300 K are plotted in Fig. 1(b). They are frequency dependent and, as is illustrated for the $CaNdAlO_4$ sample (see also Ref. [4]), decrease at lower temperatures. All the tested materials are highly anisotropic, as is shown for the (110) oriented $LaGaO_3$ and $BaLaGa_3O_7$ samples.

4. High-T_c Films

Bi-based (BSCCO) films were fabricated using a dc-magnetron sputtering technique previously developed in our laboratory [6]. The BSCCO-on-$CaNdAlO_4$ films of two different stoichiometries have been deposited from single $Bi_2Sr_2Ca_1Cu_2O_x$ and $Bi_2Pb_2Sr_2Ca_{2.15}Cu_{3.3}O_x$ targets. The $Bi_2Sr_2Ca_1Cu_2O_x$ film was subsequently annealed for 5 minutes at 900 °C in air, and furnace-cooled down to room temperature within 2.5 hours, while the $Bi_2Pb_2Sr_2Ca_{2.15}Cu_{3.3}O_x$ film was annealed at 869±0.5 °C for one hour. The resistance versus temperature characteristics of the above films are shown in Fig. 2. We observe sharp, single-step 80-K or 110-K transitions, indicating a dominance of either the $Bi_2Sr_2Ca_1Cu_2O_x$ or $(BiPb)_2Sr_2Ca_2Cu_3O_x$ phase in the films, as was verified by the X-ray diffraction. In a separate experiment [4], a 100-nm-thick, superconducting Y-Ba-Cu-O film was *in situ* deposited on a heated $CaNdAlO_4$ substrate. The film was highly epitaxial and exhibited zero-resistivity at 89 K.

Persistent attempts have been made to fabricate post-annealed BSCCO and Y-Ba-Cu-O films on $BaLaGa_3O_7$ and $SrLaGa_3O_7$ substrates. Unfortunately, the films were nonsuperconducting, apparently due to an observed severe diffusion of the Ba and Ga elements during the annealing process [3].

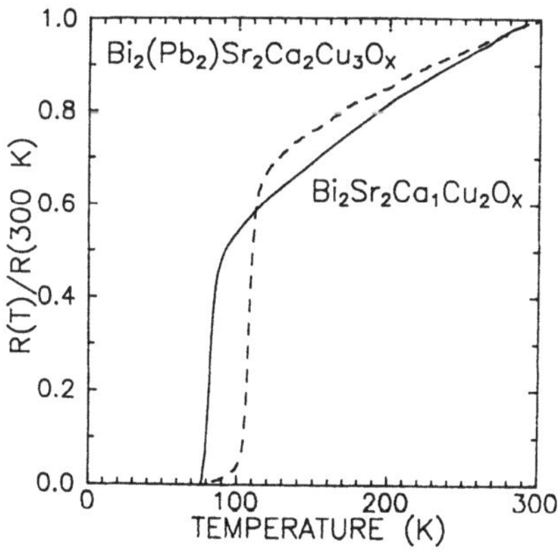

Fig. 2. Resistance vs. temperature dependence for the $Bi_2Sr_2Ca_1Cu_2O_x$ and $Bi_2Pb_2Sr_2Ca_{2.15}Cu_{3.3}O_x$ films, deposited on the $CaNdAlO_4$ substrate.

5. Conclusions

We have investigated the use of the single-crystal $CaNdAlO_4$ perovskite and tetragonal $BaLaGa_3O_7$ and $SrLaGa_3O_7$ crystals, as substrates for thin-film fabrication of high T_c superconductors and found that $CaNdAlO_4$ currently represents one of the best choices for microwave and far-infrared applications. The most unique feature of this material is that it combines chemical stability with a relatively good lattice match and very good dielectric properties at sub-THz frequencies. We believe that other materials belonging to the ABC_3O_7 and $ABCO_4$ classes are equally promising, and we intend to continue our research on their applications for high-T_c thin-film deposition.

References

[1] For a general review see e.g.: **Science and Technology of Thin-Film Superconductors** *ed. by* R. D. McConnell and S. A. Wolf (Plenum Press, New York, 1989).

[2] M. Berkowski, J. Fink-Finowicki, A. Pajączkowska, K. Mazur, and J. Sass, *Progr. in High Temp. Superconductivity,* vol. 24 (World Scientific Publ., May 1990).

[3] P. Gierłowski, W. Kula, M. Berkowski, A. Dąbkowski, R. Sobolewski, P. A. Saltykov, G. Jung, and A. Pajączkowska, *Progr. in High Temp. Superconductivity,* vol. 24 (World Scientific Publ., May 1990).

[4] M. Berkowski, A. Pajączkowska, P. Gierłowski, S. J. Lewandowski, R. Sobolewski, B. P. Gorshunov, G. V. Kozlov, D. B. Lyudmirsky, O. I. Sirotinsky, P. A. Saltykov, H. Soltner, U. Poppe, Ch. Buchal, and A. Lubig, *Appl. Phys. Lett.* - submitted for publication, and references therein.

[5] A. A. Volkov, Y. G. Goncharov, G. V. Kozlov, S. P. Lebedev, and A. M. Prokhorov, *Infrared Phys.,* **25**, 369 (1985).

[6] P. Gierłowski, S. Zarembiński, W. Kula, and R. Sobolewski, in: **Weak Superconductivity**, *ed by.* S. Benacka and M. Kedro (NOVA Scientific Publ., New York, March 1990).

On Incommensurate Phases in High-T_c Compounds

P.H.M. van Loosdrecht[1] *and A. Janner*[2]

[1]Research Institute for Materials, University of Nijmegen
[2]Institute for Theoretical Physics, University of Nijmegen, Toernooiveld, 6525 ED Nijmegen, The Netherlands

Abstract. A brief overview of incommensurate crystal structures, their classification and physical properties is given. The incommensurability in high T_c compounds is discussed for (a) the incommensurate structure of the $Bi_2Sr_2Ca_nCu_{n+1}O_{2n+6-\delta}$ and some related compounds (substitution of Pb for Bi or Yb for Ca and the Tl-based compounds), and (b) the incommensurate anti-ferromagnetic spin structure in $La_{2-x}Sr_xCuO_{4-\delta}$ which, in these compounds, seems to be related to the superconductivity.

1. Introduction

Incommensurate crystals lack lattice periodicity in at least one direction, but still have a perfect long-range ordered structure. Incommensurate phases are found in many of the compounds showing high T_c superconductivity. Some examples are the Bi-based compounds with generic formula $Bi_2Sr_2Ca_nCu_{n+1}O_{6+2n}$ and the $La_{2-x}Sr_xCuO_4$ compounds. The incommensurability is probably related to the rather complex structure of the high T_c superconductors (HT_cS).

The purpose of this paper is twofold. In the first part (sects. 2-4) a brief overview of incommensurate crystal structures is given. For an extensive review on incommensurability in crystals see Janssen & Janner [1]. The second part (sects. 5-6) is concerned with the incommensurate properties of some of the high T_c compounds.

2. Examples of incommensurate crystal structures

Incommensurability is a very general phenomenon, encountered in many physical systems. In this section some examples of incommensurate crystal structures are given. (a) Displacively modulated crystals. In these compounds the atomic positions are shifted by a modulation wave with a period incommensurate with respect to the average lattice. Examples are K_2SeO_4 and the Bi-based HT_cS [2,3]. (b) Charge and spin density wave systems. In these materials the charge or spin density wave can be incommensurate with respect to the lattice. Usually this results in a incommensurable crystal structure of type (a). Examples are $TaSe_2$ (charge density wave) and $La_{2-x}Sr_xCuO_4$ [4] (spin density wave). (c) Occupation modulation. This is the case when the probability for

Electronic Properties of High-T_c Superconductors and Related Compounds
Editors: H. Kuzmany · M. Mehring · J. Fink

occupation of a site by a certain atom is incommensurately modulated. An example is $NaNO_2$. (d) Mono-layers. Mono-layers adsorbed on a substrate can have an ordering into a two dimensional lattice incommensurate with the underlying substrate lattice. An example is a mono-layer of W adsorbed on Mo. (e) Intergrowth systems. These are systems consisting of two (or more) subsystems with mutually incommensurate lattice periodicities. An example is $Hg_{3-\delta}AsF_6$ (f) Quasi-crystals. These crystals show a diffraction pattern with a point symmetry incompatible with lattice periodicity (i.e. icosahedral). Yet, the diffraction peaks are sharp, indicating a long range order. An example is shechtmanite (Al-Mn).

3. Structure description for a modulated crystal

Ordinary crystals can be described with a density function having a Fourier decomposition $\rho(\mathbf{r}) = \sum_{\mathbf{k}\in\mathbf{\Lambda}^*} \hat{\rho}(\mathbf{k})e^{i\mathbf{k}\cdot\mathbf{r}}$, where **k** can be indexed by three integer indices (h,k,l) and runs over the reciprocal lattice $\mathbf{\Lambda}^*$.

In incommensurate crystals this relation no longer holds due to the lack of lattice periodicity. Instead one can write for the Fourier decomposition $\rho(\mathbf{r}) = \sum_{\mathbf{k}'\in\mathbf{M}^*} \hat{\rho}(\mathbf{k}')e^{i\mathbf{k}'\cdot\mathbf{r}}$, where $\mathbf{k}'$ now runs over the Fourier module $\mathbf{M}^*$, and is indexed by 3+d integers (h,k,l,m,...), where d is the dimension of the modulation. For a one-dimensionally modulated structure (d=1) with modulation wave-vector **q** the module $\mathbf{M}^*$ is freely generated by $\mathbf{a}^*$, $\mathbf{b}^*$, $\mathbf{c}^*$and **q**.

Common concepts in solid state physics based on the lattice periodicity (i.e. Brillouin zone, Bloch theorem, etc.) appear to be no longer valid for incommensurate crystals. One can, however, in a simple manner restore the lattice periodicity in these compounds by embedding the module $\mathbf{M}^*$ as a 3+d dimensional lattice in a higher dimensional reciprocal space. This leads naturally to a 3+d dimensional crystal structure with lattice periodicity, having a *one to one correspondence* with that of the real crystal.

4. Physical properties of incommensurate crystals

Incommensurability has some influence on virtually every physical property of a crystal. The selection rules resulting from symmetry, for example, are governed by the superspace groups [5]. Other effects are: satellite peaks appearing in diffraction patterns, which must be indexed by 3+d integers [6,7] instead of the usual 3, satellite faces in the crystal growth forms, additional modes in optical spectra due to a coupling of $\mathbf{K}=m\mathbf{q}$ with $\mathbf{k}=0$ modes, amplitudons or phasons, changes in lineshapes of NMR and ESR resonances because of the modulation of the local site environment and finally the presence of optical rotation in crystals with a centrosymmetric average structure [8].

5. Incommensurability in high T_c superconductors

$Bi_2Sr_2Ca_nCu_{n+1}O_{2n+6-\delta}$. There have been numerous reports on the incommensurate nature of the $Bi_2Sr_2Ca_nCu_{n+1}O_{2n+6-\delta}$ (usually labeled 2 2 n n+1) family of HT$_c$S [2,3,9-14]. For the 2201 compound the incommensurate wave-vector lies in the ($\mathbf{a}^*$,$\mathbf{b}^*$)-plane (**a** is the long axis), and is found to be $\mathbf{q}_1 \approx \alpha \mathbf{a}^* + 0.21\mathbf{c}^*$. The value α is also continuously varying (thus non-rational), and is probably strongly depending on the Sr-content of the 2201 compound (values have been reported ranging from 0.33 to 1.05) [10,11]. In 2212 and 2223 the component of the incommensurate wave-vector in the $\mathbf{c}^*$ direction is the same as in 2201. The component lying in the $\mathbf{a}^*$ direction, however, is now commensurate [11,12].

The structure is found to be temperature independent, indicating that no structural phase transition to a commensurate structure takes place. The Tl-based compounds (where Bi, Sr are replaced by Tl, Ba) also exhibit a weak incommensurate structure. In this case, however, the incommensurability is only observed at low temperatures [3].

The amplitude of the modulation is the largest for the Bi-atoms in the Bi-O layers (up to 0.7 Å in 2201), and decreases for increasing n. Observations of up to fourth order satellite reflections may indicate that the modulation is not purely sinusoidal. A discussion on the possible origins of the modulation has been given by Zandbergen et al. [3].

There have been several studies on the effects of substituting Bi with Pb [11-13]. It has been shown that in 2201 the modulation disappears when Bi is partially substituted by Pb [10]. In 2212 and 2223 the substitution of Pb has a more interesting effect. In this case a second incommensurate modulation (*coexisting* with $\mathbf{q}_1$) appears with $\mathbf{q}_2 \approx 0.11 \ldots 0.16\mathbf{c}^*$, depending on the Pb-content [11-13]. At high Pb concentrations the $\mathbf{q}_1$-modulation disappears, and only the $\mathbf{q}_2$-modulation remains [12].

Casanove et al. [14] studied the influence of the substitution of Ca^{2+} by Yb^{3+} on the incommensurability in 2212. Their results show a decreasing $\mathbf{q}_1$ upon increasing the Yb-content, indicating a connection between the electronic structure and the modulation.

$La_{2-x}Sr_xCuO_{4-\delta}$. Neutron scattering experiments on $La_{2-x}Sr_xCuO_{4-\delta}$ compounds have revealed an incommensurate anti-ferromagnetic spin density wave (ISDW) for at least $0.08 < x < 0.11$ [15-17]. In this x range these compounds also exhibit superconducting behaviour. Experiments by Thurston et al. [15] on both superconducting (x=0.11) and non-superconducting (x=0.06) samples indicate that the ISDW is only present in superconducting samples, and this suggests a possible connection between the superconductivity and the ISDW. An explanation of the ISDW in terms of a Hubbard model with a nearly half filling magnetic band is given by Machida [18].

6. Concluding remarks

The incommensurate nature of many HT$_c$S has a subtle but distinct effect on the electronic and magnetic properties of these compounds. A direct relation between incommensurability and superconductivity is, however, doubtful, although in the $La_{2-x}Sr_xCuO_{4-\delta}$ compounds the ISDW seems to be a prerequisite for the superconducting state. However, simply disregarding incommensurability can possibly lead to erroneous conclusions. Quite independently, it is interesting to study the influence of incommensurability on dynamical, electronic and magnetic properties of high temperature superconductors.

References

[1]T. Janssen, A. Janner, Adv. of Phys. **36**, 519 (1987), and references therein.

[2]Y. Gao, P. Coppens, M.A. Subramanian, A.W. Sleight, Science **241**, 954 (1988).

[3]H.W. Zandbergen, W.A. Groen, F.C. Mijlhoff, G. van Tendeloo, S. Amelinckx, Physica C **156**, 325 (1988), and references therein.

[4]H. Yoshizawa, S. Mitsuda, H. Kitazawa, K. Katsumata, J. Phys. Soc. Jpn. **57**, 3686 (1988).

[5]P.M. de Wolff, T. Janssen, A. Janner, Acta Cryst. A **37**, 625 (1981).

[6]A. Janner, T. Janssen, Phys. Rev. B **15**, 643 (1977).

[7]P.M. de Wolff, Acta Cryst. A **30**, 777 (1974).

[8]H. Meekes, A. Janner, Phys. Rev. B **38**, 8075 (1988).

[9]B.C. Chakoumakos, P.S. Ebey, B.C. Sales, E. Sonder, J. Mater. Res. **41**, 767 (1989).

[10]Y. Gao, P. Lee, J. Ye, P. Bush, V. Petricek, P. Coppens, Physica C **160**, 431 (1989).

[11]O. Eibl, to appear in Physica C.

[12]L. Pierre, J. Schneck, J.C. Tolèdano, C. Daguet, Phys. Rev. B **41**, 766 (1990).

[13]M. Clin, K. El Boussiri, J. Rezpski, D. Morin, J. Schneck, Revue Phys. Appl. **25**, 13 (1990).

[14]M.J. Casanove, P. Baules, E. Snoeck, C. Roucau, Physica C **159**, 461 (1989).

[15]T.R. Thurston, R.J. Birgeneau, M.A. Kastner, N.W. Preyer, G. Shirane, Y. Fujii, K. Yamada, Y. Endoh, K. Kakurai, M. Matsuda, Y. Hidaka, T. Murakami, Phys. Rev. B **40**, 4585 (1989).

[16]R.J. Birgeneau, Y. Endoh, K. Kakurai, Y. Hidaka, T. Murakami, M.A. Kastner, T.R. Thurston, G. Shirane, K. Yamada, Phys. Rev. B **39**, 2868 (1989).

[17]G. Shirane, R.J. Birgeneau, Y. Endoh, P. Gehring, M.A. Kastner, K. Kitazawa, H. Kojima, I. Tanaka, T.R. Thurston, K. Yamada, Phys. Rev. Lett. **63**, 330 (1989).

[18]K. Machida, Physica C **158**, 192 (1989).

Substitutional and Oxygen Doping in $YBa_2Cu_3O_y$

H. Kuzmany[1], *T. Pichler*[1], *M. Matus*[1], *A. Hanninger*[2], *P. Rogl*[2], *and S. Pekker*[3]

[1]Institut für Festkörperphysik, Universität Wien, Strudlhofgasse 4, A-1090 Wien, Austria and
Ludwig Boltzmann Institut für Festkörperphysik, Kopernikusgasse 15, A-1060 Wien, Austria
[2]Institut für Physikalische Chemie, Universität Wien, Währingerstr. 42, A-1090 Vienna, Austria
[3]Institute of Physics, Hungarian Academy of Science, Budapest, Hungary

Abstract. We report on *Cu/Ga* and *Y/Ca* substitutional and oxygen deficiency doping in the system $YBa_2Cu_3O_y$. The extra electrons in *Ga* doped samples were observed to compensate oxygen doping whereas the extra holes in *Ca* doped samples supported oxygen doping. T_c decreased in both cases with substitution but increased initially with reduction in the *Ca*-doped system. A balance between c-parameter behavior and hole doping provides a wide stability range in this system.

1. Introduction

As compared to the other high temperature superconductors $YBa_2Cu_3O_y$ is very sensitive to the oxygen content. Oxygen reduction leads to a metal-insulator transition around y = 6.5. In the same concentration range an orthorhombic to tetragonal phase transition and an anomaly in the c-parameter occurs [1,2]. On the other hand the superconducting properties of this material can also be changed by substitution which may cause additional p- or n-doping (compensation). Among the possible n-dopands, *Cu/Ga* substitution is of some interest [3]. Beside the reduction of hole concentration, there is an influence on the crystal structure. *Ga* concentrations higher than 5% switch the X-ray pattern of the material to a tetragonal structure [4]. However, electron microscopy still exhibits orthorhombic microdomains. Additional p-doping may be established by *Y/Ca* substitution [5].

In order to study the nature of the doping and structural effects on the superconductivity, oxygen deficiency doping was performed in both the *Cu/Ga* and the *Y/Ca* substituted samples.

All samples were characterized by X-ray diffraction and conductivity measurements. Raman and IR measurements are reported in [6].

2. Experimental

The ceramic superconductors were prepared by solid state reactions of stoichiometric amounts of the constituents. From the well oxidized samples, defined oxygen stoichiometries were obtained by using the equilibration technique de-

Springer Series in Solid-State Sciences, Vol. 99
Electronic Properties of High-T_c Superconductors and Related Compounds
Editors: H. Kuzmany · M. Mehring · J. Fink

scribed elsewere [7]. The equilibration process was performed at 420°C for 40 hours, followed by a linear decrease of temperature to room-temperature within 10 hours.

X-ray analysis of the different samples was performed with a Huber-Guinier camara using CuK_α radiation. The lattice parameters were calculated using at least 15 reflections. The temperature dependence of the conductivity was measured between 4 and 300 K by the standard four-probe technique.

3. Results

Fig.1 shows characteristic changes of the c-parameter and T_c with oxygen concentration in unsubstituted ceramics [2]. The turning point in the c-parameter coincides with the loss in superconductivity. We have reported previously similar results for the *Ga*-doped system, though the increase of the c-parameter and the drop in T_c was more continuous for this system [3]. X-ray diffraction from the *Ga* doped system showed a well expressed line broadening for reflections around $2\Theta = 47^o$ characteristic for the O/T-transition. The broadening appeared immediately after equilibration and disappeared with a reduction down to $O_{6.7}$. Simultaneously with the reflex broadening ρ(T) became activated. In the *Ca*-doped samples one can observe a decrease of the resistance and a decrease of T_c with increasing *Ca* content. This result suggests that *Ca* replaces *Y* and thus dopes the system, even though the changes in the lattice parameters for 16 molpercent substitution were only 0.1 % for the c-axis and less than that for the a,b-axis. Reducing the *Ca* doped system resulted in a weak delay of the O/T-phase transition as shown in Fig.2(a). The behavior of the c-parameter is anomalous (Fig.2(b)). It increases first but decreases with progressing reduction. On the other hand, in the T_c vs y-diagram (Fig.3) the *Ca*-doping caused an anomalous strong delay of the T_c-quenching with reduction (arrow) as compared to the undoped system and in particular as compared to the *Ga* doped system.

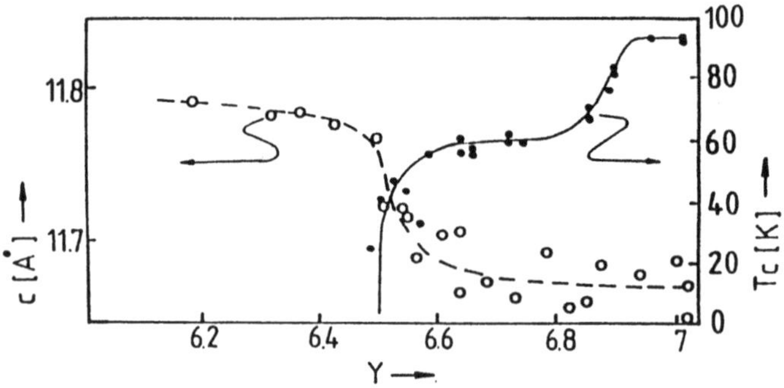

Fig.1: c-axis parameter and T_c vs y for $YBa_2Cu_3O_y$.

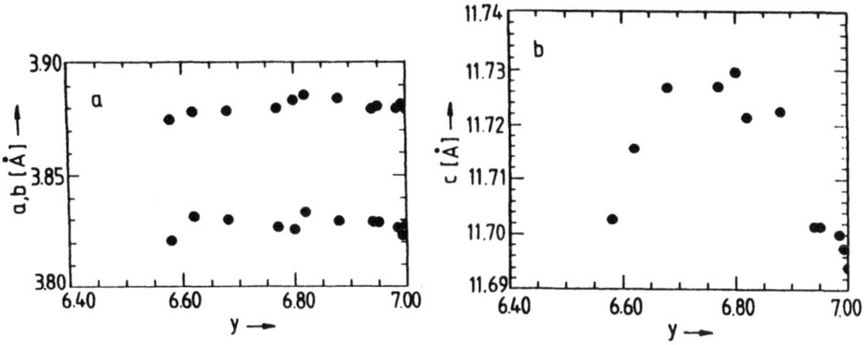

Fig.2: Unit cell parameters vs y for $Y_{0.84}Ca_{0.16}Ba_2Cu_3O_y$.

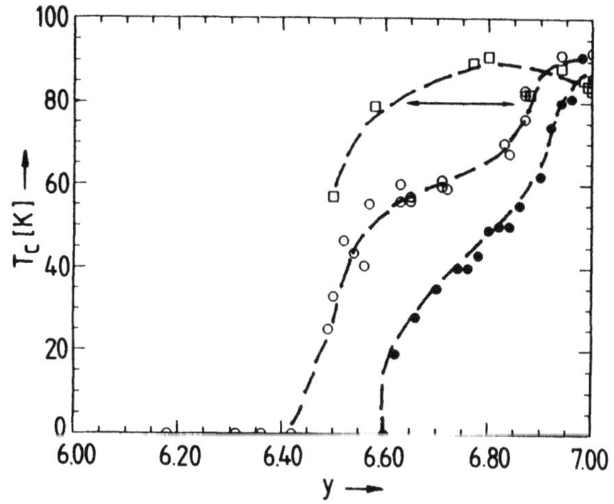

Fig.3: T_c vs y for: $YBa_2Cu_3O_y$ (○), $YBa_2(Cu_{0.94}\text{-}Ga_{0.06})_3O_y$ (●), and $Y_{0.84}Ca_{0.16}Ba_2\text{-}Cu_3O_y$ (□).

4. Discussion

The *Ga*-substituted samples behave unusually in the sense that they change to a semiconductor-like $\rho(T)$ after the annealing process, even without any detectable loss of oxygen. This effect is correlated with a characteristic line broadening in the X-ray diffraction and thus originates definitely from the bulk of the sample. Internal strains are likely to be the reason for this effect which finally leads to a localization of electron wavefunctions and thus to an activated $\rho(T)$. The strain may either originate from a rearrangement of the domain structure during the equilibration process or from point defects generated by e.g. reaction of the sample with traces of water. However, as long as internal strains and localization are considered as responsible for the activated behaviour of the carriers the basic origin of the strain is not important [6].

Concerning the relation between doping on the one side and T_c and c-parameter on the other side we observed a very well expressed c-axis anomaly in the unsubstituted samples which clearly coincides with the quenching of superconductivity. The *Ga*-substituted samples behave similarly in a sense that T_c decreases with increasing c-parameter. However, no discontinuous behaviour could be observed. The shift in the T_c vs y curve corresponds exactly to the extra electron provided by the *Ga* atom [3]. The behaviour of the *Ca*-doped samples is reversed. With increasing c-parameter, T_c increases until, for lower oxygen concentrations, T_c starts to decrease but only very slowly. The c-parameter change also reverses sign. The T_c vs y relation for these samples in general and its anomalous strong shift to lower oxygen concentrations in particular can be understood from two processes. The *Ca*-substitution provides the system with additional holes and thus compensates the loss of hole doping from the chains to the plains resulting from the increasing c-parameter with decreasing y or more generally from the decreasing oxygen concentration. Around y=6.8 T_c should drop to the lower plateau, but it does not because the c-parameter decreases as well and thus supports hole doping from the chains. Thus, c-axis behaviour and *Ca*-doping stabilize optimum conditions for superconductivity for a quite wide range of oxygen deficiency.

Acknowledgement. This work was supported by the the Fond zur Förderung der Wissenschaftlichen Forschung in Austria (grant P7059).

References

1. R.J. Cava, B. Batlogg, K.M. Rabe, E.A. Rietman, P.K. Gallagher, and L.W. Rupp Jr., Pysica C 156, (1988) 523.

2. G. Oszla'nyi, S. Pekker, Gy. Hutiray, E. Zsoldos, and L. Mihaly, Journ. of modern Physics B 1, (1988) 1305.

3. M. Matus, H. Kuzmany, B. Sumetsberger and T. Pichler, Physica C 162–164, (1989) 967.

4. G. Xiao, M.Z. Cieplak, A. Gavrin, F.H. Streitz, A. Bakhshai and C.L. Chien, Phys. Rev. Lett. 60, (1988) 1446.

5. J.J Neumeier, T. Bjørnholm, M.B. Maple, and I.K. Schuller, Phys. Rev. Lett. 63, (1989) 2516.

6. M.Matus, B.Sumetsberger, and H. Kuzmany, in this volume.

7. H. Kuznany, M. Matus, E. Faulques, S. Pekker, Gy. Hutiray, E. Zsoldos, and L. Mihaly, Physica Scripta T 25, 82 (1989).

The Effect of Cu/Zn Substitution upon Superconductivity in $Nd_{1.85}Ce_{0.15}CuO_4$

G. Hilscher[1], *S. Pöllinger*[1], *M. Forsthuber*[1], *N. Pillmayr*[1], *K. Remschnig*[2], *P. Rogl*[2], *M. Reissner*[3], *W. Steiner*[3], *and P. Knoll*[4]

[1]Institut für Experimentalphysik, TU Wien, Wiedner Hauptstr. 8–10, A-1040 Wien, Austria
[2]Institut für Physikalische Chemie, Universität Wien, Währingerstr. 42, A-1090 Wien, Austria
[3]Institut für Angewandte und Technische Physik, TU Wien, A-1040 Wien, Austria
[4]Max-Planck-Institut für Festkörperforschung, Heisenbergstr. 1, D-7000 Stuttgart 80, Fed. Rep. of Germany and Institut für Experimentalphysik der Universität Graz, Universitätsplatz 5, A-8010 Graz, Austria

Abstract. We report on the structural chemistry, dc and ac magnetic, magnetoresistance and Raman measurements of $Nd_{1.85}Ce_{0.15}(Cu_{1-x}Zn_x)O_4$ where Cu/Zn substitution suppresses superconductivity with an initial rate of -4.5K/mol% Zn.

Members of the new class of electron doped copper oxide superconductors $(R_{1-y}Ce_y)_2CuO_4$ (with R = Pr, Nd, Sm and Eu) are unique in that the charge carriers are electrons. They crystallize in the tetragonal T' - Nd_2CuO_4 structure where the copper atoms are in square planar coordination with oxygens forming the Cu-O sheets [1-3]. Since this structure exhibits only one Cu site in the Cu-O plane, Cu/Zn substitution can be explored without the complication of two different lattice sites for Cu, as is the case in YBCO. $(Nd,Ce)_2CuO_4$ is a magnetic superconductor where antiferromagnetism of the R ions coexists with superconductivity at low temperatures [2]. This makes the system attractive to study the mechanisms of superconductivity in ceramic cuprates where the magnetic and superconducting ordering temperatures are close to each other.

X-ray analysis of the samples $Nd_{1.85}Ce_{0.15}Cu_{1-x}Zn_xO_4$ proved the existence of an extended solid solubility up to values slightly larger than x= 0.05 with a random distribution of (Nd,Ce) and (Cu,Zn) on their respective lattice sites.

The superconducting fraction estimated from the comparison of the diamagnetic signal with that of a Nb superconductor of the same geometry in fields of 80 A/m (Fig.1) is 48 % , 43 % and 3-4 % for x = 0.0, 0.01 and 0.025, respectively. To determine the superconducting fraction an ac field of 80 A/m was used, since both types of samples (the pellets with 2.5 mm diameter and bar shaped samples cut from cylindrical samples

Springer Series in Solid-State Sciences, Vol. 99
Electronic Properties of High-T_c Superconductors and Related Compounds
Editors: H. Kuzmany · M. Mehring · J. Fink

of 8 mm diameter) show that at this ac field the diamagnetic response is mainly due to the screening of the grains.

The temperature dependence of the magnetization measured in fields of 0.16 mT confirms the results of the ac susceptibility measurements. The field dependence of the magnetization at 4.2 K is strongly curved for all x-values, pointing to a still existing magnetic correlation above T_N. The influence of the Cu/Zn substitution on the bulk moment at 4.2K and 6T is not significant (15.5, 16.1, 15.8 and 15.0 emu/g for x = 0.0, 0,01, 0.025 and 0.05). From the reduction of the width of the hysteresis loop a drastic decrease of the critical current density of about one order of magnitude for the 1% Zn-doped sample can be concluded. Furthermore H_{c1} and H_{c2} at 4.2 K are also reduced by one order of magnitude upon a Cu/Zn substitution of 1 mole% [4]. The measurements of the relaxation of the magnetization indicate that the number of effective pinning centers for both the weak links and the grains increase with the Zn content.

In agreement with magnetic data the resistivity mesurements show that Zn supresses superconductivity at an initial rate of 4.5 K/mol% Zn. For the sample with 2.5% Zn we still observe a finite resistivity at 4.2 K, since the superconducting fraction of about 3%, as estimated from the ac susceptibility (Fig. 1), is clearly below the percolation limit to obtain zero resistivity.

The electrical resistivity in various external fields is shown in Fig.2. The zero field resistivity curve has a sharp transition (ΔT_c (10%-90%) = 3.4K), however, broadens quickly with rising field, and after an initial drop of the resistivity a shoulder occurs even in small fields. The shoulder is most pronounced in the field range 10 mT $\leq H_{ext} <$ 1 T. We attribute this anomaly to a depression of the coupling currents, which is also observed in the ac susceptibility as the ac field amplitude is raised above 0.5 mT [4].

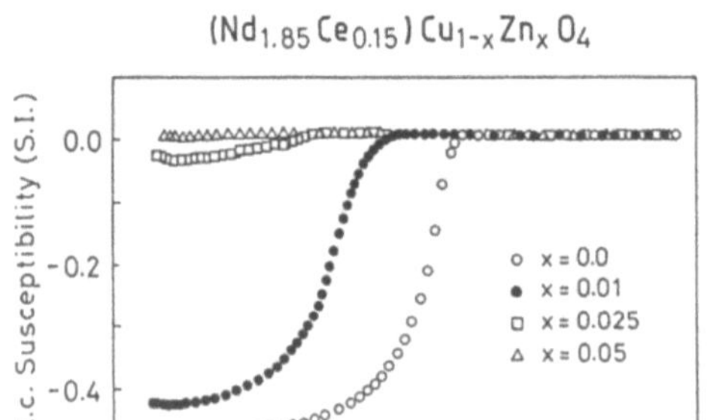

Fig.1: ac susceptibility as a function of temperature for various Zn concentrations. ac field (rms) = 80 A/m (0.1 mT), frequency = 1kHz.

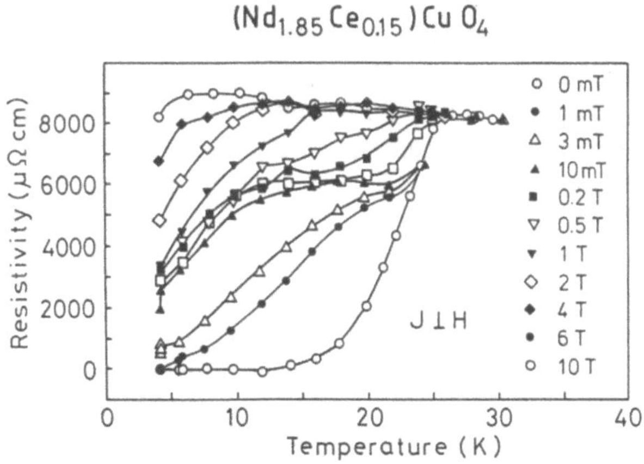

Fig.2: Resistivity as a function of temperature for various external fields, as labeled, with the transport current transverse to the field.

The overall behaviour of ρ(T,H) in sintered $(Nd,Ce)_2CuO_4$ deviates from that usually observed for ceramic high T_c materials, where the resistive transition is gradually washed out while the shift of the superconducting onset temperature is of the order of only -5% of T_c at 10T [5]. In $(Nd,Ce)_2CuO_4$, apart from the shoulder at low fields, the shift of T_c onset is -80% at 10 T. According to magnetization and heat capacity measurements [4] we suppose that magnetic correlations persist up to temperatures of about 10K - 15K. This together with the rather low superconducting transition temperature may account for the rapid suppression of the coupling currents in weak external fields.

The resistivity **below** T_c is found to be strongly current dependent. The most striking feature is that with increasing transport current the kink below T_c develops into a local minimum, and for transport currents larger than 100mA ρ(T) increases rapidly beyond values extrapolated from the normal state resistivity down to low temperatures. The behaviour described is reminiscent of ultrathin superconducting films above and below the percolation limit [6,7]. We suggest that the intergrain coupling can be suppressed not only by external fields as shown in Fig.2 but also with a transport current larger than a critical value, leading to a situation similar to that in random superconducting networks below the percolation limit [6].

The main characteristics of the Raman spectra at 230, 330, and 480 cm^{-1}, which belong to motions of Nd and oxygen (at the Wyckoff site 4d) in the z and x-y directions, remain quite stable with Zn substitution, demonstrating that the crystal structure is hardly affected, in agreement with X-ray measurements. An exception is the

480 cm^{-1} feature, which systematically varies with rising Zn concentrations. Besides an increase of Raman intensities indicating changes in the electronic properties, the frequency of the oxygen E_g vibration in the x-y plane rapidly decreases for low Zn contents, levelling off at above 2.5 mole % . This behaviour can be interpreted as a displacement of the oxygen 4d site out of its original position with increasing values of x. Due to the small variation of the X-ray structure factors with respect to the shift of the 4d oxygen, this effect is, however, far too small to be resolved with X-ray techniques.

Zn substitution has obviously a less dramatic effect upon the depression of superconductivity (-4.5 K/mol% Zn) than in YBCO [8] and LSCO [9], which is in good agreement with results recently obtained by Felner et al. [10]. Moreover we stress that our recent experiments on Cu/Co and Cu/Ni substitution yield a dramatic depression of superconductivity; for doping levels as low as 1 mole % superconductivity is already suppressed, which is in perfect accordance with the results on Cu/Ni substitution reported by Felner et al. [10]. This appears to be the reverse situation in comparison with the hole doped $(La,Sr)_2CuO_4$ with respect to Zn and 3d metal substitution. In the latter system the magnetic 3d metal substitutions have less impact upon the depression of T_c than Zn. This different behaviour of the electron and hole doped superconductors with respect to substitutions indicates that either the order parameter has a different symmetry or the larger coherence volume in $(Nd,Ce)_2CuO_4$ plays a significant role. In this context it is worth noting that the coherence length ξ of NCCO of approximately 90 Å [11] is in between the conventional superconductors (generally larger than 400 Å) and the hole doped superconductors (smaller than 20 Å). In view of the drastic difference between the effect of magnetic and non-magnetic impurities upon the suppression of superconductivity in NCCO we might expect this material to be considerably closer to the conventional BCS superconductors. In the case of a small coherence volume non-magnetic scattering should have a similar effect upon the T_c suppression as magnetic impurity scattering, while a considerably distinct effect is to be expected for a larger coherence volume .

Acknowledgement. This work was supported by the Austrian Science Foundation Fonds under projects 7056, 7060, 7061 and 7065. P.R. wishes to express his gratitude to Auer Remy Vertriebsges. Seltene Erden GmbH, Hamburg FRG, for a generous gift of rare earth oxide materials.

References

1) Y. Tokura, H.Tagaki and S.Uchida, Nature 337 (1989) 345
2) C.L.Seaman, N.Y.Ayoub, T.Bornholm, E.A.Early, S.Ghamaty, B.W.Lee, J.T.Markert, J.J.Neumeier, T.S.Tsai and M.Maple, Physica C 159 (1989) 391
3) H.Müller Buschbaum and W.Wollschläger, Z. anorg. allg. Chem. 414 (1975) 76
4) G.Hilscher, S.Pöllinger, M.Forsthuber, N.Pillmayr, K.Remschnig, P.Rogl, M.Reissner, W.Steiner, and P.Knoll, Physica C (1990)
5) A.P. Malozemoff, Physical Properties of High Temperature Superconductors I, Ed. D.M. Ginsberg, World Scientific Publishing Co., London (1989) p.72
6) B.G.Orr, H.M.Jaeger, and A.M.Goldmann, Phys.Rev.B 32 (1985) 7586
7) E.Simanek, Phys. Rev. B 25 (1982) 237
8) G.Hilscher. N.Pillmayr, R.Eibler, E.Bauer, K.Remschnig and P.Rogl, Z. Phys. B 72 (1988) 461
9) K.Remschnig, P.Rogl, E.Bauer, R.Eibler, G.Hilscher, H.Kirchmayr and N.Pillmayr, in High Tc Superconductors, Ed. H.W.Weber, Plenum Press, New York (1988) p. 99
10) I.Felner, D. Hechel and U. Yaron, preprint 1990
11) J.P.Sröbel, M.Klauda, M.Lippert,B.Hensel, G.Saemann-Ischenko, W.Gerhäuser, H.W.Neumüller, W.Ose and K.F.Renk, Physica C 162-164 (1989) 1209

Hole Density in the High-Temperature Superconductor (Bi,Pb)SrCaCuO

W. Lang, H. Jodlbauer, P. Schaffarich, and H. Kuzmany

Ludwig Boltzmann Institut für Festkörperphysik, Kopernikusgasse 15, A-1060 Wien, Austria and
Institut für Festkörperphysik, Universität Wien, Strudlhofgasse 4, A-1090 Wien, Austria

Abstract. The complex magnetic a.c. susceptibility, the electrical resistivity and the Hall effect of the high-temperature superconductor (Bi,Pb)SrCaCuO were investigated in a series of samples in which Bi was partially replaced by Pb. The hole density did not exhibit a direct relation to the Pb content of the samples but is essentially invariable in samples with similar superconducting properties but different Pb-concentration. The results indicate typical carrier concentrations for both the low-Tc (Tc=75 K) and the high-Tc (Tc=108 K) phases and are incompatible with a charge transfer from (Bi,Pb)O layers to the CuO_2 sheets induced by lead doping. This indicates that the lead addition mainly influences the synthesis of these compounds resulting in an enhanced formation of the high-Tc phase at about 20 per cent Pb-substitution.

1. Introduction

In the course of achieving higher critical temperatures in ceramic superconductors it was empirically found that an increasing number of CuO_2-layers per unit cell leads to enhanced transition temperatures. The systems $Tl_2Ba_2Ca_{n-1}Cu_nO_x$ and $Bi_2Sr_2Ca_{n-1}Cu_nO_x$ each form a series of compounds in which n=1,2,3 denotes the number of CuO_2-planes. Unfortunately it was not possible to synthesize a pure n=3 compound in the Bi-based system. Yet it was observed /1/, that a partial substitution of lead promotes higher Tc's and samples with sharp superconducting transition and zero resistance at 112 K could be prepared, containing only very small amounts of impurity and of the n=2 phase /2/.

Although the preparation condition which favours single phase materials with Tc about 110 K were studied systematically to find out optimum values for the nominal composition and the sinter parameters, it remains unclarified if the lead addition has a "doping" effect on the normal state transport properties of these compounds. In order to investigate this behaviour we performed Hall measurements on a series of ceramic samples of the nominal composition $(Bi_{1-y}Pb_y)_2Sr_2Ca_2Cu_{3.6}O_x$ with systematically varied lead substitution from y=0.0 to y=0.4.

2. Experimental

The samples were prepared by the usual solid state reaction method by sintering the ground oxides of the constituents with the nominal composition $(Bi_{1-y}Pb_y)_2Sr_2Ca_2Cu_{3.6}O_x$ (BPSCCO) for various values of y. Reaction temperature and time was 850 °C and 200 to 300 hours in air, respectively. The samples were furnace cooled afterwards. An elemental analysis confirmed the nominal composition except that an average loss of lead of about 10 % during the reaction was found.

The samples were characterized by both x-ray diffraction and magnetic field dependent a.c. susceptibility measurements. The latter technique,

Springer Series in Solid-State Sciences, Vol. 99
Electronic Properties of High-T_c Superconductors and Related Compounds
Editors: H. Kuzmany · M. Mehring · J. Fink

which was previously described in more detail /3/, yields an estimation of the superconducting volume fraction of granular compounds.

Electrical conductivity and Hall effect were measured in the standard 5-probe arrangement. The sample was cut to the desired shape by sand blasting. Contacts were made with platinum wire and platinum paste and heated overnight in flowing oxygen at 300 °C resulting in a contact resistance better than 1Ω . The density of the a.c. current was 1.75 A/cm^2 and the Hall voltage was detected by a two channel lock-in amplifier. By sweeping the magnetic field, generated by an electromagnet, repeatedly in both polarities at a fixed temperature and averaging the signal, a resolution better than 10 nV was obtained.

3. Results

The imaginary part χ'' and the real part χ' of the complex a.c. susceptibility $\chi'-i\chi''$ of a series of BPSCCO samples with various lead content y is shown in Figures 1a and 1b, respectively. Three qualitatively different behaviours can be distinguished in the susceptibility curves. The onset of superconductivity in the sample with y=0.0 mainly is at 75 K, (although a very small diamagnetic signal could be detected at temperatures up to 100 K) increasing in a second step below 50 K to χ'=-1.0. The onset of χ'' is at about 60 K, just at the same temperature at which the

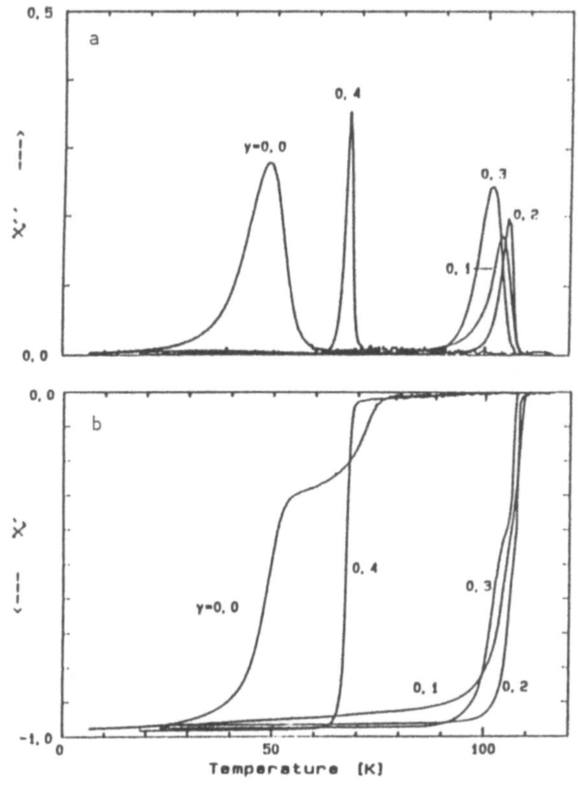

Fig.1: a) Imaginary part χ'' and b) real part χ' of the complex a.c. susceptibility $\chi'-i\chi''$ of a series of BPSCCO samples with various lead content y vs. temperature.

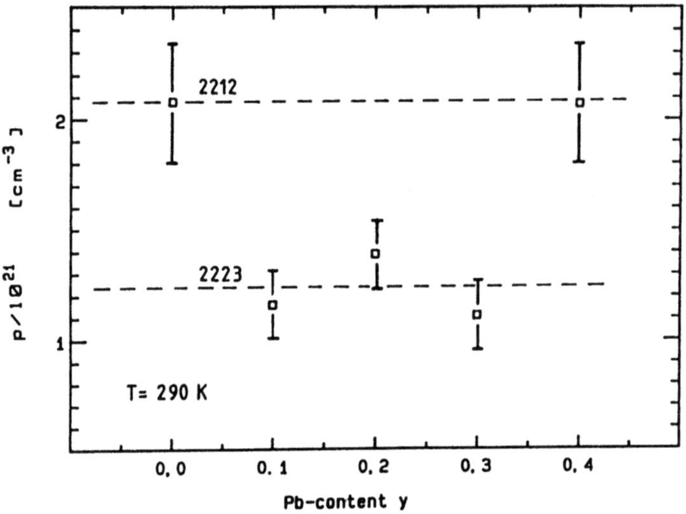

Fig.2: Hall number $p=1/eR_H$ in BPSCCO vs. Pb content y. The broken line represents mean values for the 2212 and 2223 phases, respectively.

resistivity reaches zero. The peak in χ'', associated with the breakdown of superconductivity of weak links in the sample, is at 50 K, indicating that the second phase observed is due to the intergrain coupling and is not a bulk superconducting phase. This result and x-ray diffraction demonstrate that the sample consists primarily of the 2212 phase with a small admixture of 2223 and impurity phases. The samples with y=0.1 to y=0.3 exhibit mutually a very similar behaviour with a sharp superconducting transition at 108 K, observed in both χ' and χ''. They mainly consist of the 2223 phase, where the y=0.2 sample has the best superconducting features. The sample with y=0.4 consists essentially of the 2212 phase with an already significant admixture of Ca_2PbO_4.

The characterization of the samples may be summarized in that the y=0.0 and the y=0.4 samples are for the most part 2212 phase and the samples y=0.1 to y=0.3 are 2223 phase.

The Hall voltage in all samples of BPSCCO is positive (hole like) and linear with the magnetic field B up to 1.4 T meaning a constant Hall coefficient R_H, which indicates closed Fermi surfaces. The Hall number $p=1/eR_H$, which would equal the hole concentration if we assume a simple model with one parabolic band, is plotted vs. degree of lead substitution y in Fig.2. There is no systematic variation of the Hall number p as a function of the lead content y. On the contrary, two typical values of p are observed for both the 2212 (y=0.0 and y=0.4), $p=2.1\times10^{21}$ cm^{-3}, and 2223 phase (y=0.1 to 0.3), $p=1.2\times10^{21}$ cm^{-3}, compounds.

4. Discussion

It was suggested /4/, that the BiO-layers act as an hole reservoir for the conduction process in the CuO_2-sheets but do not contribute themselves to the electronic transport. This situation is quite different to the one in YBaCuO superconductors, where the holes are supplied by the oxygen deficiency in the CuO_2-layers, which can also be concluded from the strong dependence of superconducting properties on the oxygen content in YBaCuO and the almost absence of this effect in the BPSCCO compounds. On the other hand it is evident that a substitution of Bi^{3+} by Pb^{4+} can act as a charge donor to the BiO-planes and, assuming a charge transfer to the CuO_2-layers, may change the hole concentration in these compounds. The unusual temperature dependence of the Hall coefficient R_H was explained by a multiband model with an exterior carrier source /5/. From our results, however, it can be deduced that changing valencies in the BiO planes by introducing lead has no proportional effect on the transport properties in the CuO_2-sheets, which strongly discourages the opinion that the BiO layers act as a carrier source. A very similar conclusion can be deduced from measurements of the IR-reflectivity on the same set of compounds /6/. The 2212 samples exhibit an edge in the mid-IR region which can be attributed to a Drude-type contribution to the reflectivity evoked by free carriers. This feature is shifted to lower energies for the 2223 compounds in correspondence with the lower hole density p deduced from the Hall effect measurements.

In summary, our results support the suggestion /7/ that the lead addition is an important factor during the sinter process and enhances the formation of the high-Tc (2223) phase at about 10 to 30 per cent Pb-substitution. A variation of the transport and superconducting properties due to the lead substitution seems not to be caused by a charge transfer from (Bi,Pb)O planes to the CuO_2 sheets but by the formation of different crystallographic phases.

Acknowledgement

This work was supported by the Fonds zur Förderung der wissenschaftlichen Forschung, Austria, grants P7068 and P7059

References

/1/ L.Ganapathi, S.Krishna, K.Murthy, R.Vijayaraghavan, C.N.R.Rao; Solid State Commun. 67, 967 (1988)
/2/ H.Kuzmany, P.Schaffarich, J.Geisselbrecht, T.Pichler, W.Lang, H.Jodlbauer, W.Steiner, M.Reissner; Physica C 162-164, 1219 (1989)
/3/ W.Lang, W.Markowitsch; J. Less Common Metals 151, 147 (1989)
/4/ J.M.Tarascon, W.R.McKinnon, P.Barboux, D.M.Hwang, B.G.Bagley, L.H.Greene, G.W.Hull, Y.LePage, N.Stoffel, M.Giroud; Phys.Rev. B38, 8885 (1988)
/5/ D.Y.Xing, C.S.Ting; Phys.Rev. B38, 5134 (1988)
/6/ W.Markowitsch, W.Lang, P.Schaffarich, H.Kuzmany, K.Pressl, G.Leising; Physica C 162-164, 1235 (1989)
/7/ T.Hatano, K.Aota, S.Ikeda, K.Nakamura, K.Ogawa; Jap. J. Appl. Phys. 27, L92055 (1988)

Influence of Pb Doping on the Physical Properties of the BiSrCaCuO System

B. Stadlober[1], *P. Knoll*[1,2], *H. Jodlbauer*[3], *W. Lang*[3], *N. Brničević*[4], *R. Abt*[5], *E. Schweiger*[5], *G. Leising*[5], *and H. Kahlert*[5]

[1]Institut für Experimentalphysik der Universität Graz, Universitätsplatz 5, A-8010 Graz, Austria
[2]Max-Planck-Institut für Festkörperforschung, Heisenbergstr. 1, D-7000 Stuttgart 80, Fed. Rep. of Germany
[3]Ludwig Boltzmann Institut für Festkörperphysik, Kopernikusgasse 15, A-1060 Wien, Austria
[4]Rudjer Bošković Institute, University of Zagreb, Bijenička 54, P.O. Box 1016, YU-41001 Zagreb, Croatia, Yugoslavia
[5]Institut für Festkörperphysik der TU Graz, Petersgasse 16, A-8010 Graz, Austria

Abstract: The influence of Pb on the superconducting properties of the BiSrCaCuO system has been investigated. The Pb contribution to the sinter ceramics influences the phase composition and can lead to the pure 2223 structure with a superconducting transition at 110 K. The amount of the various phases in different samples has been quantitatively determined by magnetic susceptibility measurements. Raman experiments have been performed first on pure Bi and Pb doped single crystals in order to obtain a reliable interpretation of the observed modes. By comparing the Raman features of the sinter ceramics with the amount of different superconducting phases we conclude that Pb dominantly influences the stacking order and does not change the local environment of the CuO and BiO planes.

1. Introduction

Due to the discussion of a possible contribution [1] of phonons to the superconducting mechanism of HTS the study of vibrational spectra of the Bi-based materials becomes important for finding any common feature for all HTS materials. The Bi-Sr-Ca-Cu-O system tends to form several phases with different superconducting behavior, e.g.transition temperatures. It is known [2,3] that substitution of lead for bismuth stabilizes the superconducting phase with a transition temperature around 110 K. In this work we report on systematical studies of the influence of Pb on several physical properties of the Bi superconductors by means of Raman scattering, X-ray and magnetic susceptibility measurements. So far, details of the actual crystal structure of the BiSrCaCuO system are not well known; difficulties arise from the occurrence of twinning, stacking and superstructure [4,5,6,7]. In principle, Raman scattering can be used to investigate such structural deviations because polarization selection rules of incident and scattered light strongly reflect the symmetry of the atomic arrangement.

Springer Series in Solid-State Sciences, Vol. 99
Electronic Properties of High-T_c Superconductors and Related Compounds
Editors: H. Kuzmany · M. Mehring · J. Fink

Furthermore Raman scattering probes very sensitive electronic and vibrational properties as well as electron-phonon interaction. Therefore differences in the Raman spectrum in comparison with changes in the superconducting behavior should provide valuable information about the intrinsic properties of the BiSrCaCuO system.

2. Pure and Lead Doped Single Crystals of BiSrCaCuO

In order to determine the intrinsic Raman features we first performed micro-Raman measurements on cleaved surfaces of platelet-like, shiny single crystals of the pure and Pb doped type. The polarization microscope identified both single crystals as twinned. The composition of the BiPb single crystal was determined by electron microprobe analysis (EDAX) probed on different spots on the crystal surface which yielded the average stoichiometric formula $Bi_{2.4}Pb_{0.6}Sr_2CaCu_2O_x$ within $\pm$ 0.1 experimental error. The sharp superconducting transition at 83 K observed in the resistivity of the a-b plane supports the identification as a 2212 single crystal. The higher amount on BiPb suggests a partial substitution of other elements.

The interpretation of the Raman spectra has to consider the uncertainty in the crystal structure due to different oxygen arrangements of the Bi layers. The correlation of the factor group analysis of different, suggested space groups has been presented in a previous work [8]. From the correlation table [8] we see that modes without a contribution from the BiO layers should be quite insensitive to the uncertainty in the selected crystal symmetry. Therefore, and by comparing with lattice dynamical calculations [9], we assign the peaks at 175 cm^{-1}/190 cm^{-1} to in/out of phase z-motions of the Cu and Sr atoms (Table 1). The mode at 122 cm^{-1} which corresponds to the calculated [9] one at 87 cm^{-1} must contain z-motions of the Bi and Sr atoms. The B_{1g} mode corresponding to out of phase z-vibrations of the oxygens in the CuO_2 planes can be clearly identified as the 282 cm^{-1} peak in the pure Bi single crystal in agreement with previous measurements [10]. Taking into account the variation of oxygen distribution and content in the BiO layers we assign the sample dependent weak structures around 300 cm^{-1}, 380 cm^{-1} and 655 cm^{-1} to z-

Table 1: Mode assignment for the Bi single crystal compared with the BiPb single crystal

Bi	Bi-Pb	Symmetry	Assignment
122 cm^{-1}	-	:A_g	:α_{xx},α_{yy}, z-motions of Bi,Sr (in phase)
175 cm^{-1}	170 cm^{-1}	:A_g	:α_{xx},α_{yy},α_{zz}, z-motions of Cu,Sr (in phase)
190 cm^{-1}	190 cm^{-1}	:A_g	:α_{xx},α_{yy}, z-motions of Cu,Sr (out of phase)
459 cm^{-1}	464 cm^{-1}	:A_g	:α_{xx},α_{yy}, z-motions of O(1),O(2) (in phase)
630 cm^{-1}	627 cm^{-1}	:A_g	:α_{xx},α_{yy}<<α_{zz}, z-motions of O(3),O(4)
655 cm^{-1} ⎤	650 cm^{-1} ⎤	:A_g	:α_{xx},α_{yy},α_{zz}, z-motions of O(3),O(4),O(5);
380 cm^{-1} ⎥	weak ⎥		one intrinsic, two either defect induced
300 cm^{-1} ⎦	295 cm^{-1} ⎦		or crystal field split partners
282 cm^{-1}	see text	:B_{1g}	:α_{xy}, z-motion of O(1),O(2) (out of phase)

motions of the oxygens in the BiO layers with different contributions of the bridging (O(3)) oxygen. The mode at 459 cm^{-1} seems to correspond rather to in phase z-vibrations of the O(1), O(2) oxygens in the CuO_2 planes due to the fact that this mode is well pronounced and apparently stable on several samples. In analogy to YBaCuO the strong zz polarized mode at 630 cm^{-1} is assigned to mainly bridging oxygen vibration. For that mode a dependence on the oxygen content and arrangement of the BiO planes is expected. That expected dependence may also explain the occurrence of the 655 cm^{-1} shoulder.

In the same way a mode analysis for the BiPb single crystal has been performed. In general, no significant changes to the pure Bi crystal have been observed. Deviations in line positions as given in Table 1 seem to be more related to sample inhomogeneity than to a systematic change with Pb content. One exception might be the B_{1g} mode, which on some samples showed stronger deviations and was also observed at 375 cm^{-1}. Since the B_{1g} symmetry has very strict selection rules we cannot simply interpret such deviations as impurity or inhomogeneity effects. If Pb partially substitutes other atoms, e.g. the Ca, such a behavior of the B_{1g} mode can be explained.

3. Pb Doped Sinter Ceramics of BiSrCaCuO

In contrast to our results on Pb doped single crystals, in the ceramics Pb can lead to a stabilization of the 2223 phase with a T_c at 110K. That is clearly shown by the comparison of magnetic susceptibility measurements (Fig.1) with the 002 X-ray peak position (Fig.2). Depending on different initial composition and sintering procedures different amounts of 85K and 110K phase are formed, which has been quantitatively determined from the magnetic susceptibility measurements. The Raman experiments do not show systematic changes due to the different amounts of 2212 and 2223 phase. As presented in a previous work [8] the Raman parameters (intensity, line position and linewidth) of some particular modes which are sensitive to the CuO_2 and BiO arrangement (282 cm^{-1}, 459 cm^{-1} and 630 cm^{-1}) only vary within the experimental error. That agrees with findings based on magnetic and X-ray data because changes in the long range order (as e.g. stacking) are not expected to influence the higher frequency phonon modes. According to the absence of systematic changes of the line parameters of the higher frequency CuO_2 and BiO sensitive modes due to different phase amounts, which could be related to short range structural changes such as different oxygen content or oxygen distribution, we conclude that the shift of the transition temperature from 85 K to 110K is really due to different stacking of the CuO_2 and BiO planes.

We are indebted to R.Pfleger for providing us with single crystal material.

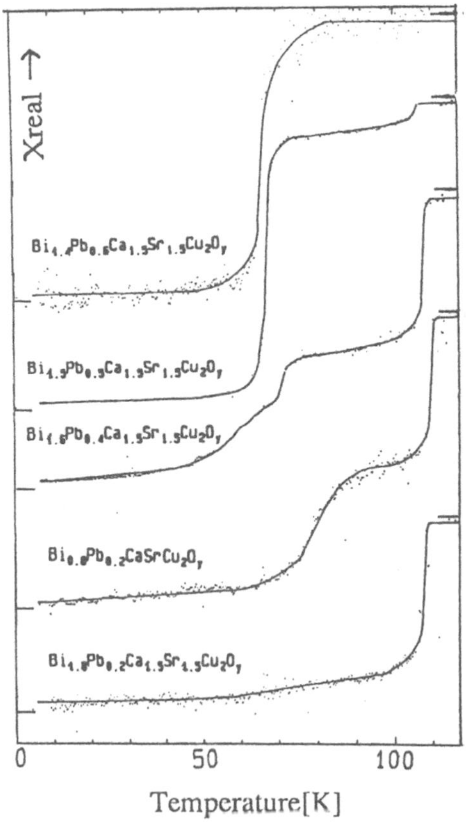

Fig.1: Absolute AC susceptibility measurements for different sinter ceramics; each slope ranges between 0 and -1

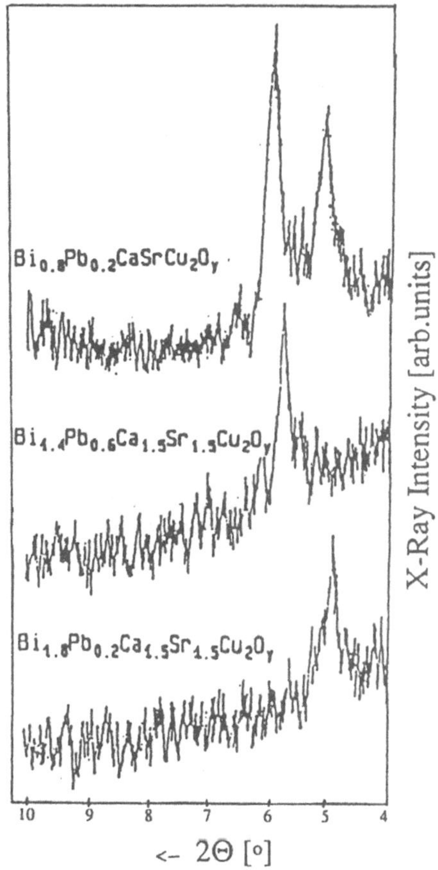

Fig.2: 002 Bragg reflection peaks of several sinter ceramics; different peak positions correspond to different c-parameters.

References:

1. R.Zeyher, G.Zwicknagl, Sol.State Comm. 66, 617 (1988)
2. C.Michel, M.Hervieu, M.M.Borel, A.Grandin et al., Z.Phys.B 68, 421 (1987)
3. H.Maeda,Y.Tanaka,N.Fukutomi,T.Asano, Jpn.J.Appl.Phys.Lett. 27 (1988) L 209
4. R.M.Hazen, C.T.Prewitt, R.J.Angel, N.L.Ross, L.W.Finger, C.G.Hadidiacos, D.R.Veblen, P.J.Heaney,P.H.Hor,R.L.Meng et al., Phys.Rev.Lett. 60, 1174 (1988)
5. T.M.Shaw, S.A.Shivashankar, S.J.LaPlaca et al., Phys.Rev.B 37, 9856 (1988)
6. R.Ramesh, G.Thomas, S.M.Green, M.L.Rudee et al., Phys.Rev.B 38, 7070 (1988)
7. J.M.Tarascon, W.R.McKinnon, P.Barboux et al., Phys.Rev.B 38, 8885 (1988)
8. P.Knoll, B.Stadlober, M.Pressl, N.Brnicevic, Physica C 162-164, 1097 (1988)
9. J.Prade, A.D.Kulkarni, F.W.de Wette, W.Kress, Phys.Rev.B 39, 2771(1989)
10. M.Cardona, C.Thomsen, R.Liu et al., Sol.State Comm. 66, 1225 (1988)

Spatial Inhomogeneity in Superconducting YBaCuO-Films

F. Kober, D. Koelle, M. Hartmann, R. Gross, and R.P. Hübener

Physikalisches Institut, Lehrstuhl Experimentalphysik II, Universität Tübingen, Morgenstelle 14, D-7400 Tübingen 1, Fed. Rep. of Germany

Abstract. Four years after the discovery of the high-T_c oxide superconductors the quality and reproducibility of the preparation of these materials has advanced considerably, and better methods for quantitatively evaluating the spatial sample homogeneity become increasingly important. Low temperature scanning electron microscopy has been extended successfully for two-dimensionally imaging the properties of high-T_c superconducting films with high spatial resolution. So far, these experiments have concentrated on YBaCuO films. The results obtained for polycrystalline and single-crystalline epitaxial films and the effects of single grain boundaries and grain boundary networks are summarized.

1. Introduction

Four years after the discovery of the high-T_c oxide superconductors the preparation of these superconducting materials has advanced considerably and better materials are now available yielding higher reproducibility between the results of different laboratories than in the beginning. Here, in addition to bulk samples and single crystals, high-T_c superconducting films have been fabricated with better and more uniform superconducting properties. Clearly, these advances in material preparation call for similarly advanced techniques for evaluating the spatial homogeneity of the superconducting properties such as the critical temperature T_c and the critical current density j_c.

Recently, low temperature scanning electron microscopy (LTSEM) has been extended and successfully applied for investigating the superconducting quantities T_c and j_c in high-T_c superconducting films with high spatial resolution (approaching 1 µm at present). [1-3] These concepts are based on earlier ideas of LTSEM, where a superconducting thin-film configuration kept at low temperatures is scanned with the electron beam of a scanning electron microscope (SEM) and a proper electric response signal is recorded as a function of the coordinate point (x,y) of the beam focus on the sample surface. [4-6] For the two-dimensional imaging of the spatial distribution of the critical temperature $T_c(x,y)$ and the critical current density $j_c(x,y)$, the high-T_c film is usually current-biased and the beam-induced change $\delta V(x,y)$ of the voltage drop along the specimen is recorded as a function of the temperature maintained at the sample stage and/or the bias current. This imaging principle is shown schematically in Fig. 1. A detailed discussion of the mechanism generating the voltage signal $\delta V(x,y)$ (and also determining the spatial resolution limit) has been given elsewhere. [1,3,7] Here the essential point is the local thermal perturbation of the specimen film by the electron beam. For a typical beam power of 1 - 10 µW in an YBaCuO film near 90 K a beam-induced local temperature increment δT = 0.1 - 1 K is expected. The beam diameter is typically less than 100 nm. For improving the sensitivity of this imaging process, the beam current is modulated (typical modulation frequency = 10 kHz) and the modulated voltage signal $\delta V(x,y)$ is recorded with a lock-in detector.

Springer Series in Solid-State Sciences, Vol. 99
Electronic Properties of High-T_c Superconductors and Related Compounds
Editors: H. Kuzmany · M. Mehring · J. Fink

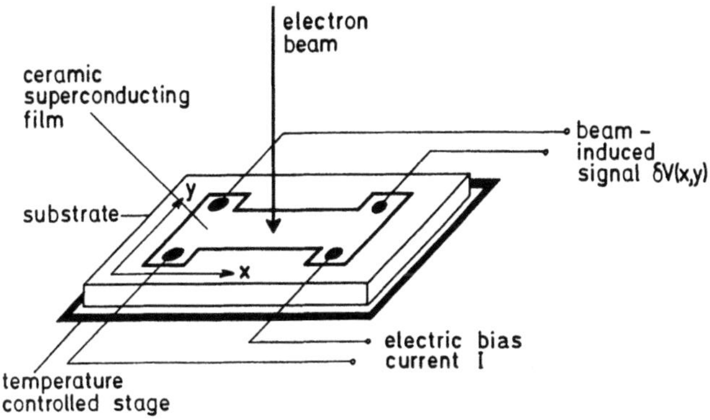

Fig. 1: Principle of LTSEM applied to high-T_c superconducting films.

Clearly, LTSEM as a functional test of the superconducting properties of high-T_c films providing high spatial resolution becomes important for evaluating the degree of sample homogeneity because of several reasons. In polycrystalline films the influence of the grain boundaries and of the orientation of the individual grains must be understood. Further, a series of sophisticated models has been proposed recently for explaining the resistive losses and the voltage current characteristics observed in high-T_c superconducting films. For example, these models include thermally assisted flux flow, giant flux creep, flux lattice melting, vortex glass aspects, and Kosterlitz-Thouless behavior. For an unambiguous clarification of the validity of these different models the degree of sample homogeneity must be established reliably. On the other hand, in the opposite direction we see sometimes the development of theoretical models trying to fit experimental results by assuming a specific spatially inhomogeneous behavior. Of course, experimental verification of this assumed inhomogeneity would be important. Last but not least, the technological applications of high-T_c superconducting films in microelectronics, presently under way, cannot evolve very far without functional tests providing high spatial resolution.

In this paper we summarize the main results obtained by LTSEM up to now for high-T_c superconducting films. So far, all of these LTSEM studies have been performed with YBaCuO films. We emphasize that we are dealing exclusively with the spatially resolved functional testing of the superconducting sample properties and not with the highly advanced techniques for materials and surface analysis available today.

2. LTSEM Studies of YBACuO Films

2.1. Polycrystalline Films

The first LTSEM studies were performed on polycrystalline films deposited by d.c.-magnetron sputtering at room temperature on $SrTiO_3$ substrates and subsequent heat treatment in an oxygen atmosphere. [8] These experiments clearly demonstrated a highly pronounced spatial inhomogeneity, the local values of the critical current density varying by more than an order of magnitude over the specimen film for constant temperature. The spatial resolution obtained was about 1 - 2 µm. Inhomogeneous behavior was observed subsequently also in polycrystalline films prepared by multilayer electron beam evaporation of Y, BaF_2, and Cu on sapphire and YSZ substrates. [3] From the latter experiments it appeared that the spatial variation of the critical current density was much more rapid for sapphire than for YSZ as substrate material. Of course, the observed strongly inhomogeneous behavior in the polycrystalline films is not surprising in view of the different orientation of the individual grains and the effect of the grain boundaries. So far, a clear correlation between variations of the local superconducting properties and of the local stoichiometry has not been worked out for polycrystalline films.

2.2 Epitaxial Films with a Single Grain Boundary

The preparation of bicrystalline films of $YBa_2Cu_3O_7$ by epitaxial growth on $SrTiO_3$ substrates [9,10] and the subsequent patterning by ablation with a focused laser beam [11] has made it possible to perform electric transport measurements across single grain boundaries. These experiments have shown that in the bicrystalline films the critical current densities are always limited by the grain boundaries, and that the critical current density of the grain boundaries decreases rapidly with grain misorientation. [9,12,13] Recently, the spatial variation of the critical current density near a single grain boundary in a bicrystalline film has been imaged two-dimensionally with a spatial resolution of one micrometer using LTSEM. [14, 15] From these images an upper limit for the effective grain boundary width of about 2 μm has been obtained. Further, spatial structures of the current flow across the grain boundary were observed, which were correlated with oscillations in the derivative of the current voltage characteristic. Apparently, these spatial structures are caused by self-excited resonances within the grain boundary region.

2.3 Epitaxial Films with Grain Boundary Networks

The combined effects of many grain boundaries on the supercurrent flow in polycrystalline high-T_c superconductors represent an important subject for experimental investigations. Recently, first experiments have been performed on epitaxial $YBa_2Cu_3O_7$ films deposited on polycrystalline $SrTiO_3$ substrates by electron beam evaporation or by laser ablation, using LTSEM. The average grain size ranged typically between 5 and 30 μm. The two-dimensional imaging confirmed that the grain boundary network limits the critical current in the sample, and that more and more grain boundaries become resistive with increasing bias current. The voltage signals $\delta V(x,y)$ change in a complex manner with the bias current and it appears that intergranular current redistribution takes place. A more detailed report on these measurements can be found elsewhere. [16,17]

2.4 Single-Crystalline Epitaxial Films

Recently, single-crystalline epitaxial $YBa_2Cu_3O_7$ films have been prepared by laser ablation [18,19] and by magnetron sputtering [20] with critical current densities exceeding 10^6 A/cm^2 at 77 K. Clearly, these advances look highly promising for technological applications of such films in microelectronics. However, adequate spatial homogeneity of the superconducting properties and the absence of local regions with weakened superconductivity are important prerequisites for the development of these applications. In this context the possible deterioration of the film quality due to the microfabrication processes (in particular in the region near the film edges) represents a subject requiring considerable attention. Here LTSEM is expected to play an increasingly important role.

The LTSEM studies performed so far on single-crystalline epitaxial $YBa_2Cu_3O_7$ films have yielded the following results:

a) Scratches on the substrate surface can cause a strong local reduction in the critical current density [21]
b) In some cases, but not always, a correlation of locally weak superconducting regions detected by LTSEM with structural features in the SEM micrograph was found. On the other hand, precipitates on the sample surface,

typical for films deposited by laser ablation, were found to have no significant effect on the local superconducting properties. [22]

c) In a c-axis oriented specimen film deposited on a (100)-oriented $SrTiO_3$ substrate by laser beam deposition (dimensions: thickness = 170 nm, width = 10 μm, length = 180 μm) with a critical current density of $5 \cdot 10^6$ A/cm^2 at 77 K, the local resistive transition at 11 - 100 μA bias current was found to occur within less than 200 mK over the whole film indicating no strong deviations from homogeneous behavior. [22]

d) LTSEM has been shown to be highly sensitive for detecting local regions with weakened superconductivity, whereas such regions can often be identified by standard surface analysis techniques (SEM, etc.) only with great difficulty, or only after the LTSEM results have indicated where to look carefully. [22]

3. Concluding Remarks

The work described above has clearly established LTSEM as an important functional test of the spatial homogeneity of high-T_c superconducting films. Whereas the spatial resolution of about 1 μm obtained so far is impressive, a further improvement of this resolution would be interesting. Here high-frequency beam modulation up to 10 - 100 MHZ appears promising. [4,23] In addition, a reduction of the beam energy, say, down to only a few keV may also yield an improvement of the spatial resolution. In this way a resolution of some tenth of a micrometer perhaps may become possible.

Another way for improving the spatial resolution may result from a combination of the signal generating concepts of LTSEM with the principle of scanning tunneling microscopy, using the current emitted from the tunneling tip for the local (thermal) perturbation of the sample. Of course, for any appreciable increase in spatial resolution (by a corresponding reduction of the size of the perturbed sample region), the cross-sectional area of the current-biased specimen must be reduced correspondingly, in order to maintain a sufficient relative amplitude of the imaging signal.

Higher spatial resolution would be particularly interesting for the experiments on grain boundaries and grain boundary networks described in Sections 2.2 and 2.3, respectively. Finally, the functional test by means of LTSEM should be combined with the standard spatially resolved techniques for structural and chemical surface analysis. However, from the experience obtained so far it has become clear, that the functional test by LTSEM is much less tedious than the spatially resolved surface analysis studies yielding local information on the structural and chemical sample properties.

Acknowledgment

Financial support of this work by the Bundesminister für Forschung und Technologie (project 13N5482) is gratefully acknowledged.

References

[1] R.P. Hübener, R. Gross, and J. Bosch: Z. Phys. B - Condensed Matter 70, 425 (1988)

[2] R. Gross, J. Bosch, H.-G. Wener, J. Fischer, and R.P. Hübener: Cryogenics 29, 716 (1989)

[3] M. Hartmann, K. Hipler, D. Koelle, F. Kober, K. Bernhardt, T. Sermet, R. Gross, and R.P. Hübener: Z. Phys. B - Condensed Matter 75, 423 (1989)

[4] J. R. Clem and R.P. Hübener: J. Appl. Phys. 51, 2764 (1980)

[5] R.P. Huebener: Rep. Prog. Phys. 47, 175 (1984)

[6] R.P. Huebener: Advances in Electronics and Electron Physics, P.W.Hawkes, ed., Academic Press, New York, Vol. 70, p. 1, 1988
[7] R.P. Huebener and R. Gross: Scanning Microscopy 3, 703 (1989)
[8] R. Gross, J. Bosch, R.P. Hübener, J. Mannhart, C.C. Tsuei, M. Scheuermann, M.M. Oprysko, and C.C. Chi: Nature 332, 818 (1988)
[9] D. Dimos, P. Chaudhari, J. Mannhart, F.K. LeGoues: Phys. Rev. Lett. 61, 219 (1988)
[10] P. Chaudhari, R. H. Koch, R.B. Laibowitz, T.R. McGuire, and R.J. Gambino: Phys. Rev. Lett. 58, 2684 (1987)
[11] J. Mannhart, M. Scheuermann, C. C. Tsuei, M.M. Oprysko, C.C. Chi, C.P. Umbach, R.H. Koch, and C. Miller: Appl. Phys. Lett. 52, 1271 (1988)
[12] P. Chaudhari, J. Mannhart, D. Dimos, C. C. Tsuei, C.C. Chi, M.M.Oprysko, and M. Scheuermann: Phys. Rev. Lett. 60, 1653 (1988)
[13] J. Mannhart, P. Chaudhari, D. Dimos, C.C. Tsuei, and T.R. McGuire, Phys. Rev. Lett. 61, 2476 (1988)
[14] J. Mannhart, R. Gross, K. Hipler, R.P. Hübener, C.C. Tsuei, D. Dimos, and P. Chaudhari: Science 245, 839 (1989)
[15] J. Mannhart, R. Gross, R.P. Hübener, P. Chaudhari, D. Dimos, and C.C. Tsuei: Cryogenics, in press
[16] J. Mannhart, R.P. Hübener, F. Kober, D. Koelle, P. Chaudhari, D. Dimos, R. Gross, A. Gupta, G. Koren, and C.C. Tsuei, 3. Bar-Ilan Conference, January 8-11, 1990, Tel-Aviv, Israel, Physica C, in press
[17] F. Kober et al., to be published
[18] B. Roas, L. Schultz, and G. Endres: Appl. Phys. Lett. 53, 1557 (1988)
[19] G. Koren, A. Gupta, E.A. Giess, A. Segmueller, and R.P. Laibowitz: Appl. Phys. Lett. 54, 1054 (1989)
[20] X.X. Xi, G. Linker, O. Meyer, E. Nold, B. Obst, F. Ratzel, R. Smithey, B. Strehlau, F. Weschenfelder, and J. Geerk: Z. Phys. B - Condensed Matter 74, 13 (1989)
[21] R. Gross, K. Hipler, J. Mannhart, R.P. Huebener, P. Chaudhari, D. Dimos, C. C. Tsuei, J. Schubert, and U. Poppe, Appl. Phys. Lett. 55, 2132 (1989)
[22] D. Koelle, F. Kober, M. Hartmann, R. Gross, R.P. Huebener, B. Roas, L. Schultz, and G. Saemann-Ischenko, Physica C, in press
[23] H. Pavlicek, L. Freytag, H. Seifert, and R.P. Hübener , J. Low Temp. Phys. 56, 237 (1984)

Part II

Transport, Tunneling, Thermal Properties

Surface Impedance of High-T_c Superconductors

G. Grüner

Department of Physics and Solid State Science Center,
University of California, Los Angeles, CA 90024, USA

Abstract. I discuss the various experiments which give information on the surface impedance $Z_S = R_S + jX_S$ of high temperature superconductors. The surface reactance X_S is related to the penetration depth λ and $X_S(T)$ is suggestive of singlet pairing. The surface resistance R_S is determined by the ac losses due to imperfections and due to thermally excited carriers, and experiments on high quality $YBa_2Cu_3O_7$ films indicate a superconducting gap which exceeds the BCS weak coupling limit. The various high frequency device applications are also briefly discussed.

1. Introduction

The surface impedance Z_S is one of the parameters which reflects the electrodynamics of the superconducting state. At the same time it is perhaps the most important technical parameter which determines the application potential of the superconductors for high frequency passive components and devices.

The surface impedance is defined as [1]

$$Z_S = \frac{E_0}{\int_0^\infty j\,dx} \tag{1}$$

where E_0 is the electric field of the surface, j the ac current in the sample and the x direction is perpendicular to the surface. Z_S is, in terms of the complex conductivity $\sigma = \sigma_1 - i\sigma_2$, given by

$$Z_S = \left(\frac{i\ \mu_0\omega}{\sigma_1 - i\sigma_2}\right)^{1/2} = R_S + iX_S \tag{2}$$

where μ_0 is the permeability of free space, R_S and X_S are the surface resistance and surface reactance, and ω the measuring frequency.

The surface reactance is given, in the superconducting state well below T_C, by

$$X_S(T) = \mu_0\omega\lambda(T) \tag{3}$$

with λ the penetration depth, and the measured $X_S(T)$ can be compared with various models of the superconducting state.

The surface resistance has been calculated in the local limit by Mattis and Bardeen [2], and $R_S(T)$ is in good agreement with experiments conducted on

Springer Series in Solid-State Sciences, Vol. 99
Electronic Properties of High-T_c Superconductors and Related Compounds
Editors: H. Kuzmany · M. Mehring · J. Fink

classical superconductors. For the high temperature superconductors, the mean free path is comparable to the coherence length ξ, and consequently finite mean free path effects are important [3], and have to be taken into account when experiments are compared with the various theories of the superconducting ground state.

2. Experimental Techniques

The majority of the experiments [4-8] are conducted by using resonant structures, with the superconducting material forming part of the structure. Both cavity endwall and cavity perturbation techniques have been utilized, they give somewhat different information on R_S and X_S. In all cases R_S and X_S are determined using a configuration where these parameters refer to electric currents flowing in the planes (if thin films are investigated). Alternatively, stripline and delay line configurations can be used with films either the central conductor or endplate of the resonator.

The various radio frequency and microwave techniques are usually applied to bulk or thin film samples, and depend on measuring relatively small changes in the value of the surface impedance in the superconducting state compared with the normal state. The sample is normally placed in the magnetic field associated with a resonant circuit and the change in the resonance frequency f_O and quality factor Q is measured as a function of the temperature. The changes normally represent a small perturbation on the resonance, and can be linearly related to the changes in the surface impedance of the sample:

$$\Delta\left(\frac{1}{Q}\right) - 2j\left(\frac{\Delta\omega_O}{\omega_O}\right) = \gamma\Delta Z_S = \gamma\left(\Delta R_S + j\Delta X_S\right) . \tag{4}$$

γ is a geometric factor associated with the dimensions of the resonator and the sample.

3. Experimental Results

The temperature dependence of λ has been evaluated using both resonant cavity [7] and delay line [8] configurations, and the results are in broad agreement with experiments utilizing low field magnetization [9] and muon spin resonance [10] techniques. All suggest singlet pairing. The expected temperature dependence of λ is different for weak and strong coupling, but the available experiments cannot convincingly determine whether strong coupling effects are important.

$R_S(T)$ measured on a ceramic, sputtered thin film and laser ablated film [1,4,5,11] is displayed in Fig. 1. One observes a finite residual resistance as $T \to 0$, and $R_S(T)$ depends also strongly on the film quality, with laser ablated films [11] — the highest overall quality — giving the lowest $R_S(T)$ values.

The large R_S for the majority of materials is due to a second phase [12] or due to grain boundaries which act like Josephson coupling between the grains [13]. In a simple model [13] of grain boundary effects on the surface impedance of thin films, the material is modeled as a network of superconducting grains

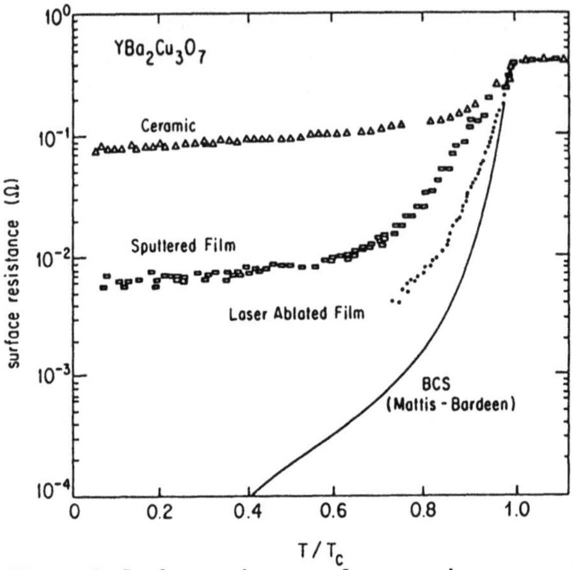

Figure 1. Surface resistance of a ceramic, sputtered thin film and laser ablated thin film of $YBa_2Cu_3O_7$. The full line is the Mattis-Bardeen expression for the surface resistance R_s. While R_N is different for different materials, in the figure all are normalized to $R_N = 0.5\Omega$, corresponding to $\rho_N = 60\mu cm$, just above T_c.

coupled by Josephson junctions, which are described using the standard resistively shunted junction model with negligible capacitance. The parameters of the model are the junction J_cR product (J_c is the junction critical current and R the junction resistance) and the effective grain size a. The parameters which enter are $L_G = \mu_o\lambda_{ab}^2$, the kinetic inductivity of the grain for current flow in the ab plane (penetration depth λ_{ab}), and $L_j = \hbar/2eJ_c$, the unit areal inductance of the junction.

For such a description of the grain boundary, the surface impedance is given by

$$Z_s = \omega\mu_o\lambda_{eff}\left[\,i + \frac{1}{2}\left(\frac{\lambda_j}{\lambda_{eff}}\right)^2 \frac{\hbar\omega}{2eJ_cR}\right], \tag{5}$$

with

$$\lambda_{eff} = \left[\lambda_{ab}^2 + \lambda_j^2\right]^{1/2}, \qquad \lambda_J = \left[\frac{\hbar}{a2eJ_c\mu_o}\right]^{1/2}, \tag{6}$$

and λ_{ab} the penetration depth in the planes. The model accounts well [13] for the increased penetration depth and increased surface resistance of sputtered superconducting films.

For laser ablated films, grain boundary effects are not important, and $R_s(T)$ [14] can be compared with the various theories [2,3] of the surface impedance. In Fig. 2 we display the experimental values (after correction for finite film thickness) [15] with a calculation based on the Mattis-Bardeen model [2], and calculations with finite mean free path effects [3] for a gap value $2\Delta/k_BT_c = 3.52$. The lower and upper sets of experimental points have been obtained by employing two differ-

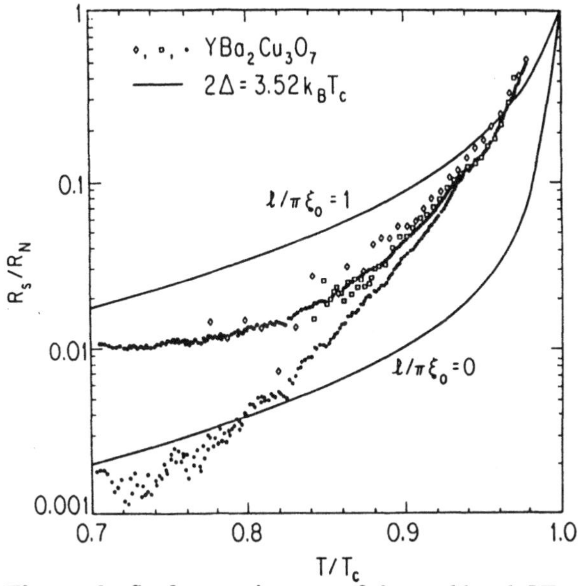

Figure 2. Surface resistance of laser-ablated $YBa_2Cu_3O_7$. The full lines are expressions, with the parameters ℓ/ξ_0 and Δ given on the figure. The curve with $\ell/\pi\xi_0 = 0$ is the Mattis-Bardeen result.

ent procedures for subtracting the low temperature residual surface resistance. They represent the lower and upper limits of the true surface resistance $R_s(T)$. It is evident that the experimental results lie below $R_s(T)$ arrived at for a realistic model, with $\ell/\pi\xi_0 = 1$ and $2\Delta = 3.5\ k_BT_c$ (corresponding to $\xi_0 = 16A$ and $\ell = 50A$, both parameters referring to the in-plane parameters). This suggests that Δ is larger than in the weak coupling limit, and a value $2\Delta \sim 5\ k_BT_c$ appears to describe our results well. Consequently, surface impedance studies suggest that $YBa_2 Cu_3O_7$ is in the strong coupling limit with Δ exceeding the weak coupling BCS value. Our results also convincingly rule out drastic deviations from a gap which opens up along the entire Fermi surface. We note that the experimental configuration employed by us leads to ac current in the CuO_2 planes, and consequently no information is gained on the ac losses for currents perpendicular to the planes.

4. Applications

Due to the low ac loss, high temperature superconductors can be used in a variety of high frequency passive devices. In high quality $YBa_2Cu_3O_7$ and Tℓ-based [16] superconducting films, R_s is significantly smaller than that of copper in the microwave spectral range. Consequently, various high frequency devices have been fabricated with performance characteristics superior to those fabricated using normal metals. Examples include microwave resonators, delay lines and filters [17]. These can be used also to study the electrodynamics of the superconducting state [8,15], and to evaluate the temperature dependence of the surface resistance and penetration depth. While such devices will most probably be the first which will be commercially available, several issues have to be resolved before commercialization takes place. Among them substrate losses, power dependences, radiation and leakage problems [18] are the most prominent.

5. Acknowledgments

This research has been supported by the Office of Naval Research and by the California Competitive Technology Program.

Most of the work reported here has been carried out by L. Drabeck and K. Holczer at UCLA; the materials were prepared by V. Venkatessan and his co-workers. We have also enjoyed a fruitful collaboration with D. Scalapino.

References

[1] C. Kittel, Quantum Theory of Solids. For a review of the early results, see: J. Carini, et al., Modern Physics Lett. B3, 5 (1989).
[2] D.C. Mattis, and J. Bardeen, Phys. Rev. 111, 412 (1958).
[3] J. Chang, and D. Scalapino, Phys. Rev. B40, 4299 (1989). J. Halbritter, Z. Phys. 266, 209 (1974).
[4] J. Carini, et al., Solid State Comm. 67, 373 (1988).
[5] N. Klein, et al. Appl. Phys. Lett. 54, 757 (1989). L. Drabeck, et al. Phys. Rev. B39, 785 (1989).
[6] D.W. Cooke, et al., Solid State Comm. 73, 297 (1990).
[7] S. Sridhar, D. Wu, and W. Kennedy, Phys. Rev. Lett. 63, 1873 (1989).
[8] S. Anlage, et al., Appl. Phys. Lett. 54, 2710 (1989).
[9] Z. Krusius-Elbaum, et al., Phys. Rev. Lett. 62, 217 (1989).
[10] D.R. Harshman, et al., Phys. Rev. B39, 851 (1989).
[11] L. Drabeck, et al., Phys. Rev. B40, 7350 (1989).
[12] K. Scharnberg, et al., (to be published).
[13] T.L. Hylton, et al., Appl. Phys. Lett. 53, 1343 (1988). T.L. Hylton, and M.R. Beasley, Phys. Rev. B39, 785 (1989).
[14] L. Drabeck, et al., Phys. Rev. B40, 7350 (1989).
[15] L. Drabeck, et al., Appl. Phys. Lett. (to be published).
[16] L.D. Chang, et al., Appl. Phys. Lett. 55, 1357 (1989).
[17] R.J. Hammond, Supercurrents 9, 66 (1989).
[18] L. Drabeck, et al., Appl. Phys. Lett. (to be published).

Detection of Flux Creep by Microwave Absorption

K.-P. Dinse, O. Dobbert, M. Rübsam, W. Goldacker, and T. Wolf**

[1]Physics Institute, Universität Dortmund, Postfach 500500,
D-4600 Dortmund 50, Fed. Rep. of Germany
*Also at: Kernforschungszentrum Karlsruhe, Postfach 3640,
D-7500 Karlsruhe 1, Fed. Rep. of Germany

Abstract. Highly structured absorption of 10 GHz microwave radiation is utilized to measure flux relaxation in weak magnetic fields. The logarithmic decay rate depends on the sample temperature and the magnetic field. The decay rate can be related to an effective pinning potential U_0. Its temperature dependence is found to scale as $U_0(t,B)=U_0\cdot(1-t^2)\cdot(1-t^4)^{1/2}/t\cdot B$ over the full temperature range for $B < 50$ mT.

1. Introduction

Observation of logarithmic flux decay in high-T_c superconductors has been confirmed by many authors [1-5]. Apart from the surprisingly large effect ("giant" flux creep), quite complex temperature dependences have been described [2]. It seemed tempting to use this relatively simple microwave absorption experiment to extract intrinsic properties of the material, i.e., pinning potential U_0 and critical current density J_c. As was pointed out by Tinkham [6], knowledge of U_0 allows an estimate of the non-resistive range of the material's behavior *independent* of its particular transition temperature T_c.

Up to now, quite contradictory data for U_0 have been reported. A comparison of the data is hampered not only by the fact that different modifications have been investigated (single crystals [2], thin films [3], ceramic material [4]) but also that different fields have been used. The field dependence of U_0 being essentially unexplored, it was practically impossible to find a relationship for the results.

Using non-resonant microwave absorption in combination with a SQUID magnetometer, we attempted to determine the

Springer Series in Solid-State Sciences, Vol. 99
Electronic Properties of High-T_c Superconductors and Related Compounds
Editors: H. Kuzmany · M. Mehring · J. Fink

flux decay rate especially at low fields and small values of $t=T/T_c$, under which conditions the "giant" flux creep is difficult to measure with conventional methods, when being restricted to small sample volumes. In particular we were interested in studying in more detail the reported creep rate maximum at approximately 30 K, which was interpreted as evidence for a distribution of pinning potentials [7].

2. Results and discussion

A particularly useful relation between U_0 and the normalized logarithmic creep rate was derived by Hagen and Griessen [7]:

$$dM/Md\ln t\big|_{t=t_b} = -(U_0/kT - \ln(t_b/\tau))^{-1}.$$

Here t_b denotes the time after the sudden change in the external field, which is used for the determination of M and $dM/d\ln t$, and τ is a phenomenological relaxation time introduced to avoid an unphysical divergence for t_b 0. Its value has been estimated [7] as in the range of 10^{-12} to 10^{-6} s. Except for fast creep rates at elevated temperatures, the $\ln(t_b/\tau)$ term can be treated as a small correction, resulting in

$$dM/M\ln t\big|_{t=t_b} = -kT/U_0.$$

Following the arguments of Tinkham [6], the temperature dependence of T/U_0 can be approximated by $t/(1-t^2)\cdot(1-t^4)^{1/2}$, invoking simple scaling laws for the thermodynamical critical field and the London penetration depth. Obviously, a monotonic increase in the normalized logarithmic creep rate with t is predicted in this simple model.

As is shown in fig. 1, even for the *unnormalized* rate, this behavior is observed with the exception of the t=0.9 value. Originating from the steep drop of M close to T_C, even this value is consistent with a monotonic increase of the normalized creep rate over the full temperature range, at variance with the reported creep rate maximum [2]. A compilation of conventional magnetization decay and microwave absorption experiments, obtained for samples field-cooled in B=24 G,

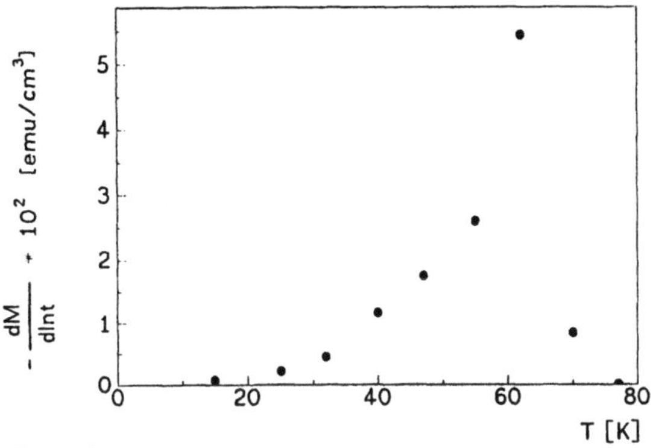

Fig. 1
Temperature dependence of the logarithmic creep rate for a single crystal of $YBa_2Cu_3O_{7-x}$. The rate is measured after cooling in zero field and application of a field of 24G parallel to the c-axis of the crystal.

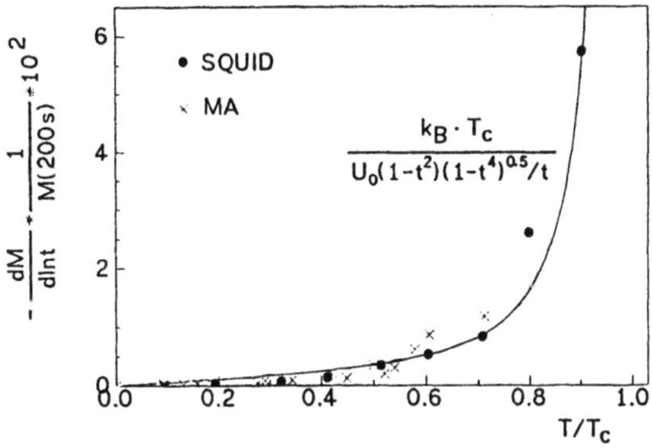

Fig. 2
Compilation of microwave absorption and conventional magnetization decay experiments for a single crystal of $YBa_2Cu_3O_{7-x}$.

is given in fig. 2. The creep rate is determined after sudden field removal. The solid line is obtained by utilizing the U(t) dependence given above. If one further assumes $U_0 \sim 1/B$ [6], the sample can be characterized by $U_0 \cdot B = 240$ eV·G for $\boldsymbol{B} \| \boldsymbol{c}$. From this value the critical current density of the sample can be calculated as $J_{c0}(0) = U_0 \cdot B / k_B \cdot C$ in the notation of ref. 6, yielding $J_{c0}(0) = 4 \cdot 10^4$ A/cm^2, well within the range expected for a twinned single crystal. Preliminary experiments with $\boldsymbol{B} \perp \boldsymbol{c}$ resulted in a critical current density for "out-of-plane" currents smaller by factor of 5, again emphasizing the close correlation of macroscopic transport currents and the flux creep behavior. Recent experiments on YBCO samples with optimized J_{c0} and correspondingly large $U_0 \cdot B$ [8] have confirmed this assumption.

Acknowledgement

This work is supported by the Bundesministerium für Forschung und Technologie under Contract No. 13N5724 5. We are indebted to Prof. E. Dormann and H. Winter, Bayreuth, for making their SQUID magnetometer available to us.

References

[1] M. Takashige, K.A. Müller and J.G. Bednorz, Jap. J. Appl. Phys. **26** suppl. 3 (1987) 1101

[2] Y. Yeshurun and A.P. Malozemoff, Phys. Rev. Lett. **60** (1988) 2202

[3] C. Rossel and P. Chaudhari, Physica C **153–155** (1988) 306

[4] M. Tuominen, A.M. Goldman, and M.L. Mecartney, Phys. Rev. **B37** (1988) 548

[5] D. Hoffmeister, O. Dobbert, K.-P. Dinse, W. Goldacker, and T. Wolf, Europhys. Lett. **8**, (1989) 369

[6] M. Tinkham, Phys. Rev. Lett **61** (1988) 1658

[7] C.W. Hagen and R. Griessen: "Thermally Activated Magnetic Relaxation in High-T_c Superconductors", in: "Studies of High Temperature Superconductors", A.V. Narlikar, Ed., Nova Science Publ., New York

[8] M. Murakami, M. Morita, and N. Koyama, Jap. J. Appl. Phys. **28** (1989) L1754

Magnetic Investigations of Grain and Weak Link Properties in $Bi_{1-x}Pb_xSrCaCu_{1.8}O_y$

M. Reissner[1], *R. Ambrosch*[1], *W. Steiner*[1], *H. Kuzmany*[2], *P. Schaffarich*[2], *R. Stroh*[3], *and W. Wruss*[3]

[1]Institut für Angewandte und Technische Physik, TU Wien, Austria
[2]Institut für Festkörperphysik, Universität Wien, Strudlhofgasse 4, A-1090 Wien, Austria and
Ludwig Boltzmann Institut für Festkörperphysik, Kopernikusgasse 15, A-1060 Wien, Austria
[3]Institut für Chemische Technologie Anorganischer Stoffe, TU Wien, Austria

Abstract. Ceramic samples with nominal composition $Bi_{1-x}Pb_xSrCaCu_{1.8}O_y$ ($0\leq x\leq 1$) were investigated by scanning electron microscopy, X-ray, electrical resistivity and d.c. magnetic measurements. For x between 0.1 and 0.2 the fraction of the 2223 phase in the samples was highest. The influence of the lead substitution on the superconducting properties of the grains of the 2212 phase is weak, but large for the weak links. The critical current density is larger in the 2223 than in the 2212 phase.

Soon after the discovery of the Bi-superconductors (2212 and 2223) by Maeda [1], it was pointed out that the 2223-modification with the higher T_c can be more easily synthesized by substituting some Bi by Pb [2,3]. To obtain additional information about the influence of Pb on the properties of both the grains and the weak links a series of ceramic samples with stoichiometry $Bi_{1-x}Pb_xSrCaCu_{1.8}O_y$ ($0\leq x\leq 1$) was prepared and investigated by scanning electron microscopy, X-ray, resistivity and d.c. magnetic measurements.

Appropriate amounts of Bi_2O_3, PbO, $SrCO_3$, $CaCO_3$ and CuO were mixed, pressed into pellets (5t, 5min), sintered twice at 850°C for 20 hours and furnace cooled. Afterwards the samples were reground, repelletized, sintered at 850°C for 334 hours and again furnace cooled. The density of the samples is about 67% of the theoretical value (table). From the intensities of the (008)-reflection for the 2212 phase and the (115)-re-

Table : Density, transition temperature and width, and fraction of superconducting volume for $Bi_{1-x}Pb_xSrCaCu_{1.8}O_y$

x		0	0.1	0.2	0.3	0.4	0.5	0.6	0.7	0.8	0.9
$\rho(g/cm^3)$		3.67	3.78	4.02	3.72	4.65	4.51	4.58	4.57	3.95	3.91
$T_c(K)^a$	2212	78				70	67	68	63		
	2223		108	108	108						
$T_c(K)^b$	2212	72	71		66	65	67	65	67	65	
	2223		104	106	106						
$\Delta T(K)^c$	2212	36	10		8	10	13	16	>67	>65	
	2223		28	16	>40						
Vsl(%) 4.2K		21	50	45	39	20	16	10	5	<1	

[a]resistivity (midpoint) [b]magnetic (onset) [c]magnetic T_c(onset)-T_c(sat)

Springer Series in Solid-State Sciences, Vol. 99
Electronic Properties of High-T_c Superconductors and Related Compounds
Editors: H. Kuzmany · M. Mehring · J. Fink

flection for the 2223 phase, fractions for the 2223 phase of 5, 50, 70, 40 and 5% were determined for x=0, 0.1, 0.2, 0.3 and 0.4, respectively. The intensities of the reflections were calibrated using volume fractions from ac-susceptibility measurements for similar samples.

The microstructure of the samples was investigated by scanning electron microscopy. Up to x=0.6 platelike grains are present. For x≥0.4 grains with spherical shape are formed. Starting with x=0.6 two phases with different Pb content are present. The mean grain size (ϕ) for the 2212 phase increases smoothly up to x=0.4 and decreases with further increasing x. ϕ is 2 to 3 times larger in the samples mainly consisting of the 2223 phase (inset fig.3).

The d.c.magnetic measurements were performed on cylindrically shaped samples in fields of 0.1mT up to 6.7T using a vibrating sample magnetometer. All measurements were corrected by appropriate demagnetization factors. The superconducting transition temperatures T_c (onset of diamagnetic signal) and the width of the transition ΔT were determined from the shielding curves (table). Whereas for the samples with 2212 phase T_c is only weakly dependent on x, the width of the transition is strongly influenced. Only one transition at 106K occurs for x=0.2, within measuring accuracy. For x=0.1 and 0.3 two transitions appear, with fractions of superconducting volume of the 2223 phase (determined at 85K) of at least 22 and 16%. These are lower limits for the shielded volume at $\mu_0 H_a \approx 0.1$mT, because no corrections due to the penetration depth were made. In agreement with the X-ray data, the sample with x=0.1 contains more 2223 phase than the one with x=0.3.

In the field dependence of the magnetization at constant temperature two hysteresis loops appear due to the weak links (below 20mT) and the grains. The hysteresis for the weak links is broadest for x=0.1, becomes much smaller with increasing x and disappears for x>0.6, indicating that the formation of weak links is strongly reduced at high Pb contents (fig.1). Both the lower critical field (H_{c1w}) and the upper critical field (H_{c2w}) of the weak links are maximal between x=0.1 and 0.2 (fig.2).

From the susceptibility measured in fields higher than H_{c2w} and lower than H_{c1g} (the lower critical field of the grains) the fraction of superconducting volume V_{sl} was estimated taking into account the different densities of the samples (table). V_{sl} shows a maximum between x=0.1 and 0.2 which fits again to the result obtained by X-ray measurements.

The lower critical field of the grains is only weakly dependent on x (fig.2). The hysteresis for the grains are broadest for x=0.2. Within Bean's critical state model [4] upper limits for the critical currents can be estimated. In this model for a sample with cylindrical shape the critical current density is given by [5] $J_c=30\Delta M/D$ (ΔM width of the hysteresis, D diameter of superconducting region). To deduce the critical current density of the grains (J_c), D is assumed to be identical with the grain size ϕ (inset fig.3). For the sample with x=0.1 J_c is higher than for the compounds with x>0.3 (fig.3a), indicating that the critical current density in the 2223 phase is higher than in the 2212 phase, a fact which was also observed by Lee et al. [6] from measurements at 77K on differently prepared samples. The low J_c for x=0.2 and 0.3 samples, which also contain the 2223 phase, might be due to an overestimation of the grain size used in the calculation (inset fig.3), since measurements of the relaxation of the magnetization exhibit a lower creep rate (defined by $S=|d\ln M/d\ln t|$) for all 2223 containing samples, pointing to stronger pinning and therefore also to higher critical current densities. The small amount of the 2223 phase visible in the X-ray data for x=0 may be responsible for the higher J_c value in this sample, too.

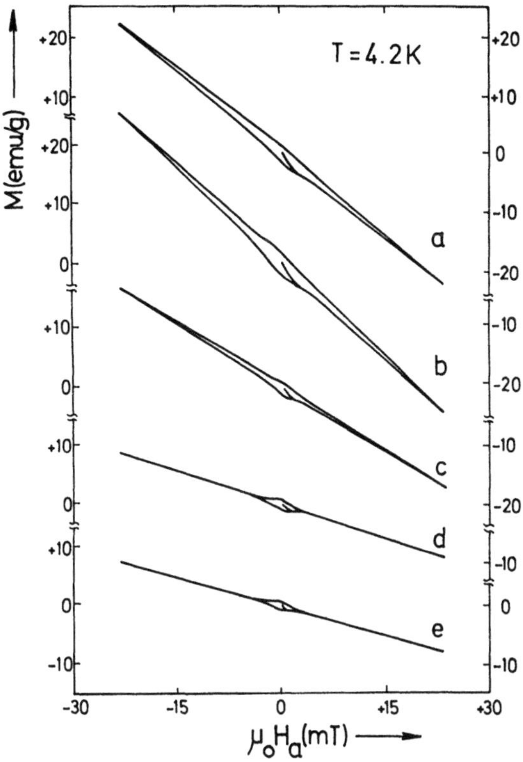

Fig. 1: Hysteresis loops in the weak link regime of $Bi_{1-x}Pb_xSrCaCu_{1.8}O_y$ for x=0.1 (a), 0.2 (b), 0.3 (c), 0.4 (d) and 0.5 (e).

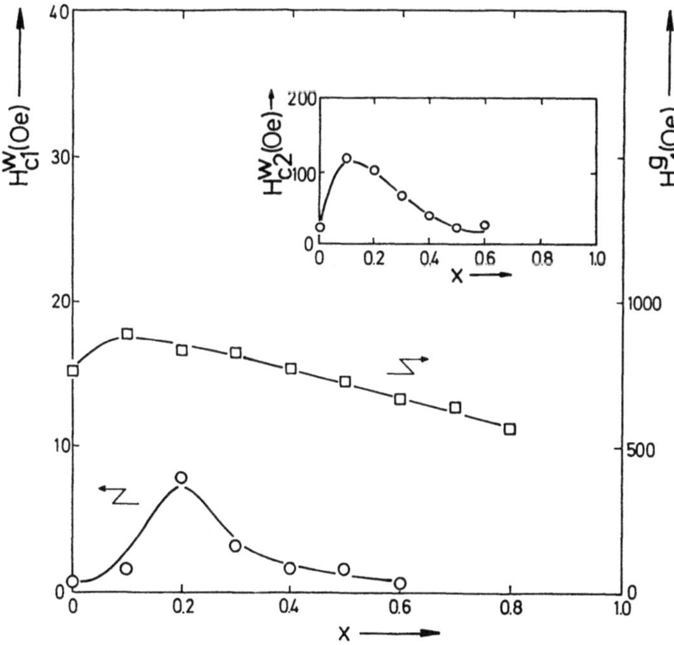

Fig. 2: Lower critical field of the grains (□) and the weak links (○) for $Bi_{1-x}Pb_xSrCaCu_{1.8}O_y$ at 4.2K. Inset: Upper critical field of the weak links at 4.2K.

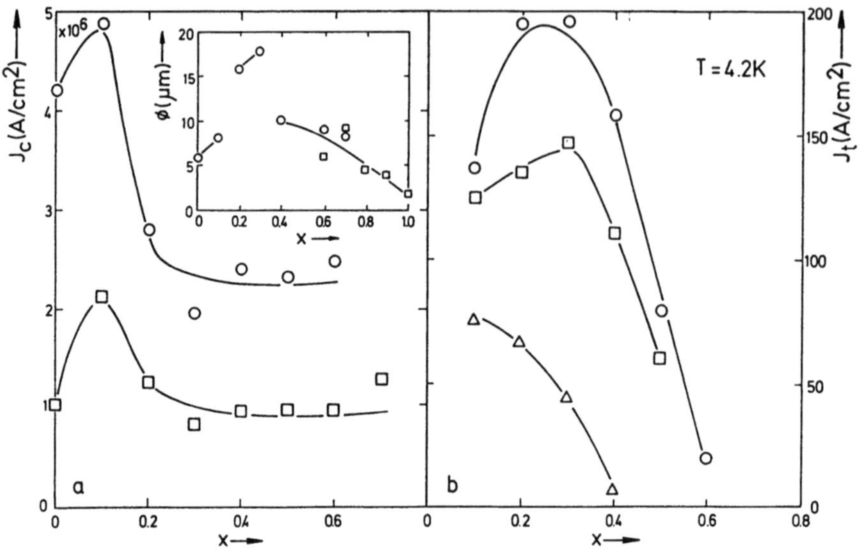

Fig.3: (a) Critical current density in the grains at 1T (O) and 6T (□), (b) critical current density in the weak links estimated from remanent magnetization (O), for fields where dM/dH=0 (□) and for 10mT (Δ). Inset: Diameter of platelike (O) and spherical (□) grains.

The same analysis can be carried out for the weak link hysteresis, leading to the critical current density within the weak links J_t, which should be comparable to directly measured transport current densities. In this analysis D is equal to the outer diameter of the sample. Contrary to the critical current density in the grains, J_t is strongly influenced by the lead substitution (fig.3b).

To summarize, with the chosen preparation conditions the grain size is much larger for the 2223 than for the 2212 phase. Both T_c and J_c are only weakly influenced by the lead substitution in the 2212 phase. The same is valid for H_{c1g}. However, the width of the transition, which reflects the homogeneity of the grains, and also the weak link properties (H_{c1w}, H_{c2w} and J_t) depend strongly on x. The influence of the substitution of lead for Bi is weak in the grains and large on the weak links. High lead content destroys both superconductivity in the grains and the weak links.

Work supported by the Austrian FFWF (grant # P7061,P7059 and P7057).

References:

[1] H. Maeda, Y. Tanaka, M. Fukutomi and T. Asano, Jpn. J. Appl. Phys. 27(1988)L209.
[2] M. Takano, J. Tokada, K. Odo, H. Kitaguchi, Y. Miura, Y. Ikeda, Y. Tamii and H. Mazaki, Jpn. J. Appl. Phys. 27(1988)L1041.
[3] S. A. Sunshine, T. Siegrist, L. F. Schneemeyer, D. W. Murphy, R. J. Cava, B. Batlogg, R. B. van Dover, R. M. Fleming, S. H. Glarum, S. Nahahara, R. Farrow, J. J. Krajewski, S. M. Zahurak, J. V. Waszczak, J. H. Marshall, P. Marsh, L. W. Rupp, Jr. and W. F. Peck, Phys. Rev. B38(1988)893.
[4] C. P. Bean, Rev. Mod. Phys. 36(1964)31.
[5] W. A. Fickz and W. W. Webb, Phys. Rev. 178(1969)657.
[6] H. K. Lee, K. W. Lee, K. Park, N. M. Huang, O. K. Oh, J. S. Kim, K. H. Yoo, Y. B. Kim, C. S. Kim, Y. K. Cho, J. C. Park and S. I. Suck, J. Appl. Phys. 64(1989)1881.

Size Dependence of the ac Susceptibility and Critical Current Densities of $YBa_2Cu_3O_{7-\delta}$ Superconductors

M. Forsthuber[1], *G. Hilscher*[1], *and F. Ludwig*[2]

[1]Institut für Experimentalphysik, TU Wien, Wiedner Hauptstr. 8–10, A-1040 Wien, Austria

[2]Sektion Physik der Humboldt-Universität zu Berlin, O-1040 Berlin, Fed. Rep. of Germany

Abstract. Measurements of the ac susceptibility of an YBCO ceramic sample were performed for different sample sizes. To fit the experimental data, the critical state model was modified by introducing a power law to the magnetic field dependence of the critical current. We found the temperature dependence of the intergranular critical current density at zero field to be $j_{ci} \sim (T_c - T)^2$, indicating that the intergranular material of the sample behaves as a normal metal and the junctions are of SNS type.

Measurements of the ac susceptibility have widely been used to investigate high-T_c superconductors, especially to deduce critical current densities. Several studies have been performed on the temperature dependence of real and imaginary part of the ac susceptibility as a function of the amplitude of the applied magnetic ac field H^{ac} [1-3]. To explain the characteristics experimentally observed, several models have been applied in the literature. Up to now the problem has been best solved by Müller's modification of the critical state model [4]. He supposed the intergranular local critical current density to be

$$j_{ci} \approx \left[\alpha_i(T)/(\mu_{eff}\mu_o)\right] \cdot \left[\,|H_i| + H_{oi}\right]^{-1} \qquad (1)$$

[5,6]. α_i is the pinning force density for intergranular Josephson vortices; it was assumed to be temperature dependent but magnetic field independent. H_i is the local magnetic field in the intergranular region; a material dependent constant, H_{oi}, was added to avoid divergency for vanishing magnetic field.

In the present paper we check Müller's critical state model with respect to the influence of the sample size on the ac susceptibility. For simplicity we consider the intergranular contribution only, i. e. the temperature region where the intragranular phase transition has completed ($T < T_g$).

We performed measurements on one sintered YBCO pellet of single phase. The sample was subsequently rubbed down to obtain various

Springer Series in Solid-State Sciences, Vol. 99
Electronic Properties of High-T_c Superconductors and Related Compounds
Editors: H. Kuzmany · M. Mehring · J. Fink

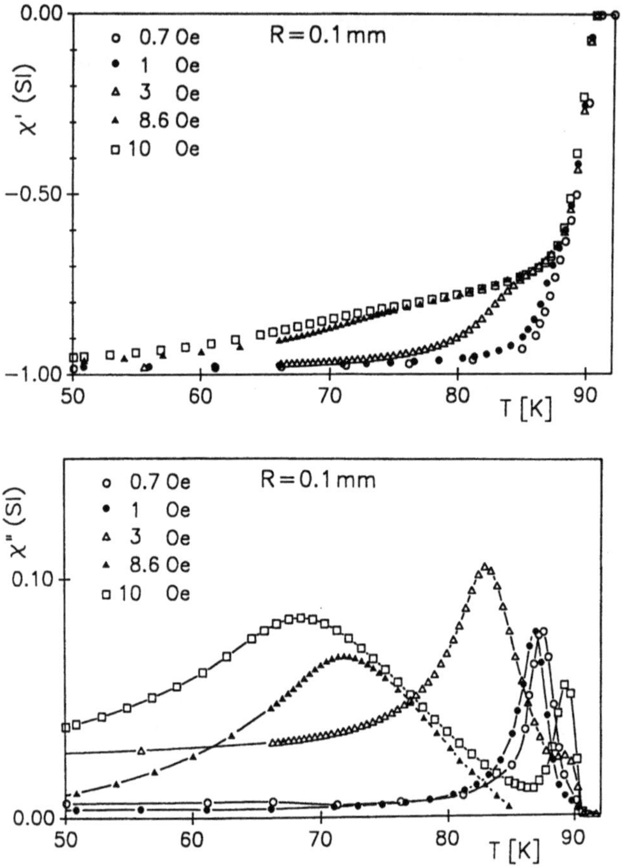

Fig. 1. Real and imaginary part of the ac susceptibility, χ' and χ'', respectively, taken at different rms values of the ac probing field, $\tilde{H}_{ex}^{ac}$, at 1 kHz, for sample 'radius' R = 0.1 mm. Lines are guides to the eye. The graph for $\tilde{H}_{ex}^{ac}$ = 8.6 Oe is omitted above 85 K for clarity.

different cylindrical geometries with respect to sample radius and thickness for the ac susceptibility measurements. We used a calibrated ac susceptometer equipped with a lock-in amplifier to measure the complex susceptibility in the temperature range between 4.2 and 100 K. All measurements were performed at different ac fields and at frequencies of 125 Hz and 1 kHz; in agreement with literature [3,7] the effect of the frequency upon $\chi''(T)$ is rather small in this frequency regime covered. The surface internal ac field peak amplitude H^{ac} was calculated from the probing field rms amplitude $\tilde{H}_{ex}^{ac}$, correcting for the demagnetization field.

Typical susceptibility measurements are shown in Fig.1 a,b for R = 0.1 mm. (2R is either the diameter or the thickness if the rotation axis

Table 1

Intergranular peak temperature, Ti, and penetration field, Hp, for various probing ac field rms amplitudes, $\tilde{H}_{ex}^{ac}$, and sample geometries. A(R) is the slope of the penetration field as described in the text.

no.	R (mm)	A(R) (Oe/K)	$\tilde{H}_{ex}^{ac}$ (Oe)	Ti (K)	Hp (Oe)
1	2.4	(-2.47)	1.0	87.2	4.67
2	1.8	-1.33	0.16	88.3	1.00
3			1.0	87.2	4.10
4			2.7	81.5	14.8
5			4.2	75.8	24.7
6	0.2	-0.69	0.10	88.7	0.154
7			1.0	87.2	1.54
8			10	73	15.6
9	0.1	-0.53	0.70	87.4	1.04
10			1.0	86.7	1.48
11			3.0	82.9	4.43
12			8.6	71.8	12.9
13			10	68.4	15.0

of the pellet is parallel or perpendicular to the applied magnetic field, respectively.) The large shift of the intergranular χ" maximum with increasing ac field amplitude is clearly visible (Fig.1 b). On the other hand, it is clear-cut that on varying the 'radius' R at constant ac field the intergranular peak is shifted to higher temperatures with increasing R, which is obvious as the temperature induced weakening of the intergranular critical current density leads to a larger penetration depth of the magnetic field. All data obtained for different sample geometries and ac fields are listed in table 1.

The maxima of χ" indicate the temperature Ti where the ac field peak amplitude equals the penetration field Hp(Ti). (Strictly speaking, $H^{ac} = q \cdot H_p$, where $q \sim 1$ is a model dependent constant.) In order to get overall information for the dependence of the intergranular peak temperature Ti upon Hp, we plotted Ti *versus* Hp (Fig.2) and found that the two quantities show a linear dependency. The 'radius' R behaves as a parameter of the straight lines, which meet at a point P^* near $T^* = 89.4$ K and $H^* = 0.50$ Oe:

$$H_p - H_o^* = A(R)\cdot(T^* - T_i). \quad (2)$$

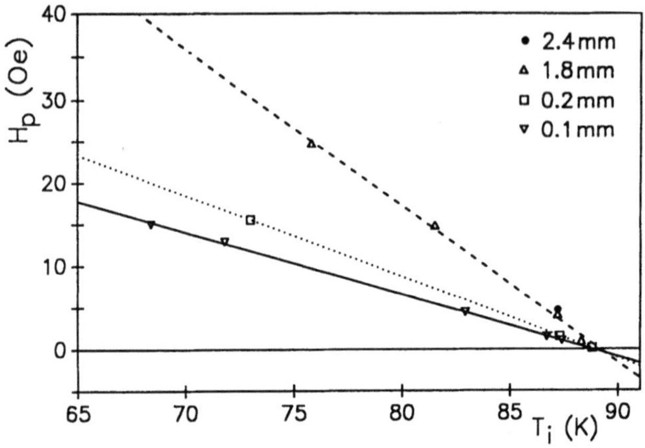

Fig. 2. Penetration field, Hp, as a function of the intergranular peak temperature, Ti, for different sample geometries. The straight lines have been obtained by least square fit.

The slopes sA of these straight lines found by least square fit are given in table 1.

As a generalizing case for eq.1 we assume a power-law model

$$j_{ci} = a(T)\cdot\left[|H_i| + H_{oi}\right]^{-b} \tag{3}$$

for the dependence of the local critical current density on magnetic field, by introducing an exponent b. Solving Ampère's law using eq.3 one obtains the penetration field

$$H_p = \left[H_{oi}^{b+1} + (b+1)a(T)R\right]^{1/(b+1)} - H_{oi}. \tag{4}$$

(compare [3]). By interpolating to 80 K using the straight lines from the least square fit (for R = 2.4 mm we extrapolated eq.2 through the data set no.1), we found a good fit for eq.4 assuming values of b = 1.64 and H_{oi} = 0.51 Oe.

So it can be concluded that the intergranular critical current density is given by

$$j_{ci} = a(T)\cdot\left[|H_i| + 0.5\text{ Oe}\right]^{-1.6}.$$

This can be explained either by a Josephson network description or by pinning effects. As for the former case, the exponent b ≅ 1.6 found experimentally is in good agreement with an Airy-like behaviour [8]. Considering the pinning, our findings indicate a weakening of pinning effects under the influence of the magnetic field obeying the relation

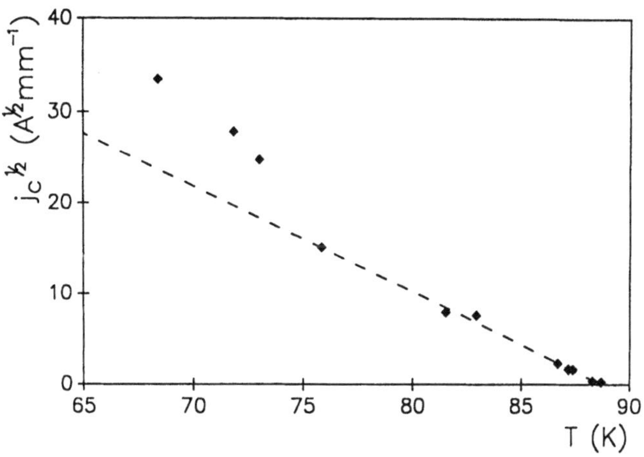

Fig. 3. Square root of the intergranular critical current density, j_{ci}, as a function of the temperature. The straight line has been obtained by a least square fit to the points above 75 K.

$\alpha_i \sim (1 + |H_i|/0.5\ \mathrm{Oe})^{-0.6}$. However, ac measurements cannot give a clear-cut decision between the Josephson network and the pinning interpretation.

Using our results, we calculated a(T) for different temperatures and estimated the temperature dependence of the intergranular local critical current density, $j_{ci}(T, H = 0)$, in zero field. In Fig.3 the square root of the critical current density is plotted against temperature, showing a linear dependency at higher temperatures. Since the critical current of SNS junctions obeys a quadratic law $j_{SNS} \sim (T_c - T)^2$ near T_c [9], we conclude that the intergranular material is essentially normal conducting. It is a crucial point that the experimental data presented here are valid for the same sample, which has only been reduced in size (i. e., rubbed smaller), so the relations for different sample thicknesses hold for the very same specimen, ruling out any individual deviations caused by eventual differences of the treatment of different samples. By the same argument, however, the extent to which the above results may be extended is very uncertain. Nevertheless our findings confirm the usual opinion, that the intergranular matrix shows normal metallic conductivity behaviour.

Acknowledgement

This work was supported by the Austrian Science Foundation Fonds (Fonds zur Förderung der Wissenschaftlichen Forschung) under project no. 7060.

References

[1] R.B.Goldfarb, A.F.Clark, A.I.Braginski and A.J.Panson, Cryogenics 27 (1987) 475

[2] F.Gömöry and P.Lobotka, Solid State Commun. 66 (1988) 645

[3] D.-X.Chen, J.Nogues and K.V.Rao, Cryogenics 29 (1989) 800

[4] K.-H.Müller, Physica C 159 (1989) 717

[5] K.-H.Müller, J.C.Macfarlane and R.Driver, Physica C 158 (1989) 69

[6] Y.B.Kim, C.F.Hempstead and A.R.Strnad, Phys.Rev.Lett. 9 (1962) 306; Phys. Rev. 129 (1963) 528

[7] M.Nikolo and R.B.Goldfarb, Phys. Rev. B 39 (1989) 6615

[8] R.L.Peterson and J.W.Ekin, Physica C 157 (1989) 325

[9] L.Solymar, Superconductive Tunneling and Applications, Wiley - Interscience 1972

$Y_{1-x}Gd_xBa_2Cu_3O_{7-\delta}$ and $YBa_2Cu_{3-x}Gd_xO_{7-\delta}$: Similarities and Differences

N. Brničević[1], P. Planinić[1], V. Žerjav[1], A. Dulčić[1], B. Rakvin[1], M. Požek[1], M. Forsthuber[2], G. Hilscher[2], and R. Kirchmayr[2]

[1]Rudjer Bošković Institute, University of Zagreb, Bijenička 54, P.O. Box 1016, YU-41001 Zagreb, Croatia, Yugoslavia
[2]Institut für Experimentalphysik, TU Wien, Wiedner Hauptstr. 8–10, A-1040 Wien, Austria

Abstract. Two different series of samples with the nominal composition $Y_{1-x}Gd_xBa_2Cu_3O_{7-\delta}$ and $YBa_2Cu_{3-x}Gd_xO_{7-\delta}$ ($0 \leq \underline{x} \leq 1$) were prepared by the solid state reaction. Pronounced sensitivity to Gd content is observed for $YBa_2Cu_{3-x}Gd_xO_{7-\delta}$. Besides formation of several impurity phases in this series the transformation of the orthorhombic phase to tetragonal symmetry is observed for $\underline{x} \geq 0.1$. The ac susceptibility, resistivity and field-dependent microwave absorption measurements show a rapid degradation of superconducting properties as a function of gadolinium content.

1. Introduction

Soon after discovery of the superconducting $YBa_2Cu_3O_{7-\delta}$ phase it was found that the substitution of Y by the rare-earth metal elements preserves the structural integrity of the orthorhombic phase and superconducting properties as well /1,2/. Respecting two distinct and chemically dissimilar copper sites in $YBa_2Cu_3O_{7-\delta}$ the metal elements in different oxidation states (Al^{3+}, Ni^{2+}, Zn^{2+}) were used for partial substitution of copper atoms /3/. Here we report on the comparative study of structural, magnetic, superconducting and microwave properties of the two different series of samples $Y_{1-x}Gd_xBa_2Cu_3O_{7-\delta}$ and $YBa_2Cu_{3-x}Gd_xO_{7-\delta}$.

2. Materials and Methods

Two series of samples : $Y_{1-x}Gd_xBa_2Cu_3O_{7-\delta}$ and $YBa_2Cu_{3-x}Gd_xO_{7-\delta}$ ($0 \leq \underline{x} \leq 1$) were prepared from stoichiometric amounts of Y_2O_3, Gd_2O_3, CuO and $BaCO_3$. The mixtures were treated isothermally at 860° and 900° C with intermediate grinding and pelletizing. The pellets were finally fired at 950° C and slowly cooled to room temperature (cooling rate 70°/h). The X-ray powder diffraction patterns were recorded with a Philips counter diffractometer using CuKα radiation. For electrical resistivity measurements a four-point technique was employed. The ac magnetic susceptibility was measured with a standard lock-in technique at 81 Hz with a field amplitude of 1 Oersted. The EPR spectra were recorded on a Varian E-109 spectrometer working at 9.56 Hz with 100 kHz field modulation. The same spectrometer was used for the examination of the low-field nonresonant microwave absorption.

3. Results and Discussion

As expected, significant differences in the structural, magnetic, superconducting, EPR and low-field microwave absorption properties of the two series of samples: $Y_{1-x}Gd_xBa_2Cu_3O_{7-\delta}$ and $YBa_2Cu_{3-x}Gd_xO_{7-\delta}$ were found. In the process of synthesis of $Y_{1-x}Gd_xBa_2Cu_3O_{7-\delta}$, besides the orthorhombic $(Y,Gd)Ba_2Cu_3O_{7-\delta}$ phase, the X-ray powder diffraction patterns indicate the formation of $BaCuO_2$ as the only impurity phase observed for $x \geqslant 0.2$ values. The transformation of the orthorhombic phase to tetragonal symmetry was not observed for any x value including $x = 1$. On the contrary, the samples of the system $YBa_2Cu_{3-x}Gd_xO_{7-\delta}$ were structurally very sensitive to the amount of Gd present. In this system the formation of several impurity phases was observed like Y_2BaCuO_5 (or Gd_2BaCuO_5) at $x > 0.05$, $BaCuO_2$ at $x \geqslant 0.2$ and the third unidentified phase at $x \geqslant 0.8$ (Fig. 1). Careful analysis of the X-ray powder diffractograms of these samples indicates the transformation of the orthorhombic phase to tetragonal symmetry occurring for $x \geqslant 0.1$. This could indicate that besides the formation of the impurity phases, very possibly containing Gd, simultaneously a certain amount of this element is involved in substitution of copper atoms in its Cu(1) sites in the $YBa_2Cu_3O_{7-\delta}$ structure.

It is interesting to follow the resistivity curves of the two series of samples. Severe degradation of superconducting properties starting at $x \geqslant 0.2$ was found only for $YBa_2Cu_{3-x}Gd_xO_{7-\delta}$ series. Fig. 2. shows the resistivity curves for $YBa_2Cu_{2.8}Gd_{0.2}O_{7-\delta}$ as a function of temperature in various dc magnetic fields. The initial drop is due to the superconductivity of the grains. The field dependent part of the resistivity curves reflects the intergranular properties of the compound.

The ac magnetic susceptibility is measured for both series of samples. Illustrative examples $Y_{0.6}Gd_{0.4}Ba_2Cu_3O_{7-\delta}$ and $YBa_2Cu_{2.6}Gd_{0.4}O_{7-\delta}$ are

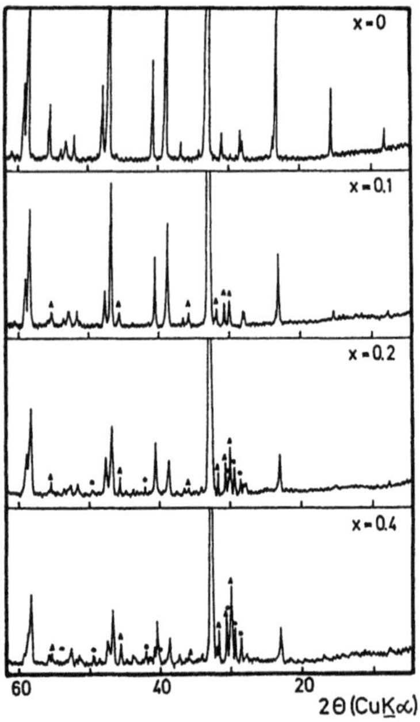

Fig. 1. X-ray powder diffraction patterns of $YBa_2Cu_{3-x}Gd_xO_{7-\delta}$ for several values of x; (▲) $(Y,Gd)_2BaCuO_5$, (●) $BaCuO_2$

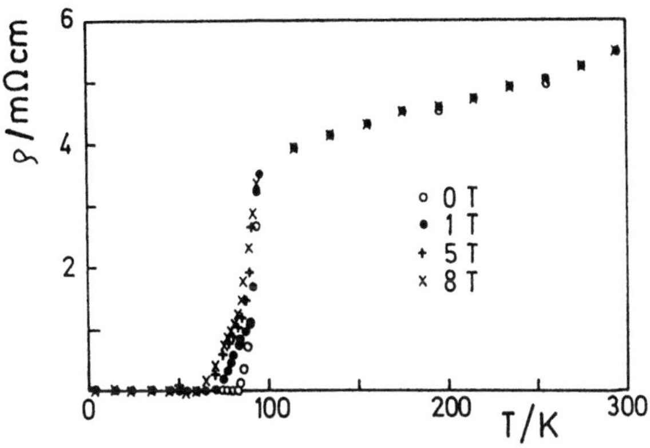

Fig. 2. Field-dependent electrical resistivity as a function of temperature for $YBa_2Cu_{2.8}Gd_{0.2}O_{7-\delta}$

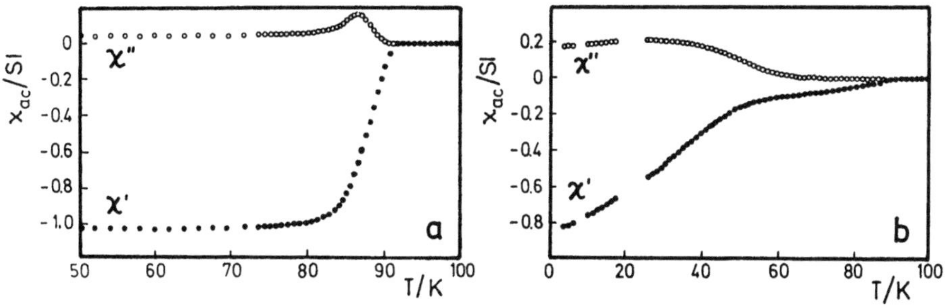

Fig. 3. Real (χ′) and imaginary (χ″) magnetic susceptibility as a function of temperature for: a) $Y_{0.6}Gd_{0.4}Ba_2Cu_3O_{7-\delta}$; b) $YBa_2Cu_{2.6}Gd_{0.4}O_{7-\delta}$

shown in Fig. 3. The onset of diamagnetism occurs at the same temperature, but the transition is broader in the latter compound. Also, the coupling between the grains, as evidenced from the peak in the imaginary part of the susceptibility, is shifted towards lower temperatures for the latter compound. In particular, at 77 K it is evident that the grains are coupled in the former compound, but not in the latter. An independent confirmation of this finding can be obtained from the analysis of the low-field microwave absorption, as shown for the two compounds in Fig. 4. The former compound exhibits a pronounced hysteresis which is due to the current induced in the course of the dc field sweep on the surface of the granular sample /4/. The latter compound shows no hysteresis at 77 K, which is an indication that practically no superconducting current flows between the grains, i.e. the grains are not coupled at this temperature.

An additional spectroscopic study of these systems can be provided by electron paramagnetic resonance (EPR). Fig. 5 shows EPR spectra of $Y_{0.6}Gd_{0.4}Ba_2Cu_3O_{7-\delta}$ and $YBa_2Cu_{2.6}Gd_{0.4}O_{7-\delta}$. The difference in the two spectra is striking. The former compound shows only a broad EPR line as expected for Gd^{3+} in $GdBa_2Cu_3O_{7-\delta}$ /5/. No detectable signal of Cu^{2+} could be found in this compound. Hence, the compound is entirely copper-based orthorhombic structure. On the contrary, the nominal $YBa_2Cu_{2.6}Gd_{0.4}O_{7-\delta}$ compound shows a strong signal similar to that found in $(Y,Gd)_2BaCuO_5$.

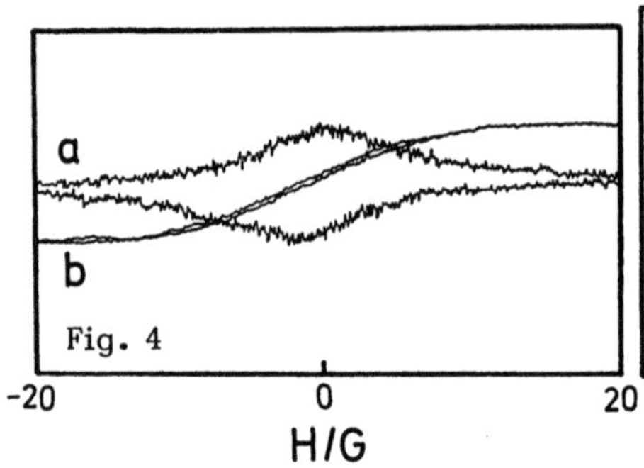

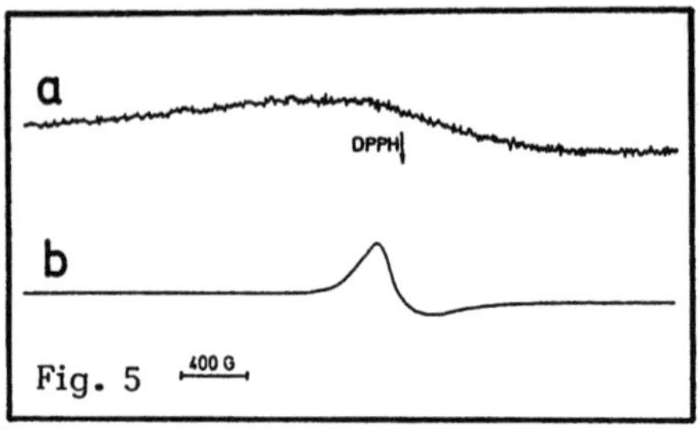

Fig. 4. Low-field modulated microwave absorption at 77 K for:
a) $Y_{0.6}Gd_{0.4}Ba_2Cu_3O_{7-\delta}$; b) $YBa_2Cu_{2.6}Gd_{0.4}O_{7-\delta}$

Fig. 5. EPR spectra of: a) $Y_{0.6}Gd_{0.4}Ba_2Cu_3O_{7-\delta}$; b) $YBa_2Cu_{2.6}Gd_{0.4}O_{7-\delta}$

The above analysis points to the conclusion that in the samples with the nominal composition $YBa_2Cu_{3-x}Gd_xO_{7-\delta}$ a copper-based orthorhombic structure is segregated forming the superconducting grains, while other nonsuperconducting phases, such as $(Y,Gd)_2BaCuO_5$ and $BaCuO_2$, are interpolated. The transformation of the orthorhombic structure to tetragonal symmetry observed for $\underline{x} \geq 0.1$ could be the consequence of the copper atoms substitution in $YBa_2Cu_3O_{7-\delta}$ orthorhombic structure in the Cu(1) sites. For small values of $\underline{x}$, the effect of the nonsuperconducting phases is mainly to reduce the coupling between the superconducting grains, while for larger $\underline{x}$ values the fraction of the superconducting phase remarkably decreases.

References

/1/ Hor P H, Meng R L, Wang Y Q, Gao L, Huang Z J, Bechtold J, Forster K and Chu C W 1987 Phys. Rev. Lett. 58 1891

/2/ Tarascon J M, McKinnon W R, Green L H, Hull G W and Vogel E M 1987 Phys. Rev. B 36 226

/3/ Jones R, Edwards P P, Harrison M R, Thanyosiri T and Sinn E 1988 J. Am. Chem. Soc. 110 6716

/4/ Požek M, Dulčić A and Rakvin B 1989 Solid State Commun. 70 889

/5/ Schwartz R N, Pastor A C, Pastor C R, Kirby K W and Rytz 1987 Phys. Rev. B 36 8858

Electron Tunneling into $Y_1Ba_2Cu_3O_7$ Thin Films

J. Geerk[1], *G. Linker*[1], *O. Meyer*[1], *Q. Li*[2], *R.-L. Wang*[3], *and X.X. Xi*[2]

[1]Kernforschungszentrum Karlsruhe, Institut für Nukleare Festkörperphysik, Postfach 3640, D-7500 Karlsruhe, Fed. Rep. of Germany
[2]University of Beijing, Beijing, People's Republic of China
[3]Chinese Academy of Science, Beijing, People's Republic of China

Abstract. We report on the fabrication and measurements of sandwich-type tunnel junctions on the HTSC materials of the 1-2-3 type. We obtained gaplike structures with high reproducibility corresponding to a gap near 16 meV and or near 30 meV depending on the fabrication procedure of the tunnel barrier. Both results agree well with tunnel data from other groups.

1. Introduction

Since the discovery of the high T_c superconductors (HTSC) the most important research from the scientific point of view is to get insight into the superconducting mechanism of these new materials. The most sensitive probe for the microscopic properties of a superconductor is electron tunneling. This experimental method primarily delivers the superconducting density of states which then can further be used to test theoretical models and to determine important parameters such as the energy gap and the electron-electron coupling spectrum.

Most of the experimental difficulties of this method arise from its sampling depth, which is of the order of the superconductor coherence length, i.e. 20-50 Å for most of the interesting superconductor compounds. For the 1-2-3 material group, e.g. $Y_1Ba_2Cu_3O_7$, from critical field measurements a coherence length of about 3 Å is deduced in the direction of the c-axis whereas for directions parallel to the a-b plane about 30 Å are found. Thus the a-b plane coherence length is comparable to that of conventional superconductor compounds and has encouraged us to study the HTSC using experimental procedures analoguous to those which proved successful for the conventional superconductors. In this paper we present tunneling results on $Y_1B_2Cu_3O_7$ films prepared on $SrTiO_3$ single crystals with (110) orientation. The junctions are prepared with natural oxide barriers as well as with artificial tunnel barriers and the results are compared to those of other groups. We note that this paper is essentially devoted to junctions of the sandwich-type or, more generally expressed, to junctions with a tunneling area of macroscopic size, e.g. 1 mm^2. Compared to junctions of the point contact type the macroscopic size junctions, formed either by evaporated counter electrodes or by squeezing a piece of a soft metal like Pb or In onto the

Springer Series in Solid-State Sciences, Vol. 99
Electronic Properties of High-T_c Superconductors and Related Compounds
Editors: H. Kuzmany · M. Mehring · J. Fink

tunnel barrier, seem to deliver more reliable results, have easier to handle tunneling resistances and can be raised to higher temperatures to control if the observed structures are really due to superconductivity.

2. Experimental

The thin films of $Y_1Ba_2Cu_3O_7$ were prepared in situ by sputtering from an inverted cylindrical magnetron device. After deposition the vacuum vesssel was vented with oxygen up to a pressure of 100 Torr. The samples were slowly cooled down till 450°C to achieve full oxygen stoichiometry. The films were deposited on (110) oriented $SrTiO_3$ substrates, where they grow with the mixed/(110)/(013) directions perpendicular to the film plane as determined by X-ray and channeling experiments. Their zero resistance appeared typically near 86 K and their critical current density at 77 was near 3×10^5 A/cm^2. It should be noted that with (100) $SrTiO_3$ substrates we get routinely 2×10^6 A/cm^2 at the same deposition conditions.

A naturally grown oxide which can serve as tunnel barrier was obtained just by cooling the samples down to RT in the oxygen atmosphere. More details on junction fabrication, counter electrodes and junction evaluation can be found in ref. 1. A substantial part of this paper is dedicated to more new results, which we obtained with an artificial tunnel barrier. These barriers were prepared in the following way: to avoid the growth of a natural oxide the oxygen was pumped away to 10^{-1} Torr on cooling down the sample from 450°C to RT. At RT a thin amorphous overlayer of 20 Å thickness of the 1-2-3 material was deposited onto the sample. This overlayer has good isolating properties and served as tunnel barrier.

3. Results and Discussion

The results obtained on junctions with the natural oxide tunnel barrier are described in detail in ref.1. Briefly summarized we have presented dI/dV data on these junctions which showed a gaplike feature with an amazing reproducibility demonstrated on as much as 15 junctions in that publication. In addition to this gaplike feature, which corresponds to an energy gap of about 16 meV, on most junctions, except a few, additional structures were found at higher voltages. For comparison with the tunnel data of other groups, we take in the following the tunnel data of a junction with and without these additional structures. These data were normalized to the featureless background determined in the normal state. It is emphasized, that this normalization only serves to demonstrate more vividly the structures the junctions show at low temperatures and is not a normalization in the traditional sense as outlined in more detail in ref. 1. In Fig. 1 we show these tunneling data in comparison with dI/dV data of two other groups. The upper trace was obtained by Takeuchi et al.[2] on sintered samples of $Y_1Ba_2Cu_3O_7$ on

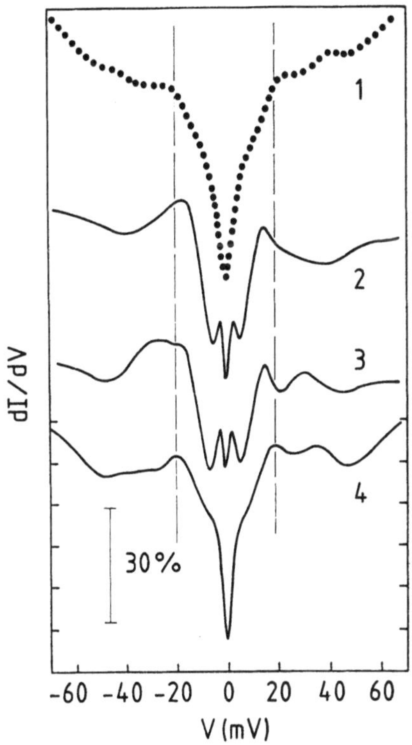

Fig. 1
Comparison of dI/dV data of tunnel junctions on Y-Ba-Cu-O from different groups. Trace 1: Takeuchi et al., Trace 2,3: Geerk et al., Trace 4: Gurvitch et al.

which a tunnel junction was formed by squeezing a piece of Pb on it at low temperatures.

The more recent data of Gurvitch et al. [3] were taken on junctions of the sandwich type prepared on single crystals of $Y_1Ba_2Cu_3O_7$. Also, these authors outlined a high reproducibility of their results and in addition the comparably very good quality of their junctions with respect to non-tunneling currents at zero bias. We observe in Fig. 1 that all the data show a gaplike structure which corresponds to a gap of about 20 meV for the upper curve, to a gap of 19 meV for the data of Gurvitch et al. and to a gap of about 16 meV for our data. The smaller gap of our thin film diodes is possibly due to a reduced quality of the tunnel barrier in the case of natural oxides. Furthermore, we observe on all the junctions at zero bias a gaplike dip which would correspond to a gap of 3 to 5 meV and above 25 meV additional structures which look asymmetric especially for the data of Gurvitch et al. On the whole all the structures look very similar and we regard them to be due to the same physical effect.

In ref. 1 we have compared our data with those of Fournel et al. [4] which like those of Gurvitch et al. were taken from a sandwich type junction prepared on single crystal $Y_1Ba_2Cu_3O_7$. Their data showed a gaplike structure with no additional structures above the gap energy and it corresponds to a gap of the very high energy of 30 meV. As we knew

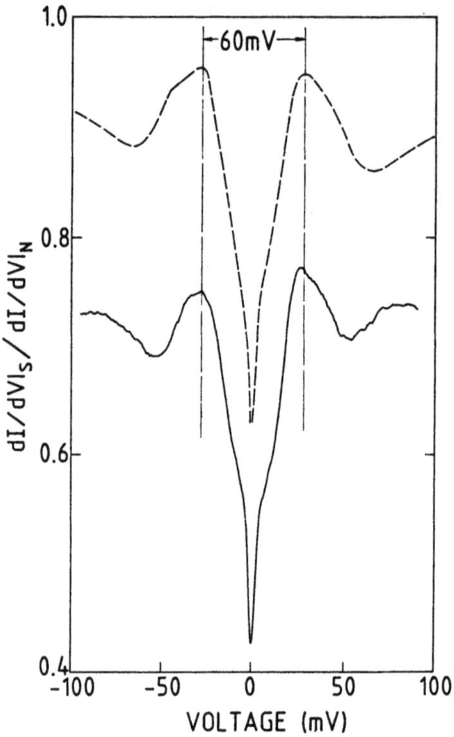

Fig. 2
Comparison of normalized dI/dV data. Solid line: $Y_1Ba_2Cu_3O_7$ thin film with artificial tunnel barrier (this work). Dashed line: Data from Fournel et al. on single crystal $Y_1Ba_2CuO_7$.

from tunneling experiments into conventional superconductors that the tunnel data can be quite sensitive to the way the tunnel barrier is formed we fabricated the junctions with artificial barriers as described above. Fig. 2 shows the normalized tunnel data taken from such a junction together with those of Fournel et al. The vertical scale is valid for both curves, which are vertically shifted for clarity. The data show a very close agreement both in shape and strength of the structures.

We summarize, that for the case of the natural oxide barriers our thin film tunnel data of $Y_1Ba_2Cu_3O_7$ show a gaplike feature with additional structures at higher energies. These structures can be regarded to be essentially confirmed later by Takeuchi et al. and most recently by Gurvitch et al. For a sample with a T_c of 90 K the energy gap has a maximum value near 20 meV. For the case of an artificial tunnel barrier we find a gaplike structure with no additional structures above the gap voltage and an energy gap near 30 meV. We consider these results to strongly confirm those of Fournel et al. A common feature of all the structures shown above is that the essential part of the structure concerns only 20-30 % of the whole tunneling current.

REFERENCES

1. J. Geerk, X.X. Xi and G. Linker, Z. Phys. B **73** (1988) p. 329
2. I. Takeuchi, J.S. Tsai, Y. Shimakawa, T. Manako, and Y. Kubo, Physica C **158** (1989) p. 83-87
3. M. Gurvitch, J.M. Valles, A.M. Cucolo, R.C. Dynes, J.P. Garno, L.F. Schneemeyer, and J.V. Wasczak, Phys. Rev. Lett. **63** (1989) 1009
4. A. Fournel, I. Oujia, I. Sorbier, Europhys. Lett. **6** (1988) p. 653

Model for Field-Dependent Microwave Absorption in Granular High-T_c Superconductors

M. Požek[1], *A. Dulčić*[1,2], *and B. Rakvin*[1]

[1]Rudjer Bošković Institute, University of Zagreb, Bijenička 54, P.O. Box 1016, YU-41001 Zagreb, Croatia, Yugoslavia
[2]Faculty of Natural Sciences and Mathematics, University of Zagreb, P.O. Box 187, YU-41001 Zagreb, Croatia, Yugoslavia

Abstract. The model for loss mechanism in field dependent microwave absorption at low magnetic fields is developed and compared with experimental signals in the first and the second harmonic detection.

1. Introduction

Microwave absorption is an important experimental technique for the study of superconductors. Measurements in the absence of a dc magnetic field show the quasiparticle response at various temperatures and can be used to study the gap properties. In another class of microwave measurements one observes the dependence of the absorption on the external magnetic field. They can be used for the study of the magnetic properties of superconductors.

Several research groups have observed magnetic field dependent microwave absorption in high-T_c superconductors /1-3/. The absorption increases as the external field is applied. When an alternating field is superimposed on the dc field, some peculiar nonlinear effects can be observed /4,5/. When magnetic field modulation and lock-in detection is used, one may expect that the output from the lock-in amplifier would be the first derivative of the absorption spectrum. However, in granular superconductors this is not the case. A pronounced hysteresis is observed in the first harmonic signal for small modulation amplitudes (fig. 1.a), while it is reduced at larger modulation amplitudes (fig. 1.b) /4,5/. On the contrary, in the second harmonic detection there is no hysteresis at low modulation amplitudes (fig. 1.c), but a small hysteresis appears at larger modulation amplitudes (fig. 1.d) /5/.

The aim of the present paper is to show that both experimentally puzzling features of the signals can be deduced from a single microscopic mechanism of absorption. Recently, we have proposed a model which explains successfully the lineshapes of the first harmonic signals /6/. Here we extend this analysis to the second harmonic of the modulation frequency.

2. Model

A granular superconductor forms a network of weak links e.g. Josephson junctions. In many cases one can represent a granular superconductor by an effective weak medium which reproduces its overall properties at low magnetic fields. A weak pinning in the medium brings about a

Springer Series in Solid-State Sciences, Vol. 99
Electronic Properties of High-T_c Superconductors and Related Compounds
Editors: H. Kuzmany · M. Mehring · J. Fink

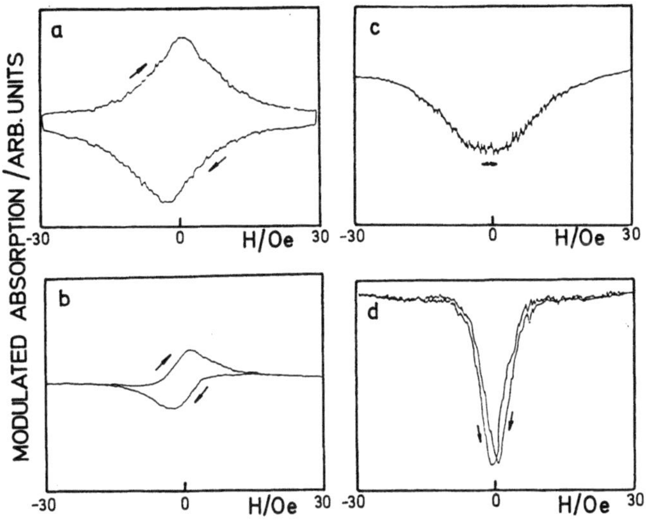

Fig. 1 Modulated absorption of a ceramic superconductor $YBa_2Cu_3O_7$ at T = 77 K: a) first harmonic (100 kHz) detection, modulation amplitude H_M = 0.01 Oe; b) first harmonic, H_M = 0.1 Oe; c) second harmonic (50 kHz) detection, H_M = 0.1 Oe; d) second harmonic, H_M = 1.6 Oe.

boundary current I_o whose sense depends on the direction of the magnetic field sweep.

The simplest microscopic model is to observe the microwave absorption of a single Josephson junction embedded in the effective medium. The differential equation reads

$$\frac{C\hbar}{2e}\frac{d^2\varphi}{dt^2} + \frac{1}{R}\frac{\hbar}{2e}\frac{d\varphi}{dt} + I_c \sin\varphi = I_o + I_M \cos\omega_M t + I_{mw} \cos\omega_{mw} t$$

where C, R and I_c are the capacitance, normal resistance and critical current of the junction, respectively. The modulation and microwave fields induce currents which are superimposed on I_o.

We assume that I_o is limited by weaker junctions in the medium, and observe a representative junction with $I_c > I_o$ ($I_o = I_c \sin\varphi_o$). Next, we consider that the response of the superconductor is fast on the scale of the applied modulation frequencies so that modulation can be treated adiabatically

$$I_c \sin(\varphi_o + \varphi_M) = I_o + I_M \cos\omega_M t \ .$$

Both I_c and I_o should decrease with the applied external field. Instead of a typical diffraction like dependence, we take a smooth envelope as an average over a distribution of the junction size (fig. 2). The applied external field includes the dc field and the modulation field, so that we consider the quantity $I_c(H+H_M \cos\omega_M t)$.

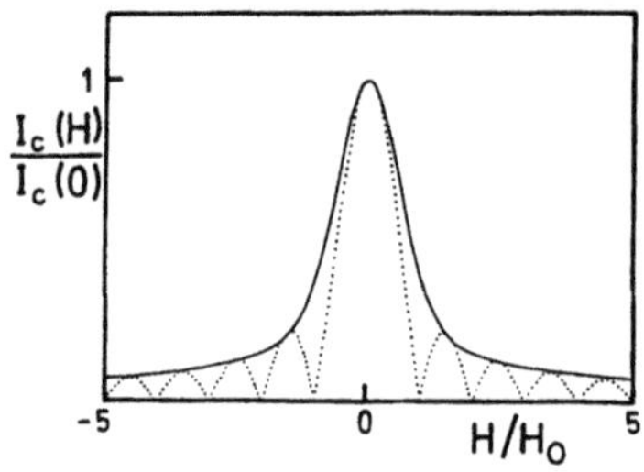

Fig. 2 The diffraction pattern and its envelope used as a reduction factor for the critical currents in Josephson junctions.

3. Results

Solutions can be found for weak microwave fields, i.e. when small oscillations are induced in the phase

$$\varphi(t) = \varphi_o + \varphi_M(t) + \varphi_{mw}(t), \qquad \varphi_{mw} << \varphi_o .$$

The differential equation can be linearized in ω_{mw} and then solved. The field modulation causes nonlinear response of the microwave absorption. The solutions for the first and the second harmonic are

$$S_M = \frac{((1/2)I_{mw}^2 R)^{1/2}}{((1/R)(\hbar/2e)\,\omega_{mw})^2} \cdot \frac{1}{(1+\eta)^{3/2}} \left[- I_c \frac{dI_c}{dH} H_M + I_M I_o \right]$$

$$S_{2M} = \frac{1}{4} \frac{((1/2)I_{mw}^2 R)^{1/2}}{((1/R)(\hbar/2e)\,\omega_{mw})^2} \cdot \frac{1}{(1+\eta)^{3/2}} \left[I_M^2 - \left(\frac{dI_c}{dH}\right)^2 H_M^2 - \right.$$

$$\left. - I_c \frac{d^2 I_c}{dH^2} H_M^2 + \frac{3}{1+\eta} \cdot \frac{1}{((1/2)(\hbar/2e)\,\omega_{mw})^2} \left(I_o I_M - I_c \frac{dI_c}{dH} H_M \right)^2 \right]$$

where $$\eta = \frac{I_c^2 \cos^2 \varphi_o}{((1/R)(\hbar/2e)\,\omega_{mw})^2} .$$

Reversing of the field sweep changes sign of $I_o = I_c \sin \varphi_o$ and hysteresis appears.

The first harmonic signal has a sweep independent (reversible) part and a part which changes sign when the field sweep is reversed. The former dominates at large modulation amplitudes, while the latter yields a pronounced hysteresis at smaller modulation amplitudes (figs. 3.a and 3.b.). The second harmonic signal has many terms, most of which are independent of the field sweep. At higher modulation amplitudes one expects a small hysteresis from a sweep dependent term (figs. 3.c and 3.d).

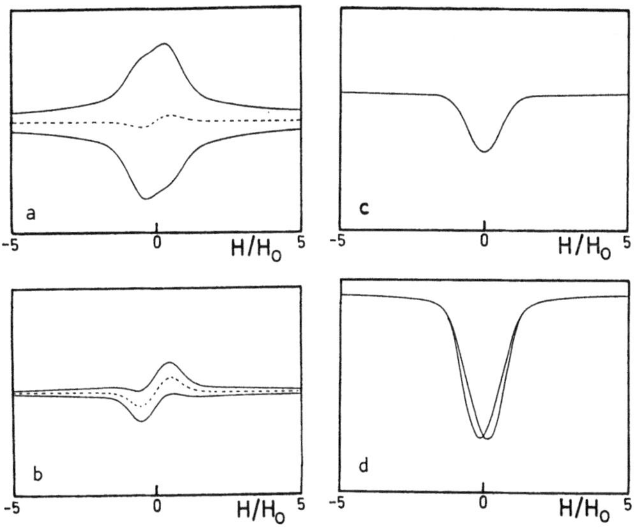

Fig. 3 Modulated absorption signals calculated in forward and reverse field sweeps: a) first harmonic (S_M) for the low modulation amplitude; b) first harmonic for high modulation amplitude; c) second harmonic (S_{2M}) for low modulation amplitude; d) second harmonic for high modulation amplitude. Dotted line represents sweep independent term S_M.

In conclusion, the analytical solutions of the present model contain the basic features which were observed experimentally in the microwave signals.

References

1. K.W. Blazey, K.A. Müller, J.G. Bednorz, W. Berlinger, G. Amoretti, E. Buluggiu, A. Vera and F.C. Matacotta, Phys. Rev. **B 36**, 7241 (1987).
2. M. Perić, B. Rakvin, M. Prester, N. Brničević and A. Dulčić, Phys. Rev. **B 37**, 522 (1988).
3. E.J. Pakulis and T. Osada, Phys. Rev. **B 37**, 5940 (1988).
4. M. Stalder, G. Stefanicki, M. Warden, A.M. Portis and F. Waldner, Physica **C 153–155**, 659 (1988).
5. B. Rakvin, M. Požek, M. Paljević and N. Brničević, Solid State Commun. **70**, 729 (1989).
6. A. Dulčić, B. Rakvin and M. Požek, Europhys. Lett. **10**, 593 (1989).

The Tunneling Gaps of $YBa_2Cu_3O_x$ High-T_c Superconductors

A. Plecenik and Š. Beňačka

Institute of Electrical Engineering, Slovak Academy of Sciences,
842 39 Bratislava, Czechoslovakia

Abstract. The preparation and measurement of tunneling properties of sandwich-type junctions and point contact junctions with a $YBa_2Cu_3O_x$ base electrode and a Pb or Nb counter electrode are reported. In all cases a natural oxide of the base electrode was used as the tunnel barrier. The highest energy gap $\Delta(0) = 21$ meV we ascribe to a 60 K $YBa_2Cu_3O_x$ phase with $2\Delta/kT_c \approx 8$, and lower energy gaps to oxygen deficient parts of high-T_c superconductor. For different junctions the relatively large spread in the small energy gap $\Delta(0) = 1$–4 meV can be explained by proximity effect enhancement of the critical temperature of the counter electrode, or by a large critical temperature dispersion of the HTSC. The absence of 90 K phase near the barrier is jusified by oxygen deficiency due to creation of counter electrode oxides.

1. Introduction

The most sensitive method for investigation of surface properties of new high-T_c superconductors (HTSCs) is electron tunneling spectroscopy. Because of the small coherence length of HTSCs the quality of the surface plays a crucial role in tunneling results. In many papers large dispersion in energy gap values as well as $2\Delta(0)/kT_c$ ratio have been reported [1]. The energy gap for 123 materials is very sensitive to different methods of tunnel barrier preparation [2]. With a natural oxide barrier a gap-like structure was reproducibly observed at about 20 meV [3,4]. Much better results were obtained on thin film structures with an artificial tunnel barrier: $\Delta(0) = 30$ meV, $2\Delta/kT_c \approx 8$ for the $Ba_2Cu_3O_x$ 90 K phase [5]. Optimistic results were reported on epitaxial multilayer structures [6].

In this paper we present results obtained on sandwich-type and point-contact-type junctions with a Pb or Nb counter electrode and a natural oxide tunneling barrier prepared on the surface of the HTSC. The differential characteristics exhibit one, two, and sometimes even three gap-like features between 10 and 21 meV, which may be ascribed to the HTSC. Further, a small energy gap opening below 9 K, with values of $\Delta(0)$ from 1 to 4 meV, is evident in tunnel structures.

Springer Series in Solid-State Sciences, Vol. 99
Electronic Properties of High-T_c Superconductors and Related Compounds
Editors: H. Kuzmany · M. Mehring · J. Fink

2. Experimental

The majority of our tunnel structures were realized on $YBa_2Cu_3O_x$ randomly oriented granular thin films, prepared in situ by vacuum co-deposition [7] in the presence of oxygen (using two individual thermal sources and one electron gun source). The films were deposited on $SrTiO_3$, Al_2O_3 and MgO single crystal substrates and poly-Al_2O_3 substrates. The onset of the phase transition from the normal to the superconducting state was at $T_{con} = 90\,K$, and the zero resistance critical temperature T_{c0} was typically 80–85 K. The thickness of films was 500–700 nm for randomly oriented thin films and about 100 nm for preferentially oriented films with the c-axis perpendicular to the substrate. The films were in situ superconducting, without a post-annealing process. The substrate temperature did not exceed 650°C. The critical current density at 4.2 K and zero magnetic field was higher than $10^4\,A/cm^2$. The cross-type tunneling junctions were prepared by vacuum deposition of Pb film through a metallic mask on a clean surface of a HTSC film. The tuneling area (0.1–0.3 mm^2) was defined by SiO isolation mask. The point contact junctions (with tunneling area of $\approx 0.1\,mm^2$) were prepared by means of Pb or Nb tips located on the surface of thin film or bulk HTSC.

The differential voltage–current characteristics (dU/dI) versus voltage (U) of all types of contacts exhibit similar features. The energy gap values of $YBa_2Cu_3O_x$ were near 20 meV but in some cases more complicated structure was observed in this region. Figure 1 illustrates differential tunneling characteristics of three different point contacts with bulk (curves a, b) and thin film base electrodes (curves c, d, e), at temperatures below and above 9 K.

3. Results

In the lowest voltage region a small energy gap with typical value $\Delta(0) = 3\,meV$ was observed in the overwhelming majority of contacts. This energy gap is dependent on temperature and disappears at about 9 K. It was interpreted [2] as an energy gap of the upper electrode with an enhanced gap value due to the proximity effect. Because this energy gap moves toward lower voltages with increasing magnetic field, and because it does not disappear in magnetic fields up to 1.4 T [8], we believe that it belongs to a superconductor with high critical magnetic field. The scatter of $\Delta(0) = 3\,meV$ gap values depends on the technological conditions of tunnel structure preparation, and is zero above 9 K.

In the voltage range 15–20 meV the gap-like structures are ascribed to HTSCs. For different tunnel junctions the value of $\Delta(0)$ was around 20 meV. Similar results were published in [5]. With increasing temperature the gap structure moves toward lower voltages. The temperature dependence of the

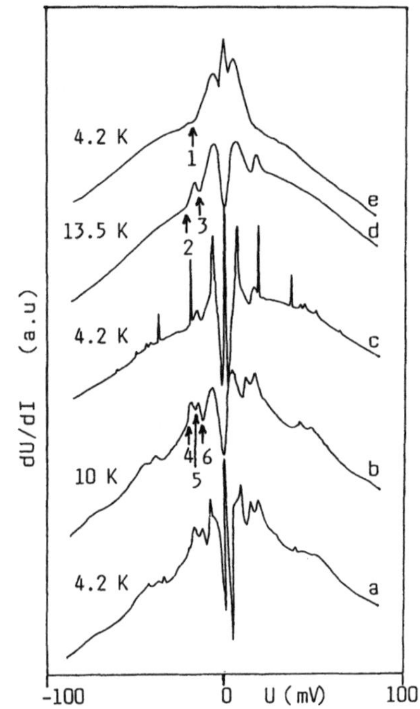

Fig. 1. Differential resistance dU/dI versus voltage U for $YBa_2Cu_3O_x$/Pb point contact junction with bulk YBacuO base electrodes (curves a, b), and for $YBa_2Cu_3O_x$/Pb (curves c, d), and $YBa_2Cu_3O_x$/Nb (curve e) with thin film $YBa_2Cu_3O_x$ electrodes. Different temperatures are denoted in the figure

gap structure in Fig. 1, curve d, we described in [9]; it was in good agreement with the theoretical dependence for strongly coupled superconductors [10] (when one considers the local suppression of superconductivity near the barrier surface). In some samples we observed a more complicated structure of differential characteristics (Fig. 1, curve d, points 2, 3 and curve b, points 4, 5, 6). These structures may be explained as multigap structures originating from superconducting phases, with different T_c and so with different Δ, localized near the barrier surface.

Finally, we present our view of the structure of the junction prepared directly on the unoxidized surface of the superconductor. Below 9 K the tunneling structure behaves as $S_1 - I - S_2 - S_3$ with S_1 as the counter electrode (Pb), S_2 as $YBa_2Cu_3O_x$ with $T_c \approx 9$ K and S_3 as the YBCO base electrode. Above 9 K the $N_1 - I - N_2 - S_2$ tunneling structure occurs, where N_1 is the counter electrode in the normal state and N_2 is a very thin film of $YBa_2Cu_3O_x$ in the normal state due to the oxygen deficiency arising during the formation of Pb oxide on the Pb/YBaCuO interface. This view may explain the variety of energy gaps observed in experiments. The above conclusions confirm preliminary results of XPS spectroscopy [11] of oxide grown on a clean $YBa_2Cu_3O_7$/Pb boundary. Following the *Geerk* results [5], i.e. $2\Delta/kT_c \approx 8$ for the 90 K phase, in our structures only the 60 K phase is present, evidently due to the oxygen deficit on the tunneling surface of $YBa_2Cu_3O_x$.

4. Conclusions

The differential tunneling characteristics for a natural oxide tunnel barrier exhibit a small energy gap with a typical value of $\Delta \approx 3\,\mathrm{meV}$, and gap-like structures corresponding to one, two, or three energy gaps present in the HTSC, with values of gaps in the range 10–20 meV. This may be explained by assuming that the multigap structure in the HTSC originates from superconducting phases with different critical temperatures near the tunneling barrier surface. The deficiency of oxygen in the HTSC arises as a result of counter electrode oxidation. This also explains better results achieved with more stable artificial barriers.

References

1. R. Srinivasan: *Quasiparticle Tunneling in High Temperature Superconductors*, Studies of HTSC, Vol. 5, ed. by A.V. Narlikar, in press
2. S. Beňačka et al.: Solid State Commun. **68**, 753 (1988)
3. M. Gurwitch et al.: Phys. Rev. Lett. **63**, 1008 (1989)
4. S. Beňačka et al.: IEEE Trans. Magn. **25**, 2583 (1989)
5. J. Geerk et al.: Physica C **162–164**, 837 (1989)
6. U. Poppe et al.: Solid State Commun. **71**, 569 (1989)
7. Š. Chromik et al.: IEEE Trans. Magn. **25**, 2484 (1989)
8. S. Takács et al.: Physica C **162–164**, 1045 (1989)
9. A. Pleceník et al.: *Weak Superconductivity*, Proc. 5th Int. Symp. on Weak Superconductivity, Smolenice Castle, 1989 (Nova Science, Commack, NY 1990)
10. D.J. Thouless: Phys. Rev. **117**, 1256 (1960)
11. P. Mikušík et al.: In preparation

Superconducting Pair Density from Specific Heat Measurements

J.W. Loram and K.A. Mirza

IRC in Superconductivity, Madingley Road, Cambridge CB3 0HE, UK

Abstract. We describe high precision measurements of superconducting specific heat anomalies in several high T_c series, concentrating particularly on the information these provide on the superconducting pair density. From the series $YBa_2(Cu_{1-x}M_x)_3O_7$ with M=Zn we derive the electronic specific heat coefficient $\gamma(x,T)$ from 1.6K to 300K, and from M=Co obtain possible evidence for CuO chain superconductivity in stoichiometric YBCO. In the series $La_{2-x}Sr_xCuO_4$ and $Y_{1-x}Ca_xSr_2Cu_2Tl_{.5}Pb_{.5}O_7$ the anomaly height and sharpness (pair density and phase coherence) are very much more sharply peaked with x than is T_c

1. Introduction

In this paper we show our specific heat data on several superconducting series examined in sufficient detail to reveal significant trends. In considering these data the following points should be borne in mind.

(a) The size of the specific heat jump $\Delta C(T_c)$ at T_c provides an approximate but unambiguous measure of the zero-temperature superconducting pair density n* from the relation $n^* \sim \Delta C(T_c)/3k_B$ ($\sim 3.10^{21} \cdot \Delta C(T_c)$ cm^{-3} if $\Delta C(T_c)$ is in mJ/g.at K). n* is largest for $YBa_2Cu_3O_7$ $\sim 10.10^{20}$ cm^{-3} and is in the range $1\text{-}3.10^{20}$ cm^{-3} for the optimum members of other systems. Note that these values are significantly lower than the carrier density n ($\sim 2\text{-}6.10^{21}$ cm^{-3}).

(b) The superconducting condensation energy density $U \sim \Delta C(T_c).T_c$ is determined by the size and temperature of the anomaly. Since $U = H_c^2(0)/8\pi$ and $H_{c2} = \sqrt{2}\kappa H_c$, where H_c and H_{c2} are the thermodynamic and upper critical fields, the size of the anomaly is of obvious technical importance.

(c) The mean field steps at T_c seen in conventional superconductors are seldom observed for the present systems. This is due to fluctuations in the order parameter which are important when the number of pairs in a coherence volume $n^*\Omega \sim 1$. Taking $\Omega \sim 20 \times 20 \times 5 = 2000$ Å^3 we have $n^*\Omega \sim 2$ for YBCO and ~ 0.2 for LSCO (cf ~ 1000 for Al). Critical fluctuations are revealed by a sharpening of the anomaly and a tail above T_c.

Springer Series in Solid-State Sciences, Vol. 99
Electronic Properties of High-T_c Superconductors and Related Compounds
Editors: H. Kuzmany · M. Mehring · J. Fink

(d) A sharp specific heat anomaly requires phase coherence over large volumes containing many pairs. The phase coherence length l_ϕ may be limited by spin flip scattering, chemical inhomogeneity or an applied magnetic field, and a rounding of the transition will be observed if $l_\phi < 5\text{-}10\ \xi_0$ for the present systems.

2. Results and Discussion.

Our ceramic samples were prepared by conventional solid state reaction taking particular care over homogeneity and annealing procedures. To resolve the small specific heat anomalies we employ a high precision differential technique (~1:10^4 resolution) where we measure directly the difference in specific heat between each sample and a suitable reference, the reference being chosen to have a similar phonon spectrum to the sample (1). In favourable cases only rather small corrections have to be made for the residual difference of phonon and low temperature magnetic terms. This has proved to be particularly successful for the series $YBa_2(Cu_{1-x}Zn_x)_3O_7$ with $YBa_2(Cu_{.93}Zn_{.07})_3O_7$ as reference, the very small change in phonon specific heat on Zn doping providing a unique opportunity to determine the electronic coefficient $\gamma(x,T) = C^{el}/T$ over a wide temperature range (2). Zn has been shown (3) to substitute for both plane Cu(2) and chain Cu(1) and has little effect on the Hall number n_H and the thermopower (4). The results for $\gamma(T)$ for this series are shown in Fig.1, uncertainty being greatest below 20K. The curves, which show no further change up

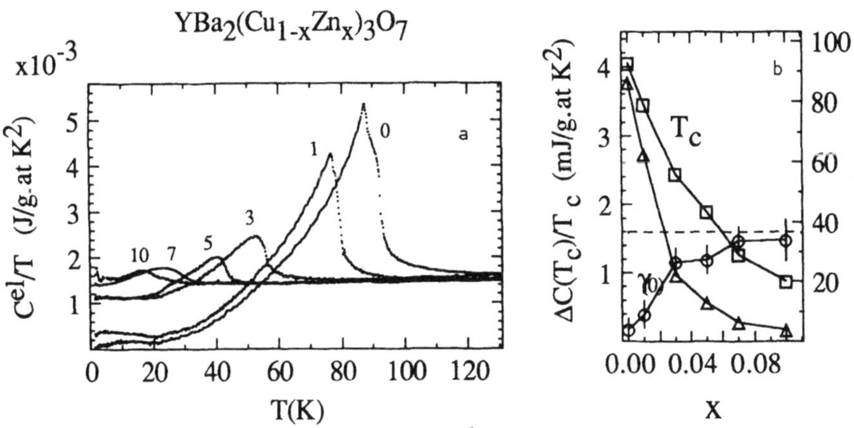

Fig 1. (a) Electronic specific heat C^{el}/T vs T for $YBa_2(Cu_{1-x}Zn_x)_3O_7$. Labels show 100x. The zero of the ordinate axis has an uncertainty ± .15 mJ/g.at K^2.
(b) T_c (□), low temperature electronic coefficient $\gamma(0)$ (o), and step height $\Delta C(T_c)/T_c$ (Λ) vs x.
The broken line shows γ_n=1.6mJ/g.at.K^2.

to 300K, yield a normal state value γ_n=1.6 mJ/g.at.K^2 independent of x. This value of γ_n agrees with that estimated for undoped YBCO by Junod (5) from C -C^{ph}, and is consistent with the Pauli susceptibility (6) . From γ_n we obtain a density of states per spin $g(E_F)$=4.4 states/eV.fu , independent of x, and thus E_F~ 0.12 eV for 1 hole per fu. $g(E_F)$ is at least a factor four larger than band structure values, indicating strong correlation effects for the Fermi surface electrons. The measured ratio $\Delta C(T_c)/T_c(\gamma_n - \gamma(0)) \sim 2.65$ is substantially larger than the value 1.43 for a weak coupling BCS superconductor and corresponds to a gap parameter $2\Delta(0)/k_BT_c$~4.8. For undoped YBCO we find $\gamma(0)$= .15±.15 mJ/g.at.K^2, which is not inconsistent with $\gamma(0)$=0.

The anomalies at T_c are sharp for x=0 and x=.01 showing long range phase coherence with a well-defined fluctuation regime above T_c .Increased rounding and broadening for x≥.025 (3) demonstrate limited phase coherence when the mean separation of Zn ions falls below ~ 13Å. The linear decrease in T_c with x and the rapid reduction in superconducting entropy with correlated increase in $\gamma(0)$ are qualitatively consistent with magnetic pair breaking and suggest that the substitution of magnetic Cu^{2+} by non-magnetic Zn produces strong magnetic scattering. The finite $\gamma(0)$ for x ≥ .01 results from excitations extending down to zero energy due to lifetime broadening of the gap edge quasiparticle peak. On this interpretation the reduced superconducting volume fraction $f_s = 1 - \gamma(0)/\gamma_n$ should be viewed not as separate normal and superconducting regions but as fluctuations between paired and unpaired states.

The results shown in Fig.1 are inconsistent with the condensation at T_c of preformed bose pairs. Taking n~ 1 hole/fu for this system and n* ~ n/2 this model predicts entropies $S^{el}(T_c)$ independent of T_c and anomaly heights $\Delta C(T_c)/T_c \sim 1000/T_c$ mJ/g.at K^2 which are clearly not observed. The results provide strong support for a Fermi liquid model of k-space pairing which correctly predicts the observed entropy $S^{el}(T_c) \propto T_c$, $C^{el} \propto T$ for T>T_c and from the anomaly heights $n^* \sim \Delta(0)g(E_F) \sim n.k_BT_c/E_F << n$. We estimate that around 20% of the Fermi sphere is paired in $YBa_2Cu_3O_7$ and about 7% in $La_{1.85}Sr_{.15}CuO_4$.

We turn next to the system $YBa_2(Cu_{1-x}Co_x)_3O_7$. Trivalent ions such as magnetic Co or Fe or non-magnetic Ga or Al are thought to substitute primarily for chain Cu(1) (7). In these systems T_c remains constant up to x ~ .01 - .03 before decreasing at higher x rather slowly for non-magnetic dopants and more rapidly for magnetic ions (7). Fig.2 shows the superconducting anomalies for Co concentrations 0 ≤ x ≤.07 (8). T_c initially increases from 92.5K at x=0 to 94K at x=.015 before falling linearly with x above x=.028. In contrast to T_c there is an immediate rapid decrease in size of the anomaly (pair density) with x up to .028 followed by a region of constant step height. Similar behaviour for low x has been observed for Fe^{3+} (9) and Cr^{3+} (10) doping. The initial rise in T_c is especially surprising in view of this sharp fall in pair

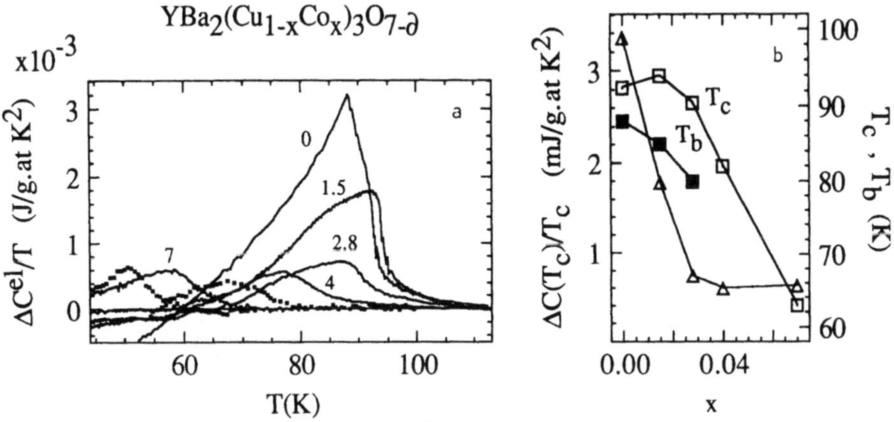

Fig 2. (a) Connected points: $\Delta C^{el}/T$ for $YBa_2(Cu_{1-x}Co_x)_3O_7$ where $\Delta C^{el}=C^{el}-C_n^{el}$. Labels show 100x.
Disconnected points: $\Delta C^{el}/T$ for $YBa_2Cu_3O_{7-\delta}$ with δ=.2 (T_c~74K) and δ=.5 (T_c~ 54K).
(b) Anomaly height (Δ) and transition temperatures T_c (□) and T_b (■) against x for $YBa_2(Cu_{1-x}Co_x)_3O_7$.
We tentatively associate T_c and T_b with the onset of plane and chain superconductivity respectively.

density and the very rapid decrease in the Hall carrier density n_H with x (11). The specific heat behaviour on Co doping closely mirrors the effect of oxygen depletion in $YBa_2Cu_3O_{7-\delta}$. Junod et al.(12) find that the anomaly exhibits a rapid decrease in magnitude with little change in T_c for $0 \le \delta < 0.15$ similar to our results for x≤.028. We include in Fig.2 our measurements for δ=.2 and .5 with T_c = 74K and 54K respectively, and these are consistent with the trend for Co doping with x ≥ .04. It is therefore interesting to speculate whether there is a common origin for the behaviour of both systems.

We propose (8) that the large additional superconducting entropy in stoichiometric YBCO is due to a more 3D superconducting order parameter extending into the CuO chains but which is rapidly destroyed by trivalent ion substitution or oxygen depletion which disrupt the chains. Koch et al.(13) conclude from the anisotropic plasma edge that the hole density on the chains is double that on the planes for δ=0 and thus the low value of $\gamma(0)$ ($\le$ 10% γ_n) found above suggests a superconducting energy gap for both planes and chains. Possible evidence for two distinct but coupled order parameters may be seen in the shape of the anomalies for the two different x=0 samples in Figs.1 and2. These show a sharp symmetrical peak at T_b~88K signalling the onset of long range phase coherence, and a rapid second order-like drop with mid-point at T_c =92.5K. Doping the chains with Co strongly suppresses the 88K peak leaving small broad bumps at T_b~85K and 80K for x=.015 and .028 respectively, but has a much weaker effect on the step at T_c.

We tentatively attribute T_c and T_b to the onset of superconducting order on the planes and chains respectively. It is significant that the rather constant anomaly size for $x \geq .04$ or $\delta \geq .2$ is similar in magnitude to that in oxide superconductors without a chain structure and we suggest that in this regime superconductivity is limited to the planes. If this interpretation of the specific heat results is correct it is clear that chain superconductivity in YBCO more than doubles the condensation energy U.

Figs.3 and 4 show the progression of the specific heat anomaly as a function of x for $La_{2-x}Sr_xCuO_4$ (14) and the new 108K superconducting system $Y_{1-x}Ca_xSr_2Cu_2Tl_{.5}Pb_{.5}O_7$ (15). The Tl system is tetragonal but otherwise has the $YBa_2Cu_3O_7$ structure with $Tl_{.5}Pb_{.5}$ replacing Cu(1). Both systems progress with increasing x directly from a semiconducting to a superconducting phase (at x ~ .06 and 0.5 respectively). Above an optimum composition x_0 (.16 and .8) T_c decreases although the normal state conductivity continues to increase. Antiferromagnetic order in the parent compound La_2CuO_4 is rapidly destroyed by Sr doping but short range magnetic correlations survive into the superconducting region, decreasing in range with x. The magnetic behaviour of the Tl system has not yet been established but we observe low temperature specific heat anomalies in the semiconducting region closely similar to those associated with spin glass order in LSCO (14). Antiferromagnetic order has been observed in the related compound $YBa_2Cu_2TlO_7$ and it is therefore very likely that the magnetic phase diagram of the Tl system is similar to that of LSCO and other oxide superconductors.

Figs.3 and 4 reveal striking similarities between the two systems. In each case the anomaly increases rapidly in area and sharpness as x approaches x_0, reflecting a corresponding increase in pair density and range of phase coherence. However T_c (defined as the mid

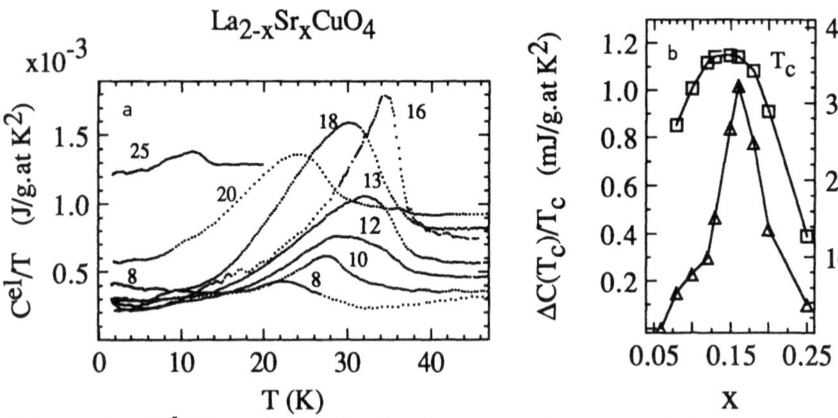

Fig 3. (a) C^{el}/T for $La_{2-x}Sr_xCuO_4$. Labels show 100x.
(b) T_c (□), and anomaly height $\Delta C(T_c)/T_c$ (Δ) vs x

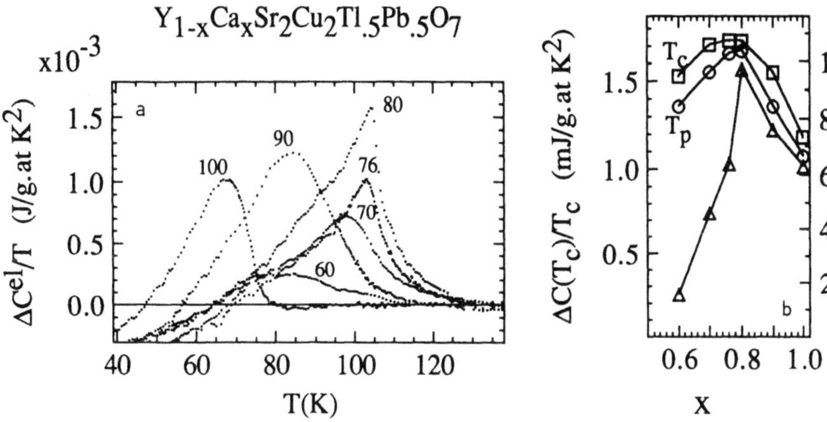

Fig 4. (a) $\Delta C^{el}/T$ for $Y_{1-x}Ca_xSr_2Cu_2Tl_{.5}Pb_{.5}O_7$ where $\Delta C^{el}=C^{el}-C_n^{el}$. Labels show 100x.
(b) T_c (□), peak temperature T_p (o) and step height $\Delta C(T_c)/T_c$ (Δ) vs x.

point of the transition) increases only a little and the onset temperature is almost independent of x. The picture that emerges is one of small fluctuating regions with $T_c \approx T_{c,max}$ growing in volume with x and finally merging to form a fully coherent superconducting state over a narrow composition range around x_0, in agreement with the conclusions of Harshman et al.(16). Phase frustration by static or slowly varying magnetic correlations must play a significant role in the region $x<x_0$. The immediate fall in T_c and entropy for $x>x_0$ is still an unsolved problem, but might be understandable if the pairing interaction is of magnetic origin since this will be progressively screened by an increasing carrier density.

We have examined the progression of the specific heat anomalies along four superconducting series, placing particular emphasis on the information contained in the size and shape of the anomaly. Our measurements suggest a possible explanation in terms of chain superconductivity for the uniquely large anomaly in YBCO (by a factor 2-3), and reveal in two systems that the pair density and phase coherence optimise over a surprisingly narrow composition range. Specific heat is a bulk property and is insensitive to microstructure, surface effects or traces of superconducting phase which may dominate the behaviour of other properties. This contributes to the systematic behaviour observed for our ceramic samples and the fact that specific heat measurements on ceramic and single crystal samples give similar results. We have noted that the entropy (pair density) varies much more rapidly with composition than does T_c and as such is a more sensitive indicator of the superconductivity of a system than Tc from both the practical and fundamental point of view.

Acknowledgements. We are indebted to P.A. Freeman, R.S. Liu, J. Osborne and J.L. Tallon for providing the samples. They are co-authors in final publications of the results.

References.

(1) J.W.Loram: J.Phys E16 (1981) 367
(2) J.W.Loram,K.A.Mirza,P.F.Freeman: L.T.19, Physica B (1990).
(3) G.Roth et al: Physica C162-164 (1989) 518
(4) J.Clayhold: Private communication.
(5) A.Junod: "Physical Properties of High Temperature Superconductors 2",
Ed.D.M.Ginsberg. World Scientific (1990).
(6) B.Batlogg "High Temperature Superconductivity:Los Alamos Symposium 1989 Ed. K.Bedell. Addison Wesley (1990)
(7) J.T.Markert et al.: "Physical Properties of High Temperature Superconductors 1", Ed.D.M.Ginsberg, World Scientific (1989) 265-337.
(8)J.W.Loram,K.A.Mirza,P.F.Freeman,J.L.Tallon: L.T.19, Physica B (1990).
(9) A.Junod,A.Bezinge,D.Eckert,T.Graf,J.Muller: Physica C152 (1988) 495
(10) S.Kim, R.A.Fisher,N.E.Phillips,J.E.Gordon: Physica C162-164 (1989) 494
(11) J.Clayhold et al.: Phys. Rev B39 (1989) 777
(12) A.Junod,D.Eckert,T.Graf,G.Triscone,J.Muller: Physica C162-164 (1989) 482
(13) B.Koch,H.P.Geserich,Th.Wolf: Solid State Commun. 71 (1989) 495
(14) J.W.Loram,K.A.Mirza,J.Osborne: Physica C162-4 (1989) 498.
(15) J.M.Liang,R.S.Liu,Y.T.Huang,S.F.Wu,P.T.Wu,L.J.Chen: Physica C165 (1990) 347.
(16) D.R.Harshman et al.: Phys. Rev. Lett. 63 (1989) 1187.

Part III

High-Energy Spectroscopies

Electron Energy-Loss Studies on High-T_c Superconductors

J. Fink, N. Nücker, H. Romberg, M. Alexander, G. Roth, and P. Adelmann

Kernforschungszentrum Karlsruhe, Institut für Nukleare Festkörperphysik, Postfach 3640, D-7500 Karlsruhe, Fed. Rep. of Germany

Abstract. An introduction to the electronic structure of high-T_c superconductors and their investigation by high-energy spectroscopies is given. In particular we review electron energy-loss studies on the electronic structure of the high-T_c superconductors. Energy-loss spectra in the low-energy range ($0.1 \leq E \leq 50$ eV) yield information on intra- and interband transitions and on collective excitations of the charge carriers. Absorption edges of the O 1s and Cu 2p core levels provide information on the symmetry and the character of empty states close to the Fermi level.

1. Introduction

In order to clarify the mechanisms for high-T_c superconductivity in the cuprates it is essential to understand their unique electronic structure in the normal state. The undoped parent compounds, e.g. La_2CuO_4, exibit semiconducting properties because of an opening of a gap in the half filled antibonding Cu 3d-O 2p band due to strong correlations on the Cu sites. This leads with the present values of parameters determining the electronic structure to a charge-transfer insulator [1]. The valence band is formed predominantly from O 2p states with some admixture of Cu 3d states. The conduction band is mainly composed of Cu 3d states and some admixture of O 2p states. The amount of admixture in the valence and conduction bands is a measure of covalency in the CuO_2 layer. Upon p- or n-type doping, electrons are removed from or added to the CuO_2 layers, respectively, and at higher doping levels a metallic state is formed exhibiting high-T_c superconductivity. There are many models describing the electronic structure in the extreme dilution limit of a single hole in the CuO_2 layer. On the other hand it is not clear whether the dilute limit is related to the state which is formed at higher doping levels exhibiting high-T_c superconductivity. Possibly a completely new state is formed which cannot be described in terms of dilute limit models. At present several models for the doped case in the non-dilute limit are being discussed. They are illustrated for the p-type and n-type doped cases in Fig. 1 together with a picture for the charge-transfer insulator. In a rigid-band-like model for the charge-transfer insulator, holes in the valence

Springer Series in Solid-State Sciences, Vol. 99
Electronic Properties of High-T_c Superconductors and Related Compounds
Editors: H. Kuzmany · M. Mehring · J. Fink

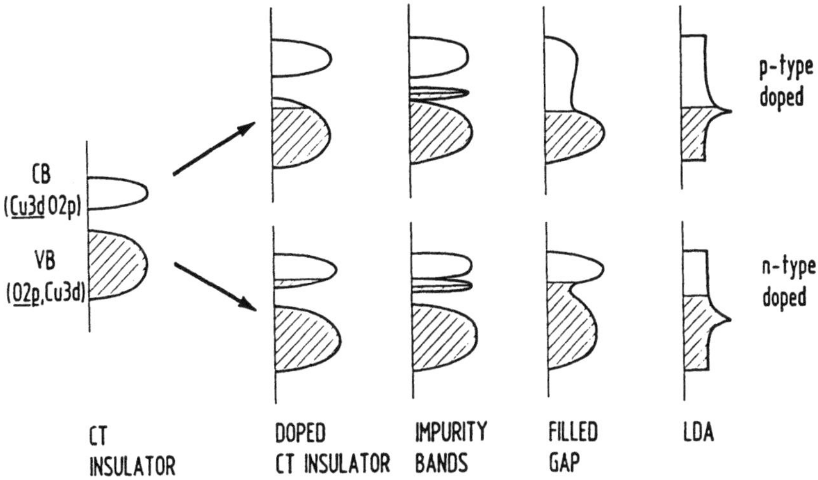

Fig. 1 Illustrations of the electronic structure of the charge transfer insulator and various models for p- and n-type doped systems.

band (with mainly O 2p character) and electrons in the conduction band (with mainly Cu 3d character) will be formed upon p- or n-type doping, respectively. Other models predict the existence of impurity bands or a filled gap [2]. Finally we show in Fig. 1 the antibonding CuO band as derived from band structure calculations in the local density approximation (LDA) with the position of the Fermi level depending on the amount of p- or n-type doping. Since the electronic structure of the doped CuO_2 layer and the insulator-metal transition is rather unclear at present, experiments on the electronic structure are urgently required. High-energy spectroscopies are a tool from which important information on the electronic structures can be obtained.

2. High-Energy Spectroscopies

In Fig. 2 we illustrate the various high energy spectroscopies currently applied to the high-T_c superconductors. The various methods yield information on occupied (UPS, XPS, XES) and unoccupied (BIS, XAS, EELS) states while AES provides information on a two hole state. Since it is very difficult to prepare surfaces of the high-T_c superconductors which represent bulk properties, it is important to consider the typical depth D from which information is obtained. Several methods (UPS, XPS, BIS) are extremely surface sensitive and therefore it is difficult (although feasible) to obtain reliable information on the electronic structure of high-T_c superconductors. Others measure bulk properties (EELS in transmission, XAS in transmission, XES) and are therefore particularly suitable for investigations of cuprates. It is also interesting to compare the energy

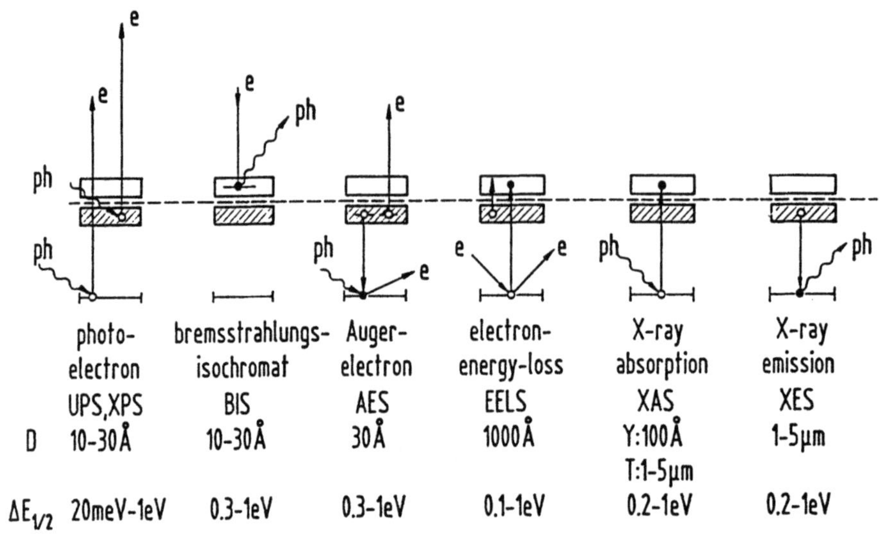

Fig. 2 High-energy spectroscopies. D: typical depth from which information is obtained. Y: partial yield method for XAS; T: XAS in transmission. $\Delta E_{1/2}$ typical energy resolution.

resolution of the various methods. At present, only ultraviolet photoemission (UPS) and EELS in reflection (not shown) have enough energy resolution to resolve differences in the electronic structure between the normal and superconducting state. All other methods can only provide information on the normal state properties.

In the following we report on results derived from electron energy-loss spectroscopy in transmission using electrons with a primary energy of 170 keV. The loss function $\mathrm{Im}[-1/\varepsilon(q,\omega)]$ depending on the frequency and momentum dependent dielectric function $\varepsilon(q,\omega)$ is measured. For loss energies smaller than ~50 eV, information on plasmons and intra- and interband transitions is obtained. At higher energies, core-level excitations yield information on the local density of unoccupied states, provided that the interaction with the core hole is small.

3. Results

Typical loss functions of cuprates [3,4] show a broad feature around 20 eV caused by the plasmon of all valence electrons. Superimposed on this plasmon are low-lying core-level excitations and interband transitions. In the undoped parent compounds a gap of about 1.7 eV is realized. In the doped compounds, a second well-pronounced plasmon between 1 and 2 eV is realized. This plasmon is due to collective oscillations of the charge carriers formed upon doping. In Fig. 3 we show measurements of the charge carrier plasmon in $Bi_2Sr_2CaCu_2O_8$ as a function of momentum transfer q parallel to the CuO_2 layers.

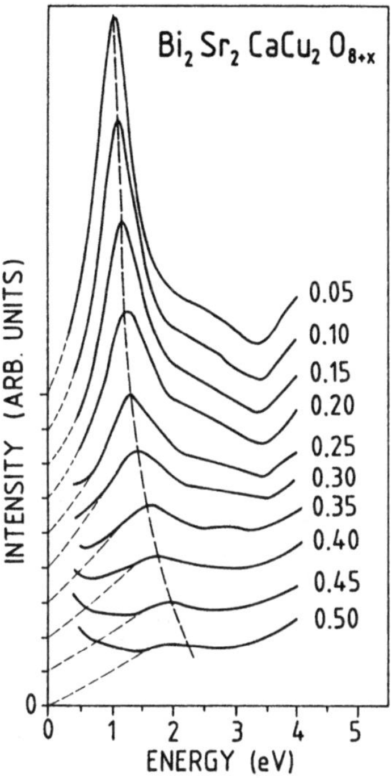

Fig. 3 Charge carrier plasmon of $Bi_2Sr_2CaCu_2O_8$ as a function of momentum transfer.

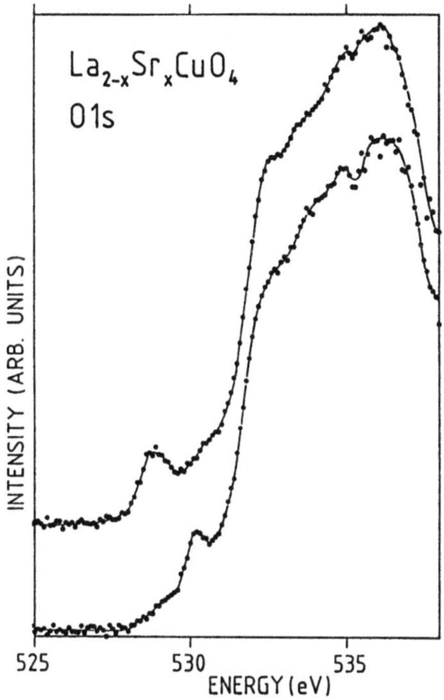

Fig. 4 O 1s absorption edges of La_2CuO_4 (lower curve) and $La_{1.85}Sr_{0.15}CuO_4$ (upper curve).

This plasmon dispersion shows the in-phase collective motion of charge carriers on the CuO_2 (and BiO) layers for different wavelengths. If we plot the plasmon energy against the momentum transfer, the data can be well represented by a relationship $E_p(q) = E_p(0) + (\hbar^2/m)\alpha q^2$. Using the random phase approximation, the dispersion coefficient is given by $\alpha = 3/8[mv_F/E_p(0)]$. Compared to plasmons in a 3-dimensional system, the prefactor for a 2-dimensional system is 3/8 instead of 3/10. Using these relations, the Fermi velocity can be calculated to be $v_F = 0.51 \times 10^8$ cm/sec. This value is very close to $v_F(\text{II}) = 0.49 \times 10^8$ cm/sec which was calculated using LDA band-structure calculations [5]. On the other hand this perfect agreement should not be stressed. On a low energy scale ($E \leq 100$ meV) there may be a coupling of the charge carriers to various excitations which cannot be resolved on the high energy scale ($E \sim 1$ eV) in the range of the plasmon energy. This situation is similar to the plasmon of Si at ~16 eV which is hardly influenced by the gap of about 1.1 eV.

Next, we review results from core excitations providing information on the character and symmetry of unoccupied states. In Fig. 4 we compare O 1s absorption edges, i.e. transitions from the O K shell into unoccupied O 2p states, for La_2CuO_4 and p-type doped $La_{1.85}Sr_{0.15}CuO_4$.

In Fig. 5 we show similar spectra for Nd_2CuO_4, n-type doped $Nd_{1.85}Ce_{0.15}CuO_4$, reduced $Nd_{1.85}Ce_{0.15}CuO_{4-\delta}$ (superconducting), and Nd_2O_3. As can be seen directly from a comparison of the Nd_2CuO_4 and Nd_2O_3 spectra the excitations above 531 eV are mainly due to transitions into O 2p states hybridized with Ln (Ln = La,Nd) 5d and 4f states. In La_2CuO_4, the absorption edge at ~530 eV is assigned to transitions into O 2p states admixed to the conduction band (see Fig. 1). There is some spectral weight below this edge starting at 528.2 eV, i.e., about the gap energy below the conduction band. These transitions are due to states in the gap probably formed upon p-type doping by excess O in La_2CuO_4. This indicates that the Fermi level in this compound is very close to the valence band. With increasing Sr doping, a new threshold at lower energy is formed and the intensity at ~530 eV decreases [6]. When moving from the antiferromagnetic insulator to the superconducting metal, probably the gap is filled mainly by conduction band states pulled to lower energy. In addition the Fermi level is shifted more and more into the valence band thus forming holes having predominantly O 2p character. The measurements are compatible with the "filled gap" model, shown in Fig. 1 but not with impurity band models assuming narrow bands. The formation of holes on O upon p-type doping is also seen in other p-type doped systems such as, e.g., $YBa_2Cu_3O_{7-\delta}$ [7], and $Bi_2Sr_2Ca_{1-x}Y_xCu_2O_8$ [8]. The observation of a pre-peak appearing in Nd_2CuO_4 and the n-type doped compounds leads to the question whether in these compounds also hole states on O must be assumed. The new measurements on $La_{2-x}Sr_xCuO_4$, where clearly two pre-peaks were observed, and measurements on insulating CuO and $(Ca,Sr)CuO_2$, where also a pre-peak is realized, indicated that this pre-peak in the T' phase compounds is due to transitions into the O 2p states admixed to the conduction band. This is further supported by systematic measurements of the O 1s threshold energy (E_{TH}) for various undoped and doped cuprates. These threshold energies are plotted in Fig. 6 versus the Cu-O distance in the CuO_2 planes. In addition we have plotted the gap energy versus Cu-O distance for various insulating cuprates as derived from optical measurements [9]. The shift in threshold energy for the insulating compounds is due to both a change in the gap energy and a chemical shift of the O 1s level. There is a linear relationship between E_{TH} and the Cu-O distance for the undoped systems (except $(CaSr)CuO_2$) indicating that in all these comopunds the threshold is caused by transitions into the conduction band. The decrease of E_{TH} with increasing Cu-O distance can be explained by the fact that at larger Cu-O distances (weaker covalency) there are more O 2p electrons at the O sites. This leads to a better shielding of the nuclear charge and thus reduces the binding energy of the O 1s level. It is interesting to note that the threshold energy for several p-type doped systems is about 1.5 eV lower (the gap energy E_g~1.7 eV) than that for the undoped systems.

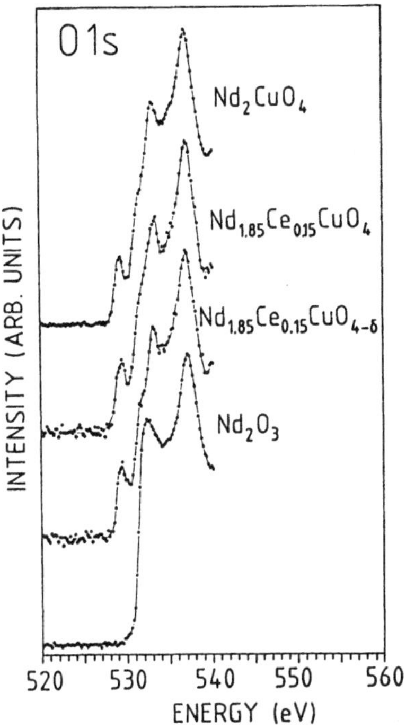

Fig. 5 O 1s absorption edges of Nd_2CuO_4, $Nd_{1.85}Ce_{0.15}CuO_4$, superconducting $Nd_{1.85}Ce_{0.15}CuO_{4-\delta}$ and Nd_2O_3.

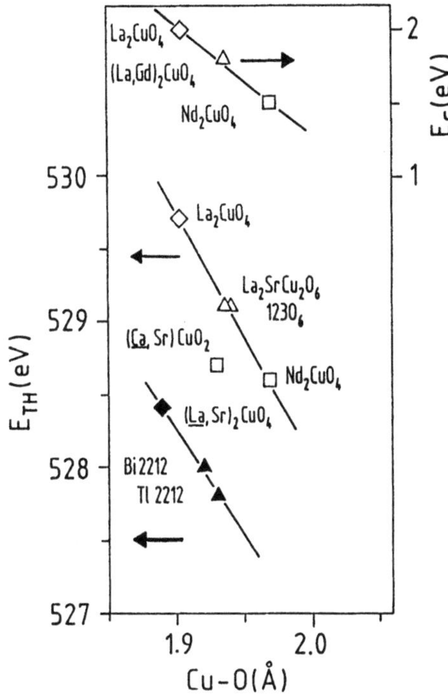

Fig. 6 Plot of the gap energy E_g and the O 1s threshold energies E_{TH} for various undoped (open symbols) and doped (closed symbols) cuprates.

The pre-peak in the O 1s edge of Nd_2CuO_4 hardly changed upon n-type doping. This indicates that the O 2p admixture to the conduction band is almost unchanged. On the other hand, in the Cu 2p edge a reduction of Cu 2p-3d transitions by about 20% is observed [10]. This indicates a filling of Cu 3d states in the conduction band forming charge carriers having probably Cu 3d character. The measurements are compatible with the "doped charge transfer insulator" model or the "filled gap" model shown in Fig. 1 but not with the impurity band model.

References

[1] A. Fujimori and F. Minima, Phys. Rev. B **30**, 957 (1984); J. Zaanen, G.A. Sawatzky, and J.W. Allen, Phys. Rev. Lett. **55**, 418 (1985); S. Hüfner, Z. Phys. B **61**, 135 (1985)

[2] P. Horsch, W.H. Stephan, K. v. Szczepanski, M. Ziegler, and W. von der Linden, Physics C **162-164**, 783 (1989); P.C. Pattnaik and D.M. Newns, Phys. Rev. B **41**, 880 (1990); C. Melo and S. Doniach, Phys. Rev. B **41**, 6633 (1990)

[3] N. Nücker, H. Romberg, S. Nakai, B. Scheerer, J. Fink, Y.F. Yan, and Z.X. Zhao, Phys. Rev. B **39**, 12379 (1989)

[4] H. Romberg, N. Nücker, J. Fink, Th. Wolf, X.X. Xi, B. Koch, H.P. Geserich, M. Dürrler, W. Assmus, and B. Gegenheimer, Z. Phys. B **78**, 367 (1990)

[5] L.F. Mattheiss and D.R. Hamann, Phys. Rev. B **38**, 5012 (1988); L.F. Mattheiss, private communication

[6] H. Romberg, M. Alexander, N. Nücker, P. Adelmann, and J. Fink, this volume

[7] N. Nücker, J. Fink, J.C. Fuggle, P.J. Durham, and W.M. Temmerman, Phys. Rev. B **37**, 5158 (1988)

[8] J. Fink, N. Nücker, H. Romberg, M. Alexander, S.Nakai, B. Scheerer, P. Adelmann, and D. Ewert, Physica C **162**, 1495 (1989)

[9] Y. Tokura, T. Arima, S. Koshihara, T. Ido, S. Ishibashi, H. Takagi, and S. Uchida, Proc. of the Second Intern. Symp. on Superconductivity, Springer Verlag, to be published

[10] M. Alexander, H. Romberg, N. Nücker, P. Adelmann, and J. Fink, this volume

Electronic Structure of Transition Metal Compounds

S. Hüfner

Fachbereich Physik, Universität des Saarlandes,
D-6600 Saarbrücken, Fed. Rep. of Germany
and Institut de Physique, Université de Fribourg,
CH-1700 Fribourg, Switzerland

Abstract. The analysis of high energy spectroscopy on transition metal compounds is reviewed. It is concluded that the states responsible for the superconductivity in the cuprate compounds probably have a considerable dispersion.

The detection of superconductivity in perovskite cuprates /1/ has renewed a long-standing interest in the electronic structure of the transition metal compounds. Here we shall give a brief account of the present insight into this electronic structure as obtained by high energy probes like photoemission spectroscopy (PES), electron energy loss spectroscopy EELS or X-ray absorption spectroscopy among others.

An analysis of the final state structure in PES core level spectra of copper compounds lead to the realization /2,3/ that in copper compounds the lowest transitions leading to charge fluctuations (measured e.g. in optical spectra as the optical gap energy E_g) are of the charge transfer type. This is seen from Fig. 1, where for the Cu dihalides good agreement is found between the calculated and measured satellite intensity (E(sat)) provided the experimental optical charge transfer gap E_g is inserted into the calculations.

This observation can be generalized and the 3d transition metal compounds can be classified into those where the gap is a transition between the ligand states (L) and the first empty metal d state (charge transfer compounds) and those where the gap connecting the highest occupied and lowest unoccupied state is of the Mott type (Fig. 2) /4/.

The question where in the row of the 3d transition metal compounds the transition occurs from the Mott case to the charge transfer case is not yet decided. Veal and Paulikas /5/ proposed that the charge transfer regime holds for all 3d

Springer Series in Solid-State Sciences, Vol. 99
Electronic Properties of High-T_c Superconductors and Related Compounds
Editors: H. Kuzmany · M. Mehring · J. Fink

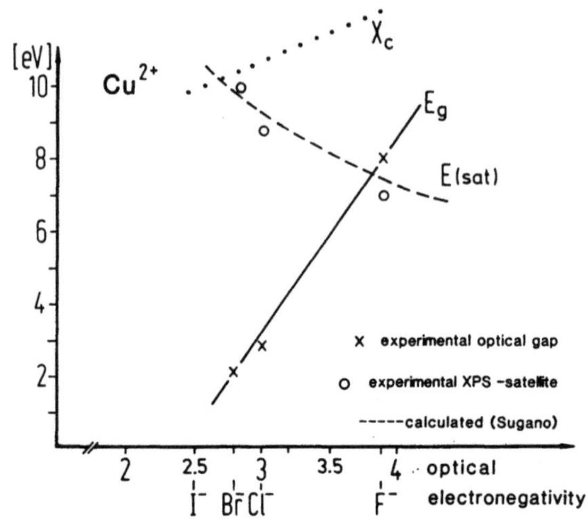

Fig. 1. Analysis of the main line satellite distance (E(sat)) in the XPS spectra of Cu dihalides in terms of the model of Asada and Sugano. E_g is the optical gap energy, X_c the lowering of the Cu energy levels by the core hole.

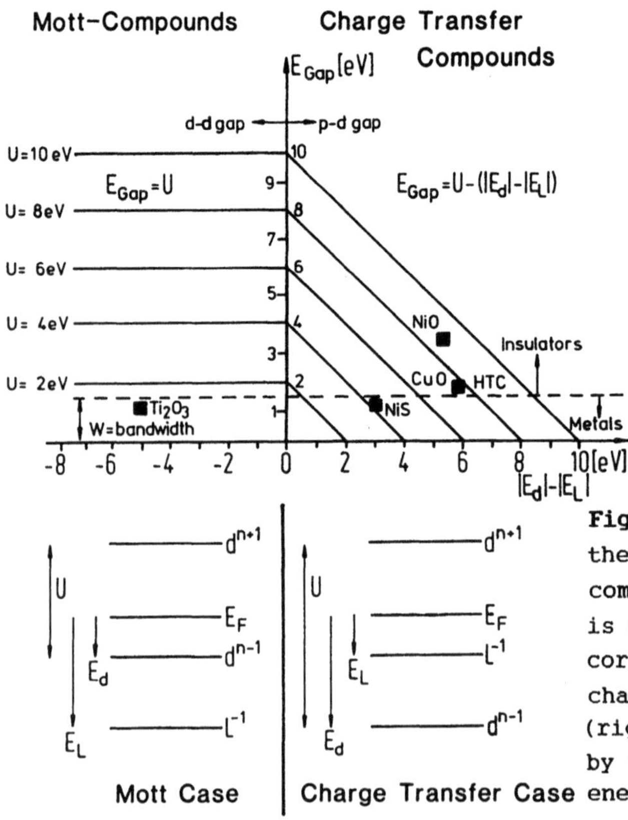

Fig. 2. Diagram showing the regions for Mott compounds (left side, gap is given by the d-d correlation energy) and charge transfer compounds (right side, gap is given by the p-d transition energy).

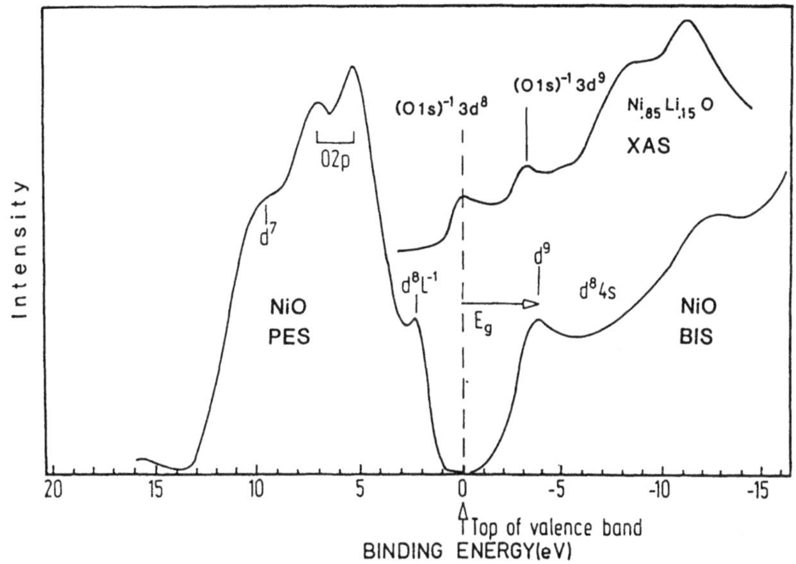

Fig. 3. Combination of PES, BIS and XAS data for NiO and hole doped NiO which is $Ni_{1-x}Li_xO$. There is an energy difference of 2 eV between the itinerant holes at E_F and the localized holes d^8L^{-1}.

transition metal ions. This conclusion is based on an analysis of the exchange splitting of 3s core levels. Kinsinger et al. /6/ come to a different conclusion namely that for $n_d \leq 5$ the 3d transition metal compounds are of the Mott type, for $n_d = 8$ and 9 they are of the charge transfer type and for $n_d = 6$ and 7 the wave functions are so strongly mixed that a distinction between the two cases seems not to be meaningful.

The knowledge accumulated so far on the electronic structure of the 3d transition metal compounds can be used to analyse spectroscopic data on a prototype material namely NiO. The experimental data /7-10/ are shown in Fig. 3. We only want to stress one point concerning the data around the Fermi energy: NiO always contains Ni defects /11/, which lead to holes at the top of the valence band and these holes pin the Fermi energy. These holes have e.g. been detected by XAS /9/. These mobile holes seem to be different from the localized holes d^8L^{-1}, which are the final state of a PES experiment, and which have a binding energy of ~2 eV /7/. The valence band states can be measured by BIS (Bremsstrahlung Isochromate Spectroscopy) /8/ or XAS /9/.

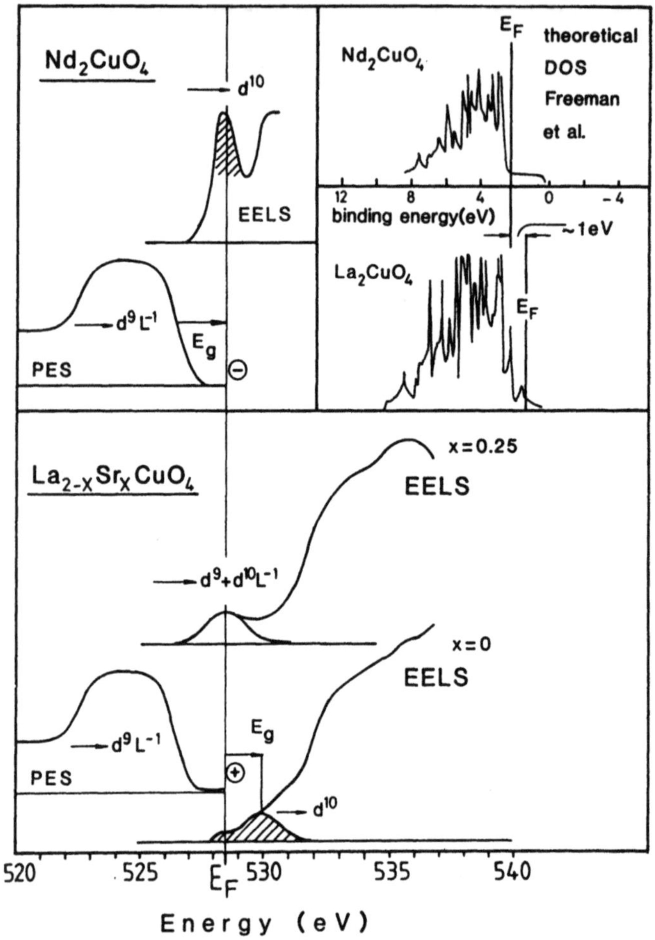

Fig. 4. Combination of PES and EELS data for Nd_2CuO_4 (electron superconductor) and $La_{2-x}Sr_xCuO_4$ (hole superconductor). The insert shows the theoretical density of states for Nd_2CuO_4 and La_2CuO_4.

In the same way as we have analysed the spectroscopic data for NiO, these for the high temperature superconductors can be tentatively understood /12/. As an example the combination of PES, EELS and XAS data /13-16/ for $La_{2-x}Sr_xCuO_4$ and Nd_2CuO_4 is given in Fig. 4. From this compilation it seems appealing /17/ that in $Nd_{2-x}Ce_xCuO_4$ the conductivity is by electrons and that the empty states measured by EELS /16/ or XAS /18/ are of the $Cu3d^{10}$ type with a strong Cu4s hybridization. For instructive purposes the theoretical densities of states /19/ of La_2CuO_4

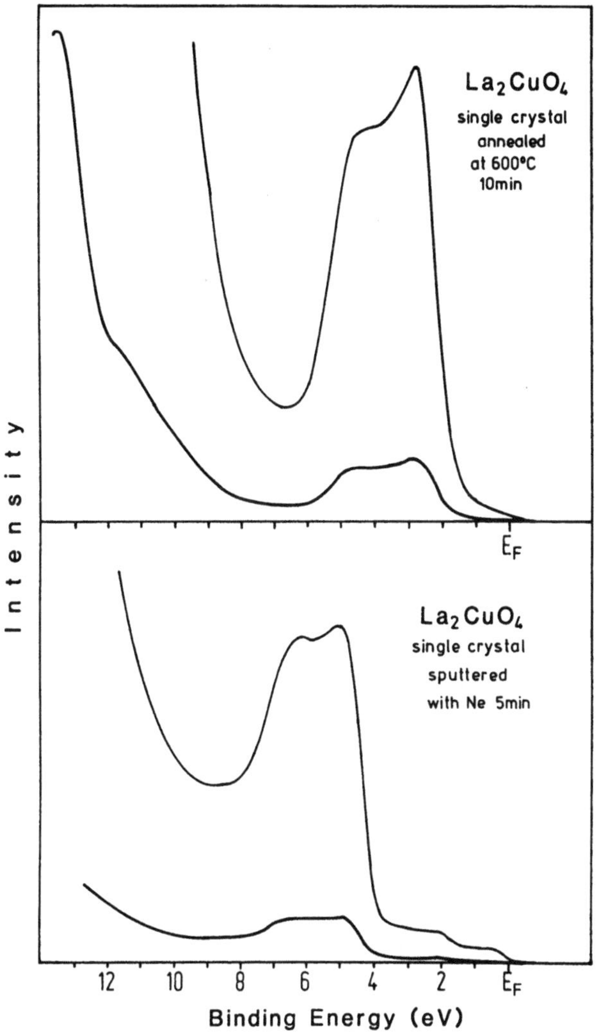

Fig. 5. Comparison of the PES spectra of La_2CuO_4 (hole conductor) and sputtered La_2CuO_4 (electron conductor). There is a shift between the two spectra of ~2 eV corresponding to the optical gap.

and Nd_2CuO_4 are shown as an insert in Fig. 4. It should be emphasized that there seems to be a difference between Nd_2CuO_4 doped with electrons (see Fig. 4) and La_2CuO_4 doped with electrons (which can be achieved by sputtering) as is demonstrated in Fig. 5) /21/.

We finally emphasize that the bands of the high temperature superconductors have probably considerable dispersion. This can be seen from the density of states measurement of Cu_2O

/22/ which indicates a bandwidth of 3 eV. A similar magnitude of the dispersion for the oxygen copper bands is obtained from a study of the bands in the p(2x1)O-Cu(110) structure /23/ where also large dispersions are observed.

Acknowledgments. This work was supported by the Deutsche Forschungsgemeinschaft and the Bundesministerium für Forschung und Technologie. The author thanks the Institut de Phyique de l'Université de Fribourg for its hospitality.

References

/1/ Bednorz J.G. and Müller K.A.: Z.Physik **B64**, 189 (1986)
/2/ Asada S. and Sugano S.: J.Phys.Soc.Japan **41**, 1291 (1976)
/3/ Hüfner S.: Solid State Commun. **47**, 943 (1983)
/4/ Hüfner S.: Physik **B61**, 135 (1985); Z.Physik **B58**, 1 (1984); Solid State Commun. **53**, 707 (1985)
/5/ Veal B.W. and Paulikas A.P.: Phys.Rev. Letters **51**, 1995 (1983)
/6/ Kinsinger V., Steiner P. and Hüfner S.: Solid State Commun., accepted for publication
/7/ Hüfner S. and Riesterer T.: Phys.Rev. **B33**, 7267 (1986)
/8/ Hüfner S., Osterwalder J., Riesterer T. and Hulliger F.: Solid State Commun. **52**, 793 (1984)
/9/ Kuiper P., Kruzinga G., Ghijsen J., Sawatzky G.A. and Verwey H.: Phys.Rev.Letters **62**, 221 (1989)
/10/ Hüfner S., Steiner P. and Sander I.: Solid State Commun. **72**, 359 (1989)
/11/ Kröger F.A.: J.Phys.Chem.Solids **29**, 1889 (1989)
/12/ Hüfner S.: Solid State Commun., accepted for publication
/13/ Steiner P., Hüfner S., Jungmann A., Kinsinger V. and Sander I.: Z.Physik **B74**, 173 (1989)
/14/ Nücker N., Fink J., Renker B., Ewert D., Politis C., Weijs P.J.W. and Fuggle F.C.: Z.Physik **B67**, 9 (1987)
/15/ Grassmann A., Ströbel J., Klauda M., Schlotterer J. and Saemann-Ischenko G.: Europhys.Letters **9**, 827 (1989)
/16/ Nücker N., Adelmann A., Alexander M., Romberg H., Nakai S., Fink J., Rietschel H., Roth G., Schmidt H. and Spille H.: Z.Physik **B75**, 421 (1989)

/17/ Tokura Y., Takagi H. and Uchida S.: Nature **337**, 345 (1989)

/18/ Takahashi T., Katayama-Yoshida H. and Matsuyama H.: Z.Physik **B78**, 343 (1990)

/19/ Freeman A.J., Yu J. and Fu C.L.: Phys.Rev. **B36**, 711 (1987)

/20/ Massida S., Hamada N., Yu J. and Freeman A.J.: Physica C, **157** (1989)

/21/ Courths R. and Weirich M., private communication

/22/ Steiner P., Hüfner S., Jungmann A., Kinsinger V. and Sander I.: Z.Physik **B74**, 173 (1989)

/23/ Courths R., Cord B., Wern H., Saalfeld H. and Hüfner S.: Solid State Commun. **63**, 619 (1987)

Gap Measurements of High-T_c Superconductors by High Resolution Electron Spectroscopy

Y. Petroff

LURE, Bât. 209 D, Université Paris-Sud, F-91405 Orsay, France

The determination of the gap Δ and the value of $2\Delta/KT_C$ are very important for the understanding of the origin and nature of superconductivity in the new class of high-T_C superconductors. It has been shown recently that photoemission is a very good tool to measure these parameters. In this paper we present and discuss results obtained by different groups using high resolution photoemission and energy loss spectroscopy.

1 - Introduction

Since about 15 years ago angle resolved photoemission has been extensively used to obtain information on the band structure of solids[1]. This is due to the fact that without many adjustable parameters it is rather easy to obtain the dispersion of the bands $E(\vec{K})$. The situation is more complicated with the high T_C superconductors : the large number of atoms per unit cell gives so many bands that it is impossible to separate them. However, near the Fermi level, using high resolution (< 40meV) very important information can be obtained.

Suppose that we have a metal with a constant density of states, N(O), in the normal state ; in the BCS model the density of states in the superconducting state is given by

$$N(E) = N(O)\frac{E}{\left[E^2 - \Delta^2\right]^{1/2}} \quad \text{for } /E/>/\Delta/ \qquad (1)$$

$$= 0 \quad \text{for } /E/</\Delta/$$

This is illustrated in fig 1. The important point to notice is that the states which in the normal state are within a distance $/\Delta/$ of the Fermi surface are <u>moved out</u> and <u>piled up</u> ouside the energy gap region.

Springer Series in Solid-State Sciences, Vol. 99
Electronic Properties of High-T_c Superconductors and Related Compounds
Editors: H. Kuzmany · M. Mehring · J. Fink

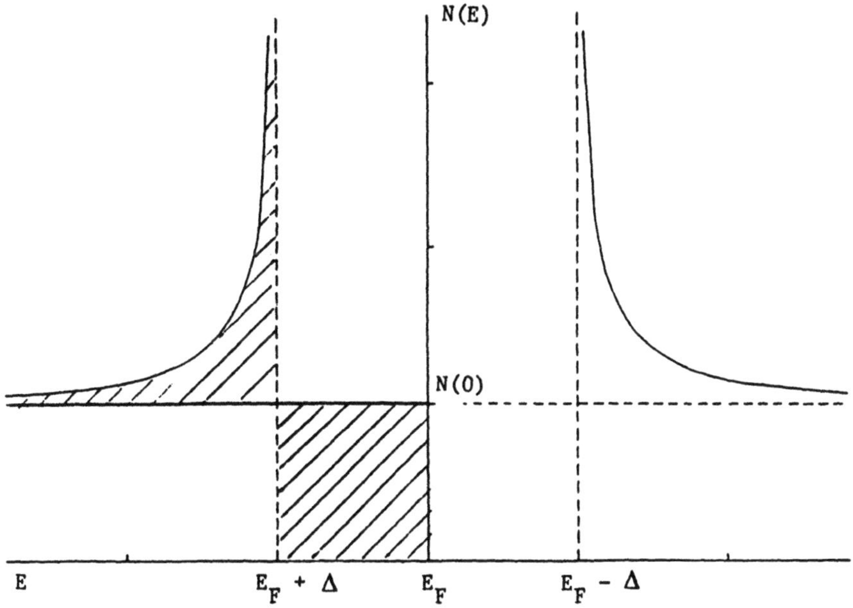

Fig. 1 - Opening of the gap in the BCS model

We would like to point out that only X-ray photoemission (XPS) measurements can determine the density of states. At low photon energy ($h\omega \sim$ 10 - 50eV) even angle integrated photoemission does not give the density of states due to strong different cross-section variations for the s, p, d, f... levels.

2 - Photoemission results

A - Angle integrated photoemission measurements

Imer et al. [2] were the first to report angle integrated photoemission experiments on single crystals of $Bi_2Sr_2CaCu_2O_8$ with a resolution of 20 meV. The spectra, taken in the normal state (T = 105K) and in the superconduction state (T = 15K), are represented in fig 2.

The most striking feature in this figure is the appearance of a peak near the Fermi level. The normal state curve for T = 105K has been simulated by assuming a linearly decreasing density of states convoluted with a Gaussian instrumental line of 20 meV (FWHM) and multiplied by the Fermi function. For $T < T_c$ the BCS density of quasiparticle excitations in the superconducting states has been calculated using equation (1). The best fit is obtained for

$$\Delta = 30 \pm 5 \text{ meV}$$

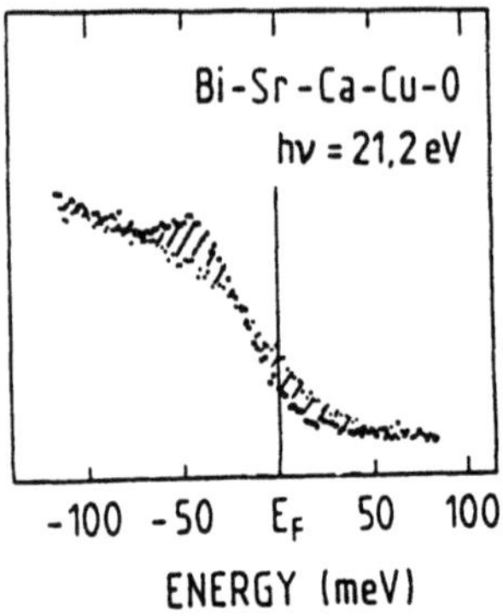

Fig. 2. Photoemission spectra taken at T = 15K (•) and T = 105K(.) [from reference 2]

corresponding to a reduced gap of $2\Delta/KT_C = 8 \pm 1.4$. We notice that the number of new states appearing in the peak correspond to the ones which have disappeared in the gap (hatched area).

B - Angle resolved photoemission measurements

Olson et al.[3] have made a very nice study with good angle (2°) and energy (30 meV) resolution. This has allowed them to isolate the behaviour of the states in a single band near the Fermi level. Fig 3 shows the angle resolved photoemission curves taken at different angles in the direction $\overline{\Gamma M}$ at temperature above (90K) and below T_C (20K). The opening of a gap is is again evident.

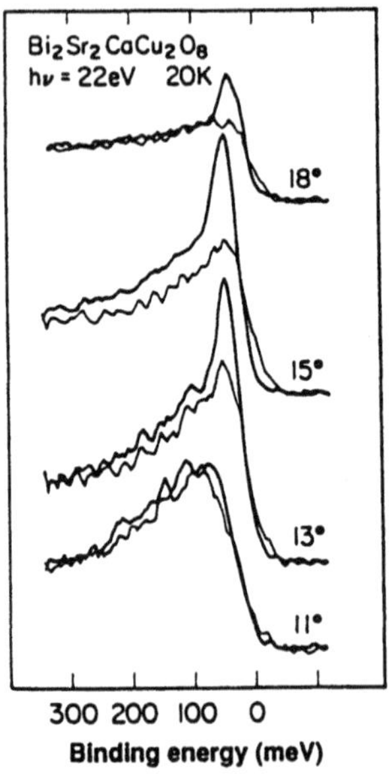

Fig. 3 - Angle resolved energy distribution curves for a surface of $Bi_2Sr_2CaCu_2O_8$ cleaved and measured at 20K (heavier lines) with the 90K spectra superimposed with lighter lines (from reference 3).

At 11°, where the band is well below E_F, there is only a slight sharpening of the band with reduced temperature but no gap opening.
To model the spectra one can start with the simplest case, the 90K spectrum at 18°. A Lorentzian is used to represent the intrinsic spectrum and a linear function for the region at greater binding energy. The Lorentzian and linear term are then multiplied by a 90K Fermi function and convoluted with the instrument function. Finally the Lorentzian and linear term can be used to fit the 20K spectrum by multiplying their sum by a BCS density of states and a 20K Fermi function.
A good fit to the edge is obtained for Δ = 24 meV giving $2\Delta/KT_C \sim 7$.

C - Comparison between the results of angle integrated and angle resolved photoemission studies.

. Fig. 3 shows that Δ is different at 13°, 15° and 18°. The measured value at 15° is 34 meV different from 24 meV at 18°. This should not be taken as evidence for gap anisotropy but rather as an inadequacy of the model when the peak moves away from E_F.
However, we should point out that the Δ value obtained by angle resolved photoemission is strongly dependent on the angle resolution of the analyzer [4].

. Angle integrated photoemission measurements are overestimating Δ because they average, for example, the curves at 18°, 15°, 13° and 11° of Fig. 3, in addition to other symmetry directions, giving a larger value to the one near E_F (18°).

. There are discrepancies between angle integrated and angle resolved data. In the results of Imer et al.[2] the states appearing in the peak below T_C, (Fig. 2), are taken from the gap as expected in the BCS model. The situation is very different in the results of Olson et al.[3]. As shown in Fig. 3 at 18° a very small amount of states comes from the gap, at 13° and 15° it is even worse. This contradicts the BCS model and therefore the value obtained for Δ is meaningless.
These results have been used by Anderson [4] in favor of a model in which one separates the spin and charge and where the superconducting state takes most of the intensity for the superconducting peak from lower energies and not from the gap region. However at present the results of Olson et al.[3] disagree with both models.

. The photoemission results show that Δ is almost temperature independent between 15 and 70K. Above these temperatures the signal to noise ratio is too poor for any definitive conclusion.

. Olson et al.[5] have, by taking angle-resolved photoemission spectra at several points on the Fermi surface, shown that the gap Δ remains constant. This result does not support d-wave coupling theory[6].

. It is not possible to rule out the existence of two gaps (for example with Δ = 30 meV and Δ = 10 meV).

3 - Electron-energy-loss spectroscopy results

High resolution (7meV) electron-energy-loss experiments on $YBa_2Cu_3O_7$ and $Bi_2Sr_2CaCu_2O_8$ have been reported by Demuth et al.[7]. The spot focusing (~ 70 μm diameter) has proved crucial as it has allowed the study of small flat regions of the surfaces. In general, it has been found that about 1/3 of the 1:2:3 surface areas examined show superconductivity compared with only about 5-10 % for 2:2:1:2.

Fig. 4 shows a series of loss spectra for 2:2:1:2 taken after cleavage at 22K. As is typical of all 2:2:1:2 cleaves, the features depend on the beam location on the sample. The least resistive spectrum (dotted line) corresponds to a resistivity of 20 μΩcm, the dashed line to 30 μΩcm and the most resistive spectrum (solid line) to 1000 μΩcm.

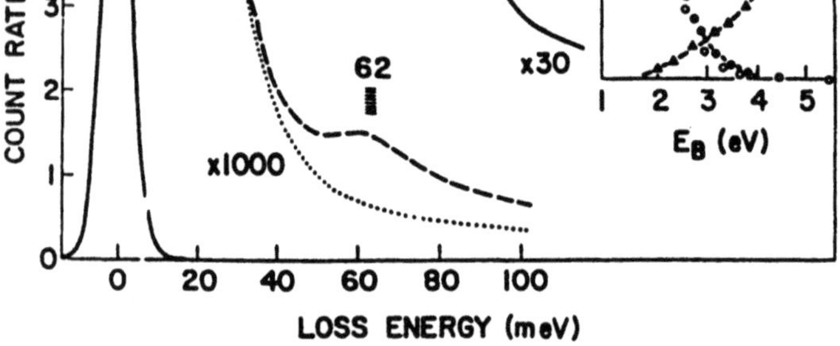

Fig. 4 Electron-energy-loss-spectra for 2:2:1:2 obtained at 22K. Inset : the low resistivity termination (circles) with a different specular reflectivity than the high-resistivity termination (triangles). [from reference 7]

The features at 55 and 82 meV are two vibrations modes corresponding most closely to the infrared modes involving oxygen vibrations out of the Bi-O planes and between the Bi and Cu atoms, respectively. The more-metallic surfaces shown in Fig. 4 strongly screen the incident fields and prevent the bulk vibrations from being observed. The asymmetric feature with its maximum at ~ 62 meV is only seen at low temperatures and corresponds to the superconducting gap.
For the range of T_C's studied on both 1:2:3 and 2:2:1:2 materials the average value of $2\Delta/KT_C$ is 7.8 ± 0.3, in good agreement with the photoemission data.

The temperature dependence indicates a non-BCS behaviour.

4 - **Conclusion**

Many more experiments, especially in the field of angle resolved photoemission, are necessary before one is able to give a definitive answer to the origin and the value of the observed gap.

REFERENCES

[1] see for example P. Thiry, D. Chandesris, J. Lecante, C. Guillot, R. Pinchaux and Y. Petroff - Phys. Rev. Lett. 43, 82 (1979)

[2] J.M. Imer, F. Patthey, B. Dardel, W.D. Schneider, Y. Baer, Y. Petroff, A. Zettl. Phys. Rev. Lett. 63, 102 (1989)

[3] C.G. Olson, R. Liu, A.B. Yang, D.W. Lynch, A.J. Arko, R. S. List, B.W. Veal, Y.C. Chang, P.Z. Jiang and A.P. Paulikas - Science 245, 731 (1989)

[4] P.W. Anderson and Y. Ren (unpublished)
P.W. Anderson (unpublished)

[5] C.G. Olson, R. Liu and D.W. Lynch, R.S. List, A.J. Arko, B.W. Veal, Y.C. Chang, P.Z. Jiang and A.P. Paulikas (to be published)

[6] However, we have to point out that the Δ value reported (18 meV) is different from the one given in reference 3

[7] J.E. Demuth, B.N.J. Persson, F. Holtzberg and C.V. Chandrusekar - Phys. Rev. Lett. 64, 603 (1990)

Electronic Structure of $Bi_2Sr_2CaCu_2O_8$ by Angle-Resolved Photoemission

R. Claessen[1], *T. Buslaps*[1], *J. Fink*[2], *G. Mante*[1], *R. Manzke*[1], *and M. Skibowski*[1]

[1]Institut für Experimentalphysik, Universität Kiel, Olsenhauserstr. 40, D-2300 Kiel 1, Fed. Rep. of Germany

[2]Kernforschungszentrum Karlsruhe, Institut für Nukleare Festkörperphysik, Postfach 3640, D-7500 Karlsruhe, Fed. Rep. of Germany

Abstract. Angle-resolved photoemission spectroscopy (ARPES) on $Bi_2Sr_2CaCu_2O_8$ single crystals is shown to provide valuable information on the applicability of a Fermi liquid picture to the normal state of high-T_c superconductors. Spectra taken along high symmetry directions show band dispersions qualitatively similar to the results of band structure calculations, though strongly renormalized due to correlation effects. Fermi-Dirac-like cutoffs in the spectra indicate the existence of a Fermi surface whose shape agrees with the theoretical prediction.

1. Introduction

A detailed knowledge of the normal state electronic structure is an important prerequisite for a fundamental understanding of high-T_c superconductivity. The question has been posed, whether the charge carriers can be conventionally described in a Fermi liquid picture, to which some new kind of pairing mechanism has to be added in order to account for the high critical temperatures, or whether a theory on high-T_c superconductivity has to start from a non-Fermi liquid state above T_c as described e.g. by an extended Hubbard Hamiltonian [1]. Recently, the anomalous properties of the normal state have been adressed in a phenomenological way [2] and the existence of a"marginal" Fermi liquid has been discussed.

Angle-resolved photoemission spectroscopy (ARPES) is a very suitable experimental technique to obtain detailed information on the electronic structure, because ARPES measures essentially the occupied part of the momentum- and energy-resolved density of states, i.e. in the case of a conventional metal the **k**-resolved band structure. The information depth probed by ARPES is restricted due to the strong interaction of the escaping photoelectron with the remaining electronic background to the order of 10 A. Therefore, high quality surfaces are required, which are well-defined and clean on an atomic scale. This has shown up to be a particular problem with the

Springer Series in Solid-State Sciences, Vol. 99
Electronic Properties of High-T_c Superconductors and Related Compounds
Editors: H. Kuzmany · M. Mehring · J. Fink

1-2-3 superconductors, which suffer a severe surface degradation due to loss of oxygen [3], so that the surface consists primarily of the insulating phase. In contrast, the Bi-based high-T_c superconductor proved to be much more suitable for ARPES investigations, not only because sufficiently large crystals can be grown from this material, but also because the surfaces obtained by cleavage of the crystals remain chemically stable both at room-temperature and under ultra-high vacuum conditions.

Here we report on our latest ARPES results on single crystals of $Bi_2Sr_2CaCu_2O_8$ [4] concerning the detailed electronic structure near the Fermi energy and the shape of the Fermi surface. The data were obtained by using synchrotron radiation from the DORIS II storage ring at HASYLAB, Hamburg.

2. Results and Discussion

We have performed ARPES at room-temperature for various angles of photoelectron emission along the ΓX and $\Gamma\overline{M}$ direction of the Brillouin zone of the face-centred quasi-tetragonal crystal structure. The spectra taken at a photon energy of 18 eV are displayed in Fig. 1. An overall energy resolution of 80 meV was applied, the acceptance angle of the analyzer was less than 0.5°.

The original ARPES data (represented by dots in Fig. 1) around normal emission (corresponding to the Γ point) show a slight emission structure at about 200 meV binding energy accompanied by a broad intensity background which can possibly be interpreted in terms of incoherent spectral weight. Along the ΓX direction this peak moves towards the Fermi level as the angle of emission ϑ is increased, until at $\vartheta=7.5^\circ$ we observe a Fermi edge in the spectrum, defined by the coincidence of the turning point of the emission onset with the independently determined Fermi energy and its width being equal to the width of a Fermi-Dirac distribution function convoluted with the experimental spectrometer function. Both criteria can be estimated with an accuracy of about 10 meV. At an emission angle of 10.5° the peak and the Fermi edge have disappeared. Thus, a band appears to disperse upwards through the Fermi energy between $\vartheta=7.5^\circ$ and 10.5° corresponding to about 1/3 ΓX, in reasonable agreement with the occurrence of a band in the unoccupied energy region at about 1/2 ΓX observed by angle-resolved *inverse* photoemission (ARIPES) [5]. Starting again at 200 meV for $\vartheta=0^\circ$ this structure also disperses upwards along the $\Gamma\overline{M}$ direction (right panel of Fig. 1) until it

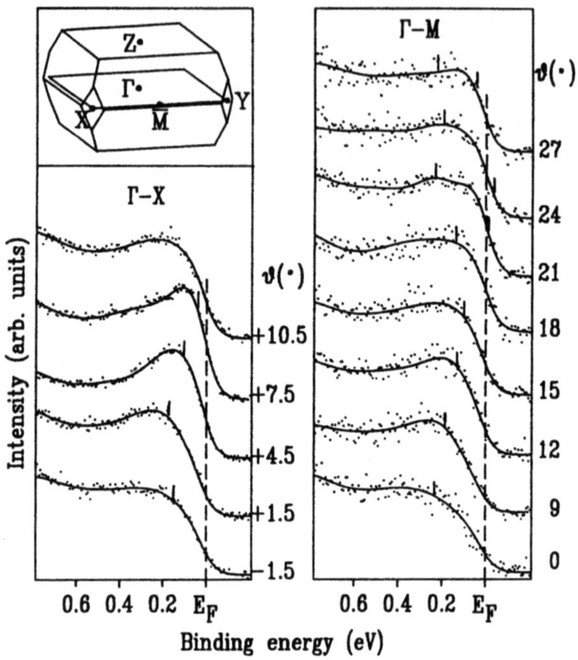

Fig. 1: Angle-resolved photoemission spectra of $Bi_2Sr_2CaCu_2O_8$ along the ΓX (left panel) and $\Gamma\bar{M}$ (right panel) direction taken with 18 eV photon energy at 300 K. The solid lines result from a fit of the orginal spectra (see text). The position of the peaks near E_F are indicated by ticks. The inset shows the Brillouin zone.

becomes rather difficult to unambiguously interpret the spectra for $\vartheta > 15^\circ$. This may be understood from the presence of several bands near E_F at the $\bar{M}$ point as predicted by band structure calculations (e.g. [6]). However, for $\vartheta = 21^\circ$ (i.e. near $\bar{M}$) a clear Fermi-Dirac cut-off is observed, again indicating a crossing of a band through E_F.

In order to extract more quantitative information from the data, in particular from the $\Gamma\bar{M}$ spectra, we have applied a numerical fitting procedure for the low binding energy features by carefully taking into account the broad background intensity visible in the spectra (details are described in Ref. 4). The result of these fits are shown as solid lines in Fig. 1, the energy location of the fitted structures is denoted by tick marks. Along the $\Gamma\bar{M}$ direction we obtained improved convergence for $\vartheta \geq 21^\circ$ by considering an additional Gaussian line profile. Though appearing very slightly above E_F for $\vartheta = 21^\circ$ and 24°, we interpret the occurrence of this additional peak as due to a band touching the Fermi level from above.

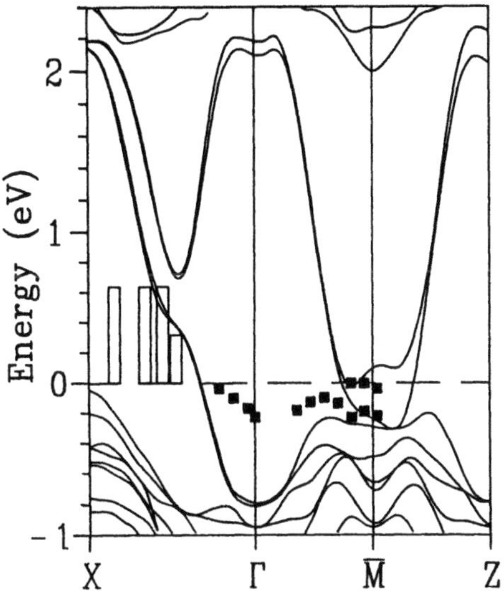

Fig. 2: Experimental band structure of $Bi_2Sr_2CaCu_2O_8$ close to the Fermi level compared with the LDA calculations of Krakauer and Pickett [6]. The small squares denote occupied states from ARPES and the large bars empty states from ARIPES (from Ref. 5), their height representing the experimental resolution. The smaller bar denotes the first observation of a band at the Fermi level in the ARIPES spectra.

We have converted the fit results for the lowest binding energy peaks into an experimental $E(k_{\parallel})$ band structure, which is shown in Fig. 2 together with the LDA calculation of Krakauer and Pickett [6] and the ARIPES results of Ref. 5. Combining both ARPES and ARIPES results one can clearly identify the O(1)-derived conduction band along ΓX, whose unoccupied part constitutes the holes responsible for the superconductivity in $Bi_2Sr_2CaCu_2O_8$. By temperature-dependent ARPES we have observed the opening of the superconductivity gap in this band at that point in **k**-space where it intersects the Fermi level, i.e. directly on the Fermi surface [7]. While the point of intersection agrees favourably with the LDA calculation, the experimental band width below E_F of about 0.2 eV is in strong disaccordance with the theoretical value of about 0.8 eV, indicating a large effective mass ($m^*/m \approx 4$) of the charge carriers. Along $\Gamma\overline{M}$ the O(1) band disperses slightly upwards to about 1/2 $\Gamma\overline{M}$ and downwards again up to the $\overline{M}$ point, while the additional band mentioned above appears at E_F near the zone boundary. The qualitative behaviour of the measured bands along $\Gamma\overline{M}$ is similar to the predicted one, but again with a

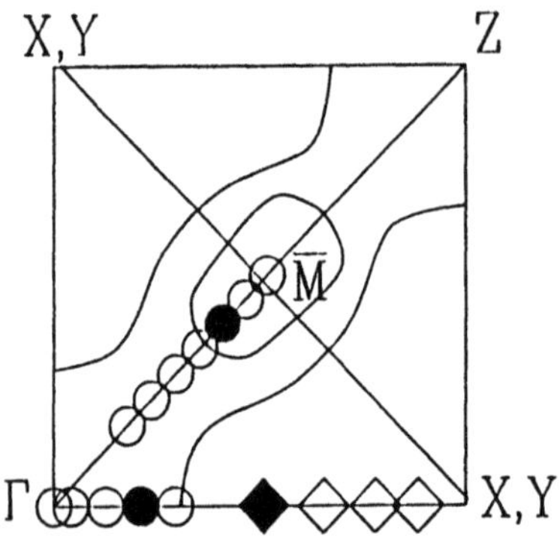

Fig. 3: Basal plane of the Brillouin zone including the Fermi surface as calculated in Ref. 6. Circles and rhombs denote the points in **k**-space measured by ARPES and ARIPS [5], respectively. Solid circles denote the observation of a Fermi-Dirac cutoff in ARPES, the solid rhomb marks the first clear appearance of a band above E_F observed by ARIPES.

strongly renormalized dispersion. The discrepancy between experimental and calculated band widths points to strong electronic correlation effects, which the local density approximation is obviously not able to properly account for.

Despite the disagreement concerning the band dispersions we find for the crossing points of the bands through the Fermi level a remarkable coincidence between experiment and theory. This can also be seen from Fig. 3, where we have plotted the basal plane of the Brillouin zone including the points in **k**-space at which we have taken ARPES and ARIPES spectra, in comparison to the calculated Fermi surface of Ref. 6. Again, one can clearly observe the intersection of the hole-like Fermi surface along ΓX given by the Fermi-Dirac cut-off in ARPES (solid circle) and the first appearence of clear intensity at E_F in ARIPES (solid rhomb). The relatively wide regime in **k**-space for the experimental crossing may be attributed to the difficulty of an unambiguous identification of a Fermi-Dirac cut-off in ARIPES due to the low energy resolution of 640 meV. The data points along $\Gamma\bar{M}$ provide evidence for the existence of an electron pocket around the $\bar{M}$ point. Thus, concerning the shape of the Fermi surface we find for $Bi_2Sr_2CaCu_2O_8$ a similar agreement between ARPES data and LDA band structure calculations as in the recent work of Olson et al. [8]. It is interesting to note that also in other highly correlated systems such as heavy fermion systems the Fermi surface seems to be well reproduced by LDA band structure calculations [9].

3. Conclusion

By a detailed evaluation of ARPES spectra taken on $Bi_2Sr_2CaCu_2O_8$ single crystals we found the qualitative behaviour of the conduction band in good agreement with theory, while the band widths appear to be drastically reduced compared with the predictions of LDA calculations, pointing to strong electronic correlation effects. However, from the observation of bands crossing the Fermi level and corresponding Fermi-Dirac cutoffs in the spectra we can confirm the existence of a Fermi surface and find its shape in reasonable agreement with that calculated by LDA theory.

In the recently suggested phenomenological model of a "marginal" Fermi liquid [2] the quasiparticle character near the Fermi level vanishes, so that the spectra shown here would have to be interpreted entirely in terms of incoherent spectral weight. Such an interpretation is somewhat contradictory to the results of temperature-dependent photoemission, in which the spectral intensity below T_c can be very well described by a BCS-type quasiparticle excitation spectrum [7, 10, 11].

Thus, from photoemission experiments there seems to be some support for a Fermi liquid-like description of the high-T_c compound $Bi_2Sr_2CaCu_2O_8$, though P.W. Anderson [12] has recently noted that his RVB-model is capable of explaining the **k**-dependence of the photoemission spectra.

References

[1] Proc. 2nd International Conf. on High Temperature Superconductors and Materials and Mechanism of Superconductivity (Stanford, 1989) Eds. W. Harrison, N. Phillips, and R. Shelton, special issue of Physica C 162-164 (1989)

[2] C.M. Varma, P.B. Littlewood, S. Schmitt-Rink, E. Abrahams, and A.E. Ruckenstein, Phys. Rev. Lett. 63, 1996 (1989)

[3] R.S. List, A.J. Arko, Z. Fisk, S.-W. Cheong, S.D. Conradson, J.D. Thompson, C.B. Pierce, D.E. Peterson, R.J. Bartlett, N.D. Skinn, J.E. Schirber, B.W. Veal, A.P. Paulikas, and J.C. Campuzano, Phys. Rev B 38, 11966 (1988)

[4] G. Mante, R. Claessen, T. Buslaps, S. Harm, R. Manzke, M. Skibowski, and J. Fink, submitted to Z. Phys. B

[5] R. Claessen, R. Manzke, H. Carstensen, B. Burandt, T. Buslaps, M. Skibowski, and J. Fink, Phys. Rev. B 39, 7316 (1989)

[6] Henry Krakauer and Warren E. Pickett, Phys. Rev. Lett. 60, 1665 (1988)
[7] R. Manzke, T. Buslaps, R. Claessen, and J. Fink, Europhys. Lett. 9, 477 (1989)
[8] C.G. Olson, R. Liu, D.W. Lynch, R.S. List, A.J. Arko, B.W. Veal, Y.C. Chang, P.Z. Jiang, A.P. Paulikas, submitted to Phys. Rev. B
[9] G.G. Lonzarich, J. Magn. Magn. Mat. 76&77, 1 (1988)
[10] J.-M. Imer, F. Patthey, B. Dardel, W.-D. Schneider, Y. Baer, Y. Petroff, and A. Zettl, Phys. Rev. Lett. 62, 336 (1989)
[11] C.G. Olson, R. Liu, A.-B. Yang, D.W. Lynch, J. Arko, R.S. List, B.W. Veal, Y.C. Chang, P.Z. Jiang, and A.P. Paulikas, Science 245, 731 (1989)
[12] P.W. Anderson, preprint

The Influence of Hybridization on the Spin and Charge Distribution in High-T_c Superconductors

H. Eskes and G.A. Sawatzky

Materials Science Center, Department of Solid State and Applied Physics, University of Groningen, Nijenborgh 18, 9747 AG Groningen, The Netherlands

We discuss the importance and consequences of strong Cu3d-O2p hybridization in the high T_c superconductors. We show that the spectral weights found in various high energy spectroscopies require strong hybridization effects. The strong redistribution of spectral weights as a function of hole doping observed in X-ray absorption is shown to be explained by strong hybridization effects. The role of covalency effects on the ligand field splittings, charge and spin density redistribution, the superexchange interaction as well as in the stabilization of the Zhang-Rice singlet is discussed.

Introduction.

Since the discovery of the high T_c superconductors there have been two quite contrasting opinions about the amount of covalency in these compounds. On the one hand, some NMR and neutron scattering experiments seem to indicate that the Cu atom is almost purely ionic (Cu^{2+}-d^9). On the other hand, band structure calculations and the various kinds of spectroscopic experiments seem to tell that these materials are strongly covalent, and that Cu is something between Cu^{2+} and Cu^{1+} (or even Cu^{3+}). In this article we will present arguments in favour of the covalent picture. It is divided roughly in three parts. In the first place we will describe how the photoemission data can be used to obtain estimates of parameters occurring in model Hamiltonians for the high T_c's. Then, using these parameters, the ground state for the insulating anti-ferromagnetic compound and the first ionization state (or the states which are occupied upon hole doping) are studied.

1. Parameter estimates from spectroscopic data.

Many of the models describing the high T_c superconductors involve a set of model parameters. Although these model Hamiltonians are used to describe the low energy properties (order kT), the parameters which occur in these Hamiltonians, like Coulomb interactions and hopping matrix elements, are high energy parameters (>1 eV). Because the low energy physics for these models often strongly depends on the relative value of these parameters it is certainly important to know their magnitudes.

One way to estimate such interactions is by means of LDA based density functional calculations for the valence and conduction bands. The bands thus obtained can then be compared with tight binding bands leading to estimates of the one particle matrix elements (on site energies and inter site hopping terms). The on site energy difference (between Cu-3d and O-2p) however requires some re-interpretation in terms of a configuration interaction rather then a molecular orbital or mean field picture. Because in a model Hamiltonian the Coulomb interactions are taken into account explicitly, the actual value of this charge transfer energy will in general be larger than the one estimated from the one electron band structure, where these interactions are already partly included in a mean field way.

Springer Series in Solid-State Sciences, Vol. 99
Electronic Properties of High-T_c Superconductors and Related Compounds
Editors: H. Kuzmany · M. Mehring · J. Fink

Table 1.

Parameter estimates (in eV) taken from LDA calculations (ref.1 and 2), compared with those obtained from fits to spectroscopic data. U_d is the repulsion between two holes in the $d_{x^2-y^2}$ orbital, U_p between two holes in a p_x orbit. A, B, C are the three Racah parameters describing the d^8 multiplets on Cu. F_0, F_2 are the two Slater integrals describing the p^4 interactions on oxygen. pdσ..ppπ are Slater Koster transfer integrals. $t_{pp} = \frac{1}{2}(pp\sigma - pp\pi)$. $t_{pd} = \frac{1}{2}\sqrt{3}\, pd\sigma$. Values in brackets are the ones used in the calculations.

	MMS (ref.1)	HSC (ref.2)	experimental
Δ	1.3-3.3	2-4 (3.6)	2.75-3.75 (3.5)
pdσ	1.85	1.5	1.5 (t_{pd} = 1.3)
pdπ	-0.75		-0.7
ppσ	-1.0	t_{pp} = 0.65	$t_{pp} \simeq 0.65$
ppπ	0.3		
U_d	8.5	9-11 (10.5)	8.8 (A = 6.5 B = 0.15 C = 0.58)
U_p	4.1-7.3	3-8 (4)	6 (F_0 = 5 F_2 = 6)
U_{pd}	0.6-1.3	0.7-1.7 (1.2)	< 1 (0)

These explicit Coulomb interactions can also be estimated from LDA. This is done in a somewhat more indirect way by constraining occupation numbers on the different atoms, and determining the ground state energy as a function of the occupation number. The values so found by several authors [1,2] are listed in table 1.

Here we will describe an alternative way to arrive at parameter estimates. Our strategy is to compare spectroscopic data with impurity and small cluster calculations. Apart from data for the high T_c's we also look at the model compounds CuO and Cu_2O. The actual calculation involves the (exact) diagonalization of an extended "11 band" Hubbard model including all Cu-3d and O-2p orbitals, which is necessary in order to compare with the measurements. Because the number of many particle states grows very rapidly with the number of Cu sites, the maximum number of Cu atoms in practice is limited to only two. We have studied the effect of the smallness of the clusters by comparing the different XPS spectra (up to 17eV binding energy) as calculated from an impurity approach and from a CuO_4, Cu_2O_8 and Cu_2O_7 cluster respectively, using the same set of parameters [3]. The spectra were all found to be very similar and differences occur only in the smaller detail (0.5-1 eV). Although the actual number of parameters is quite large (see table 1), most of them were fixed by either band structure calculations, spectroscopic experiments other than photoemission or by symmetry arguments as described elsewhere [4,5]. In the comparison with photoemission only the four most important ones were taken as free parameters, namely the charge transfer energy Δ, the Coulomb energy on the copper (Racah A parameter), the Cu-O hopping parameter t_{pd} (or pdσ) and the oxygen 2p bandwidth (mainly determined by ppσ).

It is generally accepted that the added holes (going from the insulating to the superconducting materials) are of predominantly oxygen character, while adding electrons to the insulating materials mainly results in filling the holes in the Cu 3d band, placing the high T_c superconductors in the charge transfer region of the ZSA [6] diagram. The energy difference between the two processes is directly proportional to the energy difference between the Cu 3d and the O 2p states. Thus the charge transfer energy Δ can be determined from (but is not equal to) the energy gap, known to be about 1.5 to 2 ev, as determined from, for instance, a combined photoemission and inverse photoemission measurement [7] or optical spectroscopy [8]. Here one has to be a bit careful because the calculated energy gap will depend on the size of the cluster used, due to "banding" effects [9].

A resonant photoemission experiment using photons with an energy around 74 eV gives a clear fingerprint of configurations with two holes on the Cu atom. From this resonance the value of the on site Coulomb interactions (or better $U_d-\Delta$) can be estimated. In this estimate it is important to realize that the on site Coulomb and exchange multiplets are distributed over an energy range of about 7 eV [4]. This is clearly seen in an Auger measurement for the Cu metal [10]. Thus, although the monopole part of the interaction is quite strongly screened as compared with the free atom, the actual splittings, described by the Racah B and C parameters are almost unchanged and we take the free ion values for those.

Because of the low density of oxygen holes, the photoemission spectra of CuO or the high T_c's are not very sensitive to the value of the repulsions on the oxygen site (U_p). This makes it hard to estimate U_p in this way. For the compound Cu_2O the whole procedure is more straightforward. In this case the ground state does not have one hole per Cu atom, like in the A.F. compounds, but instead Cu_2O is a full band insulator (Cu^+-d^{10}, O^{2-}-p^6). In order to see the effects of the on site hole-hole repulsions, two electrons have to be removed locally and at the same time, i.e. by means of an Auger process (Cu $L_{23}M_{45}M_{45}$ Auger leads to a $3d^8$ and O $KL_{23}L_{23}$ leads to a $2p^4$ partially filled shell). From fits to these spectra the Coulomb repulsion on the Cu site is found to be about 1 eV larger than for CuO. An explanation of this somewhat larger value is the two fold surrounding of the Cu atom in Cu_2O [7]. Because of this low coordination number the screening of the interactions can be expected to be less than in the four fold surrounded Cu in CuO. From the $KL_{23}L_{23}$ Auger on oxygen we estimate $U_p \sim 6.5$ eV, which is by the same reasoning expected to be about 0.5-1 eV smaller for the high T_c's. Besides U_d and U_p, the inter-site Cu-O repulsion (between d^9 and p^5) can be estimated from the position of the satellite in the $KL_{23}L_{23}$ spectrum [11]. This value is found to be smaller than 1 eV, indicating very small values for U_{pd} in the high T_c materials.

In the photoemission spectrum the relative intensity of the two main features is very sensitive to the value of the hybridization matrix element t_{pd}. Suppose the A.F. compound consists of a purely d^9 configuration with no admixture of holes on the oxygen sites. Then after a sudden removal of a 3d electron from the Cu atom (which is the case for the high photon energy XPS experiment) one ends up with a d^8 configuration. Because of the large value of U_d, one would expect that the XPS experiment only shows a significant intensity above say 7 eV ! That in reality the main part of the intensity for 3d removal is found between 1-7 eV binding energy shows that the materials are strongly covalent, resulting in a shift of d spectral weight from d^8 to $d^9\underline{L}$ character states.

2. Character of the wavefunction : Ground state (A.F. compounds).

As pointed out above, the intensity distribution of the XPS spectrum alone proves the strong covalency or hybridization of the Cu d and O 2p states. The fact that the 3d electron removal spectral weight is so strong at threshold, which is generally accepted to be a $d^9\underline{L}$ like state, has in the past often caused an incorrect identification of these states as being d^8. The reason for the strong shift in d spectral weight is two-fold. First the ground state is strongly hybridized so that, say, in a CuO_4 cluster its wave function would be of the form

$$|\Psi_G\rangle = \sqrt{n_d}\,|d^9\rangle + \sqrt{(1-n_d)}\,|d^{10}\underline{L}\rangle \qquad (1)$$

where n_d is the number of 3d holes. The d count would then be $10 - n_d$. From fits to experiments we find roughly $n_d \simeq 0.7$ yielding a Cu 3d count of 9.3, quite far from the ionic value.

We immediately see that d electron removal from the state described by (1) yields both d^8 states and $d^9\underline{L}$ states. This provides one reason for the d spectral weight in the $d^9\underline{L}$ region. More important still is the constructive interference encountered when considering the phases of the wavefunctions. The final state, reached by the removal of an electron from the state (1), is of the form

$$|F,i\rangle \simeq \alpha_i\,|d^8\rangle + \beta_i\,|d^9\underline{L}\rangle + \gamma_i\,|d^{10}\underline{\underline{L}}\rangle\ . \qquad (2)$$

The lowest energy state for a charge transfer solid ($U > \Delta$, $\Delta > t_{pd}$) is one for which $\beta_i > \gamma_i \gg \alpha_i$. The d electron removal spectrum then has the spectral weight

$$(\sqrt{n_d}\,\alpha_i + \sqrt{(1-n_d)}\,\beta_i)^2 = n_d\,\alpha_i^2 + (1-n_d)\,\beta_i^2 + 2\sqrt{n_d}\sqrt{(1-n_d)}\,\alpha_i\,\beta_i .$$

The signs of α_i and β_i are the same as for the ground state if i refers to the low energy states, so the last term adds to the intensity. However for the high energy d^8 like states the sign of α_i and β_i must be opposite and the last term reduces the intensity. This effect is, as it turns out, very large for the values of U, Δ and t_{pd} found experimentally. One should note that it is quite absurd to try to determine the amount of Cu d or O 2p character from a photon energy dependent experiment without considering these large interference effects.

The notation $\underline{L}$ (ligand hole) in equation (1) above involves taking a linear combination of a hole on the four neighboring oxygens surrounding the Cu atom. Furthermore this combination also has to have the right symmetry (also x^2-y^2) with respect to the position of the Cu atom. Thus

$$|GS\rangle = \sqrt{0.7}\,|d^9_{x^2-y^2}\rangle + \sqrt{0.3}\cdot 1/\sqrt{4}\ \{\,|p_{1,\sigma}\rangle + |p_{2,\sigma}\rangle + |p_{3,\sigma}\rangle + |p_{4,\sigma}\rangle\,\}$$

$$= \sqrt{0.7}\,|d^9_{x^2-y^2}\rangle + \sqrt{0.08}\ \{\,|p_{1,\sigma}\rangle + |p_{2,\sigma}\rangle + |p_{3,\sigma}\rangle + |p_{4,\sigma}\rangle\,\} \qquad (3)$$

now $p_{1,\sigma}$ denotes a hole on one of the neighboring oxygens which points towards the Cu atom and has the right sign. Thus, despite the strong covalency, the number of holes on a given oxygen site originating from one of its two neighboring coppers is only 8 % , which is difficult to detect with neutrons.

Another interesting effect of the strong hybridization is the ligand field splitting of the Cu 3d orbitals. In square planar symmetry the shift of the various Cu 3d orbitals is shown in fig.1 with the parameters of table 1. The strongest shift is for $3d_{x^2-y^2}$ orbital, which also causes this to be the ground state. The other symmetries are only slightly shifted causing a large splitting of about 1.3 eV between these and the ground state. Also shown in fig.1 is the degree of covalency of the various states. From this splitting one can estimate the anisotropic temperature independent van Vleck contribution to the paramagnetic susceptibility and the orbital contribution to the NMR Knight shift [12].

Perhaps the most important effect of the strong covalency is the superexchange interaction which is known to be abnormally large in CuO and in the high T_c's [13]. In the clusters containing two Cu atoms, the two spins in the ground state (one per Cu) are found to form a singlet. The energy difference with the first excited state, which is a triplet, is just the superexchange interaction J between the Cu spins. Using the parameter

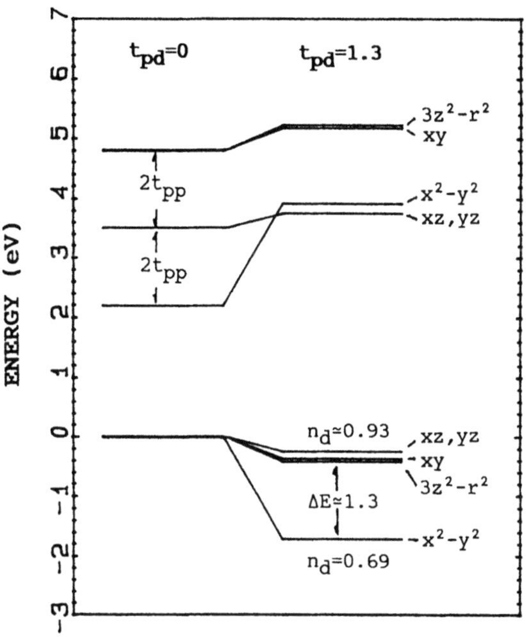

Figure 1. Ligand field energy shifts for the Cu-3d and O-2p orbitals, before and after turning on the Cu-O hybridization.

set from table 1, J is found to be 0.24 eV, somewhat larger than the experimental value of about 0.13 eV. This value is however quite sensitive to the actual parameters, and inclusion of a Cu-O inter-site interaction and exchange term in the calculation reduces this value to 0.16. All of the above effects can be explained by a strong Cu 3d-O 2p hybridization, and one would be hard pressed to explain them assuming a 3d count of much less than 9.3 .

3. Character of the wavefunction : Electron removal (addition) states.

As already mentioned, adding electrons to the A.F. compounds, as in the Nd compounds, will mainly result in filling the $d_{x^2-y^2}$ orbitals on the Cu site. In a CuO_4 cluster, using the simple basis set, the final state would then be $|d^{10}\rangle$. It should however be noted that this state can be reached from the ground state with both d electron and O 2p electron addition. In fact the intensity of this structure as seen by O_{1s} XAS or EELS will be determined by $(1-n_d)$ i.e. the covalency in the ground state. This effect was described in detail and demonstrated by Kuiper et al. for Li doped NiO, as well as for the high T_c superconductors [14]. The O_{1s} pre edge peak seen in the electron doped material [15] is therefore most likely due to the upper Hubbard band as pointed out also by Takahashi et al.[18].

It is rather interesting to see what happens to the intensity of this peak upon hole doping. In the latter case there is a state lower in energy than the upper Hubbard band of $d^9\underline{L}$ character which in fact hybridizes with the upper Hubbard band. As in XPS this causes a severe shift in spectral weight in the O_{1s} XAS spectrum towards the threshold peak and the upper Hubbard band seems to disappear. Also this was demonstrated in $Li_xNi_{1-x}O$ [14] and for $La_{2-x}Sr_xCuO_4$ and $YBa_2Cu_3O_{7-\delta}$. To demonstrate this effect we show in fig.2 the calculated O_{1s} XAS spectrum for a Cu_2O_7 cluster [3] with two holes (undoped) and with three holes (50 % hole doping) showing the strong decrease in intensity of the upper Hubbard band for the latter case. The experimental observation of this effect again demonstrates the strong covalency in these systems.

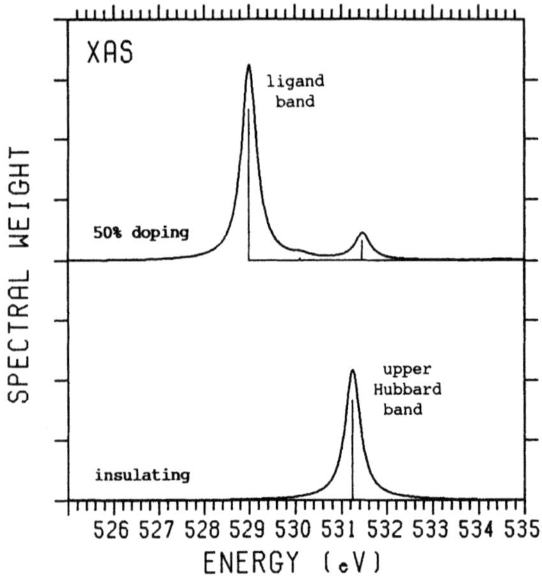

Figure 2.

O_{1s} XAS spectrum for a Cu_2O_7 cluster, starting from the two hole (bottom curve) and three hole ground state (top curve), using the parameters from table 1.

The first electron removal state is of quite a different origin. Using the parameters of table 1, this state is found to be a singlet state (Zhang Rice singlet) of A_g symmetry (D_{4h} point group). This state was found by us [5] to be the lowest energy state also upon inclusion of the O 2p band width and was demonstrated to be quite stable for the parameters of table 1. The wave function is of the form

$$|el.rem\rangle \simeq \sqrt{0.09}\,|d^8\rangle + \sqrt{0.68}\,|d^9\underline{L}\rangle + \sqrt{0.23}\,|d^{10}\underline{\underline{L}}\rangle \ . \qquad (4)$$

At first glance, looking at the occupation numbers of the different configurations, it seems that this state can be reached from the ground state by the addition of a ligand hole ($\underline{L}$), except for the small amount of $|d^8\rangle$ character which can not be obtained this way. One might then conclude that the d electron removal spectral weight for this state is almost zero. This is not so since as argued above the d electron spectral weight at threshold is quite large, due to interference effects. An actual calculation of the spectral weights yields 1.6 for the ratio between $\underline{L}_{x^2-y^2}$ electron removal and $d_{x^2-y^2}$ electron removal, thus they are both of the same order of magnitude. This shows that the hopping in a lattice of such a Zhang-Rice singlet can be accomplished both via the motion of a d hole as well as a ligand hole. (The hopping in a lattice can be thought of as an annihilation of a particle at one site and the creation at another site and therefore can be related to the spectral weight in photo- and inverse photoemission spectroscopy.)

The ideas of some that the oxygen holes (induced by doping) could move independently of the Cu holes (or the holes already present) as in some two fluid models can not possibly be correct since these are intimately connected as shown above. Similarly the suggestion by some that the holes induced by doping are of pure oxygen character is based on a simplified picture with neglect of the O-Cu hybridization.

The large Cu-O hybridization in the hole doped case also causes a very large exchange splitting of the Zhang-Rice singlet from the corresponding triplet state. In a CuO_4 cluster calculation this splitting is found to be about 3.5 eV. However from this one should not conclude that the Zhang-Rice singlet is the only state of importance. A full CuO_4 cluster calculation shows that a triplet state of B_1 symmetry (a product of x^2-y^2 and $3z^2-r^2$ symmetry orbitals) lies only about 1 eV above the Zhang-Rice state. This energy difference is lowered further if one also takes into account the apex oxygen [17] or if one considers an embedded cluster [12].

In the CuO_4 cluster the x^2-y^2 and $3z^2-r^2$ symmetry orbitals do not mix via one particle matrix elements of the Hamiltonian, because of the four fold symmetry. In the real solid this is no longer the case and there will be a certain amount of admixture of $3z^2-r^2$ character in the low lying electron removal states. Suppose we look at a $\vec{k}$ value of $(\pi,0)$ in the two dimensional CuO_2 plane. In this case the plane looks like a collection of chains in the x direction. This is easily seen because the p_σ orbitals in the y direction have the wrong symmetry to couple to the Cu d orbitals, and are thus isolated. For these 1D chains the x^2-y^2 and $3z^2-r^2$ orbitals now have the same symmetry, leading to a finite amount of $3z^2-r^2$ occupation. For a Cu_2O_7 cluster the amount of $d_{3z^2-r^2}$ occupation per Cu site is found to be about 10 % [3]!

In summary we have used parameters obtained from the comparison between high energy spectroscopies and model Hamiltonians to discuss the influence of hybridization on various observables. We conclude that a strong Cu-O hybridization both in the undoped and doped cases is consistent with high energy spectroscopy results as well as quantities like a large ligand field splitting, an anisotropic magnetic susceptibility, the NMR Knight shift or the large superexchange interaction and it causes a stabilization of the Zhang-Rice singlet.

Acknowledgement.

This investigation was supported by the Netherlands Foundation for Chemical Research (SON) and the Netherlands Foundation for Fundamental Research on Matter (FOM) with financial support from the Netherlands Organization for the Advancement of Pure Research (NWO).

References

1. A.K.McMahan, R.M.Martin, and S.Satpathy, Phys.Rev.B **38**, 6650 (1988).
2. M.S.Hybertsen, M.Schlüter, and N.E.Christensen, Phys.Rev.B **39**, 9028 (1989).
3. H.Eskes, G.A.Sawatzky, and L.F.Feiner, Physica C **160**, 424 (1989). H.Eskes and G.A.Sawatzky, in preparation.
4. H.Eskes, L.H.Tjeng, and G.A.Sawatzky, Phys.Rev.B **41**, 288 (1990).
5. H.Eskes and G.A.Sawatzky, Phys.Rev.Lett. **61**, 1415 (1988).
6. J.Zaanen, G.A.Sawatzky, and J.W.Allen, Phys.Rev.Lett. **55**, 418 (1985).
7. J.Ghijsen, L.H.Tjeng, J.van Elp, H.Eskes, J.Westerink, G.A.Sawatzky and M.T.Czyzyk, Phys.Rev.B **38**, 11322 (1988).
8. S.Tajima et al., J.Opt.Soc.Am.B **6**, 475 (1989).
9. M.S.Hybertsen, E.B.Stechel, M.Schlüter, and D.R.Jennison, unpublished;
10. E.Antonides, E.C.Janse, and G.A.Sawatzky, Phys.Rev.B **15**, 1669 (1977).
11. L.H.Tjeng, J.van Elp, P.Kuiper, and G.A.Sawatzky, unpublished.
12. J.F.Annett, R.M.Martin, A.K.McMahan, and S.Satpathy, Phys.Rev.B **40**, 2620 (1989).

13. G.Shirane et al., Phys.Rev.Lett. **59**, 1613 (1987); B.X.Yang et al., Phys.Rev.B **38**, 174 (1988); **39**, 4343 (1989); J.B.Forsyth et al., J.Phys.C **21**, 2917 (1988); S.Sugai et al., Phys.Rev.B **38**, 6436 (1988).
14. P.Kuiper, G.Kruizinga, J.Ghijsen, G.A.Sawatzky, and H.Verweij, Phys. Rev.Lett. **62**, 221 (1989); P.Kuiper et al., Phys.Rev.B **38**, 6483 (1988).
15. N.Nücker, P.Adelmann, M.Alexander, H.Romberg, S.Nakai, J.Fink, H.Rietschel, G.Roth, H.Schmidt, and H.Spille, Z.Phys.B **75**, 421 (1989).
16. T.Takahashi, H.Katayama-Yoshida, and H.Matsuyama, Z.Phys.B, in print.
17. A.Fujimori, Phys.Rev.B **39**, 793 (1989).

Local Partial Electronic Structure of High-T_c Superconductors and Related Oxides by X-Ray Emission Spectroscopy

F. Burgäzy[1], *H. Jaeger*[1], *C. Politis*[2], *P. Lamparter*[1], *and S. Steeb*[1]

[1]Max-Planck-Institut für Metallforschung, Seestr. 92, D-7000 Stuttgart 1, Fed. Rep. of Germany

[2]Kernforschungszentrum Karlsruhe, INFP, D-7500 Karlsruhe, Fed. Rep. of Germany

Abstract. The distribution of O 2p and Cu 3d states of $YBa_2Cu_3O_x$ with different oxygen content x, $Bi_2Sr_2CaCu_2O_{8+\delta}$, and other copper oxides is determined from O K_α and Cu L_α X-ray emission spectra. In contrast to the non-superconducting oxides the superconducting materials show O 2p states close to the Fermi energy and Cu 3d states at higher binding energies. This behaviour is also confirmed within the Y-Ba-Cu-O system. The results are compared with those of band structure calculations.

1. Introduction and Experimental

X-ray emission spectroscopy (XES) is a useful tool to study the local partial electronic structure. Since valence band X-ray spectra arise from transitions from the valence band to a core level, they are subject to electric dipole transition rules, which allow the spectroscopy of the partial electronic structure. The participation of a core-level related to one specific element results in the local character of the spectra. Therefore, with O K_α and Cu L_α spectra the O 2p and Cu 3d states can be probed separately.

The spectra were recorded with a JEOL JXA-733 electron microprobe operated at 10 kV for O K_α and at 4 kV for Cu L_α radiation. The X-ray spectrometer was equipped with a clinochlore crystal for the O K_α ($\Delta E < 0.7$ eV) and with a TAP crystal for the Cu L_α spectra ($\Delta E \approx 0.9$ eV). The clinochlore crystal was calibrated with the Cr L_ι and Mn L_ι lines and the TAP crystal with the Cu L_α line of metallic copper [1]. A description of the sample preparation is given in [2] and [3].

Since the XES spectra are measured on an X-ray energy scale, they need to be assigned to a binding energy scale. This was done with the help of X-ray photoemission spectra (XPS) of the involved core levels [4]-[8]. Since this procedure implies the same energetic position of the core levels with XES and XPS it has to be assumed that with XES the core hole is screened by the valence electrons before the transition occurs.

In compounds such as the high-T_c superconductors, where one element is located at crystallographically different positions, there may be slight differences in the core level positions. This leads to a shift of the corresponding partial spectra relative to each other and thus to a slightly distorted picture of the density of states [3]. Nevertheless, since the differences in the core level positions of the investigated compounds are so small that they cannot even be resolved with core level XPS, a qualitative comparison of XES with the density of states is quite possible. The determination of the centers of gravity of the partial densities of states via the XES spectra is even less erroneous because the assignment to the binding energy scale was done with the XPS core level spectra whose maxima represent the averaged core level positions.

2. Local Partial Electronic Structure

Fig. 1 shows the O K_α and Cu L_α XES spectra of the investigated copper oxides compared with the corresponding UPS spectra from [5],[6], and [12]. Though part of the subject has already been investigated elsewhere [9]-[11] we will present all of our results for an overall comparison of the O 2p and Cu 3d centers of gravity. With Cu_2O the O 2p and Cu 3d states are widely separated. While the main part of the O 2p states is located

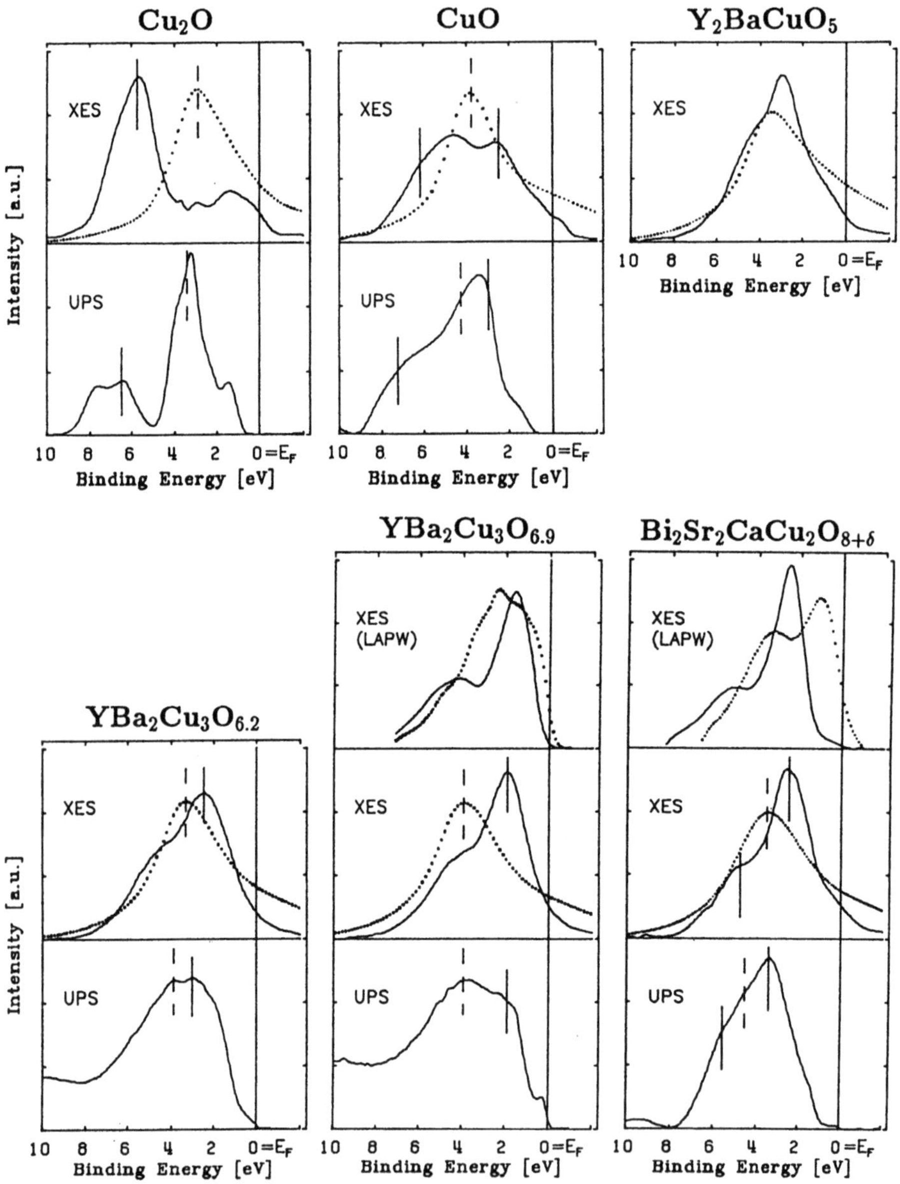

Fig 1. Measured (XES) and calculated (XES(LAPW)) X-ray emission spectra (—— : O K_α and ······· : Cu L_α) compared with UPS spectra (UPS) from [5],[6], and [12].

at high binding energies, the Cu 3d states are found rather close to the Fermi energy E_F. The comparison with the UPS spectrum allows an easy assignment of the Cu and O related features in the UPS spectrum. In contrast to Cu_2O, the Cu 3d states of CuO are positioned within the region of the broad two-peaked O 2p band. The big overlap of the two spectra renders the assignment of the features of the UPS spectrum more difficult. In Y_2BaCuO_5 there is an almost complete overlap of the Cu 3d and O 2p states. Therefore, from this point of view a large hybridization between the Cu 3d and O 2p states cannot be ruled out.

Now the important question will be addressed whether there are significant changes in the electronic structure of the superconductor $YBa_2Cu_3O_{6.9}$ when only the oxygen content is decreased to $O_{6.2}$ in order to make the compound non-superconducting. The most obvious difference between the superconducting compound with $O_{6.9}$ and the other copper oxides consists of O 2p states at low binding energies with the dominant peak at about 2 eV. Compared to the O 2p states the Cu 3d states are located at higher binding energies. The assignment of the peaks in the UPS spectrum to those of the XES spectra is in good agreement with the results obtained by varying the UPS excitation energy [8] [12]; there it can be seen that the major oxygen-related features extend around 2 eV binding energy and the copper-related features around 4 eV. Although there is only a relatively small difference in the oxygen content, non-superconducting $YBa_2Cu_3O_{6.2}$ shows a distinctly different electronic structure. Compared to $YBa_2Cu_3O_{6.9}$ the O 2p states are moved to higher and the Cu 3d states to lower binding energies. This results in a narrowing of the valence band which explains in a convincing way the narrowing of the UPS spectra observed recently by Arko et al. [12] (see Fig. 1).

$Bi_2Sr_2CaCu_2O_{8+\delta}$ is the second high-T_c superconductor investigated in the present work. The measured O K_α and Cu L_α spectra look similar to those of $YBa_2Cu_3O_{6.9}$. Again, the O 2p states are at lower binding energies than the Cu 3d states, although the separation between them is less than in $YBa_2Cu_3O_{6.9}$. Furthermore, the Cu L_α spectrum of $Bi_2Sr_2CaCu_2O_{8+\delta}$ is broader than the one of $YBa_2Cu_3O_{6.9}$. The assignment of the features of the almost structureless UPS spectrum is in good agreement with the results of photoemission spectra generated with different excitation energies [5].

Fig. 2 shows the mean positions of the O 2p and Cu 3d states of the investigated oxides. There the mean position of the O 2p states is given by the center of gravity of the O K_α spectrum. However, the mean position of the Cu 3d states had to be taken as the peak position of the Cu L_α spectrum because satellites on the high energy side of the spectrum do not allow a determination of the mean position via center of gravity. With Fig. 2 it becomes evident that the compounds of the Y-Ba-Cu-O and Bi-Sr-Ca-Cu-O systems show O 2p states at lower binding energies than the two simple copper oxides Cu_2O and CuO. The two superconducting compounds have the lowest mean O 2p binding energies; the higher the transition temperature of the compound, the lower the mean O 2p binding energy, and the larger the separation of the mean Cu 3d binding energy. In order to extend this observation upon other copper oxides,further investigations are planned.

In the case of Cu_2O the comparison of integrated densities of states obtained by band structure calculations [13] with the experimentally obtained XES spectra yields a striking agreement. With CuO, the broad oxygen-related density of states obtained from band structure calculations [14] is qualitatively in agreement with the experimental O K_α spectrum while the calculated position of the Cu states is at lower binding energies than the one observed experimentally with the Cu L_α spectrum. These discrepancies may arise from the $3d^9$ configuration of the Cu states in CuO which leads to a two-hole-

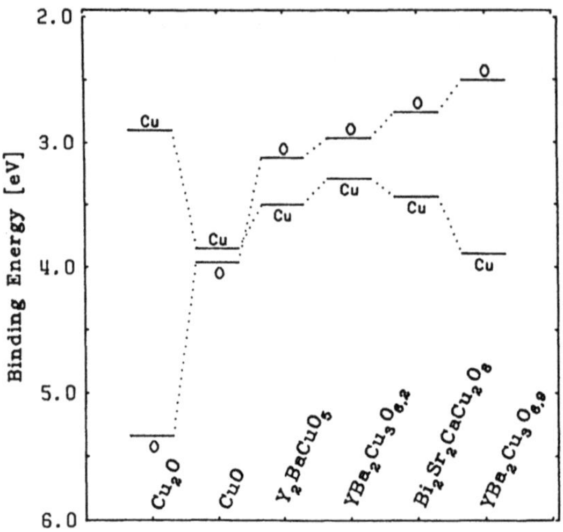

Fig 2. Mean positions of O 2p and Cu 3d states of various copper oxides.

system in the XES and UPS final state. Strong on-site Coulomb interaction between the localized 3d states in this case may influence the experimental spectra. For $YBa_2Cu_3O_{6.9}$ and $Bi_2Sr_2CaCu_2O_{8+\delta}$ XES spectra have been calculated [15][16] based on the results of LDA-LAPW band structure calculations [17] [18]. These calculated spectra are included in Fig. 1 with the label XES(LAPW). The comparison with the experimental spectra reveals good agreement for the O K_α spectra concerning not only their shape but also their positions. This is not the case with the Cu L_α spectra where theory and experiment clearly disagree. The calculated Cu L_α spectra are much broader and show more structure than the measured spectra. This discrepancy with the Cu L_α spectra does not have to be due to failure of the ground state band structure calculations but rather may arise from XES. In [19] the Cu L_α spectrum of the CuO_2 planes is calculated taking into account electron correlations with the Anderson impurity model and treating XES as a coherent second order optical process. Thereby the main X-ray transition path involves only states of e_g and b_{2g} symmetry. This leads to a spectrum resembling much more the narrow and structureless spectra measured in this work. Despite the differences in Fig. 1, the fact that the calculated Cu L_α spectrum of $Bi_2Sr_2CaCu_2O_{8+\delta}$ is broader than the one of $YBa_2Cu_3O_{6.9}$ is well confirmed by the experiment. Also the observed change in the band structure of $YBa_2Cu_3O_{6.9}$ with decreasing oxygen content (see Fig. 1) is qualitatively in agreement with LDA-calculations in [14], where the dominant Cu contributions show up at lower binding energies in the O_6-compound than in the O_7-compound, while the number of O states at lower binding energies decreases.

References

1. "International Tables for X-Ray Crystallography", Vol. IV, Kynoch Press (1974)
2. F. Burgäzy, H. Jaeger, K. Schulze, P. Lamparter, S. Steeb : Z. Naturforsch. **44a**, 180 (1989)
3. F. Burgäzy, C. Politis, P. Lamparter, S. Steeb : Z. Naturforsch. **44a**, 780 (1989)
4. P. Steiner, V. Kinsinger, I. Sander, B. Siegwart, S. Hüfner, C. Politis, R. Hoppe, H. P. Müller : Z. Phys. B **67**, 497 (1987)

5. P. Steiner, S. Hüfner, A. Jungmann, S. Junk, V. Kinsinger, I. Sander, W. R. Thiele, N. Backes, C. Politis : Physica C **156**, 213 (1988)
6. P. Steiner, S. Hüfner, A. Jungmann, V. Kinsinger, I. Sander : Z. Phys. B **74**, 173 (1989)
7. P. Steiner : private communication
8. P. Steiner, S. Hüfner, A. Jungmann, V. Kinsinger, I. Sander : "High T_c Superconductors - Electronic Structure", A. Bianconi, A. Marcelli (eds.), Pergamon Press (1989)
9. J.-M. Mariot, V. Barnole, C. F. Hague, G. Vetter, F. Queyroux : Z. Phys. B **75**, 1 (1989)
10. C. F. Hague, V. Barnole, J.-M. Mariot, C. Michel, B. Raveau : "High T_c Superconductors - Electronic Structure", A. Bianconi, A. Marcelli (eds.), Pergamon Press (1989)
11. E. Z. Kurmaev, V. I. Nefedov, L. D. Finkelstein : Int. J. Mod. Phys. B **2**, 393 (1988)
12. A. J. Arko, R. S. List, R. J. Bartlett, S.-W. Cheong, Z. Fisk, J. D. Thompson, C. G. Olson, A.-B. Yang, R. Liu, C. Gu, B. W. Veal, J. Z. Liu, A. P. Paulikas, K. Vandervoort, H. Claus, J. C. Campuzano, J. E. Schirber, N. D. Shinn : Phys. Rev. B **40**, 2268 (1989)
13. P. Marksteiner, P. Blaha, K. Schwarz : Z. Phys. B **64**, 119 (1986)
14. K. T. Park, K. Terakura, T. Oguchi, A. Yanase, M. Ikeda : J. Phys. Soc. Jap. **57**, 3445 (1988)
15. J. Redinger, A. J. Freeman, J. Yu, S. Massidda : Phys. Lett. A **124**, 469 (1987)
16. P. Marksteiner, S. Massidda, J. Yu, A. J. Freeman, J. Redinger : Phys. Rev. B **38**, 5098 (1988)
17. S. Massidda, J. Yu, A. J. Freeman, D. D. Koelling : Phys. Lett. A **122**, 198 (1987)
18. S. Massidda, J. Yu, A. J. Freeman : Physica C **152**, 251 (1988)
19. S. Tanaka, K. Okada, A. Kotani : J. Phys Soc. Jap. **58**, 813 (1989)

Polarized Cu-L_{III} XAS: Relation Between the Symmetry of the d-Holes and the Critical Temperature

P. Lagarde and A.M. Flank

LURE, Bât. 209D, Université Paris-Sud, F-91405 Orsay, France

Abstract. High resolution polarized Cu L_3 x-ray absorption spectra of single crystals and of highly oriented pellets of high T_c superconductors have been measured by using synchrotron radiation at Super-ACO. We have investigated the energy splitting Δ between the in-plane and out-of-plane 2p->3d transition as well as the intensity of the E//z component versus the overall 3d absorption and we find that these quantities are correlated with the critical temperature of the superconducting systems.

1. Introduction

It is well established that the holes induced by doping are mainly in the oxygen derived valence band i.e. ligand holes giving the $3d^9\underline{L}$ configuration [1 - 5]. In previous papers we have shown that the $3d^9\underline{L}$ states induced by doping give an high energy tail or a separate shoulder to the copper white line. In both $La_{2-x}Sr_xCu_2O_4$ x=0.1 and $Bi_2Sr_2Ca_1Cu_2O_8$ this $\underline{2p}3d^{10}\underline{L}$ shoulder in the range 1-2 eV above the white line appears for the E⊥z polarization indicating the presence of $\underline{d}(b_1)\underline{L}(b_1)$ states. The presence of holes out of the CuO_2 plane is found in the high T_c range of the $YBa_2Cu_3O_{7-\delta}$. In this case, the E⊥z spectrum of the T_c=60K sample shows an asymmetric high energy tail indicating the presence of a $3d^9\underline{L}$ configuration with $\underline{d}$ unoccupied orbitals in the CuO_2 plane $\underline{d}_x\underline{L}$ but most of the weight of the 1.5 eV shoulder observed in the E//z spectra of the T_c~80K is formed when the transition temperature T_c changes from the low range (50-60K) to the high T_c range (80-90K). All these results are in agreement with EELS investigation of the oxygen K-edge [4].

We have then addressed our interest to the variation induced by doping of the unoccupied Cu-d-states responsible for the main white line in X-ray absorption. The electronic configuration of the Cu ions in the antiferromagnetic insulating compounds is expected to be the Cu $3d^9$

Springer Series in Solid-State Sciences, Vol. 99
Electronic Properties of High-T_c Superconductors and Related Compounds
Editors: H. Kuzmany · M. Mehring · J. Fink

configuration, where the d hole is in the Cu $3d_{x^2-y^2}$ orbital of b1 molecular symmetry in the CuO_4 square plane cluster. The tilting of the elongated octahedron of the $La_{2-x}Sr_xCuO_4$ and the distortions of the CuO_5 square pyramids of $Bi_2CaSr_2Cu_2O_8$ induce a mixture of the $3d_{z^2-r^2}$ (a1 molecular symmetry) or of the $3d_{xz}$ (e molecular symmetry) configurations with the $3d_{x^2-y^2}$ one in the ground state [6]. There are several experimental indications that the holes are not present only in the σ bond between the $3d_{x^2-y^2}$ and the $O(2p_x)$ or $O(2p_y)$ in the CuO_2 layers. For instance the holes induced by doping in $YBa_2Cu_3O_{7-\delta}$ (T_c=80K system) are present in the apical oxygen O(4) of Cu(2) which is also in the linear Cu(1)-O chains. Evidence for d-d excitations in $YBa_2Cu_3O_{7-\delta}$ has been reported [7] while an early study of the polarization dependence of the "white line" in the Cu L_{III}-edge XAS of $Bi_2CaSr_2Cu_2O_8$ [3] has indicated the presence of 10%-20% of Cu 3d holes oriented in the z direction.

Some theoretical models for high T_c superconductivity such as the three band model [8,9] , the d-d excitation model [10,12] , the Kanamori model [13] , the interlayer pairing [14] require the presence of electronic states with two holes in two different orbitals. Here we have investigated the polarization dependence of the white line due to the $2p_{3/2}$->3d transition in antiferromagnetic insulating La_2CuO_4 and superconducting crystals $La_{2-x}Sr_xCuO_4$ (x=0.04 and x=0.1), in bismuth compounds of different classes with different T_c and finally in the $YBa_2Cu_3O_{7-\delta}$ system in order to investigate the presence of an electronic transition polarized in the z direction. The experimental results show a correlation between the energies and the relative weight of this z-polarized white line and T_c which could be in line with some theoretical predictions.

2. Experimental

The L_{III} x-ray absorption experiment has been carried out on the Super-ACO storage ring of the synchrotron radiation facility LURE at Orsay. A double crystal (1010) beryl monochromator with an energy resolution about 0.3 eV at 900 eV has been used. The absorption coefficient has been measured by total electron yield method with therefore a sampling depth of about 200 Å. The crystals have been cleaved with the surface normal direction z perpendicular to the CuO_2 planes while the magnetically oriented pellets had the z axis normal to the pellet surface. The polarized spectra with several incidence angles α

between the electric field of the photon beam E and the normal z to the sample surface have been recorded by rotating the sample around a vertical axis since the synchrotron radiation is linearly polarized in the horizontal plane. The procedure to obtain the relative intensities of the E//z and E⊥z white lines is the following : the "magic angle" unpolarized data is deconvoluted using the experimental data from $\alpha=10°$ and $\alpha=75°$, with the relative intensities as unknown parameters. The values obtained are then checked on two other angles. In the case of $Bi_2Sr_2Ca_1Cu_2O_8$ the z axis is parallel to the c axis while it is parallel to the b axis on lanthanum compounds. The energy calibration and the resolution of the monochromator are checked by recording the spectrum of a CuO reference sample immediately after each spectrum. Within these conditions, we think that an energy position can be obtained within an accuracy of ± 50 meV, a value which corresponds to the maximum of the shift of this reference white line over several beam injections.

3. Results

The energy separation Δ between the maxima of the two polarized white lines has been obtained by curve fitting for the insulating La_2CuO_{4-y} and for superconducting $La_{1.96}Sr_{0.04}CuO_4$ and $La_{1.9}Sr_{0.1}CuO_4$ compounds : it decreases from 230±50 meV to 80±50 meV going from La_2CuO_{4-y} to $La_{1.9}Sr_{0.1}CuO_4$. The intensity of the E//z white line is about 25% of the intensity of the E⊥z white line with an error of about 15-20%. The presence of a component of the white line for E//z polarization indicates that Cu d holes have a component of their orbital in the z direction d_z (d_z is introduced here to indicate one of the 3d orbitals : $3d_{z^2-r^2}$, d_{xz} or d_{yz} that we cannot distinguish in this experiment).

In Fig.1 the polarized white lines of $Bi_2Sr_{1.5}Ca_{1.5}Cu_2O_8$ compound are shown with their relative intensities obtained through the deconvolution procedure described above. As in the $LaCuO_4$ case, the E//z line points at lower energy and its intensity amounts to around 20 % of the total absorption intensity. The change in the relative intensities of these two components, which is given by the fitting procedure, is evidenced in Fig.2 which gives the absorption second derivative of the unpolarized spectra taken at the magic angle for three different compounds with an increasing critical temperature. We see that in going from the insulating compound to superconductor ones the relative intensity of the E//z white line increases with T_c.

In Fig.3 this energy difference Δ between the maximum of the in-plane (the xy plane of CuO_2 layers) E⊥z polarized white line and the out

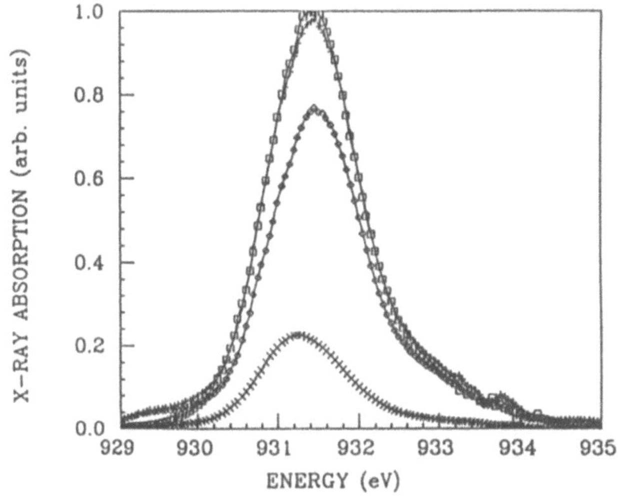

Fig.1 The white lines in both polarizations (z = crosses, xy = rhombs) with their relative intensities as well as the magic angle result (squares) in the case of $Bi_2Sr_{1.5}Ca_{1.5}Cu_2O_8$

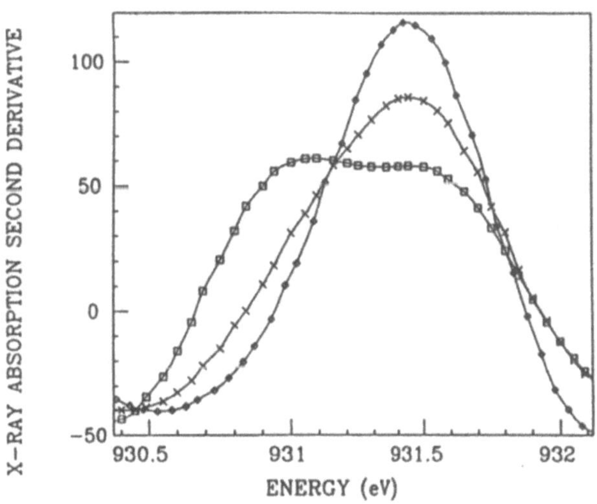

Fig. 2 Absorption second derivative of three different Bi compounds with T_c=0 (rhombs),70 (crosses) and 110K (squares) showing the increase of the intensity of the E//z component of the white line as a function of the critical temperature.

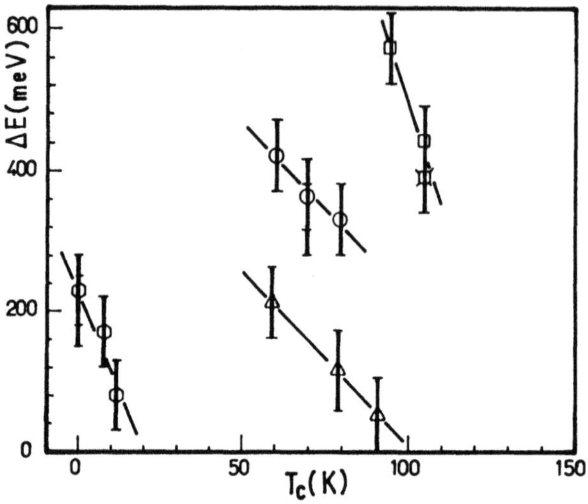

Fig. 3 Energy splitting Δ between the peaks of polarized white lines in L_3 edge XAS (E⊥z peak minus E//z peak) as function of critical temperature (T_c) for the following classes of superconductors : $La_{2-x}Sr_xCu_2O_4$ (hexagons), $YBa_2Cu_3O_{7-\delta}$ (triangles), and Bi-Sr-Ca-Cu-O (circles and squares) showing the decrease of this splitting versus Tc.

Table I

Sample	T_c(K)	ΔE(meV)	I_z(%)
$Bi_2Sr_2YCu_2O_{8+\delta}$	0	200	8
$Bi_2Sr_2CaCu_2O_{8+\delta}$	61	420	10
$Bi_2Sr_{1.5}Ca_{1.5}Cu_2O_{8+\delta}$	70	365	16
$Bi_2Sr_2CaCu_2O_{8+\delta}$	82	330	
$Bi_2Sr_2Ca_2Cu_3O_{10+\delta}$	95	570	45
$Bi_{1.8}Pb_{0.2}Sr_2Ca_2Cu_3O_{10+\delta}$	105	430	

of plane E//z polarized line for various compounds is plotted versus the critical temperature of each sample. We observe that within each class of systems the $\Delta(T_c)$ is a decreasing function of T_c. For instance the value of Δ reaches its maximum, in the series $La_{2-x}Sr_xCuO_4$, for the insulating x=0 system. In this compound as well as in the $YBa_2Cu_3O_{7-\delta}$ one, the doping induces an important change in the d unoccupied states by reducing the energy separation between the unoccupied states with components along z and that with components along the xy plane and increasing the relative weight of the z component. Table I gathers all these results on Bi compounds.

4. Conclusion

In conclusion we have found by polarized L_{III} XAS a correlation between the energy separation of the two white lines (for E//z and E⊥z) and T_c, as well as a correlation between T_c and the relative intensity of the E//z line compared to the overall L_{III} absorption. These results have to be compared with the predictions of the d-d excitation model which considers two copper hole excited states with the symmetries $3d_{x^2-y^2}$ and $3d_{3z^2-r^2}$ separated by an energy Δ closely related to the critical temperature. In such a model, T_c should change at the same time with this energy splitting Δ and with the hybridization between the two hole states which is measured in our experiment by their relative intensities following the relation $T_c = \Delta \exp(-1/\lambda_c)$ where λ_c is a function of Δ and of the degree of hybridization. While the exact relationship between these calculations and our experimental results is not yet clearly established, we have here a direct proof that a theory of such high T_c superconductors must imply the existence of two types of holes with different symmetries on the copper sites.

Recently band structure calculations of CuO_2 planes in tight-binding approximation have been carried out [15]. Depending on the oxygen-oxygen hopping parameter the authors find two Cu d-bands with the $3d_{x^2-y^2}$ and $3d_{3z^2-r^2}$ symmetries crossing the Fermi level, a result again in agreement with the XAS ones.

References

1 A. Bianconi, A. Congiu Castellano, M. De Santis, P. Rudolf, A.M. Flank, P. Lagarde and A. Marcelli *Sol. State Commun.* 63, 1009-1013 (1987)

2 A. Bianconi, J. Budnick, A.M. Flank, A. Fontaine, P. Lagarde, A. Marcelli, H. Tolentino, B. Chamberland, C. Michel, B. Raveau and G. Demazeau *Phys. Lett. A* 127, 285-291 (1988)

A. Bianconi, M. De Santis, A. Di Cicco, A.M. Flank, A. Fontaine, P. Lagarde, H. Katayama-Yoshida, A. Kotani and A. Marcelli . *Phys. Rev.* B 38, 7196-7299 (1988)

3 A.M. Flank, P. Lagarde, A. Bianconi, P. Castrucci, A. Fabrizi, M. Pompa, H. Katayama-Yoshida and H., Castelani Proc. Ninth Conf. on VUV Physics, *Physica Scripta* (1990) in press

4 J. Fink, H. Nücker, H. Romberg and S. Nakai . *High T_c superconductors : Electronic structure*, Pergamon Press, London p. 293 (1989) edited by A. Bianconi and A. Marcelli) *and references cited therein.*

5 G. Kaindl, D.D. Sarma, O. Strebel, C.T. Simmons, U. Neukirch, R. Hoppe and H.P. Muller . *Physica* C 153-155, 139-140 (1988)

6 M. Bacci *Jpn. J. Appl. Phys.* 27, L1699-L1701 (1988)

7 H.P. Geserich, G. Scheiber, J. Geerk, H.C. Li, G. Linker, W. Assmus and W. Weber. *Europhysics Letters* 6, 277-282 (1988)

8 C. Castellani, C. Di Castro and M. Grilli. *Physica C* 153-155, 1659-1660 (1988)

9 C. Castellani, C. Di Castro and M. Grilli Proc. of the Conf. "Towards the theoretical understanding of high T_c superconductors" Trieste, July 1988 edited by S. Lundqvist, E. Tosatti, M.P. Tosi and Yu Lu, *Int. Journal of Modern Physics* 1, 659-665 (1988)

10 D.L. Cox, M. Jarrell, C. Jayaprakash, H.R. Krihnamurthy and J. Deisz *Phys. Rev. Lett.*, 62 , 2188 (1989)

11 M. Jarrell, H.R. Krishnamurthy and D.L. Cox, . *Phys. Rev.* B 38, 4584-4587 (1988)

12 W. Weber *Z. Phys. Condensed Matter* 70, 323 (1988)

13 T. Nishino, M. Kikuchi and J. Kanamori *Solid State Commun.* 68, 455-458 (1988)

14 S. Kurihara *Physica C* 153-155, 1247-1248 (1988)

15 M. Grilli, C. Castellani and C. Di Castro (to be published)

Calculation of Photoionization Cross Sections and Optical Conductivity for Models of High-T_c Superconductors

G. Wendin[1] *and Z. Crljen*[2]

[1]Institute of Theoretical Physics, Chalmers University of Technology, S-41296 Göteborg, Sweden
[2]Rudjer Bošković Institute, University of Zagreb, Bijenička 54, P.O. Box 1016, YU-41001 Zagreb, Croatia, Yugoslavia

We model $YBa_2Cu_3O_7$ as a collection of free atoms and use a local-density based random phase approximation (LDRPA) to calculate photoionization cross sections in the 10-160 eV photon energy range. We compare with experimental results for the optical conductivity and propose interpretations of major spectral features.

In this paper we shall focus attention on the large scale variation of the O 2p, Cu 3d, Ba 5p,4d and Y 4p ionization cross sections in $YBa_2Cu_3O_7$, as well as on the relative strengths of various resonances, like O 2s-2p, Cu 3p-3d, Ba 5p-5d, Y 4p-4d, and Ba 4d-4f. In particular, we shall compare with experimental results for the optical conductivity from electron energy loss spectroscopy (EELS) [1-3].

We model the superconductor in the simplest possible way as a collection of free atoms and use a local density based random phase approximation (LDRPA; LDA+ RPA) to calculate total and partial photoionization cross sections [4]. Previous experience has shown that this is a reasonable approximation for describing the overall behaviour of photoionization cross section of solids in the shallow core region [5-7].

There are some obvious limitations of an atomic model for a solid:

(i) There are no molecular bonds, no hybridization, no interference effects, no charge transfer, no conduction charge carriers: this gives much too low polarizability at low excitation energies. In particular, the prominent O 2p-Ba 5d charge transfer excitations are not included.

(ii) The configurations to be used for Cu and O are not clear. We have chosen the neutral atomic ground state configurations Cu $3d^{10}4s$ and O $2p^4$ for calculating potential, charge density, and bound state and continuum wave functions. Using these orbitals we have then *in the cross section calculations* assumed that charge transfer results in the configurations Cu $3d^9$ and O $2p^{6-x}$, where we have chosen x=0.1 to model typical O 2p hole concentrations.

(iii) The LDA ionization energies of the atomic ground state configurations are sometimes quite different from those of the solid. We have adjusted these energies in the case of O 2s and Cu 3p excitation and ionization.

Descriptions of the LDRPA method can be found in Refs. [5-7] . The photoabsorption cross section is given by $\sigma_{abs}(\omega) \sim \omega \, |\langle\varepsilon| r \, \varepsilon_b^{-1}(r;\omega)|i\rangle|^2$ where $\varepsilon_b^{-1}(r;\omega)$ describes induced (local) field effects due to screening within the ion cores. ω is the photon energy, $|i\rangle$ is the initial LDA state with energy ε_i and $\langle\varepsilon|$ is the final LDA state at a continuum energy $\varepsilon=\omega+\varepsilon_i$.

Springer Series in Solid-State Sciences, Vol. 99
Electronic Properties of High-T_c Superconductors and Related Compounds
Editors: H. Kuzmany · M. Mehring · J. Fink

The LDA is totally inadequate for describing the Ba 5p,4d and Y 4p ionization cross sections. The corrections to the Cu 3d and O 2p cross sections are moderate (but significant). However, the O2p/Cu3d ratio is essentially not changed from the LDA to the LDRPA, and agrees well with the previous Herman-Skillman results by Yeh and Lindau [8] and with the experimental ratio by Ming Tang et al. [9] for O2p and Cu3d derived features in the valence band spectrum.

Figure 1 shows the calculated photoabsorption cross section $\sigma_{abs}(\omega)$ of $YBa_2Cu_3O_7$ in the 10-160 eV photon energy range. The oxygen and copper cross sections are essentially structureless apart from the weak O2s-2p and Cu 3p-3d resonances around 20 eV and 75 eV respectively. Barium and yttrium, on the other hand, show strong giant dipole resonances around 20 eV (Ba 5p-5d), 30 eV (Y4p-4d) and 110 eV (Ba4d-4f).

Agreement with photoyield measurements above ω=70 eV [9] is reasonably good, as expected. The localization of the LDA Ba4d-4f excitation to within the ion core makes the excitation spectrum insensitive to perturbations by molecular or solid state environments [10]. The 4d-4f $^3D_1$ peak does not shift between different compounds, metal and atom. The overall spectral shape of the Ba 4d-εf $^1P_1$ giant dipole resonance is determined in the core region and does not change on the average. However, scattering in the final states (interference, molecular structure, Ba5d-O2p hybridization) leads to pronounced structure. Note, however, that the double peak structure is probably not due to molecular effects, because these features exist already in the free atom and in the metal: the pronounced peak around 120 eV probably may be associated with a $4d^{-1}5p^{-1}5d^2$ double excitation [11].

In the 10-70 eV region there are, to our knowledge, no experimental photoabsorption or photoyield data. We shall instead compare with results for the optical conductivity derived from EELS [1,2] measurements via Kramers-Kronig analysis.

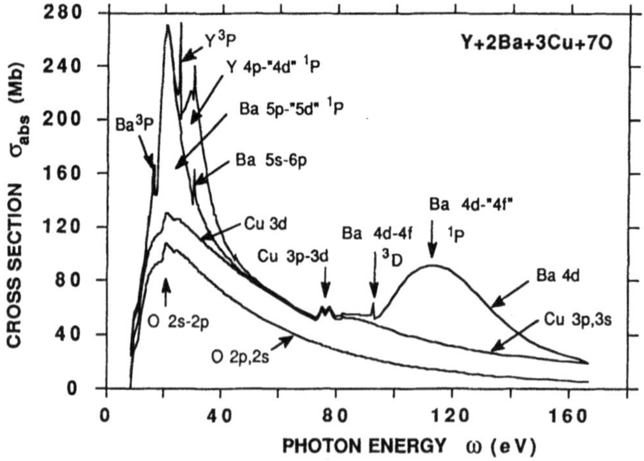

Fig. 1. Total LDRPA photoionization cross sections, accumulated in the form of 7O, 3Cu+7O, 2Ba+3Cu+7O, and Y+2Ba+3Cu +7O.

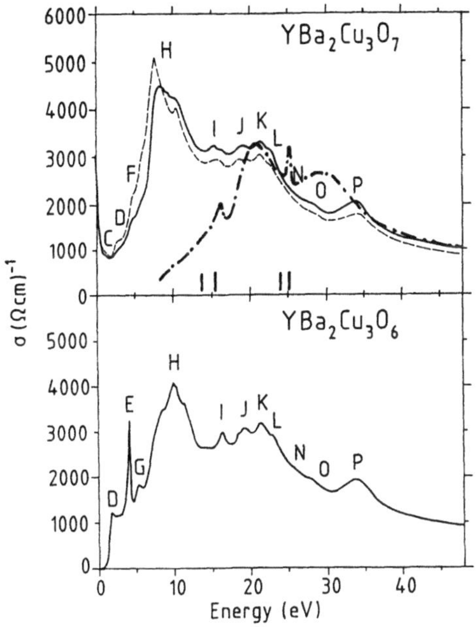

Fig. 2. Optical conductivity for $YBa_2Cu_3O_7$ and $YBa_2Cu_3O_6$ [1]. —— : film. - - - : single crystal. – . – . – : Present theory, arbitrarily normalized to experiment at 50 eV.

The optical conductivity is given by $\sigma(\omega) = \omega\, \varepsilon_2/4\pi$, where ε_2 is the imaginary part of the macroscopic dielectric function. For a collection of atoms with density (N/V), $\sigma(\omega)$ ~ (N/V) $\sigma_{abs}(\omega)$. Therefore, as long as we disregard the formation of new states, like molecular bonds and conduction states, the conductivity will be proportional to the sum of atomic photoabsorption cross sections.

Figure 2 shows the photoabsorption cross section normalized around 50 eV to the experimental optical conductivity derived from EELS [1]. The comparison in Fig. 2 demonstrates that the absorption spectrum above 15-20 eV may essentially be described within an atomic picture. The character of the spectrum has already been indicated in Fig. 1. O 2p-εd and Cu 3d-εf continuum excitations form a background with a broad maximum around 20-25 eV (O 2p-εs and Cu 3d-εp transitions dominate close to threshold, around 10 eV, but rapidly fall off at higher photon energies. The Cu 3d-4p excitation has not been included; it will show up as a spike below 6 eV). About half of the absorption strength in the 20-30 eV range is due to Ba 5p-εd ("5d") and Y 4p-εd ("4d") giant dipole resonances, showing up as peaks around 20 eV (J,K,L) and 34 eV (P) in the experimental spectrum. The incorrect theoretical position (30 eV) of the Y 4p-εd ("4d") resonance might possibly be connected with the high ionicity of yttrium (Y^{3+}) in the solid (the calculations refer to atomic configurations, and no shifts have been made for Ba and Y).

The calculations furthermore suggest that the weak structure (I) around 16 eV may be connected with a Ba 5p-5d ($^3P_1$) structure arising from spin-orbit splitting of the giant dipole resonance (intermediate coupling: 5p spin-orbit splitting smaller than the 5p-5d exchange integral) [6,7]. This indicates that in the $YBa_2Cu_3O_7$ superconductor, the Ba 5p-5d $^3P_1$ excitation is quite well localized (atomic-like): strong hybridization, like in metallic Ba, would wipe out this peak. In the same way the broad structure around 28 eV (O) could be an Y 4p-4d $^3P_1$ peak, as suggested by the calculation, but strongly influenced by hybridization.

In Fig. 2 we have marked the positions of the Ba 5p and Y 4p core levels, as determined experimentally from PES [12], defining excitations to the Fermi level (charge-transfer screened core holes). The Ba 5p-5d ^3P and Y 4p-4d ^3P excitations are both above the respective ionization thresholds and may hybridize and decay into the continuum. The large width of the Y 4p-4d ^3P peak might then possibly be due to its relatively large energy above the Fermi level. The position of the Y ions, intercalated between the CuO_2 planes, might also play a role.

The fine structure on top of the Ba 5p-d cross sections around 20 eV (J,K) is not understood. It might be due to final state molecular structure (BaO cluster). It might also be due to spin-orbit splitting [1] of non-charge-transfer screened Ba 5p levels.

As expected, the atomic model breaks down completely below 15 eV excitation energy. The atomic model contains roughly the correct number of electrons per unit cell, but partly occupuing too compact orbitals. Molecular bonds in BaO and CuO clusters will be more delocalized, with spectral weight at low frequencies corresponding to charge transfer excitations. One can therefore expect oscillator strength to be removed from higher energies and transferred to low energies. In Fig. 2 we only show theory normalized to experiment. However, preliminary theoretical results for the absolute conductivity (involving the density of cells) suggest that the theoretical curve should be multiplied by a factor of ~2. This puts too much oscillator strength above 20 eV and essentially nothing below 15 eV, in line with what is required by the oscillator strength sum rule.

Finally it should be mentioned that we have results for the partial cross sections of individual shells in the neighbourhood of the resonances, giving information about the orbital character of the emitted electrons. In the description of resonances, we have included Auger decay of the core hole in the LDRPA scheme [4,6,7]. A resonance therefore becomes characterized by an Auger decay width Γ_A and an autoionization (recombination) width Γ_a. In the case of the Cu 3p-3d and O 2s-2p resonances, the Auger width is a good deal larger than the recombination width, i.e. $\Gamma_A \gg \Gamma_a$. The resonances therefore mainly branch to two-hole-one-electron states, i.e. resonant satellites, and show up as window resonances (anti-resonances; dips) [4] in the single-hole cross sections.

This general behaviour is already fairly well established for the Cu 3p-3d resonance [3,12,13]. In the O 2s-2p case, we note that the experimental resonant PES results of Arko et al. [13] show a *clear dip around 22 eV* in the intensity of an O 2p peak centred about 100 meV below the Fermi level. We suggest that this might be evidence for an O 2s-2p resonance in the YBaCuO superconductor. However, this interpretation remains tentative because molecular types of interference effects may give rise to peaks and dips, and hybridization may change the resonance behaviour of main lines and satellites.

References:

1. H. Romberg, N. Nücker, J. Fink, Th. Wolf, X.X. Xi, B. Koch, H.P. Geserich, M. Dürrler, W. Assmus and B. Gegenheimer, Z. Phys. B **78**, 367 (1990)
2. N. Nücker et al., Phys. Rev. B **39**, 12379 (1989)
3. G. Wendin, Physica Scripta **T27**, 31 (1989), and references therein
4. G. Wendin and Z. Crljen, Physica C **162-164**, 1327 (1989); this paper gives a preliminary discussion of some of the present results.
5. G. Wendin, in New Trends in Atomic Physics, eds. G. Grynberg and R. Stora, Les Houches Session XXXVIII, 1982 (Elsevier, 1984)
6. G. Wendin, Phys. Rev. Lett. 53 (1984) 725
7. G. Wendin, in Giant Resonances in Atoms, Molecules and Solids, NATO ASI Series B, vol. B151, eds. J.-P. Connerade, J.-M. Esteva and R.C. Karnatak (Plenum Press, New York, 1987), p.171
8. J.J. Yeh and I. Lindau, At. Data Nucl. Data Tables **32**, 1 (1985)
9. K.-L. Tsang et al., Phys. Rev. B **37**, 2293 (1988); R.A. Rosenberg and C.-R. Wen, Phys. Rev. **37**, 9852 (1988); M. Onellion et al., Phys. Rev. B **36**, 819 (1987); Ming Tang et al., Phys. Rev. **38**, 897 (1988)
10. P. Rabe, K. Radler and H.-W. Wolff, in Vacuum Ultraviolet Radiation Physics, eds. E.E. Koch, R. Haensel and C. Kunz, Vieveg/Pergamon, 1974, p. 247
11. G. Wendin, ibid, p. 225
12. P. Thiry, et al., Europhys. Lett. **5** , 55 (1988)
13. A.J. Arko et al., Phys. Rev. B **40**, 2268 (1989)

Evidence for Jahn-Teller Distortions in Bi-Sr-Ca-Cu-O High-T_c Superconductors

J. Röhler and A. Larisch

II. Physikalisches Institut, Universität zu Köln, Zülpicherstr. 77, D-5000 Köln 41, Fed. Rep. of Germany

Abstract. We have measured the extended absorption fine structure beyond the Cu-*K* x-ray absorption edge in "2212" and "2223" Bi superconductors. The temperature dependences of geometrical distortions in the CuO_5 and CuO_4 configurations were extracted from the "beat" analysis of the interfering nearest-neighbor signals. The base bonds, Cu-O11, O12 differ by ≅ 0.115 Å ("2212") and 0.105 Å ("2223"). The lengths of the apex Cu-O2 bonds are 2.53 Å ("2212") and 2.37 Å ("2223") at 300 K and decrease at $T \lesssim T_c$ by 0.015 Å and 0.05 Å, respectively.

1. Introduction

The detailed geometry of the oxygen configurations coordinating the Cu^{2+} sites in the high-T_c cuprates ought to provide insight into their electronic structure and possible electronic excitations. Cu^{2+} ($3d^9$) is a Jahn-Teller (JT) active ion lifting the double degeneracy of the E_g ground state by a distortion of the oxygen environment. For example, the axial elongation of the O_6 octahedra or O_5 pyramids splits the Eg doublet into two states with $3d_{x^2-y^2}$ and $3d_{3z^2-r^2}$ symmetry. In an isolated pyramidal geometry JT excitations were shown to be possible within the E_g manifold ($3d_{x^2-y^2} \rightarrow 3d_{3z^2-r^2}$) and from E_g- into T_{2g}-states ($3d_{x^2-y^2} \rightarrow 3d_{xy,yz}$). Both types of excitations show up as characteristic distortions [1]. In a corner-linked CuO_5 network, however, cooperative effects must be taken into account [2]. Detailed knowledge of the nature of the distorted Cu^{2+} environments in the high-T_c cuprates might be important for *d-d* excitation models, postulating the pairing of the O *p* conduction holes occurs via excitations within the E_g manifold [3].

The basal plane of the CuO_5 pyramids in the "123" rare earth cuprates is orthorhombically distorted (≅ 2%). Here the lower-than-tetragonal site symmetry of Cu2 clearly originates from the formation of O-Cu1-O chains along the crystallographic *b*-axis. However, one may argue that the JT character of Cu^{2+} stabilizes the rhombic distortion and thus plays an structural role. For example, single ion

Springer Series in Solid-State Sciences, Vol. 99
Electronic Properties of High-T_c Superconductors and Related Compounds
Editors: H. Kuzmany · M. Mehring · J. Fink

$3d_{x^2-y^2} \rightarrow 3d_{3z^2-r^2}$ excitations asymmetrically stretch the in-plane oxygens, favouring an orthorhombic environment [1].

The Bi cuprates have strongly modulated structures, which are nearly tetragonal (≅ 0.2% orthorhombicity). The modulations are largest in the Bi-O double layers and weakest in the CuO_5 polyhedra. Diffraction data of the "2212" phase give no unique picture of the distorted Cu^{2+} environments. Refinements in the average space groups Am*aa* or A2*aa* result in up to four different Cu-O1 base bonds [4]. Also results obtained from different experimental groups deviate significantly from each other [5]. It is not clear whether these discrepancies result from the choice of an inappropriate space group of the average structure, compositional variations, cation vacancies or different oxygen concentrations. Little is known about the temperature dependences of the Cu-O bondlengths and even less about distortions of the CuO polyhedra in the "2223" phase [6].

We have used the extended absorption fine structure (EXAFS) beyond the Cu-*K* x-ray absorption edge to probe *locally* the average distortions of the CuO polyhedra. In *R*-space the nearest neighbor oxygens are well separated from the Ca, Sr cations. The base oxygen, apex oxygen and in-plane Cu-O-Cu scattering paths were analyzed. The latter also yields information on the bond angle between the corner-linked CuO polyhedra and will be reported in a forthcoming paper [7].

2. Experimental details and results

The preparation of the polycrystalline "2212" phase is described in detail elsewhere [8]. From the onset of the ac-susceptibility T_C was observed at 75 K. Within the accuracy of x-ray powder diffraction a pseudotetragonal unit cell was found (space group F*mmm*) with $a = b = 5.408$ Å and $c = 30.68$ Å (300 K). No impurities of other superconducting phases or of CuO could be detected.

The polycrystalline "2223" phase was obtained from the starting composition $Bi_{1.6}Pb_{0.4}Sr_{1.6}Ca_2Cu_{2.8}O_{9.2}$. The preparation process and the detailed sample characterization are described in [9]. From the maximum of the specific heat anomaly $T_C = 106.9$ K. Powder x-ray diffraction gave the lattice parameters (spacegroup F*mmm*) $a = b = 5.405$ Å and $c = 37.04$ Å (300 K). Inclusions of additional phases were 3-4% "2212" phase, ≤ 5% CuO and ≤ 7% Ca_2PbO_4.

The absorption data were recorded with synchrotron radiation *(HASYLAB, DESY)* from carefully powdered pellets. Useful data were obtained up to $k = 16$ Å. Experimental details will be given elsewhere [7].

Normalization of the data was performed in the usual manner. 10% Gaussian windows were applied for Fourier transforming $\chi(k)k^n$ $(n = 2,3)$ into the *R*-space. The single scattering contributions from the py-

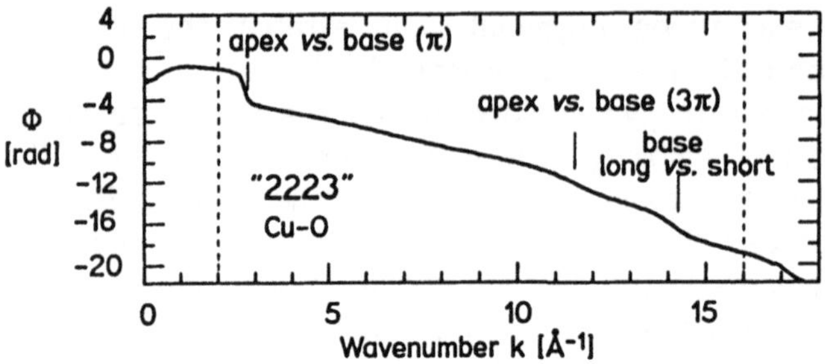

Fig. 1. Total Cu-O phase, Φ, in "2223" Bi-cuprate at 85 K for $\bar{R}$ = 1.9115 [Å]. Dashed vertical lines indicate the lower and the upper bounds of the Fourier transform from the k-space into the R-space.

ramidal oxygens ($R \leq 2.6$ Å) could be properly isolated from all other contributions within $R_{eff} = 0.9 - 2.3$ Å. We then used the "beat" analysis to extract the bondlength *differences* in the distorted CuO configurations [10]. This technique is particularly well suited for measurements of bond stretching effects. Multi-shell fitting techniques mostly suffer from a large number of correlated parameters and thus are inappropriate for the analysis of weakly distorted shells [11].

Two superimposed oscillations, corresponding to two nearest neighbors at slightly different distances, cause beats, i.e. minima in the effective amplitude and jumps in the effective phase at $k_n = (2n+1)\pi/2\Delta R_{12}$, $n = 0, 1, 2...$ Fig. 1 shows the k dependence of the nearest neighbor Cu-O phaseshift, $\Phi_{Cu-O}(k)$, in the "2223" compound. Here the O_5 pyramids contribute by 2/3, and the O_4 interlayers by 1/3 to the scattering signal. The jumps caused by the long apex bonds emerge

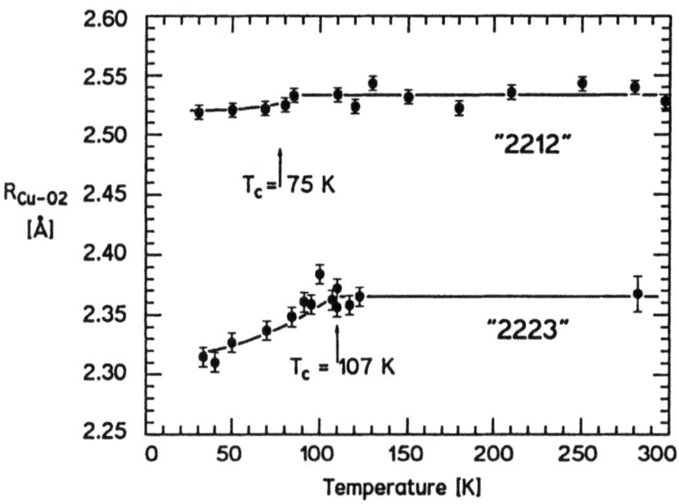

Fig. 2. Temperature dependence of the long apex bonds in Bi cuprates.

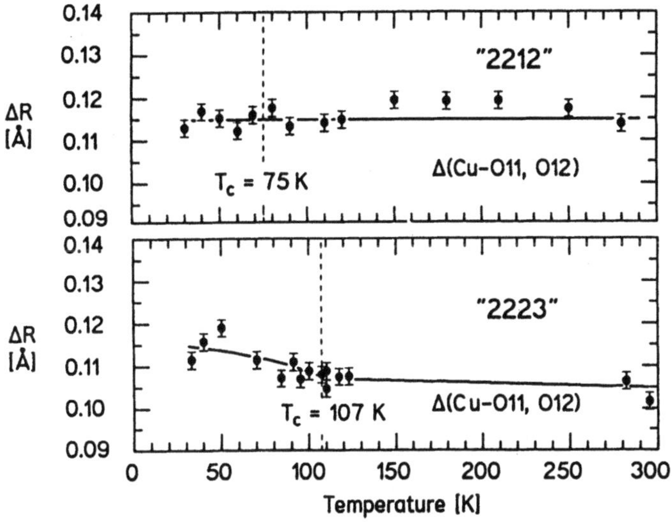

Fig. 3. Temperature dependence of the bondlength differences in the CuO_4 basal plane in "2212" (top) and "2223" (bottom) Bi cuprates.

clearly at $k_1 \cong 3$ and $k_2 \cong 11.5$ Å^{-1}. They move with temperature correlated by about 2π. On the other hand, the jump occurring at $k \cong 14.5$ Å^{-1} exhibits a completely different temperature behavior and is attributed to two different base bonds, Cu-O11 and Cu-O12. Φ_{Cu-O} extracted from the "2212" sample shows up very similarly. By comparison Φ_{Cu-O} extracted within the same transformation range from linearily coordinated Cu in Cu_2O exhibits a monotonically decreasing k dependence without jumps. Fig. 2 shows the temperature dependences of the long apex bonds. The absolute bondlengths are determined relative to Φ_{Cu-O} obtained from a CuO reference with R_{Cu-O} = 1.958 Å. Fig.3 (top and bottom) shows the temperature dependences of the bondlength differences $\Delta R_{Cu-O11,O12}$ in the basal planes.

3. Discussion

Our data yield two different Cu-O11, O12 base bonds differing by $\cong$ 0.1 Å. This distorted planar Cu-O configuration must be fitted into the well-established pseudotetragonal symmetry of the Cu sublattice. One can achieve it by: *i.* rotating the asymmetrically stretched oxygen squares (Cu is centrosymmetric) in the a-b plane according to an "antiferro"-type ordering scheme, and by *ii.* puckering the planes along the *a*- or *b*-axis. In the latter case Cu no longer takes centrosymmetrical sites in the CuO_5 basal plane. The analysis of the EXAFS Cu-O-Cu multiscattering paths supports the existence of ordered CuO_4-planes: the angle formed by the in-plane Cu-O-Cu bonds deviates from 180^0 by 5^0±3 [7]. From our data alone one cannot decide whether "antiferro"-type or puckering-type ordering occurs. The existence of longe range modulations along the *b*-axis and the diffraction data [4] support the puckering type of ordering and hence *in-plane stretching* of the base oxygens. Such a type of distortion might be stabilized by $3d_{x^2-y^2}$ $(E_g) \rightarrow 3d_{xy,yz}$ (T_{2g}) excitations [1].

Acknowledgement. We thank the staff of *HASYLAB* for their hospitality.

References

[1] M. Bacci, Chem. Phys. **104**, 191 (1986)
[2] L.F. Feiner and J. Zaanen, these proceedings
[3] W. Weber, Z. Phys. **B70**, 323 (1988) ,and these proceedings
[4] P. Bordet, J.J. Capponi, C. Chaillout, J. Chenavas, A.W. Hewat, E.A. Hewat, J.L. Hodeau, M. Marezio, J.L. Tholence, and D. Tranqui, Physica **C156**, 189 (1988)
[5] A. Sequeira, H. Rajagopal, and J.V. Yakhmi, Physica **C157**, 519 (1989)
[6] W. Carrillo-Cabrera and W. Göpel, Physica **C161**, 373 (1989)
[7] A. Larisch and J. Röhler, in preparation
[8] F. Seidler, Diplomarbeit Universität zu Köln (1989) unpublished
[9] W. Schnelle, E. Braun, H. Broicher, H. Weiß, H. Geus, S. Ruppel, M. Galffy, W. Braunisch, A. Waldorf, F. Seidler, and D. Wohlleben, Physica **C161**, 123 (1989)
[10] G. Martens, P. Rabe, N. Schwentner, and A. Werner, Phys. Rev. Lett. **39**, 1411 (1977)
[11] P. P. Lottici, G. Antonioli and F. Licci, Physica **C152**, 468 (1988)

Electron Energy-Loss Studies on n-Type Doped Superconductors

M. Alexander, H. Romberg, N. Nücker, P. Adelmann, and J. Fink

Kernforschungszentrum Karlsruhe, Institut für Nukleare Festkörperphysik, Postfach 3640, D-7500 Karlsruhe, Fed. Rep. of Germany

Abstract. The electronic structure of n-type doped $Ln_{1.85}M_{0.15}CuO_{4-\delta}$ (Ln = Pr,Nd,Sm; M = Ce,Th), $Nd_2CuO_{3.8}F_{0.2}$ and their semiconducting parent compounds has been studied by high-energy electron energy-loss spectroscopy (EELS) in transmission. Core-level spectroscopy on O 1s and Cu 2p levels yields information on the partially unoccupied density-of-states near the Fermi level. We observe a pre- edge structure in the O 1s spectrum of all compounds under investigation similar to those observed in the p-type doped superconductors. We assign that pre-preak to transitions into the conduction band mainly consisting of Cu 3d with a slight admixture of O 2p states. This pre-peak was found to be almost independent of the dopant concentration and polarized in the CuO_2-plane. In the Cu $3d^9$ configuration we found a slight admixture (~10%) of Cu $3d_{3z^2-r^2}$ to the Cu $3d_{x^2-y^2}$ hole states similar to the p-type doped superconductors. Qualitatively we recognized a reduction of the excitonic Cu $2p_{3/2}$ line, indicating that the introduced electrons go onto the Cu sites, thus producing charge carriers having predominantly Cu 3d character.

1. Introduction

Since their discovery by Tokura et al. [1] the n-type doped superconductors $Ln_{2-x}M_xCuO_{4-\delta}$ (Ln = Pr,Nd,Sm; M = Ce,Th) and $Nd_2CuO_{4-x}F_x$ have attracted wide-spread interest among the high-T_c community. Upon doping, a trivalent rare earth (RE) ion is replaced by a formally tetravalent one, and threfore, it is argued that electrons are introduced onto the CuO_2 planes. Several authors [2,3,4] using X-ray absorption (XAS) or X-ray photoemission spectroscopy (XPS) concluded that monovalent copper is formed. Others [5,6] using the same spectroscopic methods obtain the result that no Cu^{1+} is formed. Allen et al. [7] using resonant photoemission spectroscopy proposed the appearance of a new band in the bandgap being responsible for the metallic character of $Ln_{1.85}M_{0.15}CuO_{4-\delta}$.

Springer Series in Solid-State Sciences, Vol. 99
Electronic Properties of High-T_c Superconductors and Related Compounds
Editors: H. Kuzmany · M. Mehring · J. Fink

2. Results and Discussion

In Fig. 1 we show the O 1s edges of CuO, $Ca_{0.86}Sr_{0.14}CuO_2$, Pr_2CuO_4, $Pr_{1.85}Ce_{0.15}CuO_{4-\delta}$, $Pr_{1.85}Th_{0.15}CuO_{4-\delta}$, Nd_2CuO_4, $Nd_{1.85}Ce_{0.15}CuO_{4-\delta}$, $Nd_{1.85}Th_{0.15}CuO_{4-\delta}$, $Nd_2CuO_{3.8}F_{0.2}$, Sm_2CuO_4, and $Sm_{1.85}Ce_{0.15}CuO_{4-\delta}$. In all spectra of the n-type doped superconductors and their semiconducting relatives we observed a pre-peak structure with a threshold at about 528.6 eV. This pre-peak does not change upon doping within our experimental error. Though the similarity of this pre-peak for n- and p-type doped systems, suggested the same interpretation in terms of holes in the valence band [8] we now attribute this spectral feature in the n-type doped system to transitions into the conduction band (upper

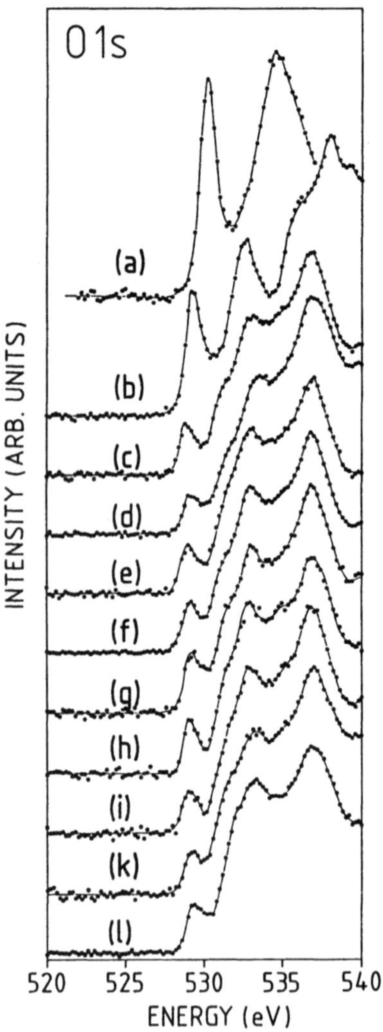

Fig. 1 O 1s edges of CuO (a), $Ca_{0.86}Sr_{0.14}CuO_2$ (b), Pr_2CuO_4(c), $Pr_{1.85}Ce_{0.15}CuO_{4-\delta}$(d), $Pr_{1.85}Th_{0.15}CuO_{4-\delta}$(e), Nd_2CuO_4(f), $Nd_{1.85}Ce_{0.15}CuO_{4-\delta}$ (g) $Nd_{1.85}Th_{0.15}CuO_{4-\delta}$ (h), $Nd_2CuO_{3.8}F_{0.2}$ (i), Sm_2CuO_4 (k) and $Sm_{1.85}Ce_{0.15}CuO_{4-\delta}$(l).

Hubbard band). By examination of the O 1s spectra of CuO and $Ca_{0.86}Sr_{0.14}CuO_2$ a similar spectral feature can be detected. In these two insulating compounds no holes in the valence band are expected and the pre-peak is assigned to transitions into the conduction band.

In Fig. 2 the Cu $2p_{3/2}$ edges of Nd_2CuO_4, $Nd_{1.85}Ce_{0.15}CuO_{4-\delta}$, $Nd_{1.85}Th_{0.15}CuO_{4-\delta}$, $Nd_2CuO_{3.8}F_{0.2}$, Sm_2CuO_4, and $Sm_{1.85}Ce_{0.15}CuO_{4-\delta}$ are shown. Qualitatively a slight reduction of the intensity of the excitonic like main peak upon doping is detected. The weight of this peak is related to the density of empty Cu 3d states. Therefore, we believe that the electrons introduced mainly go onto the Cu sites, thus producing charge carriers having predominantly Cu 3d character. In addition a slight shift upon doping (~0.2 eV) of the main line towards higher energies is detected.

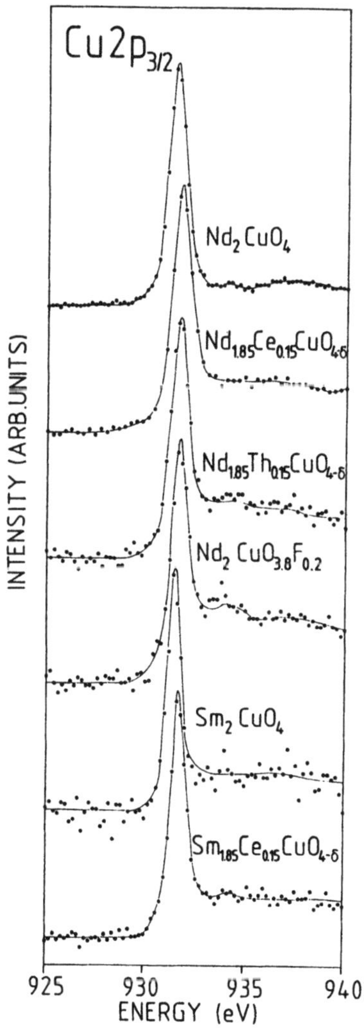

Fig. 2 Cu $2p_{3/2}$ edges of Nd_2CuO_4 $Nd_{1.85}Ce_{0.15}CuO_{4-\delta}$, $Nd_{1.85}Th_{0.15}CuO_{4-\delta}$, $Nd_2CuO_{3.8}F_{0.2}$, Sm_2CuO_4, and $Sm_{1.85}Ce_{0.15}CuO_{4-\delta}$.

This may be attributed to the enhanced screening capability of free charge carriers in the doped compounds. The enhancement of the spectral weight at the high energy side of the white line in the doped compounds is remarkable but still unclear and may be attributed to a hybridization of Cu 4s or 4p with Cu 3d states..

Fig. 3 shows O 1s spectra of single crystalline Nd_2CuO_4 both with momentum transfer **q** parallel and perpendicular to the CuO_2-plane. In addition we show the O 1s spectrum of polycrystalline Nd_2O_3.The polarization dependent spectra caused by transitions into empty O $2p_{x,y}$ states clearly exhibit the pre-peak structure to be polarized in the CuO_2-plane. This may be another hint towards the interpretation of transitions into the conduction band mainly consisting of Cu $3d_{x^2-y^2}$ and a slight admixture of O $2p_\sigma$ states.

In Fig. 4 we show the polarization dependent Cu $2p_{3/2}$ edges of single crystalline Nd_2CuO_4. It is clearly seen that the Cu 3d holes are mainly polarized in the CuO_2-plane, thus in a Cu $3d_{x^2-y^2}$ orbital. A slight admixture of states with Cu $3d_{3z^2-r^2}$ symmetry is seen.

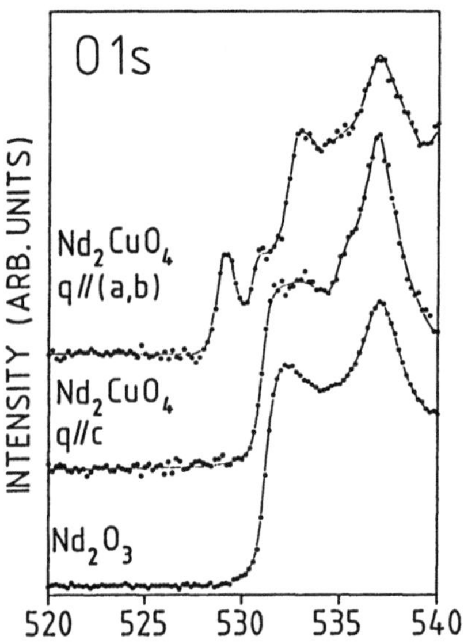

Fig. 3 O 1s edges of the single crystalline Nd_2CuO_4 compound with momentum transfer q parallel and perpendicular to the CuO_2-plane. In addition the O 1s edge of polycrystalline Nd_2O_3 is given.

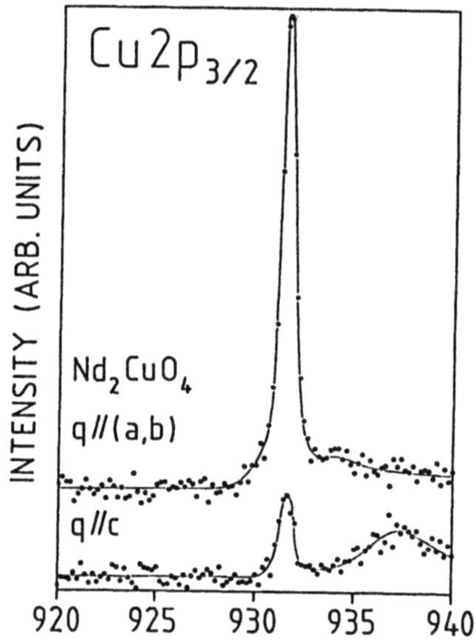

Fig. 4 Cu $2p_{3/2}$ edges of the single crystalline Nd_2CuO_4 with momentum transfer q parallel and perpendicular to the CuO_2-plane.

Summary

The present results do not favor the existence of holes with O 2p character in the valence band. By the inspection of the O 1s edges of $Ln_{2-x}M_xCuO_{4-\delta}$ we could not resolve significant changes of the partial density-of-states with O 2p symmetry in the conduction band. The electrons introduced upon doping mainly go onto the copper sites, thus filling empty Cu 3d states.

References

[1] Y. Tokura, H. Takagi, S. Uchida, Nature **337**, 345 (1989)
[2] S.M. Tranquada, S.M. Heald, D.R. Moodenbaugh, G.Liang, M. Croft, Nature **337**, 720 (1989)
[3] A. Grassmann, J. Ströbel, M. Klauda, J. Schlötter, G. Saemann-Ischenko, Europhys. Lett. **9**, 827 (1989)
[4] H. Ishii, T. Koshizawa, H. Kutaura, T. Hanyu, H. Takai, K. Mizguchi, K. Kume, I. Shiozaki, and S. Yamaguchi, Jpn. J. Appl. Phys. **28**, L1952 (1989)
[5] A. Fujimori, Y. Tokura, H. Eisaki, H. Takai, S. Uchida, E. Takayama-Muromachi, submitted to Phys. Rev. B
[6] E.E. Alp, S.M. Mini, M. Ramanathan, B. Dabrowski, D.R. Richards, and D.G. Hinks, Phys. Rev. B **40**, 2617 (1989)
[7] J.W. Allen, C.G. Olson, M.B. Maple, J.S. Kung, L.Z. Liu, J.-H. Park, R.O. Anderson, W.P. Ellis, J.T. Markert, Y. Dalichaouch, and R Lue, in print
[8] N. Nücker, P. Adelmann, M. Alexander, H. Romberg, S.Nakai, J. Fink, H. Rietschel, G. Roth, H. Schmidt, H. Spille, Z. Phys. B **75**, 421 (1989)

Does the Cu 4s Orbital Play a Role in Electron Doping of the $Sm_{2-x}Ce_xCuO_{4-\delta}$ Compound?

C.F.J. Flipse [1], *G. van der Laan* [2], *A.L. Johnson* [3], *and K. Kadowaki* [4]

[1] Surface Science Research Center, University of Liverpool, Grove Street, P.O. Box 147, Liverpool L69 3BX, UK
[2] Daresbury Laboratories, Daresbury, UK
[3] Department of Chemistry, University of Cambridge, UK
[4] Natuurkundig Laboratorium, University of Amsterdam, The Netherlands

Abstract. The oxygen 1s and copper 2p X-ray absorption spectra of the semiconductor, superconductor and metallic $Sm_{2-x}Ce_xCuO_{4-\delta}$ have been measured.Cerium doping does not change the density of unoccupied O 2p- Cu 3d molecular orbital states, but shows a decrease in the O 2p - Cu 4s molecular orbital states with a_1 symmetry.The Cu 2*p* absorption spectra show a decrease in Cu $3d^9$, although a strong Cu $3d^{10}$ peak is not observed. As a result of the strong mixing between the O 2*p* and Cu 4*s* states of a_1 symmetry, a composite state of Cu 3*d*, Cu 4*s* and O 2*p* orbitals can be formed.

1. Introduction

The discovery of "electron-doped" copper oxide superconductors $Ln_{2-x}Ce_xCuO_{4-\delta}$ with Ln = Pr, Nd or Sm has stimulated much discussion about the theories of high-T_c oxide superconductors. One of the key questions to be answered is whether these materials are really electron doped, or hole doped just as the $YBa_2Cu_3O_{7-\delta}$ compounds.

Cu-1s X-ray absorption spectroscopy (XAS) [1] and Cu 2*p* X-ray photoemission spectroscopy (XPS) [2] confirm that an electron fills the Cu 3*d* band and creates a Cu $3d^{10}$configuration after Ce doping.

On the other hand, Fujimori *et al.* [3] and Alp *et al.* [4] claim that there is no evidence for filling of Cu 3*d* orbitals by doping and Nücker *et al.* [5] concludes from O 1*s* electron energy loss spectroscopy (EELS) that there is evidence for holes at the oxygen sites after Ce doping.

Springer Series in Solid-State Sciences, Vol. 99
Electronic Properties of High-T_c Superconductors and Related Compounds
Editors: H. Kuzmany · M. Mehring · J. Fink

2. Results and discussion

Fig. 1 shows a pre-edge feature at 529.4 eV which has a width of about 1 eV. This feature, but much more intense, is also found in CuO [6] and can be attributed to oxygen $2p$ states which are covalent mixed with metal $3d$ states [7]. The structure above 531 eV is due to O $2p$ states which are covalent mixed with Cu $4s$ and $4p$ states. From an augmented-localized-spherical wave calculation for a CuO_4^{6-} cluster, the peaks at 534 eV and 537.5 eV can be ascribed to a $2a_{1g}$ and $3e_u$ molecular orbital, respectively [6]. The Sm compound shows a decrease in the $2a_{1g}$ peak compared to the $3e_u$ one after Ce doping. This indicates a decrease in the unoccupied density of states for the $2a_{1g}$ molecular orbital with O $2p$ and Cu $4s$ p character in the superconductor.

In table I we compare the intensities of the main absorption line relative to the continuum in the Cu 2p absorption spectrum [8]. The table shows that the

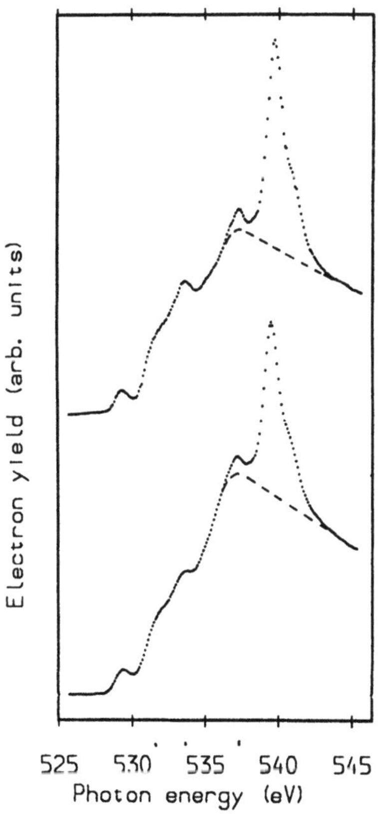

Figure 1:
The O $1s$ absorption spectra:

upper curve: $Sm_2CuO_{4-\delta}$
lower curve: $Sm_{1.85}Ce_{0.15}CuO_{4-\delta}$.

The dashed line gives the demarcation with the Sm $3d_{5/2}$ absorption structure, which comes from the second order contribution of the monochromator.

Table I: Energy position (eV) and intensity (relative to the continuum) of the main line in the Cu $2p$ absorption spectrum [8].

Compound		Energy	Intensity
CuO		931.20	12.8
$Sm_2CuO_{4-\delta}$	(semiconductor)	931.35	8.4
$Sm_{1.85}Ce_{0.15}CuO_{4-\delta}$	(superconductor)	931.46	7.3
$Sm_{1.83}Ce_{0.17}CuO_{4-\delta}$	(metal)	931.44	6.2

contribution of the $\underline{2p}3d^{10}$ final state peak at ~931 eV is reduced in the superconductor phase compared with the semiconductor. In the metallic compound the main peak is even more reduced. This indicates that the $3d^9$ character decreases with doping.

The Cu 2p absorption data do not show a contribution at 933 eV. This is an indication that defects, due to hole doping when samarium is substituted by cerium, are absent [9] . If electron doping by Ce causes filling of the Cu 3d band, one would expect a peak attributed to a $d^{10} \rightarrow \underline{2p}3d^{10}4s$ transition, which in Cu_2O is at 934 eV [10]. This peak is also absent in the spectra.

3. Conclusions

The Cu $2p$ absorption spectra indicate that the ground state of the samarium cuprate semiconductor is a mixture of $3d^9$ and $3d^{10}\underline{L}$ states. In the superconductor the $3d^9$ character is reduced, but the filling of the Cu $3d$ band is incomplete. The O $1s$ spectra indicate that after Ce doping the number of holes in the O $2p$ - Cu $3d$ molecular orbital is unchanged, and the hole concentration in the $2a_{1g}$ molecular orbital, which mixes the O $2p$ and Cu $4s$ p states, is decreased. The formation of a composite state of Cu 3d, Cu 4s and O 2p orbitals has been predicted by cluster calculations [11].

References

[1] J. M. Tranquada, S.M. Heald, A.R. Moodenbaugh, G. Liang, M. Croft Nature 337, 720 (1989)

[2] A. Grassman, J. Ströbel, M. Klauda, J. Schlötterer and G. Saemann-Ischenko, Europhys. Lett. 9, 827 (1989)

[3] A. Fujimori, Y. Tokura, H. Eisaki, H. Tkagi, S. Uchida, E. Takayama-Muromachi, Phys. Rev. in press

[4] E.E. Alp, S.M. Mini, M. Ramanathan, B. Dabrowski, D.R. Richards and D.G. Hinks, Phys. Rev. B40, 2617 (1989)

[5] N. Nücker, P. Adelmann, M. Alexander, H. Romberg, S. Nakai, J. Fink, H. Rietschel, G. Roth, H. Schmidt, H. Spille, Z. Phys. B 75,421 (1989)

[6] M. Grioni, M.T. Czyzyk, F.M.F. de Groot, J.C. Fuggle and B.E. Watts, Phys. Rev. B 39, 4886 (1989)

[7] P. Kuiper, G. Kruizinga, J. Ghijsen, G.A. Sawatzky and H. Verweij, Phys. Rev. Lett. 62, 221 (1989)

[8] C.F.J. Flipse, G. van der Laan, A.L. Johnson,K. Kadowaki, to be published in Phys. Rev. B

[9] P. Kuiper, G. Kruizinga, J. Ghijsen, M. Grioni, P.J.W. Weys, F.M.F. de Groot, G.A. Sawatzky, H. Verweij, L.F. Feiner and H. Petersen, Phys. Rev. B 38, 6483 (1988)

[10] G. Kaindl, O. Strebel,A. Kolodziejczyk, W. Schäfer, R. Kiemel, S. Lösch, S.Kemmler-Sack,R. Hoppe,H.P. Müller,D. Kissel,Physica B 158,44 (1989)

[11] T. Tohyama, Y. Ohta and S. Maekawa, Physica C 158, 525 (1989)

Hole Doping in $RE_2NiO_{4+\delta}$

M. Medarde[1,2], *X. Batlle*[3], *X. Granados*[3], *X. Obradors*[2], *J. Rodriguez*[4], *J. Fontcuberta*[3], *M. Vallet*[5], *J. Gonzalez*[5], *J. Alonso*[5], *M.J. Sayagues*[5], *J.L. Martinez*[4], *and A. Fontaine*[2]

[1]Institut de Ciencia de Materials de Barcelona, CSIC, c/ Marti i Franques, s/n, 08028 Barcelona, Spain
[2]LURE, CNRS, F-91405 Orsay, France
[3]Departament Fisica Fonamental, Facultat Fisica, Universitat Barcelona, Diagonal 647, 08028, Barcelona, Spain
[4]Institut Laue-Langevin, 156X Centre de Tri, F-38042 Grenoble, France
[5]Departamento Quimica Inorganica, Facultad Quimica, Universidad Complutense Madrid, 28040 Madrid, Spain

Abstract. We report a preliminary study of the electronic structure and the magnetic and transport properties of some hole doped $RE_2NiO_{4+\delta}$ oxides. Ni K-edge X-Ray Absorption (XAS) measurements of $Pr_2NiO_{4+\delta}$ single crystals allow us to clearly identify the $3d^9\underline{L}$ and $3d^8$ σ and π final states and to evaluate the covalent mixing. Magnetic susceptibility of $Nd_2NiO_{4+\delta}$ and $La_{2-x}Sr_xNiO_{4+\delta}$ reveals a Curie-Weiss behaviour (CW) above a characteristic temperature T^*, which depends on hole concentration. Below T^*, localization of mobile holes leads to a strong Ni^{2+}-O^- interaction, thus frustrating the otherwise AF Ni^{2+}-O^{2-}-Ni^{2+} interactions. Resistivity and Seebeck coefficient data provide additional support for this picture.

1. Introduction

Immediately after the discovery of high temperature superconductors, the relevance of hole doping in these new materials was recognized. Probably, one of the most remarkable characteristics of the prototype $La_{2-x}Sr_xCuO_4$ oxide is that the hole doping introduced by the Sr substitution takes place in an orbital of O:2p character. Not so much attention has been devoted to the closely related Ni-based oxides. Experimental features of La_2NiO_4-type oxides similar to those of Cu ones are i) the existence of an antiferromagnetic parent compound and ii) the strong dependence of both the 3D AF ordering temperature and the spin wave stiffness constant /1/ (and, consequently, the effective intralayer antiferromagnetic interaction) on the oxygen stoichiometry. In a recent paper Kuiper et al. /2/ have shown that in NiO holes introduced by Li doping reveal a predominant

Springer Series in Solid-State Sciences, Vol. 99
Electronic Properties of High-T_c Superconductors and Related Compounds
Editors: H. Kuzmany · M. Mehring · J. Fink

O(2p) character, in close similarity with Cu oxides. The study of the electronic-related properties of Ni oxides having the same crystal structure as the superconducting Cu ones may give valuable information about the relevance of the normal state properties in the appearance of high temperature superconductivity.

In this paper we report a variety of experimental data (polarized XANES, XANES, magnetism and transport) obtained from some $RE_{2-x}Sr_xNiO_{4+\delta}$ (RE = La, Pr, Nd) oxidized samples (δ>0).

2. Results and Discussion

2.1 Polarized XAS: $Pr_2NiO_{4+\delta}$

Ni K-edge absorption of a $Pr_2NiO_{4+\delta}$ single crystal was measured using an EXAFS conventional spectrometer. Fig. 1 shows polarized XAS spectra taken at different angles ϑ (ϑ = 10°, 30°, 55°, 90°). In our experimental set-up the electrical field vector $\vec{E}$ is parallel to the long axis $\vec{c}$ when ϑ = 0° (π polarization), and perpendicular to it when ϑ = 90° (σ polarized). The polarization dependence of XAS peaks strongly suggests that the (A_1, A_2) and (B_1, B_2) peaks are associated, respectively, with the $1s\rightarrow4p^*(\pi)$ and $1s\rightarrow4p^*(\sigma)$ transitions. In close analogy with recent findings in NiO /2/, the splitting of an A peak into A_1 and A_2 may be interpreted in terms of a ligand-to-metal charge transfer, i.e. A_1 should correspond to $1s^2\rightarrow1s^13d^9\underline{L}4p^*(\pi)$ and A_2 to $1s^2\rightarrow1s^13d^84p^*(\pi)$ transitions. The improved screening caused by $3d^9\underline{L}$ admixture with respect to the $3d^8$

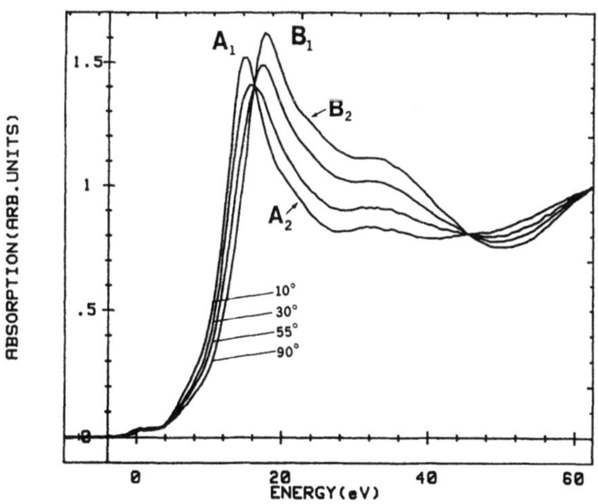

Figure 1: Ni K-edge polarized XAS of $Pr_2NiO_{4+\delta}$ single crystal at different angles ϑ.

configuration accounts for the lower energy of A_1. The B_1 - A_1 and B_2 - A_2 differences give us a value $E_\sigma - E_\pi = 2.2$ eV. This energy separation is consistent with the crystallographic data: in Pr_2NiO_4, NiO_6 octahedra are significantly elongated $(Ni\text{-}O)_c/(Ni\text{-}O)_a = 1.14$, so that $E_\sigma - E_\pi > 0$. In La_2CuO_4, where$(Cu\text{-}O)_c/(Cu\text{-}o)_a = 1.25$, this difference is found to be larger /3/.

A remarkable difference between the present results and those already reported for La_2CuO_4 is the relative intensity (A_1,B_1; A_2,B_2). In the Cu-oxide case, one single hole of d_{z^2} symmetry makes the $\vec{c}$-polarized π-transitions $1s^2 \rightarrow 1s^1 3d^{10} \underline{L} 4p^*$ more intense than the corresponding ab-polarized σ-transitions. On the other hand, similar intensities are observed for these transitions in Pr_2NiO_4, thus showing a stronger weight of the $|3d^9\underline{L}\rangle$ final configuration in both cases. Tentatively, we may interpret this result as due to the existence of two holes of d_{z^2} and $d_{x^2-y^2}$ symmetry in Pr_2NiO_4.

2.2 Magnetic Susceptibility: $Nd_2NiO_{4+\delta}$ and $La_{2-x}Sr_xNiO_{4+\delta}$

Magnetic susceptibility measurements of well-characterized $La_{2-x}Sr_xNiO_{4+\delta}$ samples were performed on a Faraday balance in the temperature range 14K-300K under a 2.4 kOe magnetic field. Magnetization measurements on a $Nd_2NiO_{4+\delta}$ sample were done by using an SQUID magnetometer in the 4.2K-300K temperature range under a 1 kOe magnetic field. The reported susceptibilities are corrected for ionic diamagnetism. Figs 2(a) and 2(b) show the inverse of the measured susceptibility for $La_{2-x}Sr_xNiO_{4+\delta}$ (x=0,1) and $Nd_2NiO_{4+\delta}$ respectively. In the high temperature range ($T^* > 100$ K) a Curie-Weiss (CW) law gives an accurate description of all the data. The Nd compound displays a higher susceptibility than La ones as a result of the contribution of the paramagnetic Nd^{3+} ions.

The most relevant fact in the context of this paper is that the magnetic moment, as deduced from the fit of the CW law for La-Sr compounds, gives $\mu_{eff} = 3.5(1)\ \mu_B$, independently of the Sr doping. The Ni^{2+} contribution to the magnetic susceptibility of $Nd_2NiO_{4+\delta}$ can be obtained after subtraction of the Nd^{3+} contribution. If we assume g = 8/11 and J = 9/2 for Nd^{3+}, the fit of $\chi_{Ni}(T) = \chi - \chi_{Nd}(T)$ at high temperatures provides a measure of the Ni magnetic moment in this compound, which results in $\mu_{eff} = 3.33(1)\ \mu_B$ for Ni^{2+}, in remarkably

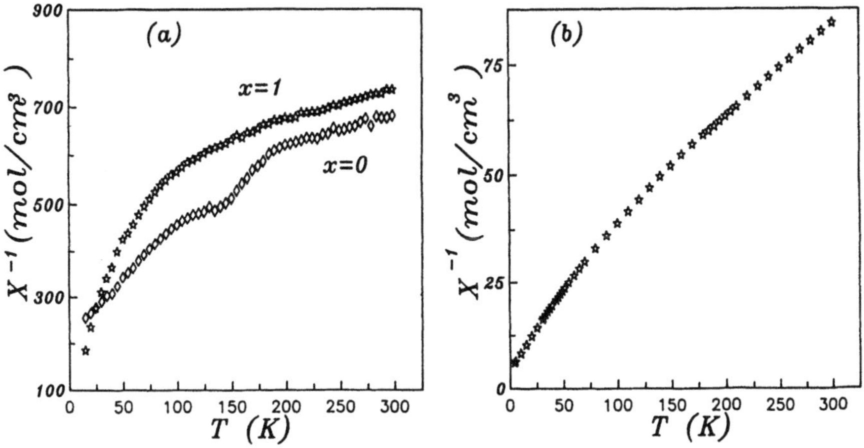

Figure 2: (a) Inverse of measured magnetic susceptibility of $La_{2-X}Sr_XNiO_{4+\delta}$ (x=0,1) as a function of temperature. (b) Inverse of measured magnetic susceptibity of $Nd_2NiO_{4+\delta}$ as a function of temperature.

close agreement with data obtained in La oxides. These values are slightly larger than the spin-only high-spin $\mu_{eff} \approx 2.82\ \mu_B$ and well below the total L+S moment $\mu_{eff} = 5.6\ \mu_B$.

The observation of a constant μ_{eff}(Ni) value for any given $(NiO)^+$ concentration seems to suggest that holes are basically formed in a band of O:2p parentage. Indeed, this is the reported behaviour from XAS experiments on $Ni_{1-X}Li_XO$ oxides /2/.

Preliminary Ni K-edge XAS measurements on the very same materials do not show any distinct feature after doping, thus providing additional support for our suggestion of the O:2p parentage of holes. Localized holes of primary O:2p character can produce a strong antiferromagnetic exchange coupling between hole spin and unpaired Ni^{2+} spins. Therefore, competition between antiferromagnetic (↑)Ni-O(↓) and (↑)Ni-O-Ni(↓) interactions would result in a frustration of the long range antiferromagnetic order in the undoped sample. Within the scope of this scheme, one would expect that the strength of the AF interactions decreases with hole doping. Also, the mechanism of AF ordering frustration via holes may be expected to be more efficient for localized O(2p) holes than for free holes. Therefore, at high temperatures when 2p holes move 'freely' in a band one should expect to observe pure paramagnetic behaviour of the Ni^{2+} $3d^8$ ions, the local AF interactions setting up as soon as holes become trapped. This state is interpreted as arising from a lower intralayer antiferromagnetic interaction in Ni oxides as compared to the equivalent Cu ones.

Accordingly, Fig. 2(a) clearly shows a CW law above a characteristic temperature T^*, which lowers on increasing the hole concentration. At $T<T^*$ deviations from the CW law can consistently be interpreted in terms of the AF interactions between localized Ni^{2+} moments and 2p holes, as discussed above. Further evidence of hole localization below T^* will come from transport data.

2.3 Transport Properties of $La_{2-x}Sr_xNiO_{4+\delta}$

Because of grain boundary effects, resistivity data alone cannot be used to get a definitive interpretation of the transport properties. However, some general trends may be safely stated. As is shown in Fig. 3, room temperature electrical resistivity greatly decreases when Sr concentration increases. The electrical resistance shows an activated behaviour at temperatures $T>T^*$, changing towards a temperature-dependent activated behaviour at lower temperatures. A typical 3D variable range hopping model may be used to fit the data. On the other hand, the Seebeck coefficient in the high temperature range can also be nicely explained in terms of an activated behaviour ($S \approx E/T + A$). Activation energies as deduced from resistivity and Seebeck measurements turn out to be similar ($\approx$ 50 meV). At low temperatures ($T \lesssim 100$ K) the Seebeck coefficient changes its slope, becoming smaller (see Fig. 4). Finally, the most heavily doped sample (x=1) displays a typical metallic dependence. The measured negative slope signals an electronic conductivity in contrast to the low doped

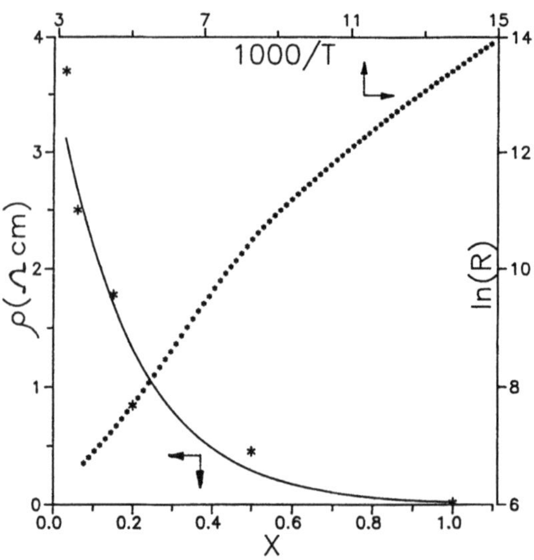

Figure 3: Resistivity vs Sr constents at room temperature and typical ln(R) vs 1/T (x=0.15) plots of $La_{2-X}Sr_XNiO_{4+\delta}$.

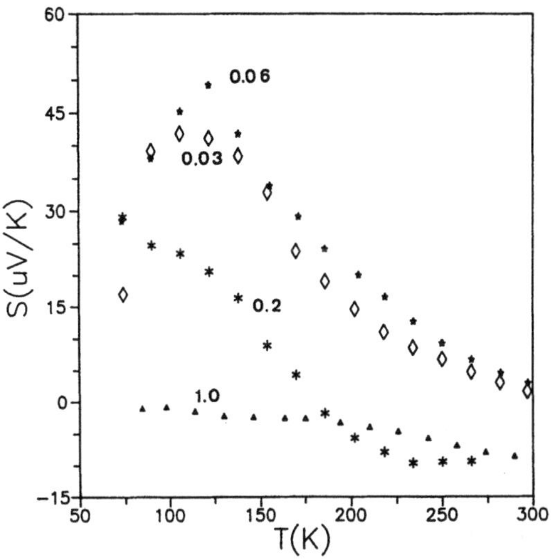

Figure 4: Seebeck coefficient vs temperature for several Sr contents of $La_{2-x}Sr_xNiO_{4+\delta}$, from x=0.06 (semiconductor-like) to x=1.0 (metal-like).

samples where a hole dominant character of the charge carriers can be deduced from these data.

The data for $T>T^*$ can be easily described in terms of a simple model in which a diffusive motion of the hole-charge carriers leads to an activated conductivity and to a temperature-dependent thermopower. The higher temperature values of S tend to saturate to a constant value which depends on the Sr contents. At $T<T^*$, the ratio of the hole concentration to the available sites increases due to hole localization near Sr^{2+} impurities, and consequently S decreases as observed. These results perfectly match magnetic data.

References

/1/ G. Aeppli, S.M. Hayden, H.A. Mook, Z. Fisk, S.W. Cheong, J.P. Remeika, G.P. Espinosa and A.S. Cooper, Phys. Rev. Lett. 62 (1989) 2052.

/2/ P. Kuiper, G. Kruizinga, J. Ghijsen and G.A. Sawatzky, Phys. Rev. Lett. 62, (1989) 221.

/3/ H. Oyanagi, K. Oka, H. Unoki, Y. Nishiyama, K. Murata and H. Yamaguchi, J. Phys. Soc. Japan 1990 (in press).

Electron Energy-Loss Studies on $La_{2-x}Sr_xCuO_4$

H. Romberg, M. Alexander, N. Nücker, P. Adelmann, and J. Fink

Kernforschungszentrum Karlsruhe, Institut für Nukleare Festkörperphysik, Postfach 3640, D-7500 Karlsruhe, Fed. Rep. of Germany

Abstract. The electronic structure of the p-type doped $La_{2-x}Sr_xCuO_4$ compound has been studied using high-energy electron energy-loss spectroscopy (EELS) in transmission. Core level spectroscopy on the O 1s level yields information on the partial unoccupied density-of-states near the Fermi level. We observe two pre-edge structures in the O 1s spectra, one of which is assigned to hole states near E_F induced upon doping, while the second structure at higher energy is assigned to transitions into the conduction band consisting mainly of Cu 3d states with a slight admixture of O 2p states.

Introduction

Among the cuprates showing high-temperature superconductivity the system $La_{2-x}Sr_xCuO_4$ is possibly one of the most suitable systems to study the electronic structure of cuprate superconductors. The undoped parent compound La_2CuO_4 is an antiferromagnetic insulator due to strong correlation effects mainly on the Cu sites, which turn the otherwise metallic compound with a half-filled band into a charge-transfer insulator [1]. A charge-transfer gap appears between the valence band with mainly O 2p character and the unoccupied conduction band (upper Hubbard band) with mainly Cu 3d character. Upon p-type doping the antiferromagnetism disappears and a transition into a metallic state appears. For $x \sim 0.15$, superconductivity is observed below $T_c \sim 35$ K. Upon further p-type doping, the metallic state persists, while T_c decreases again [2]. In spite of many theoretical and experimental studies the origin of the superconducting transition, the nature of the normal state, and the insulator-metal transition still remain unclear.

We have investigated this system by electron energy-loss spectroscopy (EELS), where transitions from occupied into unoccupied electronic states are measured. For sharp core-levels as initial states, the unoccupied local density-of-states is probed, provided that the interaction of the electron in the final state with the core hole is small. As can be shown, e.g., by the accordance of EELS and resonant BIS spectra of $Bi_2Sr_2CaCu_2O_8$ [3], this assumption seems appropriate for O 1s edges, at least in metallic systems.

Springer Series in Solid-State Sciences, Vol. 99
Electronic Properties of High-T_c Superconductors and Related Compounds
Editors: H. Kuzmany · M. Mehring · J. Fink

The polycrystalline samples were prepared as described by Torrance et al. [2] but without high-pressure oxygen annealing. Hence, our $La_{1.7}Sr_{.3}CuO_{4-y}$ sample corresponds to y~0.04, i.e., a doping equivalent to $La_{1.78}Sr_{.22}CuO_4$. From these samples slices about ~1000 Å thick were cut with an ultramicrotome using a diamond knife. The EELS measurements were performed in transmission using a dedicated 170 keV spectrometer operated at an energy resolution of 0.4 eV. The probing depth is the sample thickness (~1000 Å). Thus EELS in transmission is *not* sensitive to surface contaminations.

Results and Discussion

The O 1s absorption edges of La_2CuO_4 and $La_{1.7}Sr_{.3}CuO_4$ are shown in Fig. 1a. The dominant intensity appears above ~532 eV and remains nearly unchanged upon doping. We assign this intensity to transitions related to the LaO layers and to final states of Op character hybridized with La5d, 4f and possibly Cu states well above the Fermi level. We want to focus on the pre-edge features as shown in Fig. 1b.

In undoped La_2CuO_4 a well pronounced pre-peak (C) is observed at 530.2 eV. There is some spectral weight at lower energies starting at ~528.2 eV. Upon doping, the 530.2 eV peak C decreases and/or broadens and moves towards higher energies. The spectral weight at lower energies increases, forming a peak (V) near 528.9 eV for $x \geq 0.05$. For higher doping this peak V as well as the threshold energy (E_{th}) shifts towards lower energies. In Fig. 1c we plot the spectral weight assigned to peak V and C, respectively, as a function of x.

The pre-peak C near 530.2 eV in undoped La_2CuO_4 is assigned to transitions into the O 2p admixture of the conduction band (upper Hubbard band) having predominantly Cu 3d character. Assuming one hole state per Sr atom in pre-peak V, the amount of this admixture can be estimated between 4% and ≥12%. This value is considerably smaller than that corresponding to LDA band-structure calculation results of ~50% or that derived in a cluster approach of ~20% [4]. A similar (and, in our opinion, related) pre-edge feature was also observed in Nd_2CuO_4, $(CaSr)_2\,CuO_4$, $La_2SrCu_2O_6$, CuO and $YBa_2Cu_3O_6$ [5]. The threshold energy seems to depend systematically on the Cu-O distance.

The spectral weight at energies below this pre-edge peak C establishes unoccupied density-of-states in the charge-transfer gap expected in some theoretical approaches [6]. This spectral weight has an onset at about 1.5 eV below the conduction band edge. Since the gap as measured by optical spectroscopy is about 1.7 eV, the Fermi level is pinned just above the valence band in our "undoped" La_2CuO_4 samples. A rigid-band-like shift of the threshold energy is observed in the heavily doped compounds. We have no indication of a narrow impurity band, as the width of all the observed structures is well above our energy

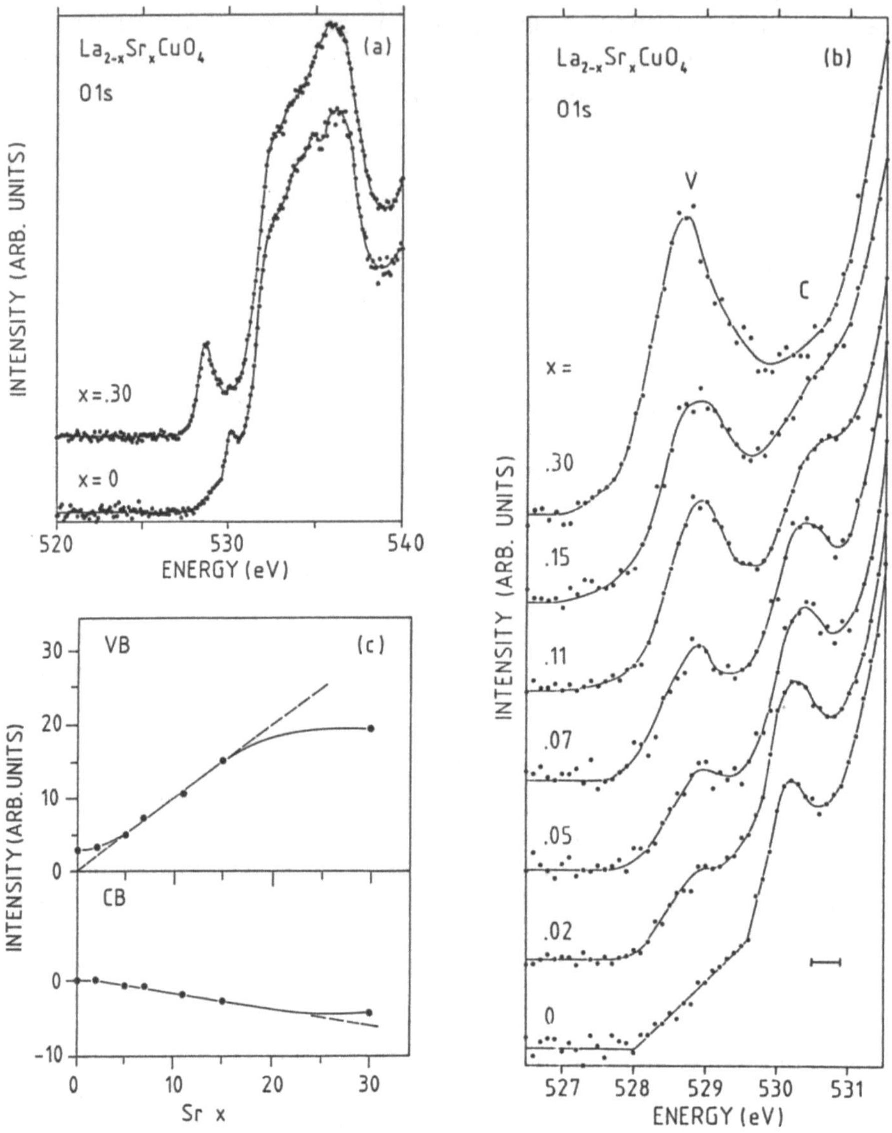

Fig. 1(a) O 1s absorption edges of La_2CuO_4 and $La_{1.70}Sr_{0.30}CuO_4$. (b) O 1s absorption edges (pre-edge structures) of $La_{2-x}Sr_xCuO_4$ for x = 0, 0.02, 0.05, 0.07, 0.11, 0.15 and 0.30. (c) The spectral weight of pre-peak V (valence band) and C (conduction band) as a function of dopant concentration.

resolution of $\Delta E_{1/2} = 0.4$ eV (marked in Fig. 1b) and the life-time broadening of 0.2 eV.

We want to mention that there exist several ideas to explain the decrease of spectral weight in the conduction band: (i) A shift of spectral weight towards lower energies which, however, may only give rise to ~1/5 of the weight observed in peak V. (ii) A change in the hybridization upon doping. (iii) A final state effect.

To summarize, we do not see any discontinuity at the insulator-metal transition in the O 2p-derived unoccupied density-of-states. The O 2p admixture to the conduction band is clearly seen and estimated to 4%-≥12%. A decrease of this feature is seen upon doping revealing changes in the electronic structure. At the same time, about 4 to 5 times more spectral weight appears as a broad feature near the low-energy side of the gap revealing hole states in the valence band being formed. No narrow impurity states are observed. In weakly doped samples the Fermi level appears near the valence band due to unoccupied density-of-states in the gap.

References

1 J. Zaanen, G.A. Sawatzky, and J.W. Allen, Phys. Rev. Lett. **55**, 418 (1985)

2 J.B. Torrance, Y. Tokura, A.I. Nazzal, A. Bezinge, T.C. Huang, and S.S.P. Parkin, Phys. Rev. Lett. **61**, 1127 (1988)

3 W. Drube, F.J. Himpsel, G.V. Chandrashekhar, and M.W. Shafer, Phys. Rev. B **39**, 7328 (1989)

4 M. Schlüter, preprint

5 J. Fink, N. Nücker, H. Romberg, M. Alexander, and P. Adelmann, these proceedings

6 P. Horsch, W.H. Stephan, K. v. Szczepanski, M. Ziegler, and W. von der Linden, Physics C **162-164**, 783 (1989); P.C. Pattnaik and D.M. Newns, Phys. Rev. B **41**, 880 (1990); C. Melo and S. Doniach, Phys. Rev. B, in print

Possible Sequence of Hole Injection in $YBa_2Cu_3O_{6+y}$ and Reverse Model of Metallization for La_2CuO_4/Nd_2CuO_4 by Impurity Hole/Electron Levels in the Band Gap

H. Tolentino[1], *A. Fontaine*[1], *A.M. Flank*[1], *P. Lagarde*[1], *J.-Y. Henry*[2], *J. Rossat-Mignod*[2], *T. Gourieux*[3], *G. Krill*[3], *and F. Studer*[4]

[1]LURE, Bât. 209D, CNRS, CEA, MEN, F-91405 Orsay, France
[2]DRF, CENG, 85X, F-38041 Grenoble Cedex, France
[3]Laboratoire de Physique des Solides/CNRS, Université de Nancy I, BP 239, F-54506 Vandœuvre-les-Nancy, France
[4]Crismat ISMRA, Bd. du Maréchal Juin, Université de Caen, CNRS, F-14032 Caen, France

X-ray Absorption Spectroscopy (XAS) of core levels has been shown to be a powerful tool to investigate the electronic structure of the new high Tc superconductors. At the Cu K-edge, the core level width is still small enough (≈1eV) to allow significant measurements of the completeness of the filling of the d band. In addition, photons of the 9 keV range allow real bulk measurements.

1-Experiment

XAS in situ measurements on $YBa_2Cu_3O_{6+y}$, powder and single crystals, at the copper K-edge, have been carried out at DCI-LURE, using the time-resolved XAS spectrometer. A full spectrum can be recorded in a single shot since the spectrometer is based on a position-sensitive detector in addition to energy-dispersive optics made of a bent silicon crystal. The key point of this experiment is the position-sensitive detector which works under high flux conditions. A full energy band pass is Bragg-reflected and focused on a 0.5 mm wide spot where the sample is kept in a furnace filled with a controlled gas ambience.The correlation of the photon energy and the reflected direction is converted into a correlation between the photon energy and the pixel number of the photodiode array. The energy resolution is limited by the Darwin width of the rocking curve to about 1 eV at 9 keV. In addition the spectrometer yields an absolute reliability of the energy scale, because of the lack of mechanical movement [1].

The oxygen stoichiometry has been swept from 6.2 to 6.95 by controlling the oxygen partial pressure and temperature [2].Three cycles which have been reversibly investigated are reported herein (fig.1). Owing to the rapidity of data aquisition (30 s) and the possibility of measuring the same crystal or powder during the whole time of the experiment, a great accuracy of the evolution of spectra is achieved.

In the one-electron picture, the Fermi Golden rule gives the probability of transition due to the dipolar interaction from the 1s initial state to the final state of the divalent Cu, which can be written as a configuration admixture $\alpha_f|1s^1...3d^9\, mp^1,...2p^6\rangle + \beta_f|1s^1...3d^{10}\, mp^1,...2p^5\rangle$, promoting the 1s electron into mp states of the excited divalent Cu. Beyond the ionic picture, even the initial state of what is called Cu^{II} in the cuprates is also an admixture

$$\alpha_i|1s^2...3d^9\, mp^0, ...2p^6\rangle + \beta_i|1s^2...3d^{10}\, mp^0,...2p^5\rangle \qquad \text{with} \quad \alpha_i^2 >> \beta_i^2,$$

which explicitly evidences that covalency gives a partial oxygen 2p character to the "d" holes. At the threshold, the final state is represented by an admixture, in a way anolog to XPS, but which takes into account that the photoelectron is just deposited in the lower empty **p** states (m=4; larger m builds a continuum). In the final state of the K-edge spectroscopy, the core hole has pulled down all the narrow bands, including the d band, allowing charge transfer from the

Springer Series in Solid-State Sciences, Vol. 99
Electronic Properties of High-T_c Superconductors and Related Compounds
Editors: H. Kuzmany · M. Mehring · J. Fink

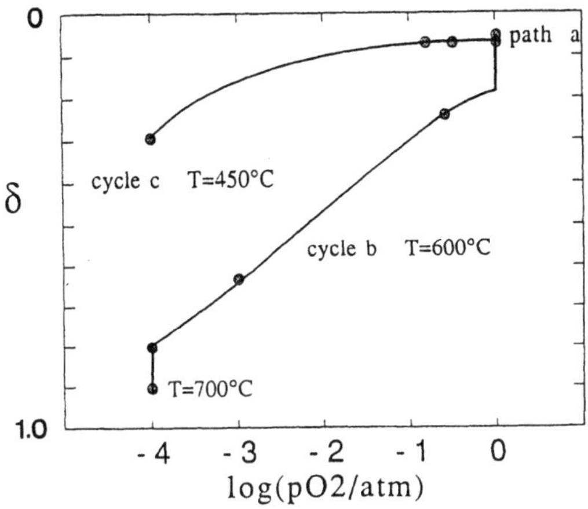

Fig.1 Stoichiometry of $Y_1Ba_2Cu_3O_y$ under different oxygen partial pressure and temperature conditions.The lines represent the paths followed during the in situ observation; the dots represents the equilibrium conditions imposed on the samples.

oxygen 2p band into the copper 3d band. As in XPS, the spectral response of the component $|1s^1...3d^{10}\,4p^1,...2p^5\rangle$ appears at lower energy (U_{cd}-$\Delta \approx 7$eV) than the response induced by $|1s^1...3d^9\,4p^1,...2p^6\rangle$ and in addition, as in XPS again, the spectral weight is transferred to the lower component , i.e. $\beta_f^2 >> \beta_i^2$.

To give an easy but simplified picture, right at the edge, the K-edge of a divalent Cu can be thought of as the product of the XPS signal convoluted by the copper 4p band. If the coordination is square planar (Nd_2CuO_4) or octahedral but distorted (La_2CuO_4) the 4p states are clearly decomposed into z, and x,y in-plane parts spread over a wide energy range. Each one-electron polarized transition $|1s^1...4p_z^1\rangle$ or $|1s^1...4p_{x,y}^1\rangle$ appears twice because of the two existing fillings (d^{10} and d^9) of the d band. The $|1s^1...3d^{10}\,4p_z^1,...2p^5\rangle$ final state gives rise to the signal at the lowest energy, its replica due to the $|1s^1...3d^9\,4p_z^1,...2p^{10}\rangle$ final state appears 7 eV above but with a spectral weight much smaller. The x,y polarized signal exhibits the same behavior but with different spectral weights because the photoelectron is emitted in the bonding plane. This is evidenced by polarization-dependent measurements on undoped single crystals such as La_2CuO_4 and Nd_2CuO_4 [3].

The situation is much simpler in the case of Cu^I as found in Cu_2O or in $YBa_2Cu_3O_6$, where the monovalent species are in a linear coordination. The final state $|1s^1...3d^{10}\,4p^1,...2p^6\rangle$ has to be differentiated only by the symmetry of the p state which is reached (non-bonding $4p_{x,y}$ if we assume the z axis along the linear O-Cu-O object). The $|1s^1...3d^{10}4p_{x,y}^1\rangle$ final state of Cu^I is located at an energy slightly below ($\approx$1.5 eV) the $|1s^1...3d^{10}\,4p_z^1,...2p^5\rangle$ final state of Cu^{II} as in Nd_2CuO_4. Therefore, integrating within a defined energy range the area under the well-defined peak due to the transition to the non-bonding $4p_{x,y}$, we have available a tool which allows quantitative measurement of the d hole creation or disappearance under oxygen uptake or removal (fig.2). A cation such as Cu^I as in Cu_2O or in $YBa_2Cu_3O_{6+y}$ is an exceptional probe in the range of K-edges because the relevant signal due to non-bonding orbitals emerges at low energy where the continuum has hardly started.

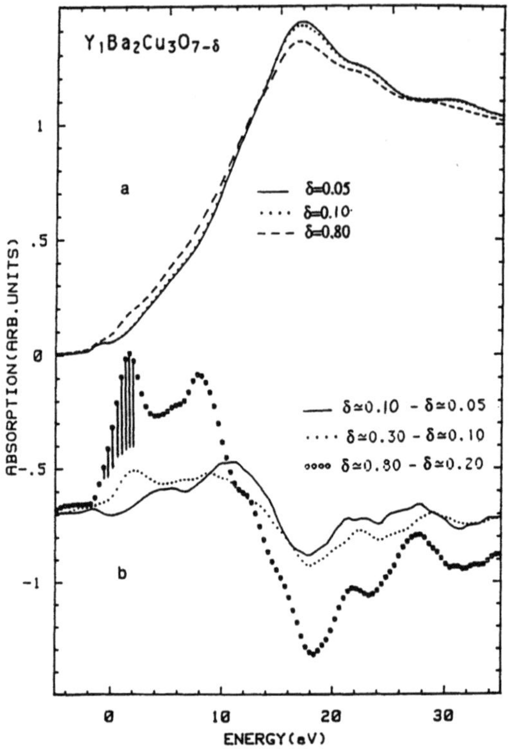

Fig.2 Evolution of the Cu K-edge spectra of $Y_1Ba_2Cu_3O_y$ as the stoichiometry is swept from y=6.95 to 6.2. The peak appearing around 1.7 eV is due to the $1s \rightarrow 4p^*_{x,y}$ transition in a $3d^{10}$ configuration.

2-Results on $YBa_2Cu_3O_{6+y}$

The required balance of the charge when $YBa_2Cu_3O_{6+y}$ takes up oxygen proceeds through two reservoirs, namely the copper d holes and the oxygen 2p holes, which are positive charges needed to compensate the 2 electrons given to the neutral oxygen added to the initial unit cell $YBa_2Cu_3O_6$ [4]. From the y-dependent in situ measurements (fig.3) we can distinguish at least three ranges in hole injection. Close to y=0 (O_6), the **d reservoir** is acting alone.Close to y=1 (O_7) it is the **2p reservoir**. In between, roughly from 6.25 to 6.75, both contributions are found. These first conclusions can be refined if we include the symmetry-dependent results given by single crystals, including those of the L_{III} spectroscopy and the oxygen K-edge spectroscopy. In addition, the antiferromagnetic properties of the insulating phase are essential to guide our views on the possible sequence of the hole injection we propose now :

I-The results on powders and single crystals show that monovalent copper exists beyond the semiconductor/metal phase transition and even above 6.75, in a small quantity. The initial oxygen uptake in the range 6-6.25 leads to injection of holes only in the copper d-band, eliminating Cu^I to produce Cu^{II} at the Cu_1 site.

II-Holes are injected in the oxygen p-band before Cu^I vanishes. When the oxygen stoichiometry is near 7, our results are consistent with the idea of hole injection of p_z symmetry, in agreement with the results of the Cu-L_{III} spectroscopy [5,6] and the O-K spectroscopy [7].

III-These results, along with the one related to magnetic ordering, suggest the following sequence of hole injection into the 2p oxygen-like band :

1 - Around 6.25 x,y-symmetry holes appear in the chains only, the system remains an insulator.

2 - Around 6.4, the planes no longer have antiferromagnetic ordering [8]. This disappearance

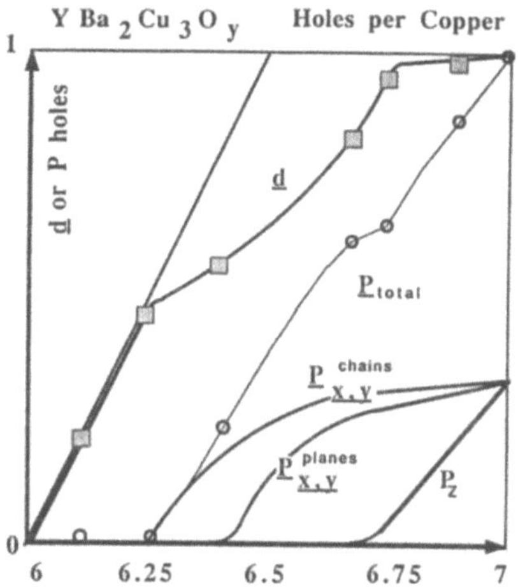

Fig.3 Measured d-hole content obtained by integrating the area within the range of the $3d^{10}$ contribution (shaded area in Fig.2) at the Cu K-edge spectrum. The total p-hole content is derived to balance the total charge given to the uptaken oxygen.

needs only a small quantity of holes (<2%) in the $(CuO_2)^{--}$ planes, as shown in $La_{2-x}Sr_xCuO_4$. These holes have x,y-symmetry. The system becomes a conductor and a superconductor below 50-60K.

3 - Finally, beyond 6.75, z symmetry holes appear injected in the 2p band of the capping oxygen. The chains become conducting and the critical temperature jumps up to 90K.

IV-Two important experimental results can be explained by this sequence :

(a) p_z holes couple the metallic $(CuO_2)^{--}$ planes, allowing a metallic conductivity both along the chains and along the c axis. This coupling transforms the 60K-superconductor system into the 3-dimensional 90K-superconductor one.

(b) If we roughly suppose that the line intensities at the O K-edge reflect the number of holes in the different symmetries, the hole concentration of x,y-symmetry in the planes is roughly 1/3, which means a hole concentration per plane of the order of 0.16, a value near the optimized one found for $La_{2-x}Sr_xCuO_4$.

3-Reverse model of metallization for La_2CuO_4 /Nd_2CuO_4

A generally accepted one-electron band picture of the cuprate insulators assumes the valence band to be of oxygen 2p character and the empty conduction band to be the upper Hubbard 3d band, which is mainly of $d_{x^2-y^2}$ symmetry. Away from the pure ionic description, Cu ^{II} inserted in this chessboard-like pattern of cuprates is better described by an admixture of configurations which takes into account the partial covalency : $\alpha_i|...3d^9,...2p^6\rangle + \beta_i|...3d^{10},...2p^5\rangle$. This means that the valence band (the bonding orbitals, using a different terminology) is the result of the hybridization of the 2p oxygen states with the copper 3d ones, which gives a partial d character to this filled band. Conversely the empty conduction band is mainly of d character but because of the hybridization the holes have an oxygen 2p component (antibonding orbitals). In other words, this one-electron band picture shows the cuprates as insulators whose gap of charge transfer type is about 2-2.5 eV. The Hubbard U repulsion, which splits the d band, is assumed much larger than Δ, the charge transfer excitation $|...3d^9,...2p^6\rangle \rightarrow |...3d^{10},...2p^5\rangle$, or $(\Delta^2+4V^2)^{1/2}$, if we take into account the covalency by V as usual.

Doping could be envisaged as an introduction of impurity levels in the gap (fig.4), the Fermi level remaining unchanged as was shown by photoemission results. In this picture the

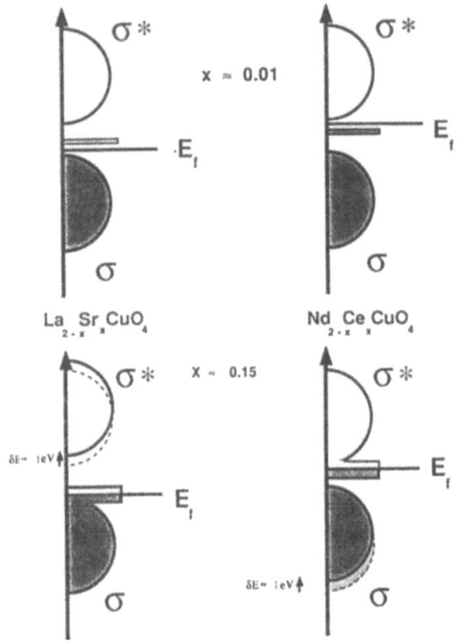

Fig.4 Schematic diagram of the one-electron band structure of square planar $(CuO_2)^{--}$. The σ and σ* bands are separated by a charge transfer gap which is increased (reduced) upon hole (electron) doping in the $La_{2-x}Sr_xCuO_4$ ($Nd_{2-x}Ce_xCuO_4$). Doping is accomplished by impurity levels in the gap.

holes due to doping form a new band, distinct and at a lower energy than that given by the 2p holes due to covalency. This has been well evidenced by Kuiper et al. [9] in the very canonical antiferromagnetic insulator NiO doped by Li substitution. In this last case the gap is too large to generate a metal at high Li concentration.

For $La_{2-x}Sr_xCuO_4$, doping holes introduced in the gap join the valence band when the concentration is large enough, and the system becomes metallic. For low x, covalent holes and doping holes tend to repel, therefore the antibonding orbitals (upper d band) are pushed to higher energy. In this process, since the original gap increases, the hybridization should decrease. The d hole component increases as the 2p empty states leaves the upper d band to feed the new impurity band. The reverse scheme for Nd_2CuO_4 should be completely symmetrical. The impurity electron band, on the top of the gap, joins the conduction band. The hybridization increases, which results in the decrease of d holes.

To prove this assumption, L_{III} spectroscopy is a direct probe which can quantitatively evaluate the amount of d holes because it takes a 2p electron and puts it into the available 3d empty states. In this spectroscopy only the component $|...2p_{3/2}^{(6)}...3d^9,...2p^6\rangle$ of the admixture of configurations $\alpha_i|2p_{3/2}^{(6)}...3d^9,...2p^6\rangle + \beta_i|2p_{3/2}^{(6)}...3d^{10},...2p^5\rangle$ of the initial state gives rise to the lorentzian-like line which corresponds to $|2p_{3/2}^{(5)}...3d^{10},...2p^6\rangle$ in the final state. The p→s channel which is dipole-allowed has too weak a spectral weight to interfere with this measurement of $\alpha_i^2=1-\beta_i^2$.

The main difficulty of such a measurement is to be able to calibrate the signal of one sample against another. This calibration has been made possible by measuring with a good statistics the response of the optics "without a copper containing sample". We carried out this measurement with a zinc foil, which yields a smooth response in the energy range of interest. This profile was periodically recorded to keep data collected during periods where the beam has been checked stable. Samples, doped and undoped, have been measured with the same careful control. Proceeding in that way, the two spectra of the doped and undoped samples overlap

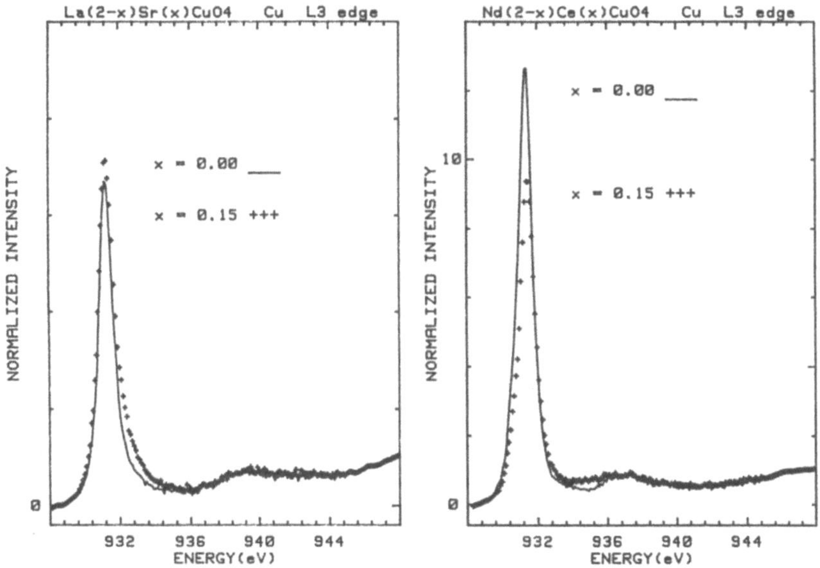

Fig.5 Cu L_{III}-edge of $La_{2-x}Sr_xCuO_4$ and $Nd_{2-x}Ce_xCuO_4$ for x=0 and x=0.15. Upon doping we observed an increase of about 15% in the integrated area intensity for $La_{2-x}Sr_xCuO4$ and a decrease of about 20% for $Nd_{2-x}Ce_xCuO4$.

nicely over a large energy range (936-950 eV) above the $3d^{10}$ line. The experiments were repeated at least four times for each system.

The experimental results, gathered in figure 5, strongly support the model proposed above. The Sr-doped sample of La_2CuO_4 exhibits an increase of the integrated intensity of the $3d^{10}$ final state line (+17%±4% for x=0.15) whereas the Ce-doped sample of Nd_2CuO_4 exhibits a decrease of the integrated intensity of the $3d^{10}$ final state line (-20%±4% for x=0.15).

The alternative way of looking at this change in d holes or, equivalently, α_i^2, which is the weight of $|2p_{3/2}^{(6)}...3d^9,...2p^6\rangle$, is to evaluate β_i^2, which is relevant to oxygen K-edge spectroscopy, because this configuration contains the 2p covalent holes. Looking qualitatively at the very complete set of data collected by EELS [10], it is easy to correlate the increase of intensity of the "line" due to the oxygen 2p-doping holes to the decrease of the intensity of the "line" due to the oxygen 2p-covalent holes for the hole-doped cuprates, namely La_2CuO_4, the 123 compounds and the bismuth cuprates. Let us recall that the hole-doping band appears at lower energy than the 2p-covalent hole band. And again the terminology 2p-covalent hole band is incomplete for the sake of conciseness:what is probed is only the 2p component of the upper Hubbard d band. On the other hand, the electron doped $Nd_{2-x}Ce_xCuO_4$ exhibits at the oxygen K-edge a slight (or no) increase of the intensity of the line due to the covalent holes. Such a clue to the increase of β_i^2 is also found indirectly by Cu K-edge spectroscopy, where the weight β_f^2 of the $|1s^1...3d^{10}\,4p_{x,y}^1,...2p^5\rangle$ component increases versus that of $|1s^1...3d^9\,4p_{x,y}^1,...2p^6\rangle$, along with a slight increase of the energy separation between the two contributions, consistent with a decrease of Δ which goes into the difference U_{cd}-Δ. This symmetry between the hole-doped cuprates and the electron-doped cuprates is satisfactory, regarding the disappearance of the magnetic order under doping.

4-Conclusion

Core hole spectroscopy often presents a lot of difficulties in being made quantitative because of the core hole effect in the final state. Because copper is the last 3d element, which leads to very unique fillings, $3d^9$ or $3d^{10}$, along with the correlated extra benefit regarding the coordination, the copper core hole spectroscopies are more quantitative than those of other elements. Such good fortune has allowed a quantitative evaluation of the linearly coordinated Cu^I. Its y-dependence is related directly to that of the doping oxygen 2p holes. Magnetic properties and results of the Cu L_{III} and the oxygen K-edge spectroscopies provide very useful information to distinguish the y-dependence of the injected p holes. In addition, the low value of the charge transfer-controlled gap of the non-doped cuprates enables metallization through creation of impurity states within the forbidden band. In NiO or nickelates, doping is not expected to lead to a metal since the gap is large.

We would like to thank the French Ministry of Research and Technology, which supports this work under the contract 87-A0609. HT is supported as a PhD fellow by a grant of CAPES from Brazil.

References:

[1] H.TOLENTINO, E.DARTYGE, A.FONTAINE, G.TOURILLON: J.Appl.Crys.21,15(1988)

[2] K.KISHIO, J.SHIMOYAMA, T.HASEGAWA, K.KITAZAWA, K.FUEKI: Jpn.J.Appl. Phys.26, L1228 (1987)

[3] H.TOLENTINO, M.MEDARDE, A.FONTAINE, G.TOURILLON, D.GUAY: in prep.

[4] H.TOLENTINO, E.DARTYGE, A.FONTAINE, T.GOURIEUX, G.KRILL, M.MAURER, M-F.RAVET, G.TOURILLON: Phys. Lett. A, 139, 474(1989)

[5] A.BIANCONI, A.CONGIU-CASTELLANO, M.DE SANTIS, P.RUDOLF, P.LAGARDE, A.M.FLANK, A.MARCELLI: Solid State Commun. 63(11), 1009 (1987)

[6] A.BIANCONI, M.DE SANTIS, A.DI CICCO, A.M.FLANK, A.FONTAINE, P.LAGARDE, H.KATAYAMA-YOSHIDA, A.KOTANI, A.MARCELLI: Phys.Rev.B 38,7(1988)

[7] N.NUCKER, H.ROMBERG, X.X.XI, J.FINK, B.GEGENHEIMER, Z.X.ZHAO: Phys.Rev.B 39(10), 6619 (1988)

[8] J.ROSSAT-MIGNOD, P.BURLET, M.J.JURGENS, L.P.REGNAULT, J.Y.HENRY, C.AYACHE, L.FORRO, C.VETTIER, H.NOEL, M.POTEL, P.GOUGEON, J.C.LEVET: J.de Phys. C8, Suppl.12, 49(1988)

[9] P.KUIPER, G.KRUIZINGA, J.GHIJSEN, G.A.SAWATZKY, H.VERWEIJ: Phys.Rev.Lett. 62, 221(1989)

[10] J.FINK, J.PFLÜGER, Th.MÜLLER-HEINZERLING, N.NÜCKER, B.SCHEERER, H.ROMBERG, M.ALEXANDER, R.MANZKE, T.BUSLAPS, R.CLAESSEN, M.SKIBOWSKI: In "Earlier and Recent Aspects of Superconductivity", by K.A.Müller, G.Bednorz, Springer Ser. Solid State Sci., Vol 90 (Springer, Berlin, Heidelberg 1990)p.377

Part IV

Hyperfine Interactions, Magnetic Properties

^{205}Tl NMR High-T_c Superconductors: Hyperfine Interactions of Localized and Delocalized Electrons

M. Mehring, F. Hentsch, and N. Winzek

2. Physikalisches Institut, Universität Stuttgart, Pfaffenwaldring 57, D-7000 Stuttgart 80, Fed. Rep. of Germany

Abstract: We summarize our ^{205}Tl NMR data obtained from Knight shift and spin lattice relacation data in $Tl_2Ba_2CuO_6$ (one layer), $Tl_2Ba_2CaCu_2O_8$ (two layers) and $Tl_2Ba_2CaCu_3O_{10}$ (three layers). A large anisotropic chemical shift is observed in all three compounds, with an isotropic Knight shift contribution $K_s = 0.1106\%$ for the TlO layer and $K_s = -0.4815\%$ for the "defect" site where Tl replaces Ca. Two different relaxation mechanisms are found: T_1 relaxation at the TlO layer sites is similar to $Cu(2)$ in $YBCO$ and in contrast Korringa type relaxation is observed for the "defect site".

1. Introduction

NMR has played an important role in determining the electronic structure and the dynamics of localized and delocalized electrons in high-T_c superconductors [1]. The NMR session at this **IWEPS 90** meeting reflects the activity in this area and highlights some of the most important observations. Most of the investigations were devoted in the past to $YBCO$ (123) and its oxygen depleted states. The lecture by Brinkmann deals with the quadrupole interaction of Cu and the spin lattice relaxation in particular with respect to the new 124 compound. Lütgemeier's lecture discusses the effect of oxygen depletion of the Cu NMR/NQR parameters. Berthier covers both ^{63}Cu and ^{17}O NMR data and gives a detailed account of the separation of Knight shifts and quadrupolar interaction in the $YBCO$ system. Alloul presents some important investigations of ^{89}Y NMR in the $YBCO$ system with variable oxygen depletion. Lüders deals with the quadrupolar interaction and demagnetizing fields as well as flux pinning as seen by zero field NQR. The La_2CuO_4 system doped with different Sr concentrations was thoroughly investigated by Rigamonti and coworkers. His lecture deals with the dynamics of doped oxygen holes in a localized antiferromagnetically coupled lattice of Cu^{++} spins.

Two theoretical lectures should be mentioned in this context. Schwarz demonstrated the power of LDA type calculations for the electric field gradient (EFG) at all nuclear sites in the $YBCO$ system. The

Springer Series in Solid-State Sciences, Vol. 99
Electronic Properties of High-T_c Superconductors and Related Compounds
Editors: H. Kuzmany · M. Mehring · J. Fink

agreement between theory and experiment is very convincing. Another theoretical lecture, namely the one by Sigmund, deals with the motion and percolation of oxygen holes in an antiferromagnetically coupled lattice of Cu^{++} spins. The calculations should be relevant for the NMR investigations in the low doping regime.

Whereas most of the NMR investigations so far have dealt with ^{63}Cu, ^{17}O and ^{89}Y in the $YBCO$ system [2-20], we have recently investigated ^{205}Tl NMR in a series of compounds, namely $Tl_2Ba_2CuO_6$ (one layer, $T_c = 85K$), $Tl_2Ba_2CaCu_2O_8$ (two layers, $T_c = 105K$) and $Tl_2Ba_2CaCu_3O_{10}$ (three layers, $T_c = 120K$) [21]. Similar investigations have been reported by others [22]. Here we concentrate on different types of observations

- chemical and Knight shift in the TlO layers
- negative Knight shift for Tl replacing Ca (defect site)
- coupling of the TlO layer via the apex oxygen to the CuO_2 layer
- antiferromagnetic fluctuations of the Cu^{++} spins as seen by the ^{205}Tl nuclei (in the TlO layer)
- Korringa-type relaxation for ^{205}Tl in the defect site where coupling to the CuO_2 layer is dominant.

2. Chemical and Knight shift

Fig.1 shows the three different ^{205}Tl NMR spectra of the three compounds discussed here [21c]. The dominant line is caused by the TlO layer nuclei, whereas the smaller line, observed only in the two- and three-layer compounds results from a defect site, where Ca is replaced by Tl. This site is located between the CuO_2 planes. The dominant line shows an average shift which is near the value expected for the Tl^{3+} oxidation state. The line is rather anisotropic, i.e. a characteristic powder spectrum is observed, which allows the principal elements of the total shift tensor to be determined. A detailed analysis has been performed and the results are summarized in Table 1 [21c].

Table 1:
Comparison of the shift values of the three investigated compounds

	$Tl_2Ba_2CuO_6$	$Tl_2Ba_2CaCu_2O_8$	$Tl_2Ba_2Ca_2Cu_3O_{10}$
δ_{iso}	3397 ppm	3108 ppm	2917 ppm
$\delta_{11} = \delta_{22}$	4174 ppm	3754 ppm	3391 ppm
δ_{33}	1844 ppm	1815 ppm	1968 ppm

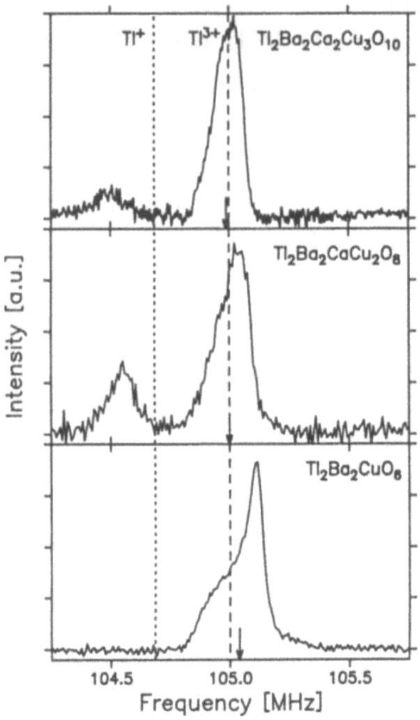

Fig.1 Room temperature ^{205}Tl NMR spectra of the three high-T_c superconductors $Tl_2Ba_2CuO_6$ (bottom), $Tl_2Ba_2CaCu_2O_8$ (middle), and $Tl_2Ba_2CaCu_3O_{10}$ (top) according to [21b]. The Tl^{3+} and Tl^{+} oxidation state reference positions are indicated. Note the appearance of the "defect line" in the Ca-containing compounds.

The total shift can be separated into the chemical shift (δ_c) and the spin dependent Knight shift (K_s)

$$\delta = \delta_c + K_s, \tag{1}$$

where we have summarized all spin independent parts, i.e. all closed shell contributions, in the "chemical shift" δ_c and all spin dependent contributions in the Knight shift K_s [21c]. The chemical shift therefore contains, e.g., the orbital Van Vleck paramagnetism, together with all the other contributions coming from the different closed shell orbitals, whereas Fermi contact terms including core polarization and dipolar interactions of the conduction electrons with the nuclei are concatenated in the "Knight shift". The reason for doing this is mainly dictated by the observation that in singlet superconductivity the spin pairing in the Cooper pairs leads to a vanishing K_s at $T = 0\,K$. This allows the separation of the two contributions in the total shift.

Moreover, the interesting part is indeed the Knight shift K_s, because of its direct relation to the spin dependent susceptibility, as

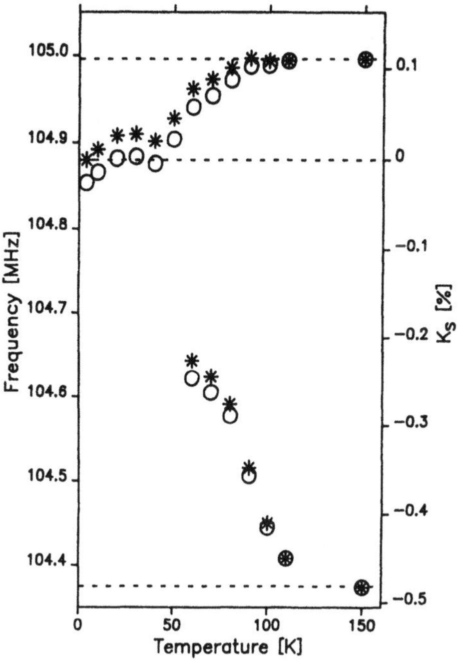

Fig.2 Temperature dependence of spectral shifts of the dominant line (top) and the defect line (bottom) [21c]. The corresponding Knight shift values are indicated on the right hand vertical axis. Uncorrected data (○) and data corrected for diamagnetic shielding (∗) are plotted. Note the negative Knight shift of the defect line.

$$K_s = \frac{\chi_s A_{zz}}{g_e \beta_e \gamma_n}, \tag{2}$$

where A_{zz} is the z-component (parallel to the externally applied field B_a) of the hyperfine tensor, g_e is the Landé factor, β_e is the Bohr magneton and γ_n is the gyromagnetic ratio of the nucleus. The spin susceptibility χ_s is given per conduction electron. It is directly related to the density of states at the Fermi energy and might contain Stoner enhancement from electron-electron correlations.

The temperature dependent shift of the spectra shown in Fig.1 is plotted in Fig.2. Note the positive Knight shift for the TlO layer nuclei and the stronger negative Knight shift for the "defect line". Below T_c both approach the $K_s = 0$ state due to spin pairing, pointing at singlet superconductivity. Both Tl NMR lines seem to be compatible with a Tl^{3+} oxidation state. In order to evaluate the shift data below T_c a correction for diamagnetic shielding had to be performed with

$$\delta = \delta_c + K_s(T) + (1 - N)\frac{M(T)}{B_a}, \tag{3}$$

where $M(T)$ is the temperature dependent diamagnetic shielding and N is the demagnetizing factor. For a granular powder we take as an average value $N = 0.4$, compared with $N = 1/3$ for a sphere. The temperature dependence of $K_s(T)$ is similar to the one proposed by Yoshida for BCS superconductors [23].

The fact that the "defect line" shows a negative Knight shift resembles the observation of a negative ^{89}Y Knight shift in the $YBCO$ system [16]. In the Tl-containing superconductors discussed here the defect Tl is at the Ca site between the CuO_2 planes and couples to the oxygen holes in a similar fashion as the Y atoms in $YBCO$. The negative Knight shift of the Tl defect site supports therefore also the σ-bonding in the CuO_2 plane as was discussed for the Y site [16]. The positive Knight shift of the TlO layer atoms on the other hand comes about by the coupling of the Tl $6s$-orbital to the p_z-orbital of the apex oxygen, which in turn is coupled to the $3d_{3z^2-r^2}$ orbital of the $Cu(2)$ site. The Knight shift K_s is then caused by a certain fraction of spin density of Cu^{++} spins in the $3d_{3z^2-r^2}$ orbital. This requires an admixture between the $3d_{3z^2-r^2}$ orbital and the $3d_{x^2-y^2}$ orbital. The observation of a positive Knight shift at the ^{205}Tl nuclei in the TlO layer is therefore a consequence of this admixture.

3. Spin lattice relaxation

Nuclear spin lattice relaxation of ^{63}Cu and ^{17}O in $YBCO$ has been investigated in detail [2-12]. It reflects the electron spin dynamics of the Cu^{++} spins as well as the oxygen hole spins. It is by no means trivial to disentangle their different contributions to the observed rate and this has initiated intense discussions. We have demonstrated that ^{205}Tl relaxation measurements [21b] are capable of observing two distinct features of the electron spin dynamics, namely

- antiferromagnetically correlated Cu^{++} spin dynamics, and

- Korringa-type (Fermi statistic) relaxation of the oxygen hole spins.

These two distinct features are shown in Figs. 3 and 4. Fig. 3 shows for comparison the temperature dependent relaxation rates for the three different Tl-containing superconductors discussed here. The prominent feature is the fairly temperature independent relaxation rate above T_c, which correlates nicely with the $Cu(2)$ relaxation. The weak temperature dependence above T_c can be explained by strong antiferromagnetic fluctuations ($q = \frac{\pi}{a}, \frac{\pi}{a}$) of the copper spins. The coupling of the ^{205}Tl nuclei to the Cu^{++} spins is mediated by the orbital connections as discussed

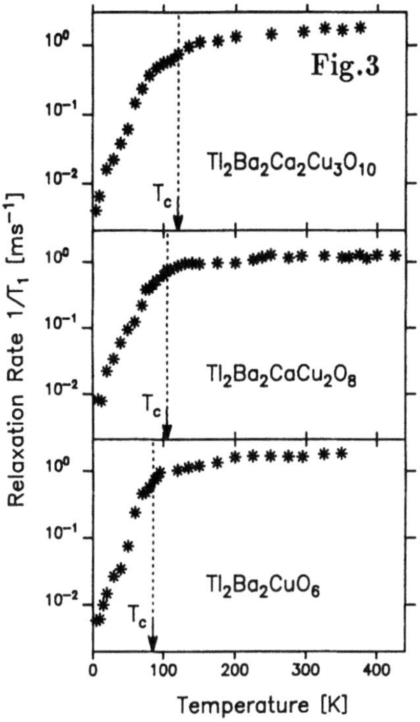

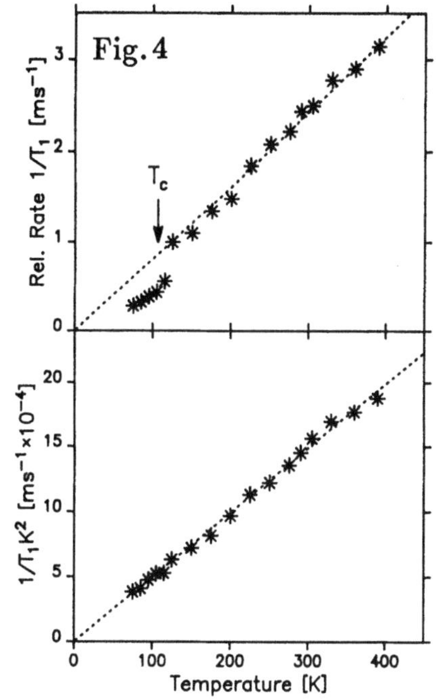

Fig.3 Korringa-type ^{205}Tl relaxation of the defect line [21b].
Top: $T_1 \sim T$ dependence above T_c. *Bottom*: with the correction due to the temperature dependent Knight shift K below T_c taken into account.

Fig.4 Comparison of the temperature dependent ^{205}Tl relaxation rates for the three different compounds discussed here [21b]. Note the flat temperature dependence and the onset of the decrease above T_c.

in section 2. The decrease of $1/T_1$ with decreasing temperature sets in already above T_c, pointing to a magnetic gap opening above T_c. Below T_c a rapid decrease of $1/T_1$ is observed, consistent with spin pairing, which affects also the Cu^{++} spins. This does not necessarily mean that the Cu^{++} spins become paired below T_c; it might also be explainable by a magnetic gap for low energies. This aspect is currently being debated in connection with recent neutron scattering results.

The defect site ^{205}Tl relaxation rate on the other hand shows a clear Korringa-type relaxation of the form

$$K_s^2 T_1 T C_0 S_K = 1, \tag{4}$$

where

$$C_0 = \frac{4\pi k_B}{\hbar}\left(\frac{\gamma_n}{\gamma_e}\right)^2$$

and S_K is a scaling factor which takes electron-electron correlations and low-dimensionality effects into account. For a simple Fermi gas we get $S_K = 1$. Interestingly enough we obtain from our Korringa analysis for the "defect line" a value of $S_K = 0.4$, which is very close to the ideal Fermi gas value. Similar results were obtained for ^{89}Y relaxation in $YBCO$. These findings might elucidate the discussion on the Fermi liquid properties of high-T_c superconductors.

The nuclear spin lattice relaxation rate $1/T_1$ caused by hyperfine interactions of the form

$$\hat{H}_{IS} = \sum_j A_j I S_j, \tag{5}$$

where the nuclear spin I couples by isotropic hyperfine interaction (the extension to anisotropic interactions is straightforward) to N electron spins at sites j with strength A_j. The corresponding relaxation rate can be expressed as [24-27]

$$\frac{1}{T_1} = \frac{1}{2} \sum_q |A_q|^2 \int_{-\infty}^{\infty} dt \langle S_q^+(t) S_{-q}^-(0) \rangle e^{i\omega_e t}, \tag{6a}$$

where $\langle S_q^+(t) S_{-q}^-(0) \rangle$ is the wavevector dependent electron two-spin correlation function with

$$S_q = N^{-1/2} \sum_j S_j \exp(-i\vec{q} \cdot \vec{r}_j)$$

and $$A_q = N^{-1/2} \sum_j A_j \exp(-i\vec{q} \cdot \vec{r}_j)$$

is the collective hyperfine interaction.Eq.(6a) may be rewritten as (for S = 1/2)

$$\frac{1}{T_1} = A_0^2 \sum_q F(q) S(q, \omega_e) \tag{6b}$$

where $$A_0 = N^{-1/2} \sum_j A_j$$

and $$S(q, \omega_e) = \frac{1}{2} \int_{-\infty}^{\infty} dt \langle S_q^+(t) S_{-q}^-(0) \rangle e^{i\omega_e t} \tag{7}$$

is the q-dependent dynamic structure factor of the electron spin fluctuations. Similar treatments have been used in the context of nuclear spin lattice relaxation in high-T_c superconductors by others. We want to point out that the characteristic frequency at which the fluctuation is observed is the electron spin Larmor frequency ω_e. The nuclear Larmor frequency ω_n is neglected, because $\omega_n \ll \omega_e$.

Moreover, we have introduced in Eq.(6b) a filtering function $F(q) = |f(q)|^2$ where

$$f(q) = \frac{\sum_j A_j e^{-i\vec{q}\cdot\vec{r}_j}}{\sum_j A_j} \tag{8}$$

The connection with the well-known Moriya expression [25] is readily obtained according to the fluctuation-dissipation theorem by replacing $S(q, w)$ by

$$S(q,\omega) = 2k_B T \ Im\left(\frac{\chi_\perp(q,\omega)}{\omega}\right) \tag{9}$$

in Eq. (6b). The filtering function $F(q)$ for the ^{205}Tl site in the TlO layer is $F(q) = 1$ independent of the wavevector $\vec{q}$. The corresponding relaxation rate is therefore dominated by the enhanced antiferromagnetic fluctuations of the Cu^{++} spins with $\vec{q} = (\frac{\pi}{a}, \frac{\pi}{a})$. In contrast the ^{205}Tl defect site replacing a Ca atom in the center between two CuO_2 planes has a strongly q-dependent filtering function. For the center of an equilateral square of electron spin sites with sides of length a, one calculates from Eq.(8) a filtering function

$$F(q) = \cos^2(q_x \frac{a}{2}) \cos^2(q_y \frac{a}{2}) \qquad (\ Ca,\ Y \text{ site}) \tag{10}$$

as for the Y site in $YBCO$ [16]. For completeness we note that for the oxygen sites in the CuO_2 plane the filtering function is calculated as

$$F(q) = \cos^2(q\frac{a}{2}) \qquad (\ O \text{ site }) \tag{11}$$

where $q = (q_x \text{ or } q_y)$ depending on the particular site. It is important to note that the oxygen site (Eq.11) and even more the Ca, Y site are "blind" for the antiferromagnetic fluctuations of the Cu^{++} spins. Their relaxation rate is dominated by fluctuations with $q \leq \frac{1}{2}(\frac{\pi}{a}, \frac{\pi}{a})$. It is therefore not surprising that the relaxation of the "defect line" shows a completely different temperature dependence. The observation of a Korringa-type relaxation for this part of q-space demonstrates the Pauli character of the dynamic susceptibility, or in other words the Fermi gas behaviour ($\sim k_B T$) of the excitations for small q-values. The importance of this finding for the "Fermi liquid"-like character near $q \simeq 0$ has to be investigated further.

The different relaxation phenomena in different parts of q-space do not necessarily advocate a "two spin fluid model" of the high temperature phase of these superconductors. Bulut et al. [26] and Monien and Pines [27] have developed a theoretical model which can explain these observations in a unified way within a correlated "one spin fluid model". Still more detailed theoretical, as well as experimental investigations are required to clarify this point.

Acknowledgments: We would like to thank A. Simon and Hj. Mattausch for sample preparation and discussions. We are indebted to R. Kremer for SQUID measurements of the magnetization. The Bundesministerium für Forschung und Technologie (BMFT-Project-Nr. 13N5577) and the Landesamt für Forschung (BW) have given financial support.

References

1) J. G. Bednorz, K. A. Müller, *Z. Phys.* **B64** (1986) 189

2a) M. Mali, D. Brinkmann, L. Pauli, J. Roos, H. Zimmermann, and J. Hullinger, *Physics lett.* **A124** (1987) 112

b) M. Mali, J. Roos and D. Brinkmann, *Physica* **C153-155** (1988)

3) H. Riesenmeier, Ch. Grabow, E. W. Scheidt, V. Müller, and K. Lüders, *Sol. State Commun.* **64** (1987) 309

4a) W. W. Warren, R. E. Walstedt, G. F. Brennert, G. P. Espinosa, J. P. Rameika, *Phys. Rev. Lett.* **59** (1987) 1860

b) R. E. Walstedt, W. W. Warren, R. F. Bell, G. F. Brennert, G. P. Espinosa, R. J. Cava, L. F. Schneemeyer, J. V. Wascazak, *Phys. Rev.* **B38** (1988) 9299

5a) Y. Kitaoka, S. Hiramatsu, K. Ishida, T. Kohara, Y. Oda, K. Amaya, K. Asayama, *Physica* **148B** (1987) 298

b) Y. Kitaoka, S. Hiramatsu, T. Kondo, Y. Oda, K. Asayama, *J. Phys. Soc. Jpn.* **57** (1988) 30

6) T. Imai, T. Shimizu, H. Yasuoka, Y. Ueda, K. Kosuge, *J. Phys. Soc. Jpn.* **57** (1988) 2280

7) E. Lippmaa, E. Joon, I. Heinmaa, V. Miidel, A. Miller, R. Stern, I. Furo, L. Mihaly, P. Banki, *Physica* **C153-155** (1988) 91

8a) M. Takigawa, P. C. Hammel, R. H. Heffner, Z. Fisk, J. L. Smith, R. B. Schwarz, *Phys. Rev.* **B37** (1988) 7944

b) M. Takigawa, P. C. Hammel, R. H. Heffner, Z. Fisk, J. L. Smith, R. B. Schwarz, *Phys. Rev.* **B39** (1989) 300

c) P. C. Hammel, M. Takigawa, R. H. Heffner, Z. Fisk, K. C. Ott, *Phys. Rev. Lett.* **63** (1989), 1992

9) H. Lütgemeier, *Physica* **C153-155** (1988) 95
10) C. H. Pennigton, D. J. Durand, D. B. Zax, C. P. Slichter, J. P. Rice, D. M. Ginsberg, *Phys. Rev.* **B37** (1988) 7944
11) I. Watanabe, K. Kumagai, Y. Nakamura, T. Kimura, Y. Nakamichi, H. Nakajima, *J. Phys. Soc. Jpn.* **56** (1987) 3028
12a) M. Horvatic, P. Segrasan, C. Berthier, P. Butaud, J. Y. Henry, M. Coach, J. P. Chaminade, *Phys. Rev.* **B39** (1989) 7332
b) M. Horvatic, Y. Berthier, P. Butaud, Y. Kitaoka, P. Segrasan, C. Berthier, H. Katayama-Yoshida, Y. Okabe, T. Takahashi, *Physica* **C159** (1989), 689
c) C. Berthier, Y. Berthier, P. Butaud, M. Horvatic, Y. Kitaoka and P. Segrasan, in "Dynamics of Magnetic Fluctuations in High-T_c Materials", eds. G. Reiter, P. Horsch and G. Psaltakis, Plenum Press (1990)
13) P. Wzietek, D. Köngeter, P. Auban, D. Jerome, J. P. Contures, B. Dubois, Ph. Odier, *Europhys. Lett.* **8** (1989) 363
14) E. Lippmaa, E. Joon, I. Heinmaa, A. Miller, R. Stern, S. Vija, Paper presented at "International Seminar on High Temperature Superconductivity", Dubna (1989) USSR
15) C. Coretsopoulos, H. C. Lee, E. Ramli, L. Reven, T. B. Rauchfuss, E. Oldfield, *Phys. Rev.* **B39** (1989) 781
16a) J. T. Markert, T. W. Noh, S. E. Russek, R. M. Cotts, *Sol. State Commun.* **63** (1987) 847
b) H. Alloul, P. Mendels, G. Collin P. Monod, *Phys. Rev. Lett.* **61** (1988) 746
c) H. Allouol, T. Ohno, P. Mendels *Phys. Rev. Lett.* **63** (1989) 1700
17a) H. Seidel, F. Hentsch, M. Mehring, J. G. Bednorz, K. A. Müller, *Europhys. Lett.* **5** (1988) 647
b) F. Hentsch, H. Seidel, M. Mehring, J. G. Bednorz, K. A. Müller, *Physica* **C153-155** (1988) 727
18) H. Lütgemeier, M. W. Pieper, *Sol. State Commun.* **64** (1987) 267
19) Y. Kitaoka, S. Hiramatsu, T. Kohara, K. Asayama, K. Oh-Ishi, M. Kikuchi, N. Koboyashai, *Jpn. J. Appl. Phys.* **26** (1987) 397
20) M. Mehring, *IBM J. of Res. and Dev.* **33** (1989) 342
21a) F. Hentsch, N. Winzek, M. Mehring, Hj. Mattausch, A. Simon, *Physica* **C158** (1989) 137

b) F. Hentsch, N. Winzek, M. Mehring, Hj. Mattausch, A. Simon, R. Kremer, *Physica* **C165** (1990) 485
c) N. Winzek, F. Hentsch, M. Mehring, Hj. Mattausch, A. Simon, *Physica* **C168** (1990) 327
d) M. Mehring, F. Hentsch, Hj. Mattausch, A. Simon, *Z. Phys.* **B77** (1989) 355
e) M. Mehring, in "Earlier and Recent Aspects of Superconductivity", eds. J. G. Bednorz and K. A. Müller, Springer Ser. in Solid-State Phys., Vol. **90**(1990)
f) M. Mehring, F. Hentsch and N.Winzek, in "Dynamics of Magnetic Fluctuations in High-T_c Materials", eds. G. Reiter, P.Horsch and G.Psaltakis, Plenum Press 1990
22a) H. B. Brom, D. Reefman, J. C. Jol, D. M. deLeeuw, W. A. Groen, *Preprint* submitted to *Phys. Rev.* (Dec. 1989)
b) K. Tompa, I.Bakonyi, P.Banki, I.Furo, S.Pekker, J. Vandlik and L.Mihaly, *Physica* **C15n2** (1988) 486
23) K. Yosida, *Phys. Rev. Lett.* **3** (1959) 325
24) F. Borsa, A. Rigamonti in "Magnetic Resonance and Phase Transitions", p. 79, Academic Press (1979)
25) T. Moriya, *J. Phys. Soc. Jpn.* **18** (1963) 516
26) N. Bulut, D. Hone, D. J. Scalapino, N. E. Bickers, *Phys. Rev.* **B41** (1990) 1797
27) H. Monien and D. Pines, *Phys. Rev.* **B41** (1990) 6297

Cu NQR and NMR Studies in Y-Ba-Cu-O Superconductors

D. Brinkmann

Physik-Institut, Universität Zürich, Schönberggasse 9,
CH-8001 Zürich, Switzerland

Abstract. This paper reviews some copper NQR (nuclear quadrupole resonance) and NMR (nuclear magnetic resonance) studies in the superconductors $YBa_2Cu_3O_{7-x}$ and $YBa_2Cu_4O_8$ performed at the University of Zürich. The presentation will emphasize the kind of information which can be obtained by NQR/NMR on static and dynamic electronic properties at an atomic scale, *e.g.* structural changes at T_c and other temperatures, electric field gradients as a function of temperature and pressure, Knight shift and spin-lattice relaxation times.

1. Introduction

Nuclear magnetic resonance (NMR) and nuclear quadrupole resonance (NQR) can provide essential information on static and dynamic properties of superconductors on an atomic scale [1]. As an example, this paper will review some studies performed by the NMR/NQR group at the Physik-Institut on two members of the Y-Ba-Cu-O family: the 1-2-3 structure $YBa_2Cu_3O_{7-x}$ (T_c = 92 K) containing single Cu1-O chains, and the 1-2-4 structure $YBa_2Cu_4O_8$ (T_c = 82 K) with double Cu1-O chains. The study of these related structures offers the possibility to derive conclusions by looking for similarities or contrasting behavior. The information obtained can serve as an input for testing models or constructing theories to explain the mechanism of superconductivity in these new materials [2].

2. Temperature and Pressure Dependence of NQR Frequencies

The frequencies ν_Q of the Cu NQR signals are of great importance because they are related to the principal components V_{ii} of the electric field gradient (EFG) tensor present at the Cu site by

$$\nu_Q = \frac{eQV_{zz}}{2h}\sqrt{1+\frac{1}{3}\eta^2} \quad . \tag{1}$$

Here, η is the asymmetry parameter defined as $\eta = (V_{xx} - V_{yy})/V_{zz}$

Springer Series in Solid-State Sciences, Vol. 99
Electronic Properties of High-T_c Superconductors and Related Compounds
Editors: H. Kuzmany · M. Mehring · J. Fink

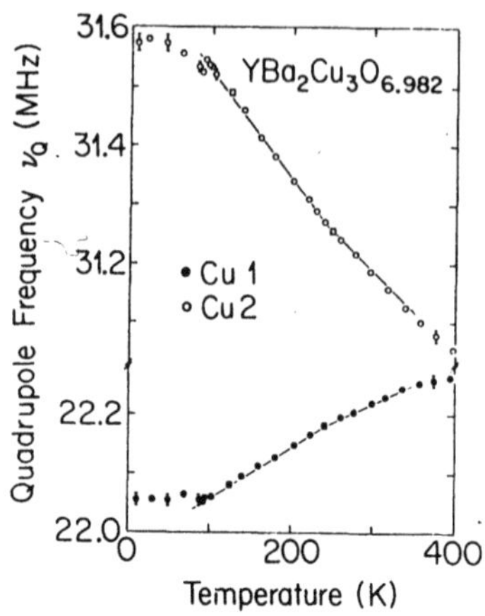

Fig. 1. Temperature dependence of the ^{63}Cu NQR frequencies in $YBa_2Cu_3O_{6.982}$ (from Ref.[5]).

where $|V_{zz}| \geq |V_{yy}| \geq |V_{xx}|$ and eQ is the nuclear electric quadrupole moment. The copper NQR spectrum in both 1-2-3 and 1-2-4 consists of four signals each because there are two Cu sites and two copper isotopes ^{63}Cu and ^{65}Cu. The experimental ratio of the NQR frequencies of the two isotopes is equal to the ratio of their quadrupole moments (1.081) for both Cu sites in 1-2-3 [3] and 1-2-4 [4]. This proves that no internal magnetic field due to any type of magnetic ordering is present at these Cu sites in the temperature range studied. Such a field would split or shift the NQR lines as it does for the Cu2 signal in antiferromagnetic $YBa_2Cu_3O_6$.

The dependence of ν_Q on temperature T and pressure P can provide detailed information on structural and electronic changes of the compound since the EFG tensor is very sensitive to small atomic displacements or modifications of charge distributions. In Figure 1 we have plotted $\nu_Q(T)$ for both Cu sites in a 1-2-3 sample with oxygen deficiency $x = 0.018 \pm 0.001$ [5]. While Riesemeier *et al.* [6] have reported a steep decrease of ν_Q at T_c, our data, at least those for Cu1, do not provide evidence for such a behavior and they are in agreement with our previous results [2]. We thus exclude a major structural transition at T_c. The change of slope in the $\nu_Q - vs - T$ dependence at T_c and at 240 K is therefore attributed to effects such as small changes of inter-atomic distances, Cu-O bond corrugations, valence state changes due to a minute electron transfer between orbitals and lattice vibrations. Perhaps Jahn-Teller excitations (see the contribution by W. Weber) should be taken into account as a conceivable origin of distinct changes of ν_Q.

In 1-2-4, the $\nu_Q(T)$ curve exhibits even more subtle changes which are outside the experimental errors and the Cu2 data above 300 K behave quite differently from those of Cu2 in 1-2-3 [4]. Present *ab initio* calculations of the EFG's in 1-2-3 [7] yield good agreement with experimental data, although for Cu2 a discrepancy of about 50 % appeared.

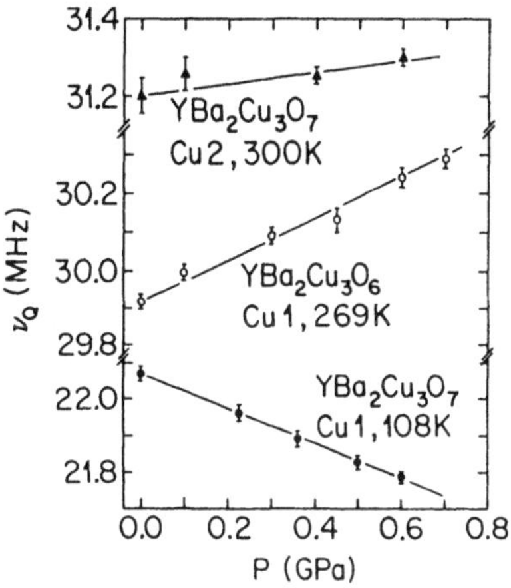

Fig. 2. Pressure dependence of the ^{63}Cu NQR frequencies for Cu sites in $YBa_2Cu_3O_7$ and $YBa_2Cu_3O_6$ (from Ref.[8]).

Thus, changes of ν_Q by a fraction of a percent cannot be accounted for by these calculations.

One therefore has to refer again to simpler methods such as the point charge model. In order to put constraints on these calculations we have measured ν_Q as a function of pressure. However, these data by themselves complementing the temperature dependent measurements, provide already valuable information. Figure 2 shows $\nu_Q(P)$ for various sites in the 1-2-3 structures [8]. Similar results were obtained for Cu2 in 1-2-4 with a slope of $+190\pm20$ kHz/GPa [9]. Such a positive frequency shift may be interpreted as the result of a slight increase of the charge carrier concentration.

It turned out that it is impossible to describe with the same set of point charges both the temperature and pressure dependence of the NQR frequencies *and* the magnitude of the EFG at both Cu sites in 1-2-3-7 [2]. Thus the present experimental data which are more accurate than previous results, still wait for an adequate interpretation.

3. Knight Shifts

By combining NQR and NMR measurements one obtains the complete tensors of both the EFG and the Knight shift present at the Cu sites. The Knight shift which is caused by conduction electrons, is defined as the displacement of an NMR line with respect to the frequency in a non-conducting reference sample. Both the EFG and the Knight shift tensors are highly anisotropic in high-T_c superconductors.

Here we will briefly report on new precise measurements of the Knight shift tensor K at both Cu sites in the normal conducting state of oriented powder samples of the 1-2-4 superconductor [10]. These data replace earlier less precise values we had obtained from non-oriented

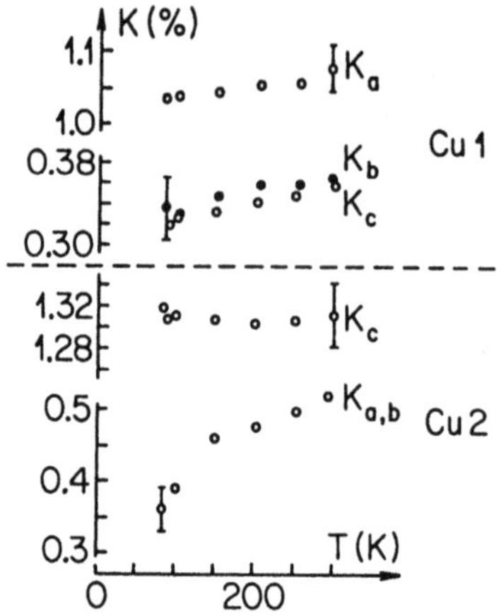

Fig. 3. Temperature dependence of the principal components of the ^{63}Cu Knight shift tensors for both Cu sites in $YBa_2Cu_4O_8$ (from Ref.[10]).

powder samples [4]. Figure 3 exhibits the temperature dependence of the various principal components of the K tensor for both Cu sites.

All components except $K_c(Cu2)$ decrease towards lower temperatures. This contrasts with the temperature independence (within the 0.02 % uncertainties) of the Knight shift components in 1-2-3 [11]. The Cu1 values in 1-2-4 are smaller than the corresponding values in the 1-2-3 structure [11,2] which contains only single Cu1 chains. The Cu2 tensors are nearly the same in both compounds.

The measurement of K provides useful microscopic information because K is proportional to the total electronic susceptibility and the hyperfine fields present at the nuclear site. The susceptibility χ contains spin, dipolar and diamagnetic contributions. We conclude that the dominant contribution to χ arises from Cu 3d spins in the $Cu - O_2$ planes [10]. By taking the orbital part of χ from data of the 1-2-3 structure [11] we can separate the contributions of Cu 3d and Cu 4s orbitals to the spin part of K. A more detailed analysis requires Knight shift measurements in the superconducting state; these experiments are in progress.

4. Spin-Lattice Relaxation Times

Measurements of the spin-lattice relaxation time T_1 microscopically probe the spin dynamics at the nuclear sites. As an example Figure 4 shows the relaxation rates of ^{63}Cu in 1-2-4 [4]. Important features of the data are: (i) the non-exponential temperature dependence of the rate below T_c which contrasts with the traditional BCS theory of NMR relaxation; (ii) the "smooth" transition of the Cu2 relaxation rate from the normal to the superconducting state without any discontinuity or abrupt change in the slope which is a striking difference to the behavior in 1-2-3 [3]. We

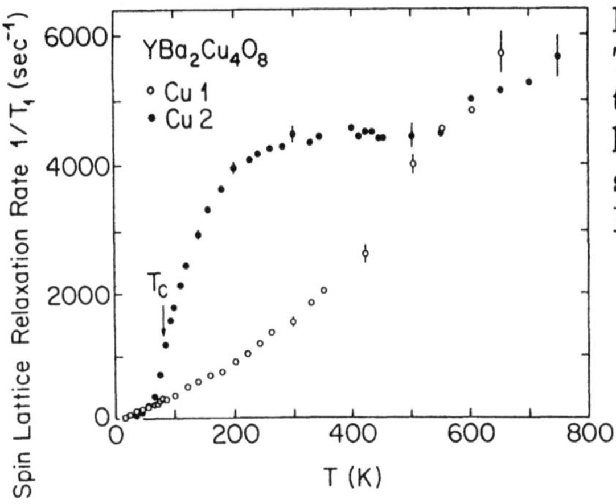

Fig. 4.
Temperature dependence of the ^{63}Cu NQR spin-lattice relaxation rates for both Cu sites in $YBa_2Cu_4O_8$ (from Ref.[4]).

have therefore tentatively described the Cu1 and Cu2 relaxation above T_c [4] in terms of a model [12] which postulates itinerant quasiparticles which interact strongly in the case of Cu2 and with no interaction for the Cu1 system. On the other hand, it seems also conceivable [10] that the strong enhancement of the Cu2 rate with respect to the so-called Korringa value [4] arises from antiferromagnetically correlated spin fluctuations at Cu2 sites. The decrease of $1/T_1$ by three orders of magnitude between 200 and 10 K then suggests the opening of a gap in the excitation spectra of these antiferromagnetic correlations. Further experimental and in particular quantitative theoretical investigations are necessary to clarify the nature of the spin dynamics.

References

1. See for instance: Proceedings of the Intern. Conf. on High Temperature Superconductors and Materials and Mechanisms of Superconductivity (M^2S – $HTSC$), Physica **153-155** (1988).

2. D. Brinkmann, Z. Naturforsch. **44a**, in press.

3. M. Mali, D. Brinkmann, L. Pauli, J. Roos, H. Zimmermann and J. Hulliger, Physics Letters **A 124**, 112 (1987).

4. H. Zimmermann, M. Mali, D. Brinkmann, J. Karpinski, E. Kaldis and S. Rusiecki, Physica **C 159**, 681 (1989).

5. H.P.Meister, M. Mali, J. Roos, H. Zimmermann and D. Brinkmann, to be published in Physica C.

6. H. Riesemeier, Ch. Grabow, E.W. Scheidt, V. Müller, K. Lüders and D. Riegel, Solid State Commun. **64**, 309 (1987).

7. C. Ambrosch-Draxl, P. Blaha, and K. Schwarz, Physica **C 162-164**, 1353 (1989).

8. Katharina Müller, M. Mali, J. Roos and D. Brinkmann, Physica **C 162-164**, 173 (1989).

9. H. Zimmermann, M. Mali, I. Mangelschots, J. Roos, D. Brinkmann, J. Karpinski, S. Rusiecki and E. Kaldis, submitted to European Materials Research Society (E-MRS) Conference 1990.

10. H. Zimmermann, M. Mali, I. Mangelschots, J. Roos, L. Pauli, D. Brinkmann, J. Karpinski, S. Rusiecki and E. Kaldis, submitted to European Materials Research Society (E-MRS) Conference 1990.

11. M. Takigawa, P.C. Hammel, R.H. Heffner, Z. Fisk, J.L. Smith and R.B. Schwarz, Phys. Rev. **B 39**, 300 (1989).

12. H. Monien, D. Pines and C.P. Slichter, preprint.

^{89}Y NMR Studies of the Electronic Properties of the 123 Compounds

H. Alloul

Laboratoire de Physique des Solides, Université Paris-Sud, F-91405 Orsay, France

Abstract. An extensive study of the ^{89}Y NMR in $YBa_2Cu_3O_{6+x}$ compounds has been performed for a broad range of temperatures and concentrations of oxygen, from the metallic to the insulating state. The ^{89}Y nuclei are found to probe the susceptibility of the CuO_2 planes through a shift of the ^{89}Y NMR associated with a coupling with the O(2p) orbitals of the planar oxygen sites. From the negative sign of the hyperfine coupling it is deduced that holes reside on Cu(3d)-O(2pσ) orbitals. Comparison of ^{17}O and ^{89}Y NMR shifts ensures that the magnetic properties of the system can be described by a single spin fluid. The ^{89}Y nuclear spin lattice relaxation time T_1 follows a Korringa law for the O_7 composition, and for $x<1$ a correlation between the variations of T_1T and K is evidenced. The ^{89}Y and ^{17}O T_1's are only weakly sensitive to the antiferromagnetic correlations on the copper sites, and the small q components of the spin fluctuations have a behaviour quite similar to that of a Fermi liquid.

1. Introduction

A determination of the mechanism of high temperature superconductivity in cuprates requires as a prerequisite a better understanding of how the local spin behaviour of the parent AF compounds [1] is modified by massive hole doping. Major questions concern of course the interaction between the doped holes and the local spins. Does the system result in a single band of correlated spins (as for the t-J model) [2], or can we consider that a band of doped holes interacts with local spins, as for local moments in metals? Some indications of the real situation can be expected from NMR measurements which directly probe the local susceptibilities [3-7]. Experiments performed in my group have demonstrated [3][4] that the ^{89}Y NMR allows one to probe the static and dynamic susceptibilities of the CuO_2 planes in $YBa_2Cu_3O_{6+x}$. A large T dependence of the spin susceptibility χ_s has been evidenced for oxygen depleted samples, and will be recalled in Sec.2. It implies some variation of the AF correlations with T, and provides an experimental capability of determining hyperfine fields. Their analysis, reported in Sec.3 allows deduction of the symmetry of the Cu(3d)-O(2p) hole orbitals. These results have stimulated further interest in ^{63}Cu and ^{17}O NMR measurements in depleted samples which have been reported since by two groups [8][9]. Comparison with our ^{89}Y data demonstrates that any susceptibility associated with an independent oxygen band is negligible and

Electronic Properties of High-T_c Superconductors and Related Compounds
Editors: H. Kuzmany · M. Mehring · J. Fink

confirms our former analysis in terms of a single band model (Sec.4). A better estimate of the ^{89}Y chemical shift gives a more accurate quantitative analysis of the correlation between the relaxation rate $(T_1)^{-1}$ and the NMR shift . In the metallic state these results can be interpreted (among others?) in a Fermi liquid model, with sizable AF electron electron correlations in order to explain the large enhancement of the ^{63}Cu relaxation rate [7] detected even in $YBa_2Cu_3O_7$ samples (Sec.5).

2. ^{89}Y NMR shift data

We report in Fig.1 the ^{89}Y NMR shift taken on various $YBa_2Cu_3O_{6+x}$ ceramic samples for x values ranging from the semiconducting to the metallic state. The position $^{89}\Delta K = {}^{89}\delta + {}^{89}K_s$ of the NMR signal involves a chemical shift contribution $^{89}\delta$ due to filled electronic shells and a spin contribution $^{89}K_s$ due to the spin susceptibility of the CuO_2 planes. The striking feature noticeable in Fig.1 is that, except for $YBa_2Cu_3O_7$, a large T variation of $^{89}\Delta K$ can be detected in all other samples. As detailed in ref.[4], $^{89}\Delta K$ scales linearly with the macroscopic susceptibility χ_m measured on clean samples. The Y site therefore probes all the T dependent part of χ_m, which can be partitioned into two terms

$$\chi_m = \chi_0(x) + \chi_1(x,T), \text{ and} \qquad (1)$$

$$^{89}\Delta K = {}^{89}\alpha\, \chi_1(x,T) + \beta(x) = {}^{89}K_s(x,T) + {}^{89}\delta(x). \qquad (2)$$

The data suggest that $^{89}K_s$ and $\alpha\chi_1$ coincide, but they might still differ by a T independent contribution as will be discussed in Sec.4. The **negative** hyperfine

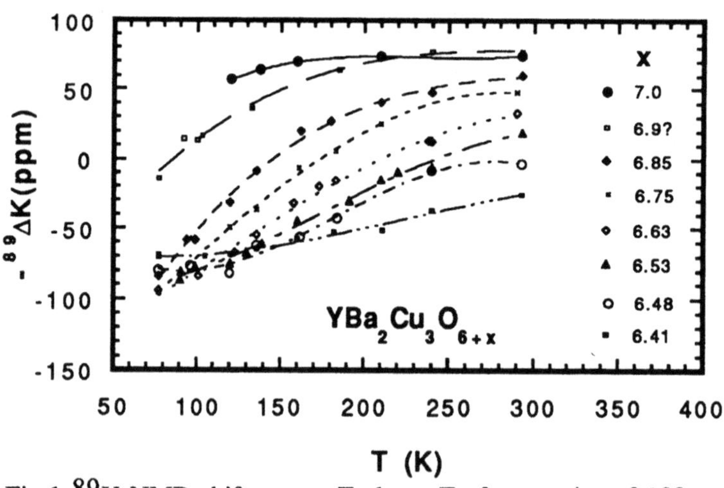

Fig.1 ^{89}Y NMR shift versus T above T_c for a series of 123 samples, from the metallic to the semiconducting state. The lines are guides to the eye. The zero of the susceptibility corresponds to $-{}^{89}\Delta K = -{}^{89}\delta = -200$ppm.

coupling $^{89}\alpha$ of the Y nuclei with the susceptibility of the CuO_2 planes is independent of x within experimental accuracy. As for $^{89}\delta$, Dupree and Smith [10] have measured ^{89}Y chemical shifts in a variety of non-magnetic yttrium oxides, which yields a rather large window $150<^{89}\delta<360$ ppm for ^{89}Y in $YBa_2Cu_3O_{6+x}$.

3. The hyperfine coupling

The ^{89}Y nuclear spin should quite generally sense the electronic spin susceptibilities $\chi_s(Cu)$ and $\chi_s(O)$ on its eight neighbouring coppers and oxygens, through two distinct hyperfine couplings $^{89}\alpha(Cu)$ and $^{89}\alpha(O)$,

$$^{89}K_s = 8\,^{89}\alpha(Cu)\,\chi_s(Cu) + 8\,^{89}\alpha(O)\,\chi_s(O). \tag{3}$$

A quick review of the ^{89}Y results might lead one to conclude that the direct hyperfine coupling $^{89}\alpha(Cu)$ is dominant. This has been done explicitly by Butaud et al. in their comparison between our ^{89}Y data for x=0.63 and their ^{17}O results taken for x=0.65 [8]. I want to insist here on the fact that the Y and Cu atomic orbitals (AO) are far enough apart to yield a negligible $^{89}\alpha(Cu)$. Indeed, as the $Cu(3d_{x^2-y^2})$ orbital has no overlap by symmetry with the Y(5s) AO, the only significant contribution to $^{89}\alpha(Cu)$ is through $Cu(3d_{x^2-y^2})$-Y(4d) hybridization. A simple quantum chemistry calculation allows one to estimate a value for $^{89}\alpha(Cu)$ about three orders of magnitude smaller than the measured $^{89}\alpha$ [11].

The dominant isotropic coupling of the ^{89}Y nuclear spins with the CuO_2 planes is then through the less distant and more extended O(2p) orbitals [3][12]. A direct overlap between the Y(5s) AO and O(2p) AO, which yields a large **positive** hyperfine field $^{89}\alpha(O)$, only occurs for oxygen orbitals pointing towards the Y nuclei [12]. Therefore the data exclude any sizable hole occupancy of these π bonding orbitals. On the other hand, O(2pσ) AO have no direct overlap with Y(5s) AO by symmetry and might then only be coupled with the Y through **a negative polarization** of the internal AO. The largest negative contribution to $^{89}\alpha(O)$ can be expected from an exchange polarization of the inner filled O(2pπ) orbitals which strongly overlap with Y(5s), induced by the hole spin density on the O(2pσ) orbitals. Our data, together with the ^{17}O NMR shift on the planar sites [6][9], favor a Cu(3d)-O(2pσ) bonding of the hole orbitals, which is the starting point for most theoretical models of the electronic structure of the CuO_2 planes.

4. One or two bands?

The linear relation found between $^{89}\Delta K$ and χ_m proves that the dominant contribution to the spin polarization on the oxygen holes scales with $\chi_s(Cu)$. This spin polarization is either induced by Cu(3d)-O(2pσ) covalency or (and) by an indirect RKKY coupling if the O(2p) holes have an independent degree of freedom. In this latter case a T independent contribution $\chi_s^h(O)$ of the O(2p) band is expected

$$^{89}K_s = {}^{89}\Delta K - {}^{89}\delta = {}^{89}K_s{}^h + {}^{89}K_s{}^t(Cu) = 8\alpha(O)[\chi_s{}^h(O) + \gamma\chi_s(Cu)]. \quad (4)$$

The total planar spin susceptibility is then $\chi_s = (1+2\gamma)\,\chi_s(Cu) + 2\chi_s{}^h(O)$, and we might therefore identify $\chi_1(x,T)$ with $(1+2\gamma)\,\chi_s(Cu)$.

To decide whether $\chi_s{}^h(O) \neq 0$, we need an independent determination of $^{89}\delta$ and of the zero of the transferred term $^{89}K_s{}^t(Cu)$. We can take advantage of the ^{17}O NMR data taken on T_c=60 K samples, as the ^{17}O probes $\chi_s(Cu)$ and $\chi_s{}^h(O)$ on different footings. Indeed, the dominant isotropic positive contribution to $^{17}K_s$ is due to the non-zero Cu(3d)-O(2s) admixture, and therefore

$$^{17}K^{iso} - {}^{17}\delta^{iso} = {}^{17}K_s{}^{iso} = {}^{17}\alpha_{2s}\,\chi_s\,(Cu). \quad (5)$$

The O(2pσ) orbitals couple to the ^{17}O nuclear spin through the traceless dipolar interaction, which yields an anisotropic contribution $^{17}K_s{}^{ax}\,(3\cos^2\theta-1)$, with

$$^{17}K^{ax} - {}^{17}\delta^{ax} = {}^{17}K_s{}^{ax} = {}^{17}\alpha_{dip}\,\chi_s(O) = {}^{17}\alpha_{dip}\,[\chi_s{}^h(O) + \gamma\,\chi_s(Cu)]. \quad (6)$$

The data reported by the two groups differ inasmuch as Takigawa [9] finds that $^{17}K^{ax}$ and $^{17}K^{iso}$ have similar T dependences, while Butaud et al.[8] find that $^{17}K^{ax}$ is nearly T independent. Comparing $^{89}\Delta K$ with the latter result would imply $\chi_s{}^h(O) > \chi_s(Cu)$, from Eq.(4), and a huge value for $^{89}\delta$ well above the 360 ppm upper limit obtained on yttrium oxides!

Such a discrepancy does not occur for the other set of data for which $^{17}K^{ax}$ scales linearly with our $^{89}\Delta K$ data for x = 0.63 as displayed in Fig.2(a). As both spin susceptibilities are expected to vanish in the superconducting state at T = 0, Takigawa et al. obtain $^{17}\delta^{ax}$ = 0.013% from their data for T $\to$0 [9]. This allows us to deduce

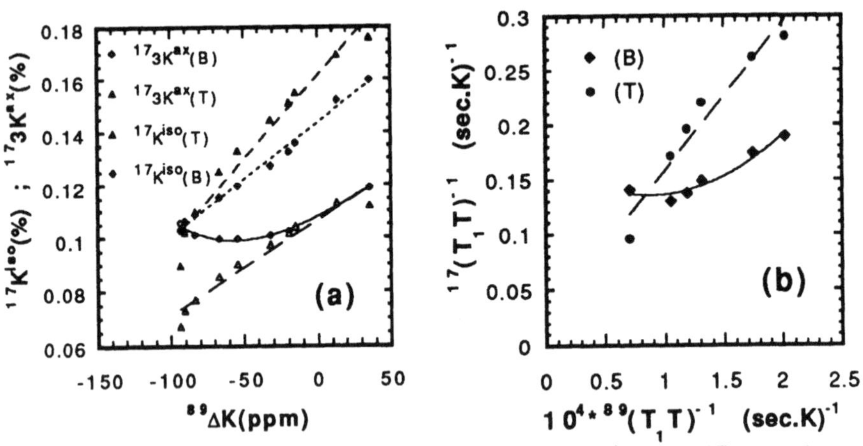

Fig.2 Plot of interpolated values of (a) $^{17}K_s{}^{ax}$ and $^{17}K^{iso}$,(b) $^{17}(T_1T)^{-1}$, versus respectively $^{89}\Delta K$ and $^{89}(T_1T)^{-1}$. While the data of ref [9] (T) always scale linearly with our results on ^{89}Y, those of ref [8] (B) display anomalies for $^{17}K_s{}^{ax}$ and $^{17}(T_1T)^{-1}$ at low T. The dotted lines are linear least squares fits, while the full lines are guides to the eye.

$^{89}\delta = 200$ ppm from Eqs. (4) and (6). An estimate of $^{89}K_s^h(O)$ can then be obtained by a comparison of $^{89}\Delta K$ and $^{17}K^{iso}$ from Eqs. (4) and (5). Although some difference between the data for $^{17}K^{iso}$ can again be seen in Fig.2(a), linear relationships between $^{89}\Delta K$ and $^{17}K^{iso}$ are obtained for both results. From $0 < {}^{17}\delta^{iso} < 0.04\%$ [6][9], we deduce $190 < {}^{89}\delta + {}^{89}K_s^h(O) < 260$ ppm, from Takigawa's result. Even with Butaud's result, $^{89}K_s^h(O)$ can hardly be negative and would correspond to a very weak susceptibility $\chi_s^h(O)$. Overall, our ^{89}Y data and the results of Takigawa can be explained coherently in a single spin fluid picture, with $\chi_s^h(O) = 0$, $^{89}\delta = 200$ppm, $^{17}\delta^{iso} = 0.035\%$, and $^{17}\delta^{ax} = 0.013\%$.

5. Nuclear spin lattice relaxation

The fluctuating components of the local field induced by the Cu spins yields a contribution to the spin lattice relaxation rate given quite generally by [13-16]

$$(T_1T)^{-1} = \gamma_n^2 \alpha^2 z^2 k_B \sum_q f^2(q) \; \mathrm{Im}\,\chi(q, \omega_n) / \omega_n. \tag{7}$$

The various nuclear sites have different symmetries in the CuO_2 lattice, which results in form factors $^{17}f^2(q) = \cos^2(q_{x(y)}a/2)$ for ^{17}O, coupled to z=2 Cu spins on the x(y) axis, $^{89}f^2(q) = \cos^2(q_x a/2)\cos^2(q_y a/2)\cos^2(q_z d/2)$ for ^{89}Y coupled to z=8 Cu spins located in two CuO_2 planes separated by a distance d.

These form factors vanish for ^{89}Y and ^{17}O at the wave vector $Q_{AF} = (\pi/a, \pi/a, 0)$ of the AF structure of $YBaCuO_6$, while it is non-zero for ^{63}Cu. From the q dependences of f(q) it can be understood that the sensitivity to a singularity at $q = Q_{AF}$ of the magnetic response of the Cu spin system decreases from ^{63}Cu to ^{17}O to ^{89}Y[14].

The results for ^{17}O differ again qualitatively as Takigawa finds that $^{17}T_1^{-1}$ decreases monotonously with decreasing T, while Butaud et al. find that $^{17}T_1^{-1}$ decreases down to 120K and then shows a maximum at about 80K, associated with an independent relaxation mechanism, as suggested by the authors. This difference is seen in the plot of Fig.2(b) where the $^{17}(T_1T)^{-1}$ data are plotted versus the ^{89}Y data for x=0.63, with T as an implicit parameter. The results of Takigawa scale linearly with ours, as can be expected from the discussion above. The absence of intercept, within experimental accuracy, agrees with the previous conclusion that any contribution from an oxygen band is negligible. On the other hand, the data of Butaud depart from a linear behaviour as the extra contribution to $^{17}(T_1T)^{-1}$ grows at low T.

As done in ref [4], we can now proceed to a quantitative comparison of the "Korringa products" $S = T_1TK_s^2 = \kappa S_0$, where $S_0 = z\hbar(\gamma_e/\gamma_n)^2/(4\pi k_B)$ is the free electron value (9.24×10^{-4}secK for ^{89}Y). If the form factors $f^2(q)$ filter completely the $q = Q_{AF}$ components of the spin susceptibility for both nuclei, κ should be the same for O and Y. For $YBaCuO_7$, we have $^{17}S = {}^{17}(T_1TK_s^{iso\,2}) \sim 1.02\times10^{-5}$ sec.K [6] and $^{89}S \sim 3.5\times10^{-4}$sec.K with $^{89}\delta = 200$ppm, which yields $^{89}\kappa / {}^{17}\kappa \sim 1.1$. Similarly from the slopes of $^{17}(T_1T)^{-1}$ versus $^{89}(T_1T)^{-1}$ and ^{17}K versus $^{89}\Delta K$ obtained in Fig.2, we deduce $^{89}\kappa / {}^{17}\kappa = 1.2$. This ratio just exceeds unity, which might indicate that AF correlations are not fully screened.

It is interesting to notice that for the composition $YBaCuO_7$ for which T_c is highest, χ_m and χ_s are T independent, and both Y and O follow a T_1T=constant Korringa-like relation with $^{89}\kappa \sim 0.38$ rather close to the free electron value, but smaller than unity. For depleted samples the correlation between T_1T and K_s is nearly T_1TK_s = constant for O [9] as well as for Y, with the actual estimate of $^{89}\delta$. This can be seen in Fig.3 where the data of ref [4] for $^{89}(T_1TK_s)$ for all our samples with $0.41 < x < 1$, taken with $^{89}\delta$ =200ppm, have been plotted versus $^{89}\Delta K$.

6. Discussion

In the usual description in which one assumes an exponential time decay τ for the spin-spin correlation, with a q independent static susceptibility, near q=0, one writes

$$\sum_q \mathrm{Im}\chi(q,\omega)/\omega = \chi(q=0)\tau \;, \quad \text{which, for} \quad K_s = z\,\alpha\,\chi(0) \quad \text{yields}$$

$$S = S_0\,(4\pi/\hbar\,\gamma_e^2)\;\chi(q=0)/\tau\,. \tag{8}$$

So the finding of a Korringa relation mainly proves that the spin susceptibility $\chi(q=0)$ and the correlation rate τ^{-1} are driven by a single energy scale (E_F in simple metals). Such a property is usually found in Fermi liquids, for instance for Kondo impurities in metals at low T (lower than the Kondo temperature T_K), for which $\chi \sim (\hbar\gamma_e)^2 / k_BT_K$, and $\hbar/\tau \sim k_BT_K$ [17].

An "extended" Korringa relation, in the presence of a T dependence of χ, is only expected if $\chi_s(T)$ results from a fine detail in the energy dependence of the density of states. If $\chi_s(T)$ is due to ferromagnetic electron correlations, Moriya and Ueda [18] have calculated $T_1TK_s^{\mu}$=constant with $\mu=1$, which is indeed observed in 3D [19]. For lower dimensionalities RPA calculations by Bourbonnais et al. [20] give $\mu=2$ for 1D, and $\mu = 3/2$ for 2D.

Here we rather consider a system in which the electronic correlations are antiferromagnetic. In such a case, one expects a decrease of the weight of the low q part of $\chi(q,\omega)$, that is a decrease of $(T_1T)^{-1}$ and $\chi(0)$ with respect to the non-interacting electron gas. If both quantities are reduced similarly, one does expect a decrease of the Korringa product with increasing AF correlation.This has been found in RPA calculations by Bulut et al. [14], and we indeed obtain $S<S_0$ for x=1.

The increase of AF correlations is probably at the origin of the large decrease of $\chi(T)$ detected in depleted samples ($\chi(T)$ is decreased by a factor of about 4 at T_c with respect to its room temperature value for the x=0.63 sample). Correspondingly the progressive decrease of T_1TK_s at low T, or with decreasing x, seen in Fig.3 can be understood as resulting from a strengthening of the AF correlations.

Most of these features may then be described within a band picture relevant for a nearly antiferromagnetic itinerant system, although a more quantitative analysis for varying x, similar to that developed by Millis et al. [16] for x=1, is still required. This might not be the only possible description, and nature could yield similar answers for completely different starting points. This has been suggested for instance by Shastry [13], who pointed out that a constant Korringa product, with nearly the

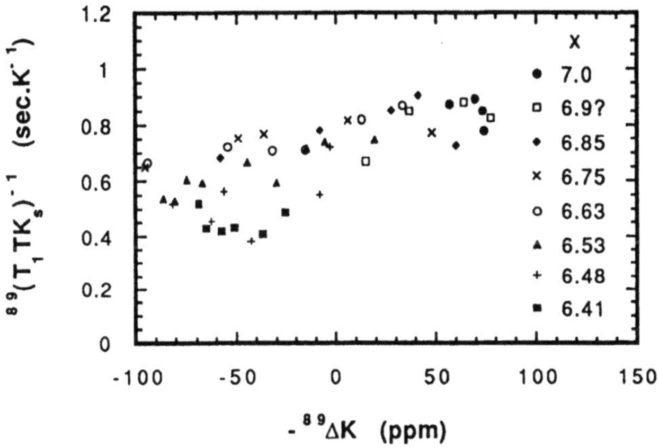

Fig.3 $^{89}(T_1TK_s)^{-1}$ versus $-^{89}\Delta K$ for a series of 123 samples, from the semiconducting to the metallic state. This quantity smoothly increases with x and $-^{89}\Delta K$, that is, with decreasingly AF correlations.

same value as for free electrons, can even be expected on the O (and Y) sites in the paramagnetic state of the 2D AF insulator! The observation of a Korringa relation for the O_7 sample, might in such a case be a pure coincidence, which would only point out that the statics and the dynamics of the spin system at low q are governed by a single energy scale. However, better experimental knowledge of the magnetic behaviour of the non-metallic samples is certainly required before concluding firmly that a Fermi liquid behaviour with AF correlations occurs in the metallic state.

I would like to acknowledge here those who have been involved in the NMR experiments, namely P.Mendels, T.Ohno, and more recently H.Casalta, as well as J.Arabski, G.Collin and J.F.Marucco, who, in addition to their critical suggestions for the physics, have been very patient in letting us benefit from all their knowledge in sample preparation, oxidation, thermal treatments, and X-ray diagrams. I have also greatly benefited from various discussions with C. and Y. Berthier, and M.Takigawa, who were kind enough to communicate their ^{17}O results prior to publication.

References

[1] R.J.Birgeneau and G.Shirane, Physical Properties of High Temperature Superconductors, D.M.Ginsberg Editor (World Scientific Publishing 1989).

[2] F.C.Zhang and T.M.Rice, Phys.Rev.B**37**, 3759 (1988).

[3] H.Alloul, P.Mendels, G.Collin and P.Monod, Phys.Rev.Lett.**61**, 746 (1988).

[4] H.Alloul, T.Ohno and P.Mendels, Phys.Rev.Lett.**63**, 1700 (1989).

[5] M.Takigawa, P.C.Hammel, R.H.Heffner and Z.Fisk, Phys.Rev.B**39**,7371(1989); M.Horvatic, Y.Berthier, P.Butaud, Y.Kitaoka, P.Segransan, C.Berthier, H.Katayama-Yoshida, Y.Akabe and T.Takahashi, Physica **C159**, 689 (1989).

[6] M.Takigawa, P.C.Hammel, R.H.Heffner, Z.Fisk, K.C.Ott, and J.D.Thompson, Phys.Rev.Lett. **63**,1865(1989).

[7] R.E.Walstedt, W.W.Warren, R.F.Bell, G.F.Brennert, G.P.Espinosa, R.J.Cava, L.F.Schneemeyer and J.V.Waszcak, Phys.Rev. **B38**,9299(1988).
[8] P.Butaud, M.Horvatic, Y.Berthier, P.Segransan, Y.Kitaoka, C.Berthier, Physica C166,301 (1990).
[9] M.Takigawa, Proceedings of Nato Workshop on the Dynamics of Magnetic Fluctuations in High Temperature Superconductors. Alghia Pelagia Greece 1989.
[10] R.Dupree and M.E.Smith, Chem.Phys.Lett.**148**, 41 (1988).
[11] F.Mila (private communication).
[12] F.J.Adrian, Phys.Rev.Lett.**61**,2148(1988); H.Alloul, ibid. **63**,690(1989); F.J.Adrian,ibid. **63**,691(1989).
[13] B.S.Shastry, Phys.Rev.Lett.**63**,1288(1989).
[14] N.Bulut, D.W.Hone, D.J.Scalapino, and N.E.Bickers, Phys.Rev.**B41**, 1797(1990).
[15] F.Mila and T.M.Rice, Phys.Rev.**B40**,11382(1989).
[16] A.J.Millis, H.Monien, and D.Pines (preprint).
[17] H.Alloul, Phys.Rev.Lett.**34**,460(1975);Physica **86-88B**,449(1977).
[18] T.Moriya and K.Ueda, Sol. State Commun.**15**,169(1974).
[19] H.Alloul and L.Mihaly, Phys.Rev.Lett.**48**,1420(1982).
[20] C.Bourbonnais, P.Wzietek, F.Creuzet, D.Jérome, P.Batail, and K.Bechgaard, Phys.Rev.Lett. **62**,1532(1989).

^{17}O NMR Investigation of the Electronic Structure of High-T_c Superconducting Oxides

C. Berthier[1], *Y. Berthier*[1], *P. Butaud*[1], *M. Horvatić*[2], *and P. Ségransan*

[1]Laboratoire de Spectrométrie Physique, Université J. Fourier Grenoble I, (LA08 CNRS), BP. 87, F-38042 St Martin d'Hères, France

[2]Institute of Physics of the University, P.O. Box 304, YU-41001 Zagreb, Yugoslavia

Abstract. We discuss the possibility of descriminating between the t-J and two band models by comparing ^{17}O and ^{89}Y NMR data in $YBa_2Cu_3O_{6+x}$. We conclude the existence of an O(2p) band weakly coupled to the Cu(2) electronic moments for $T>T_c$ in $YBa_2Cu_3O_{6.65}$.

The existence of quasi-localized electronic spins 1/2 at the Cu(2) sites (CuO_2 plane) in $YBa_2Cu_3O_{6+x}$ for x values larger than 0.4 (i.e. for oxygen concentration corresponding to high T_c superconductors) is now fairly well established from NMR experiments: this includes a careful analysis of the the nuclear spin-lattice relaxation rate (NSLRR) and the magnetic hyperfine shift (MHS) tensor **K** in $YBa_2Cu_3O_7$ [1,2] and the x dependence of the NSLRR and the MHS K_{cc}.[3-5] This was recently confirmed by neutron inelastic scattering in $YBa_2Cu_3O_{6+x}$ for x = 0.5, 0.46 and 0.52.[6-8] From this starting point, a crucial issue for the understanding of the low energy part of the electronic structure of the high T_c superconducting oxides is to determine the strength of the coupling between the electron spins of the mobile holes occupying the O(2p) orbitals and those borne by the Cu(2) atoms. Below $T_{c,}$ the temperature dependence of the Cu(2) and the O(2,3) NSLRR in $YBa_2Cu_3O_7$ [9] and that of the O(2,3) MHS tensor **K** [10] show that, in this compound at least, both types of spins are strongly coupled and should be described by a single degree of freedom. In Fig.1 for instance we show the correlation between the temperature dependence of the O(2,3) MHS K_{cc} [10] and that of the Cu(2) NSLRR as measured from NQR. [11]. Above T_c the situation is less clear. Zhang and Rice [12] have proposed that a hole introduced by doping should delocalize on the four oxygens neighbouring a Cu(2) and form a spin singlet state with the localized d-hole. Such a mechanism would justify the description of the system by a "single effective band " in the so-called t-J model.[13] Some other authors have proposed that the coupling between the two spin species should not be so strong, at least above $T_{c,}$ so that the system should be rather described by a two band model in the normal state. [14,15]

One method to discriminate between these two descriptions is to compare the temperature dependence of the isotropic and the axial part of the O(2,3) MHS tensor. These two quantities reflect the respective polarization of the O(2s) and the O(2p) orbitals. Within the t-J model, one expects [10]

$$K_{iso}^{spin} = 2f_s A^{2s} \chi_{Cu} \qquad \text{and} \qquad K_{ax}^{spin} = 2f_p A^{2p} \chi_{Cu} \,, \qquad (1)$$

where f_s (fp) are respectively the fraction of spin polarized O(2s) (O(2p)) orbitals per

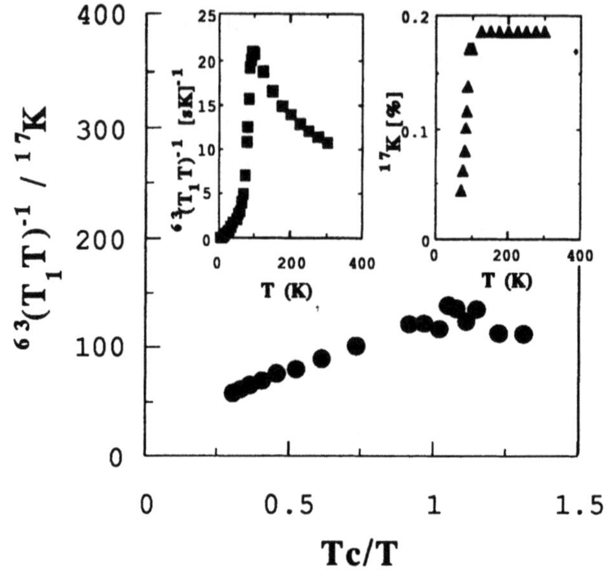

Fig.1 The temperature dependence of the Cu(2) NSLRR $(T_1T)^{-1}$ [11] and that of the O(2,3) MHS K_{cc} [10] in $YBa_2Cu_3O_7$ are shown in the left and right insets, respectively. Their ratio, which is proportional to $\chi''(q,\omega_n)/\omega_n\chi(0,0)$, is plotted as a function of T_c/T and is seen to be roughly constant below T_c.

Cu(2) moment, $A^{2s(2p)}$ are the atomic hyperfine field, and χ_{Cu} is the spin susceptibility per Cu(2) atom, whereas in the two band model,

$$K_{iso}^{spin} = 2f_s A^{2s} \chi_{Cu} \qquad \text{and} \qquad K_{ax}^{spin} = A^{2p}\,(2f_p \chi_{Cu} + \chi_{holes})\,. \quad (2)$$

In those compounds where the macroscopic susceptibility is strongly temperature dependent, like $YBa_2Cu_3O_{6.65}$ or $La_{1.85}Sr_{0.15}CuO_4$, it has been shown that the Cu(2) moments are responsible for this temperature dependence.[16,17] Thus the ratio $K_{ax}^{spin}(T)\,/K_{iso}^{spin}(T)$ must be temperature independent in the t-J model and equal to $f_p\,A^{2p}/f_s\,A^{2s}$, whereas in the two band model, this ratio is temperature dependent. We have measured the temperature dependence of $^{17}K_{iso}$ and $^{17}K_{ax}$ in the compound $YBa_2Cu_3O_{6.65}$.[17] Our results distinctly show a large difference in the temperature dependence of both quantities. These results would therefore favour the two band model, but the final interpretation relies too strongly on the separation of the experimental values K_{iso} and K_{ax} into their spin and the orbital (Van-Vleck) contributions to be definitely conclusive. The same restrictions apply to the data of Takigawa [18], which rather favour the t-J model.

We have thus proposed another method to discriminate between the t-J and the two band models, which is based on the comparison of the O(2,3) and the Y NMR data [19](including NSLRR and MHS data). Within the t-J model, the relaxation rates of the O(2,3) and Y nuclei are governed by a single dynamical spin susceptibility, that of the Cu(2) electronic moments. Neglecting the anisotropic contributions to the NSLRR one has

$$ {}^{s}(T_1T)^{-1} = {}^{s}z\left(\frac{{}^{s}A}{\hbar}\right)^2 \frac{k}{N}\sum_q {}^{s}f_q \frac{\chi''(q,\omega_n)}{\omega_n} \quad \text{and} \quad {}^{s}K_{iso} = \frac{{}^{s}z\,{}^{s}A}{\gamma_n \hbar g \mu_B}\chi_{Cu}(0,0), \quad (3) $$

where s specifies the observed nuclei (17 (89) for O (Y)), ${}^{s}A$ is the transferred hyperfine field from next-neighbour Cu spins, ${}^{s}z$ is the number of Cu first neighbours, ${}^{s}f_q$ is a form factor reflecting the symmetry of the position of the s nuclei with respect to the Cu spins, $\chi''(q,\omega_n)$ ($\chi(0,0)$) are respectively the imaginary part at q and $\omega = \omega_n$, and the nuclear Larmor frequency (the real part at q=0 and $\omega = 0$) of the dynamical spin susceptibility.
For O(2,3) nuclei, ${}^{17}f_q = (1+\cos\ q_{x(y)}a\ )$ and $z = 2$, and for yttrium nuclei, ${}^{89}f_q = (1+\cos q_z a)\ (1+\cos q_x a)\ (1+\cos q_y a)$ and $z = 8$. Both f_q vanish for the wavevector of the antiferromagnetic fluctuations $Q = (\pi/a, \pi/a)$ observed by neutron inelastic scattering. Thus, whatever the temperature dependence of the quantity $(K^2_{iso}\ T_1T)$, one has the relationship

$$ \frac{1}{{}^{17}(T_1T)} = \frac{{}^{17}z}{{}^{89}z}\left(\frac{{}^{17}\gamma_n}{{}^{89}\gamma_n}\right)^2 \eta \left(\frac{{}^{17}K_{iso}}{{}^{89}K_{iso}}\right)^2 \frac{1}{{}^{89}(T_1T)} = C\eta\ \frac{1}{{}^{89}(T_1T)}\ , \quad (4) $$

where

$$ \eta = \frac{\sum_q {}^{17}f_q\,\chi''(q,\omega_n)}{\sum_q {}^{89}f_q\,\chi''(q,\omega_n)}\ . $$

Within the two band model, one has to add the contribution of the O(2p) band (which we take to be temperature independent) to the MHS and the NSLRR of both yttrium and O(2,3) nuclei. One has for yttrium

$$ {}^{89}K_{iso} = 8\ {}^{O\text{-}Y}A\ (2f_p\chi_{Cu} + \chi_{holes}) \quad \text{and} \quad {}^{89}(T_1T)^{-1}_{exp} = {}^{89}(T_1T)^{-1}_{Cu} + {}^{89}(T_1T)^{-1}_{holes} $$

where ${}^{O\text{-}Y}A$ is the transferred hyperfine field from O(2,3) to yttrium per oxygen spin

and for O(2,3)
$$ {}^{17}(T_1T)^{-1}_{exp} = \lambda\ {}^{89}(T_1T)^{-1}_{Cu} + \mu\ {}^{89}(T_1T)^{-1}_{holes} $$

where $\lambda = C\eta$ from Eq. (4) and $\mu \approx \left(\frac{\langle \mu_B / r^3\rangle_{17}}{{}^{O\text{-}Y}A}\right)^2$.
We thus get the simple relationship

$$ {}^{17}(T_1T)^{-1}_{exp} = C\eta\ {}^{89}(T_1T)^{-1}_{exp} + (\mu - C\eta)\ {}^{89}(T_1T)^{-1}_{holes}\ . \quad (5) $$

We have plotted our experimental values of ${}^{17}K_{iso}$ in $YBa_2Cu_3O_{6.65}$ [17, 20] versus the experimental values of ${}^{89}K$ in $YBa_2Cu_3O_{6.63}$ [19] with the temperature as an implicit parameter. This procedure allows an accurate determination of the ratio ${}^{17}K_{iso}/{}^{89}K_{iso}$, which is found equal to 4.84, without the need to know precisely the orbital contribution to the MHS, since these latter are temperature independent.

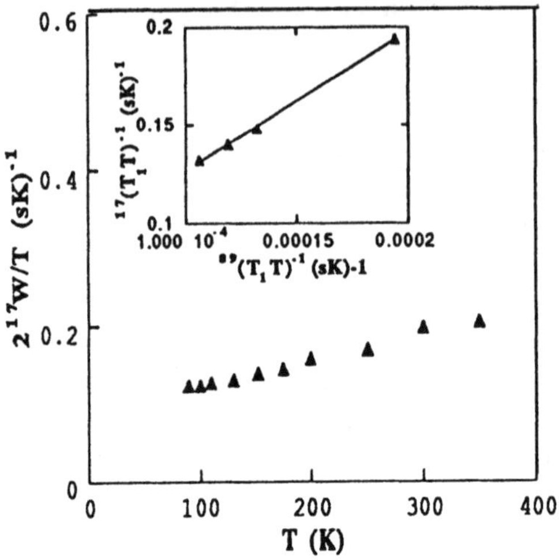

Fig.2 Temperature dependence of the O(2,3) NSLRR $^{17}(T_1T)^{-1}$ in $YBa_2Cu_3O_{6.65}$ [17]. The inset shows that $^{17}(T_1T)^{-1}$ scales linearly with $^{89}(T_1T)^{-1}$ in $YBa_2Cu_3O_{6.63}$[19]. The value of the slope indicates that the correlation length of the AF fluctuations is very short. ($\xi/a \approx 1$).

The value of the constant C in Eqs. (4) and (5) is thus found equal to 710. In Fig. 2, we have plotted the exprimental values of $^{17}(T_1T)^{-1}$ versus temperature in $YBa_2Cu_3O_{6.65}$ and the inset shows the same quantity versus $^{89}(T_1T)^{-1}$ as measured by Alloul et al. in $YBa_2Cu_3O_{6.63}$.[19] As expected in both models, the data scale linearly together, and from the slope of the linear regression, we found $C\eta$ equal to 720. This means that η **is equal to 1.** Such a quantity has been estimated by Millis et al. [21] for an RPA approximation of $\chi''(q,\omega_n)$ as a function of the correlation length ξ/a and $\chi''(Q_{AF}, \omega_n)/\omega_n\, \chi(0,0)$, and it turns out that such a result implies in the framework of their model a surprisingly short correlation length ($\xi/a \approx 1$). This plot also clearly demonstrates the existence of a contribution from an O(2p) band, which is worth at least 0.06 $s^{-1}.K^{-1}$

However, before claiming a definite conclusion we must keep in mind that we have compared ^{17}O and ^{89}Y data coming from two different samples; this type of analysis should now be applied to data collected from the same sample.

Acknowledgments

The Laboratoire de Spectrométrie Physique is "associé au Centre National de la Recherche Scientifique". This work was partly supported by the D.R.E.T. (Contract N° 89.34.045). We would like to thank H.Alloul and D.L. Cox for stimulating discussions and comments.

References

[1] C.H. Pennington, D.J. Durand, C. P. Slichter, J.P. Rice, E. D. Bukowski, and D. M. Ginsberg, Phys. Rev. **B39**, 274 (1989); ibid., 2902 (1989).

[2] S.E. Barret, D.J. Durand, C.H. Pennington, C.P. Slichter, T.A. Friedman, J.P. Rice and D.M. Ginsberg, (preprint).

[3] M. Horvatić, P. Ségransan,C. Berthier, Y. Berthier, P. Butaud, J.Y. Henry, M. Couach, J. P. Chaminade, Phys. Rev. **B39**, 7332 (1989)

[4] M. Horvatić et al., Physica C **162-4**, 265 (1989)

[5] M. Horvatić et al., Physica **C166**, 151 (1990)

[6] J. Rossat-Mignod et al., Proceedings of the International Conference on the Physics of Highly Correlated Electron Systems, Santa Fe, Sept. 89, to be published.

[7] J. Tranquada et al., (preprint).

[8] J. Rossat-Mignod et al., to appear in"Dynamics of Magnetic Fluctuations in High T_c Materials", eds. G. Reiter, P. Horsh and G.Psaltakis (Plenum Press, New York 1990).

[9] P. C. Hammel, M. Takigawa, R. H. Heffner, Z. Fisk and K. C. Ott, Phys. Rev. Lett. **63**, 1992 (1989) .

(10) M. Horvatić et al. , Physica **C159**, 689 (1989).

[11] T. Imai et al., J. phys. Soc. Jpn. **57**, 1771 (1988)

[12] F. C. Zhang and T. M Rice, Phys. Rev. **B37**, 3759 (1988).

[13] P.W. Anderson, in "Frontiers and Borderlines in Many Particle Physics" "Enrico Fermi" Course CIV, eds. R.A. Broglia and J.R. Schrieffer (North-Holland, Amsterdam 1988).

[14] N. Andrei and P. Coleman, Phys. Rev. Lett. **62**, 595 (1989) and refs therein.

[15] D.L.Cox and B.R.Trees, (preprint)

[16] Y. Kitaoka et al., Physica **C162/4**, 195 (1989)

[17] P. Butaud et al., Physica **C166**, 301 (1989)

[18] M. Takigawa, ibid. Ref. 8

[19] H. Alloul, T. Ohno, P. Mendels, Phys. Rev. Lett. **63**, 170 (1989)

[20] C. Berthier et al., ibid. Ref. 8

[21] A.J. Millis, H. Monien and D. Pines, (preprint)

Magnetic Order in 123 Compounds as Seen by Cu NMR and NQR

H. Lütgemeier

IFF – Forschungszentrum (KFA) Jülich, Postfach 1913,
D-5170 Jülich, Fed. Rep. of Germany

Abstract. The antiferromagnetic order of Cu in the 123 compounds $REBa_2Cu_3O_y$ of Y and the rare earth metals heavier than Pr which do not disturb the superconductivity, and of Pr which destroys superconductivity even at full oxygen loading is investigated by Cu NMR and NQR. In both cases magnetic order exists only at the Cu(2) sites. The EFG tensors of the Cu(1) sites with 2, 3 and 4 oxygen neighbours agree in both cases and indicate the same valence state of Y and Pr in these compounds.

1. Introduction

In the 123 compounds $REBa_2Cu_3O_y$ where RE is Y or one of the rare earth metals heavier than Pr, the superconducting (SC) and metallic behavior is destroyed and antiferromagnetic (AF) order in the Cu sublattice is induced if the oxygen concentration is reduced to $y<6.4$. An exception is Pr. Here a similar transition occurs already at full oxygen loading if more than 50% of Y is replaced by Pr [1,2]. The type of AF order, however, is not yet known in this case.

Since magnetic order at the Cu sites can be detected in nuclear resonance by the hyperfine field at the Cu sites carrying the magnetic moments, we have investigated the nuclear magnetic resonance (NMR) and quadrupole resonance (NQR) spectra of Cu in the Pr123 compounds with both high and low oxygen concentration. The experiments were performed at 4.2 and 1.2 K by conventional spin–echo technique with powder samples prepared by the ceramic technique.

In the first part of this contribution the results obtained in the normal case of Y and the heavier RE metals are summarized. In the second part, the spectra of Pr123 are presented.

2. Cu Spectra in Normal 123 Compounds

The two isotopes ^{63}Cu and ^{65}Cu with $I=3/2$ possess similar magnetic dipole and electric quadrupole moments. The electric field gradient (EFG) tensors of the Cu(1) sites ("chains") and Cu(2) sites ("planes") reflecting the charge distribution reveal different symmetry properties. They are described by the largest eigenvalues, the asymmetry parameter, η, and the directions of the symmetry axes. The EFG induces the splitting of the nuclear levels into the degenerate $\pm 1/2$ and $\pm 3/2$ states which lead to the NQR transitions in the region between 18 and 32 MHz with a doublet from both isotopes for each site [3,4]. The effect of a magnetic field, either a local hyperfine or dipolar field, or an external one, is to further split the $\pm m$ levels: at low field a NQR spectrum with Zeeman splitting is observed or at high field a NMR spectrum with quadrupole splitting. Since the splitting depends on the direction of the field with respect to the symmetry axes of the EFG tensor, the full EFG tensor can be obtained only from the NMR spectra of oriented samples.

In Y123 with $y>6.8$ two resonance lines of ^{63}Cu are found: 31.5 MHz with a symmetric EFG, and symmetry direction along the c–axis for Cu(2) and 22.0 MHz with $\eta=1$ for Cu(1) [3]. No magnetic splitting is observed for both resonances indicating that no stable magnetic moments exist at either site. In the AF state with $y<6.3$ only

Springer Series in Solid-State Sciences, Vol. 99
Electronic Properties of High-T_c Superconductors and Related Compounds
Editors: H. Kuzmany · M. Mehring · J. Fink

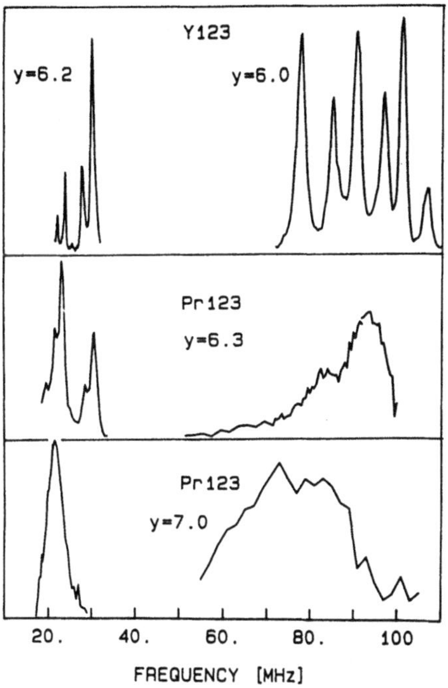

Fig. 1: Zero–field spectra of Cu in antiferromagnetic $REBa_2Cu_3O_y$ (Y123 and Pr123) measured at 1.2 K (high frequency part of the Cu(2) sites) and 4.2 K (low frequency part of Cu(1)). The oxygen concentration, y, for low values has been estimated from the relative intensities of the sub-spectra of Cu(1) with two and three oxygen neighbours at 30 and 24 MHz, respectively.

the NQR lines of Cu(1) sites with either two or three oxygen neighbours are found at 30.2 MHz (η=0) and 24.0 MHz (η=1), respectively, which again show no magnetic splitting (fig. 1). The Cu(2) resonance appears as NMR in an hyperfine field of 7.6 T (90 MHz) with quadrupole splitting which proves perpendicular orientation of the hyperfine field with respect to the c–axis (fourfold symmetry axis of the EFG) [4].

The missing magnetic splitting at Cu(1) imposes severe restrictions on the possible AF structure since the dipolar and transfered hyperfine fields from the Cu(2) sites must cancel at these sites. This holds for the structure first deduced from neutron diffraction with a stacking sequence + o − + o − of the AF Cu(2) planes [5]. Here + and − denote up and down Cu(2) spins and o the zero spin at Cu(1). Cu(1) spectra with a splitting by a field of 0.16 T have been found only at low temperature in samples containing about 1% of Fe at the Cu sites [6]. This field corresponds to the dipolar field calculated for a different spin structure with the sequence + o + − o − if a magnetic moment of 0.7 Bohr magnetons is assumed for the Cu(2) sites. The transition between both structures occurs continuously at increasing temperature with an intermediate canted spin structure, and can be seen also in the Mössbauer effect of Fe at the Cu(1) sites [7]. In the + o − + o − state,magnetic Fe ions at the Cu(1) sites are in a frustrated situation with respect to the exchange field of their Cu(2) neighbours and tend to align these parallel. This leads by a strong in–plane interaction for a reorientation of the magnetization in the whole planes.

In contrast to the Cu(1) spectra, the Cu(2) spectra are not influenced by Fe doping [8]. That means that the magnetic moments and their arrangement in the Cu(2) planes are not changed but only the relative orientation of neighbouring planes.

Exchange of Y by Nd or heavier rare earth metals introduces a magnetic sublattice into the structure which does not influence the SC behaviour. The RE moments have no influence on the Cu(1) spectra in the SC or AF states. Also in this case no local moment or field is found in the Cu(1) sites. The signals of the Cu(2) sites, the direct neighbours of the RE sites in contrast, are distorted as long as the magnetic RE moments are fluctuating above the Neel temperature of this sublattice. Only in the ordered state do the fields from the RE ions cancel at the Cu(2) sites and then the normal spectra can be observed. This is most clearly seen in Gd 123 with y=7 where the Cu(2) signal with the symmetric EFG appears only below 2.2 K [3].

3. Cu Resonance in Pr123 compounds

The lightest RE metal which can build the 123 structure is Pr. In contrast to the other RE ions, Pr suppresses the metallic behaviour and the transition to superconductivity at full oxygen loading if more than 50% of the Y is replaced. At the same Pr content AF order of Cu has been observed by Mössbauer effect studies of Fe at Cu sites [1] and by muon spin relaxation [2]. The reason for the suppression of the superconductivity by Pr in contrast to the heavier RE ions is an open question. The larger ionic radius can lead to pair–breaking by the stronger overlap of the 4f states with the electron states in the CuO_2 planes. Another reason may be that Pr enters the lattice as Pr^{4+} and fills up the holes responsible for the metallic behaviour. This model is supported by the similarity of the phase diagrams of the systems $Y_{1-x}Pr_xBa_2Cu_3O_7$ and $YBa_2Cu_3O_{7-x}$ [1,2].

The NMR spectra of Cu in Pr123 compounds have been investigated at 59 MHz for 4.2 K (fig.2). The samples with high oxygen concentration have been prepared in different laboratories. One sample has been reduced in high vacuum with final values of 10^{-8} Torr at 750°C. In samples with high oxygen concentration the typical powder spectrum for a nonsymmetric EFG with $\eta=1$ is observed (fig.2). The corresponding NQR spectrum (fig.1) contains only one broad line with the centre at 22 MHz which agrees in frequency and asymmetry parameter with the Cu(1) line of $YBa_2Cu_3O_7$. The broad linewidth indicates a much lower crystal perfection than found in the Y compounds. No NQR line is found corresponding to the symmetric EFG of the Cu(2) sites in the Y compound neither at 4.2 K nor after cooling to 1.2 K. This is a strong indication for a hyperfine field and magnetic order at the Cu(2) sites in Pr123 with high oxygen concentration y. A broad zero–field spectrum has been detected at 1.2 K in the frequency region between 50 and 100 MHz which can be explained only as the NMR of

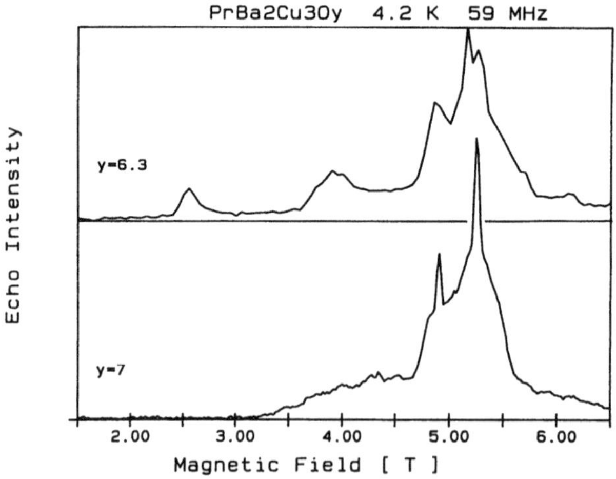

Fig. 2: NMR spectra of both Cu isotopes in the Pr123 samples of fig.1 measured by field–sweep at 59 MHz and 4.2 K, revealing a completely nonsymmetric EFG tensor for y=7 and a superposition of symmetric and nonsymmetric components at y=6.3.

both Cu isotopes in a widely distributed hyperfine field of 5(1)T, which is lower than that in AF Y123 for y=6.0. At increasing temperature, the intensity of this line disappears very fast as was described above for the Cu(2) line in Gd123.

The sample with low oxygen concentration reveals the superposition of a symmetrical EFG with a quadrupole frequency of 30 MHz and a nonsymmetric EFG in the NMR spectrum (fig.2). In the NQR spectrum, the corresponding two doublets appear at the frequencies of the Cu(1) sites with two– and threefold oxygen coordination. From the intensity ratio of these doublets we must conclude that the oxygen concentration in this sample is rather large: between 6.3 and 6.4. The attempt to remove more oxygen at higher temperature has led to a spectrum with only one line at 30 MHz but a much larger linewidth indicating a highly distorted lattice. In the high frequency region again a broad spectrum is found at 1.2 K, with the maximum at the centre frequency of the spectrum of $YBa_2Cu_3O_6$.

Thus we conclude that the EFG at the Cu(1) sites with two, three or four oxygen neighbours agrees in the 123 compounds of Y and Pr. If Pr would enter in the 4+ state in contrast to the 3+ state of Y or the other RE ions, a change of the charge distribution should occur and influence the EFG at the Cu(1) sites. Therefore these results make a different charge state of Y and Pr rather improbable. In addition the hyperfine field agrees in the AF state for the Y and Pr compounds with low oxygen concentration, whereas the hyperfine field at high y is reduced. A similar reduction of the hyperfine field is found also by muon spin relaxation [2].

4. Summary

We have discussed only the static properties of the AF state of Cu in the 123 compounds determined from the NMR and NQR spectra. The main results are:

1. No difference is found in the AF order of the Y and the PR 123 compounds.
2. Only the Cu(2) sites reveal a non–zero time average of the hyperfine field; no ordered magnetic moments exist at the Cu(1) sites. This holds also for the case of increased oxygen concentration y>6.0.
3. Paramagnetic impurities like Fe at the Cu(1) sites induce a second magnetic structure which differs only in the stacking sequence of the AF Cu(2) planes.
4. The magnetic moments of Cu(2) in the Y and Pr compounds at low y agree, the moment in the Pr compound with large y is reduced.
5. In the Y and Pr compounds the EFG tensors of the Cu(1) sites with two, three, or four oxygen neighbours, respectively, agree.

The two last results make different charge states for Y and Pr in the 123 compounds rather improbable in agreement with results from energy loss spectroscopy [9].

Acknowledgement

The author thanks I. Felner (Jerusalem), K. Fiedler (Giessen), H.D. Jostarndt (Köln), and H. Soltner and H. Bierfeld (Jülich), who made the Pr 123 samples available.

References

[1] I. Felner et al. Phys. Rev. B40 (1989) 6736.
[2] D.W. Cooke et al., Phys. Rev. Lett., submitted.
[3] H. Lütgemeier, Physica C153–155 (1988) 95.
[4] H. Yasuoka et al., Hyp. Int. 49 (1989) 167.
[5] P. Burlet et al., Physica C153–155 (1988) 1115.
[6] H. Lütgemeier et al., Physica C162–164 (1989) 1367.
[7] R.A. Brand et al., Hyp. Int. (1990) in press.
[8] L. Bottyan et al., this conference.
[9] J. Fink et al., this conference

La and Cu NQR Relaxation in $La_{2-x}Sr_xCuO_4$: Cu^{2+} Localized Spin Dynamics and the Interplay with the O(2p) Holes

A. Rigamonti

Department of Physics "A. Volta", Sezione INFN and Unità INFM-GNSM, Via Bassi n.6, I-27100 Pavia, Italy

NQR-NMR spectra and spin-lattice relaxation have been used extensively in high T_c superconductors in order to obtain local microscopic information on various magnetic properties. In the early days of high T_c superconductivity, the focus of the NQR measurements was primarily on the static properties, such as the assignments of the lines to the various nuclei (particularly in the YBCO family), the magnetic characterization and the phase diagrams, the temperature and doping dependence of the internal magnetic fields. In this line one can include the analyses of the paramagnetic and Knight shift tensors, yielding enlightening information on the hyperfine couplings and the generalized spin susceptibility, particularly when single crystals or oriented samples are available. Measurements of NQR-NMR spin-lattice relaxation driven by the fluctuating part of the nucleus-electron hyperfine magnetic interaction have been used to obtain relevant insights into the cooperative spin-dynamics of the Cu (3d) localized moments on approaching the transition to the antiferromagnetic (AF) phase, as well as around the transition temperature T_c to the superconducting (SC) phase, particularly from the NQR relaxation rate of Cu nuclei in YBCO. From NMR and/or NQR relaxation rates of nuclei not belonging to the magnetic ions (such as Y or ^{17}O) information on the quantum excitations of the Fermi liquid of O(2p) holes is also obtainable, in principle. The experimental results on the NMR-NQR relaxation have been a valuable help for the critical analysis of the generalized spin susceptibilities predicted in the framework of some theoretical models.

In this paper we are going to discuss how a proper combination of ^{139}La and ^{63}Cu NQR relaxation measurements carried out in our laboratory [1 - 6] in $La_{2-x}Sr_x CuO_4$ can be used to achieve a satisfactory clarification of many properties, such as the PA $\rightarrow$ AF transition and Cu^{2+} spin dynamics, the role of the x-doping and its effect on the AF correlation, and the interplay of the Cu localized magnetic moment and the carriers around the SC transition. In particular, attention will be focused on the Cu^{2+} spin dynamics at low frequency in the SC phase : it will be shown how the results from NQR relaxation indicate that the low-frequency spin susceptibility is dominated by the Cu (3d) states, with effective local magnetic moments which start to decrease above T_c and go to zero for $T \rightarrow 0$. As a possible consequence of the pairing process occurring in the Fermi liquid, the temperature behavior of the local r.m.s. value of the hyperfine field at the nuclear site $h_{eff}(x,T)$ (defining an effective spin $\langle S_z^2\rangle^{1/2}$) is of the form expected for the opening of a gap, with an activation energy correlated to T_c.

Springer Series in Solid-State Sciences, Vol. 99
Electronic Properties of High-T_c Superconductors and Related Compounds
Editors: H. Kuzmany · M. Mehring · J. Fink

The NQR relaxation rate due to the Cu^{2+} spin dynamics can be written

$$T_1^{-1} = A \Sigma_q \int e^{-i\omega_R t} < S^z_q (o) S^z_{-q} (t) > dt$$
$$= 2 kT A \Sigma_q \chi'' (q,\omega_R)/\omega_R \equiv A \Sigma_q S_q{}^2 J_q(\omega_R) \quad (1)$$

where the constant A, related to the strength of the hyperfine field at the nuclear site, can be measured from the spectra for $T \to 0$ in the ordered phase of La_2CuO_4 (A is q-independent for Cu NQR T_1 while for La T_1 is dominated by the component at $q=q_{AF}$). The NQR resonance frequency ω_R is much smaller than the characteristic frequencies controlling the dynamics of the collective spin components S_q. For moderate doping x (namely x<0.05), the La NQR T_1 x and T behaviors have indicated that the spin-correlation function in Eq.(1) is not controlled by the conventional thermal fluctuations but rather by "diffusion" of the charge defects introduced by the doping. In a simple model this is equivalent to saying that the normalized spectral density can be written $J_q(\omega_R) \approx \tau$, τ being an effective correlation time related to the concentration n of mobile defects; τ also controls the magnetic correlation length: $\tau = \tau_0 n^{-1} = \tau_0 \xi^2$, τ_0 being of the order of the inverse of the exchange frequency. From the experimental findings, τ and therefore n(x,T) turn out to be thermally activated: $\tau \propto n^{-1} \propto \exp [E(x)/T]$, with activation energy given by $E(x) \approx 0.85/x$ K. For the effective localized spin one has $<S_z{}^2> = (1/N) \Sigma_q (S_q)^2=1$. This simple picture of a "liquid" of mobile charge defects locally annealing the magnetic correlation with unaffected Cu^{2+} spin yields a satisfactory description for x<0.05 in a wide T range, and for larger x when $T>>T_c$. In fact, it allows one to achieve a comprehensive justification of the NQR T_1 as well as of other correlation-related properties (absolute values of the correlation length ξ, Neel temperature T_N (x), magnetic susceptibility and neutron scattering responses).

For strong doping, namely x >0.1, and $T \to T_c$, the Cu NQR T_1's evidence the breakdown of this perturbative approach, with the O(2p) carriers which affect the degree of localization of the 3d Cu electrons and the effective spin of the Cu ions. To describe these effects one could try to use tentative expressions for an x and T dependent dynamical susceptibility $\chi(q, \omega)$. A direct interpretation, which retains the main physical insights, can be given on the basis of the aforementioned heuristic picture, simply by allowing for the effective average square of the localized spin to be x and T dependent. From Eq.(1), if one still uses for $J_q(\omega_R)$ the values expected from the x and T dependence of ^{139}La NQR T_1, one can derive the temperature dependence of $<S_z{}^2>^{1/2}$, for a given x. In the framework of the same picture it is possible to prove that also the ^{139}La relaxation rate around the SC transition can be consistently fitted with the reduction of the Cu^{2+} effective localized spin moments inferred from ^{63}Cu NQR T_1. The coupling of the Cu^{2+} with the O(2p) holes can be taken into account simply by scaling the spin susceptibility which describes the contribution to the La T_1 from the carriers in the form $\chi_s(x,T)=\chi_s{}^0(<S_z{}^2>)^{1/2}$. From the analysis of the temperature behaviors of the relaxation rates, for x=0.15, 0.2 and 0.3, along the interpretative lines sketched above one derives for the local r.m.s. value of the hyperfine field at the nuclear site or, equivalently, of the effective localized spin, the equation

$$h_{eff}(x,T) \propto <S_z^2>^{1/2} \propto \exp[-\Delta(x)/T] \quad (2)$$

with a ratio $2\Delta(x)/T_c(x)$ around 3.5. One should emphasize that if the reduction of $<S_z^2>^{1/2}$ on cooling is indeeed related to the opening of a gap, it is remarkable that the effect is observed well above T_c (thus confirming analogous observation of "precursor" effects detected in YBCO; see references reported in Ref.5). One should also remark that the physical picture at the basis of the interpretative model outlined above can consistently be used in order to explain also Cu, Y and O NMR - NQR results obtained by other authors in YBCO-type systems (see Ref.5 for the references). Finally one can argue more about the spin dynamics around the SC transition if the NQR data, referring to the low-frequency range of $\omega_R \sim$ 10-100 MHz, are integrated with the findings from inelastic neutron scattering. Here the generalized susceptibility $\chi(q,\omega)$ is probed in the range 1meV $< h\nu <$ 20 meV. The normalized integrated intensity of the magnetic response function exhibits a drastic decrease on cooling for the low-energy components, while is not T-dependent for energies larger than about 15meV; the q-integrated spin is conserved (see Ref.6 for the references). The power spectrum $J_q(\omega)$ observed in neutron scattering would have to be interpreted as related to paramagnons in a strongly correlated itinerant antiferromagnetic system, without long-range order. The decrease of $<S_z^2>$ should be due to a changeover from a low frequency dynamics to a power spectrum with higher characteristic frequencies, around ω_e, far from the NQR ω_R. The decrease of T_1^{-1} on cooling around T_c can be viewed as the effect of the decrease of the low-frequency spin fluctuations induced by the correlation-breaking effect of the mobile holes. One could say that the number of active defects, controlling the low-frequency component of the spin dynamics, decreases on cooling, possibly because the spin-pairing in the liquid of the carriers inhibits the effect of the quenching on the local spin correlation. Then $h_{eff}(x,T)$ tracks the number of the carriers still active and its temperature dependence (Eq.2) reflects the BCS like pairing occurring in the Fermi liquid.

Acknowledgments

The study of the La_2CuO_4 based superconductors is being carried on in Pavia through a collaboration involving S. Aldrovandi, F. Borsa, M. Corti, A. Lascialfari and A. Rigamonti. The contributions to some of the measurements mentioned here by J. Ziolo (University of Katowice, ICTP Program for Training and Research in Italian Laboratories, Trieste) and by T. Rega (Dottorato in Fisica, University of Pavia) is gratefully acknowledged. The project is being supported by INFN (Istituto Nazionale Fisica Nucleare, Grant MSC 90-92), by INFM-GNSM (Ministero P.I., fondi 40%), and by the National Research Council of Italy (C.N.R.) under the progetto finalizzato "Superconductive and Cryogenic Technologies".

References

1) J. Ziolo, F. Borsa, M. Corti and A. Rigamonti, Physica C 153, 725 (1988).
2) A. Rigamonti, F. Borsa, M. Corti, T. Rega, J. Ziolo and G. Flor, in "Magnetic Resonance and Related Phenomena" (North-Holland, 1989) p. 611
3) F. Borsa, M. Corti, T. Rega and A. Rigamonti, Nuovo Cimento D 11, 1720 (1989); Helv.Phys.Acta 62, 671 (1989).
4) A. Rigamonti, F. Borsa, M. Corti, T. Rega, J. Ziolo and F. Waldner, in "Early and Recent Aspects of Superconductivity", Eds J. G. Bednorz and K. A. Müller (Springer-Verlag 1990), p.441.
5) A. Rigamonti, M. Corti, T. Rega, J. Ziolo and F. Borsa, Physica C 166,310 (1990)..
6) F. Borsa, M. Corti and A. Rigamonti in "HT_c Superconductivity " Proc. of SATT 3 Meeting (World Scientific Inc., 1990).

Electric Field Gradient of Ba in $YBa_2Cu_3O_y$

H. Lütgemeier[1], *V. Florentiev*[2], *and A. Yakubovksi*[2]

[1]IFF – Forschungszentrum (KFA) Jülich, Postfach 1913, D-5170 Jülich, Fed. Rep. of Germany
[2]Kurchatov Institute of Atomic Energy, Moscow, USSR

Abstract. Using samples with isotopically enriched ^{135}Ba we have been able to observe the NMR and NQR of Ba in $YBa_2Cu_3O_y$ at T=4.2 K. The pure NQR line for y=6.0 appears at 27.0(1) MHz. The field–sweep NMR spectra prove the expected symmetric EFG. A second component with a highly asymmetric EFG is found at larger y. The results confirm the calculation of the EFG by Ambrosch–Draxl et al.

1. Introduction

The electric field gradient (EFG) tensors at the different lattice sites in the 123 compound $YBa_2Cu_3O_y$ (YBCO) reflect the charge distribution in the unit cell. Since they can be measured very precisely by nuclear resonance, NMR and NQR, they present an excellent opportunity to check theories about the electronic structure in this basic HT_c superconductor and the antiferromagnetic semiconductor at reduced oxygen concentration y. The EFG tensors have been determined by NMR and NQR for all Cu sites in the compounds for y=7.0 and y=6.0 and the oxygen sites at y=7.0 and have been compiled in ref. [1]. The EFG at Y cannot be measured since no Y isotope with a quadrupole moment exists. To our knowledge, no resonance data have been published so far for Ba. The reason may be the low natural abundance of the isotopes carrying a nuclear spin I=3/2 and the small gyromagnetic ratios, $\gamma/2\pi$=4.229 and 4.731 MHz/T for ^{135}Ba and ^{137}Ba, respectively [2].

Ambrosch–Draxl et al. [1] have calculated the EFG values for all sites in YBCO by the full potential linearized augmented plane wave method using the local density approximation which leads to a good agreement for all oxygen and the Cu(1) sites. For the Cu(2) sites only the correct symmetry is obtained with an error of the eigenvalues of 50%. Thus it is very interesting to check the EFG tensor at the Ba site by measuring the Ba resonance.

2. Experiment

The samples used for the experiment have been prepared by the ceramic technique using BaO enriched with ^{135}Ba to 94%. Three samples have been used with y=6.9 (slowly cooled), 6.2 (quenched) as estimated from the NQR spectrum of Cu, and 6.0 (degassed in high vacuum). All measurements have been performed by the spin–echo technique at 4.2 K. The NMR spectra have been measured by field–sweep at fixed frequency and the NQR spectra by frequency–sweep between 15 and 32 MHz in zero field. For the spin I=3/2 only a single NQR line appears at the frequency $\nu_{NQR}=\nu_Q(1+\eta^2/3)^{1/2}$ where $\nu_Q=eQ|V_{zz}|/2h$ is the quadrupole frequency, and V_{zz} the largest absolute eigenvalue of the EFG tensor and $\eta=|V_{xx}-V_{yy}|/|V_{zz}|$ the asymmetry parameter, which in this case can be determined only in an applied magnetic field [3]. If one takes the quadrupole moment $eQ=0.18\cdot10^{-24}cm^2$ [2] of ^{135}Ba and the EFG values of the theory [1], the following NQR frequencies are obtained: 17.5 MHz with η=0.8 for y=7.0 and 24 MHz with η=0 in the case of y=6.0, which are both just in the region of the Cu NQR lines. The Larmor frequency is only

Springer Series in Solid-State Sciences, Vol. 99
Electronic Properties of High-T_c Superconductors and Related Compounds
Editors: H. Kuzmany · M. Mehring · J. Fink

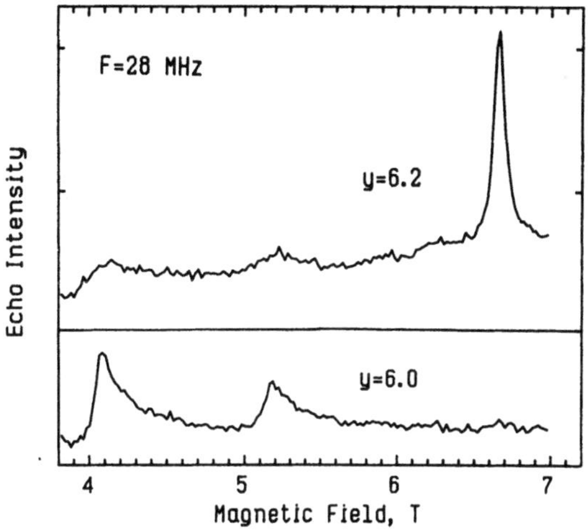

Fig. 1: NMR spectrum of ^{135}Ba in $YBa_2Cu_3O_y$ at 28 MHz and 4.2 K. The lines at 4.1 T and 5.2 T are the 3/2, 1/2 and the 1/2, −1/2 transitions, respectively, in the powder spectrum with strong quadrupole splitting by a symmetric EFG. The narrow line without a quadrupole shift corresponds to a defect site by increased oxygen concentration y.

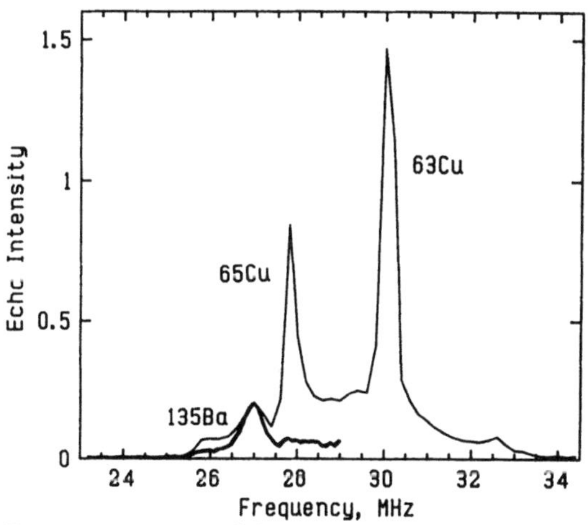

Fig. 2: Zero–field spectrum in $YBa_2Cu_3O_6$ enriched with ^{135}Ba at 4.2 K. The broad background signal may be induced by some contamination with Fe [3]. The ^{135}Ba resonance has been separated from the Cu spectrum by the different transverse relaxation times using long pulse delay (300 μs) in the echo sequence.

29.6 MHz for the largest field of 7 T. Therefore the NMR spectra of Ba are separated from the Cu resonance, but severely influenced by the quadrupole coupling.

In the antiferromagnetic sample with y=6 the transitions 3/2, 1/2 and 1/2, −1/2 in a symmetric EFG (fig. 1) could be identified in the powder spectra by the frequency dependence. The splitting corresponds to a NQR frequency of 27 MHz. The zero–field spectrum (fig. 2) with the lines of the twofold coordinated Cu(1) sites for both Cu isotopes shows a broad background signal which can be attributed to the second antiferromagnetic phase which is found in samples contaminated with about 1% Fe [3]. At 27.0(1) MHz a separate NQR line is found. The longer transversal relaxation time allows us to separate this line from the Cu resonance by using a long pulse distance in the echo sequence. The identification as the Ba resonance is possible by measuring the optimal intensity of the r.f. pulses which is much longer for Ba than for the Cu resonance due to the γ value which is smaller by a factor 2.5 compared with Cu [2].

The quenched sample showed in the zero–field spectrum broader Cu resonance lines (at 30.3 MHz for ^{63}Cu) and a weak 24 MHz line of threefold coordinated Cu(1) indicating an oxygen concentration of about 6.2. A broadened Ba NQR line was found as in the sample with y=6.0. In the field–sweep spectrum, the transitions in the symmetric EFG are found again together with a narrow line without any quadrupole shift which must be assigned to a nonsymmetric EFG with η=1.

The superconducting sample (y=6.9) showed broad NQR lines of both Cu sites which indicate a rather inhomogeneous sample. In the region between 17 and 27 MHz only one very weak Ba line was detected at 26.4 MHz which cannot contain the full Ba intensity. In the NMR spectrum, a broad line is found without any quadrupole shift. This line corresponds again to a large value of η and does not allow us to estimate the NQR frequency.

3. Discussion

For y=6, an excellent confirmation of the calculation is found in the experiments. The error is in the region of the difference of the tabulated values of the quadrupole moment of Ba [4]. At larger oxygen concentration, a part of the Ba sites loses the tetragonal symmetry when some chain sites are occupied by oxygen. The EFG tensors have been calculated for the artificial case that every other oxygen site in the chains is occupied [1]. This model reproduces the measured EFG value of the Cu(1) site with threefold oxygen coordination, but does not describe the value of η=1 for Ba. For this case the calculation with isolated oxygen ions in the chains is needed. For y=6.9 the large value of η found in the NMR spectrum agrees with the calculation, but the NQR line of the majority of the Ba sites was not yet found. An explanation may be that the EFG tensor at the Ba sites is much more sensitive to lattice distortions than at the Cu sites. Therefore samples with better perfection and larger oxygen content are needed to clarify the situation.

4. Conclusion

In spite of the fact that the local density approximation cannot describe the insulating and antiferromagnetic behaviour of the system YBCO, the EFG tensors at the Cu(1), the oxygen and the Ba sites are described correctly. The observed Ba resonance opens a new possibility to study the dynamics of the superconducting transition from the temperature dependence of the nuclear relaxation rates.

Acknowledgements

We are indebted to A.R. Kaul (Moscow State University), who supplied the samples used for this work. A.Y. and V.F. would like to thank Prof. W. Zinn for the hospitality they received at the KFA Jülich, and the DAAD for financial support.

References

[1] C. Ambrosch–Draxl et al., Hyp. Int. to be published.
K. Schwarz et al., Phys. Rev., submitted and this conference.

[2] H. Chihara and N. Nakamura, Landolt–Börnstein New Series III 20a, Springer (1987).

[3] H. Lütgemeier, Hyp. Int. to be published and this conference.

[4] Bruker Almanac (1989).

NQR Investigations on Oriented High-T_c Powder Samples

H. Riesemeier, H. Kamphausen, E.-W. Scheidt, G. Stadermann, K. Lüders, and V. Müller

Fachbereich Physik, Freie Universität Berlin, Arnimallee 14, D-1000 Berlin 33, Fed. Rep. of Germany

Abstract. For oriented $YBa_2Cu_3O_{7-\delta}$ powder samples investigations of the ^{63}Cu NQR are reported. The temperature dependence of the NQR frequency shows a non-monotonic behaviour near T_c which may be indicative for a ferroelectric or antiferroelectric transition as suggested also by ultrasonic investigations. In the superconducting state the line shape of the NQR signals depends on the magnetic field applied during the cooling procedure or after zero-field cooling which yields information about the trapped flux.

1. Introduction

For high-T_c superconductors the determination of their anisotropic properties is of particular interest. Concerning the two principal possibilities for measurement on oriented samples, single crystals and oriented powder samples, there exist specific advantages for each kind of sample .

Whereas single crystals are easier to orientate and to contact thermally, powder samples are easier to prepare with possibly more accurate crystal structure and higher oxygen content. In the special case of NMR/NQR experiments powder samples are essential due to the higher sensitivity caused by the larger effective volume. For special orientations the NQR sensitivity can be increased by a factor of about 2. In magnetic measurements the problems with the demagnetization factor may be easier to overcome by using powder samples.

In this contribution we report on NQR experiments performed on oriented powdered $YBa_2Cu_3O_{7-\delta}$ samples in connection with ac and dc susceptibility and lower critical field measurements. Besides the temperature dependence of the NQR frequency [1,2] the main point of interest is the influence of the magnetic state ("magnetic history") on the NQR line parameters. This is important for the improvement of the accuracy of NQR frequency measurements especially in connection with the temperature dependence in the vicinity of T_c. On the other hand, a line shape analysis provides information about the trapped flux.

2. Experimental

For the preparation of powder samples of $YBa_2Cu_3O_{7-\delta}$ high purity Y_2O_3, BaO, and CuO was used. The stoichiometric mixture was achieved after drying the starting materials in Ar atmosphere. Then the mixture was ball-milled in Ar atmosphere and subjected to a series of calcinations at 905°C in high purity oxygen. The final oxygen treatment was performed at 400°C and 1.2 bar for one week. The crystallites were oriented by mechanically vibrating the powder in a dc magnetic field of 8 T at room temperature. For all measurements the external field was oriented perpendicular to the c-axis of the grains. A conventional pulsed NMR spectrometer with spin-echo technique was used [3].

3. Results and discussion

Figure 1 shows the temperature dependence of the zero-field NQR frequency ν_Q of the ^{63}Cu isotope at the Cu(1) site in $YBa_2Cu_3O_{7-\delta}$ powdered samples, as obtained for different samples and different measurements. Upon cooling, ν_Q decreases nearly linearly with temperature for $T > T_c$. In the vicinity of T_c, however, a discontinuous change of ν_Q is observed. Within a very small temperature range ν_Q steeply decreases at T_c but increases again in the superconducting state. Furthermore, we observed a change in the slope of the frequency versus temperature curve near T_c: In the superconducting state ν_Q is constant. This is also reported by Brinkmann [4].

Whereas the temperature dependence above T_c may be explained by thermal vibrations and thermal expansions, the behaviour at T_c is indicative for some kind of phase transition associated with changes in the lattice parameters. One possible explanation is a freezing out of a vibrational mode of the oxygen system. This is also supported by other experiments.

From high resolution neutron powder diffraction a displacement of the chain oxygens across the chains was suggested by François et al.[5]. The O4 atoms are randomly distributed between two potential minima at about 0.1 Å to either side of the chain at $T > T_c$. In case of dynamic disorder this structure is likely to become ordered at low temperature. The oxygen displacements are then constant below T_c, forming CuO zigzag chains.

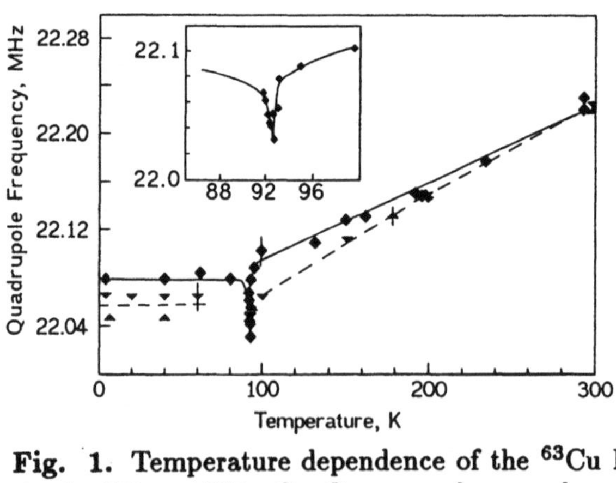

Fig. 1. Temperature dependence of the ^{63}Cu NQR frequency at the Cu(1) site in different $YBa_2Cu_3O_{7-\delta}$ powder samples

The temperature dependence of these results looks very similar to that of the NQR frequency and the same structural changes may be responsible for this behaviour. Above T_c oxygen hopping between the two possible sites of the zigzag chains leads to anharmonic oscillations and therefore to a temperature dependent NQR frequency. Below T_c the displacement is constant. This is also supported by the observation of an increased linewidth below T_c (from 170 to 250 kHz). If near T_c the ordering of the oxygen may be described by a first order phase transition (disorder-order transition), this should result in a jump of the NQR frequency. Such a change in ν_Q, however, may be small as compared to the linewidth and therefore is not necessarily observable.

From ultrasonic experiments a ferroelectric or antiferroelectric phase transition is suggested at T_c [6]. In this case the dielectric permittivity ϵ reaches high values within a very small temperature range. Since the EFG is proportional to $1/\epsilon$, also a change in ν_Q is expected.

Another example of experiments showing anomalies at T_c are EXAFS measurements. Murak et al. [7] observed in the temperature dependence of the Debye-Waller factor an anomaly at 110 K and a weak singularity at the midpoint of T_c. Finally, earlier measurements of high-resolution X-ray scattering (Horn et al. [8]) should be mentioned which show an anomalous behaviour in the temperature dependence of the orthorhomic distortion b - a near T_c.

Although there are still some experimental uncertainties for ν_Q around and below T_c, we always found the discontinuous behaviour in the temperature dependence of ν_Q. It should be mentioned, however, that there exist some experimental reasons which make precise measurements difficult. First, different samples show different behaviour. Other reasons are connected with the powder structure of the samples, e.g. the shape and average size of the grains. Since the NQR line intensity in the superconducting state is strongly influenced by the magnetic penetration depth, the intensity is considerably reduced below T_c (by about 90 % for an average grain size of 15 μm) and depends on the average grain size. Another problem is the exact determination of the sample temperature which depends on the thermal contact between the grains and the surrounding cooling chamber. Besides pseudo hysteresis effects originating from a weak thermal contact some experiments show clear evidence for structural hysteresis effects.

Other reasons which influence the NQR line shape originate from the distribution of the magnetic field inside the grains. Using conventional NMR equipment, besides the earth magnetic field additional small residual fields may act on the samples. The actual magnetic field distribution in the sample depends in a complex manner on the cooling procedure, the magnetic field strengths and the flux pinning properties. In order to learn more about these influences on the NQR line shape, i.e. the relation between trapped flux and NQR data, we started systematic investigations of the line shape under different experimental conditions. All experiments were carried out at 4.2 K in zero-field for different magnetic states: (a) zero-field cooling, (b) zero-field cooling down to 4.2 K, application of dc magnetic fields of different strengths, (c) field cooling at various field strengths.

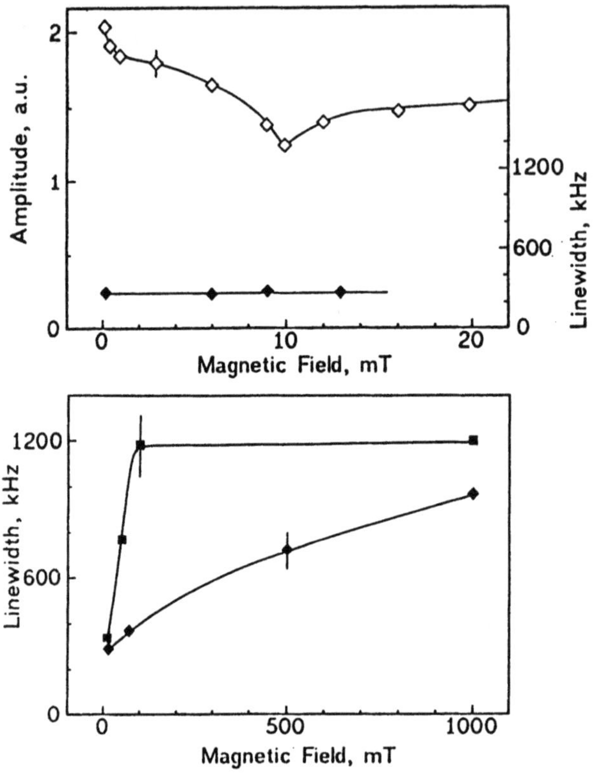

Fig. 2 and 3. Influence of the maximum magnetic field on NQR line parameters (zero-field detection)

Some of our results are presented in Figs. 2 and 3. The behaviour of the signal amplitude (maximum height) and the linewidth are plotted against the previously applied field. In the case of zero-field cooling (condition b) the linewidth keeps constant for field strengths below the lower critical field B_{c1} (Fig. 2) but increases for higher fields (Fig. 3) and is accompanied by an increase of the total line intensity. The amplitude of the signal, however, is not constant in the low field region in contrast to what is expected for the Meissner state. The amplitude decreases which may indicate the influence of the non-ideal geometry of the grains. At sharp edges the critical field is reached earlier allowing partial flux penetration. Thus some part of the center of the NQR line is shifted to other frequencies. Nevertheless, B_{c1} is clearly marked by a sharp dip (Fig. 2). Its value agrees well with those obtained by earlier dc susceptibility measurements on oriented powder samples [9].

The total intensity is enhanced in both cases (b,c) indicating that a larger amount of nuclei contributes to the NQR signal as compared to the zero-field lines (condition a). This may be explained by a larger effective penetration depth of the rf-field as compared to the London penetration depth which is caused by oscillating flux lines. The trapped flux also gives rise to the observed line broadening via Zeeman splitting of the quadrupole levels.

Acknowledgement. This work was supported by the Bundesministerium für Forschung und Technologie (F+E-Vorhaben 13N 5479/9).

References

1. H. Riesemeier, Ch. Grabow, E.W. Scheidt, V. Müller, K. Lüders, D. Riegel: Solid State Commun. **64**, 309 (1987)

2. M. Mali, D. Brinkmann, L. Pauli, J. Roos, H. Zimmermann, J. Hulliger: Phys. Lett. **A 124**, 112 (1987)

3. H. Riesemeier, J. Pattloch, E.-W. Scheidt, M. Schaefer, K. Lüders: Z. Naturforsch., in print

4. D. Brinkmann: this issue

5. M. François, A. Junod, K. Yvon, A.W. Hewat, P. Fischer: Solid State Commun. **66**, 1117 (1988)

6. V. Müller, C. Hucho, K. de Groot, D. Winau, K.H. Rieder: Solid State Commun. **72**, 997 (1989)

7. U. Murek, K. Keulerz, J. Röhler: Physica **C 153-155**, 270 (1988)

8. P.M. Horn, D.T. Keane, G.A. Held, J.L. Jordan-Sweet, D.L. Kaiser, F. Holtzberg: Phys. Rev. Lett. **59**, 2772 (1987)

9. E.-W. Scheidt, C. Hucho, K. Lüders, V. Müller: Solid State Commun. **71**, 505 (1989)

A Local Method Study of the Magnetic Transition in $YBa_2(Cu_{1-x}Fe_x)_3O_6$

L. Bottyán [1], *H. Lütgemeier* [2], *J. Dengler* [3], *S. Pekker* [1], *A. Rockenbauer* [4], *A. Jánossy* [1], *and D.L. Nagy* [1]

[1]Central Research Institute for Physics, P.O. Box 49, H-1525 Budapest, Hungary
[2]IFF – Forschungszentrum (KFA) Jülich, Postfach 1913, D-5170 Jülich, Fed. Rep. of Germany
[3]Physikalisches Institut der Universität Erlangen, Erwin-Rommel-Str. 1, D-8520 Erlangen, Fed. Rep. of Germany
[4]Central Research Institute for Chemistry, P.O. Box 17, H-1525 Budapest, Hungary

Abstract. Gd electron paramagnetic resonance, Cu(1) site nuclear quadrupole resonance and Cu(2) site zero external field nuclear antiferromagnetic resonance as well as ^{57}Fe Mössbauer spectroscopy were applied to study a magnetic transition [1] doubling the magnetic unit cell along the crystallographic *c*-axis in antiferromagnetic $Y_{1-u}Gd_uBa_2(Cu_{1-x}Fe_x)_3O_6$. The gradual rotation of the staggered magnetization of the next nearest neighbour Cu(2) planes with temperature occurs without any observable change either in the Cu(2) internal magnetic field or in the antiparallel orientation of the neighbouring Cu(2) moments along the *c*-axis. The hyperfine field distribution at Fe substituted for Cu(1) sites is attributed to Fe clustering.

1 Introduction

According to neutron diffraction results [1] coupling between both nearest and next nearest neighbour Cu(2) spins in tetragonal $YBa_2Cu_3O_6$ is antiferromagnetic (AF) with a Néel temperature of $T_N = 410$ K (structure I). Minor substitution of copper by magnetic impurities, however, results in a magnetic transition doubling the magnetic unit cell along the *c*-direction at a temperature dependent on the iron concentration (structure II). This transition is suppressed by excess oxygen [2].

The aim of this study is to reveal details of this transition by studying local magnetic and hyperfine fields using four different resonance methods such as (zero external magnetic field) Cu(2) site nuclear antiferromagnetic resonance (AFNMR), Cu(1) site nuclear quadrupole resonance (NQR), Gd^{3+} electron paramagnetic resonance (EPR) and ^{57}Fe Mössbauer spectroscopy (MS) on $Y_{1-u}Gd_uBa_2(Cu_{1-x}Fe_x)_3O_6$ samples from the *same batch* of carefully controlled oxygen and impurity doping level.

Samples with $x_{Fe} = 0$, 0.01 and $u_{Gd} = 0.0005$, 0.01 and 0.02 were prepared from high purity oxides in oxygen flow by the usual sintering method. The

Springer Series in Solid-State Sciences, Vol. 99
Electronic Properties of High-T_c Superconductors and Related Compounds
Editors: H. Kuzmany · M. Mehring · J. Fink

samples were then further heat treated in vacuum at 700° C for 12 hours. The weight loss was measured and an $O_{6.02}$ oxygen stoichiometry was found.

2. Results and discussion

NQR spectra of samples with $x_{Fe}=0$ and 0.01 at 300, 100 and 4.2 K are shown in Fig. 1.a. In the 4.2 K spectrum the splitting scales with the gyromagnetic coefficient of the two (^{65}Cu and ^{63}Cu) isotopes, therefore it is of magnetic origin and is due to a transfer magnetic field of 0.2 T from Cu(2) to Cu(1) sites [2]. The transfer field sets in at around 130 K, smoothly increases with decreasing temperature and saturates by 4.2 K. No splitting is seen in the iron-free sample. The AFNMR frequencies of samples with $x_{Fe}=0$ (structure I) and 0.01 (structure II) at 4.2 K in Fig. 1.b are equal within the experimental error, i.e. the moment on Cu(2) remains unchanged on going from structure I to structure II. The line broadening of the quadrupole split AFNMR spectrum can be attributed to random magnetic rather than quadrupole perturbation since the satellite and central lines are equally affected and can be attributed to randomly distributed Fe on Cu(2) sites.

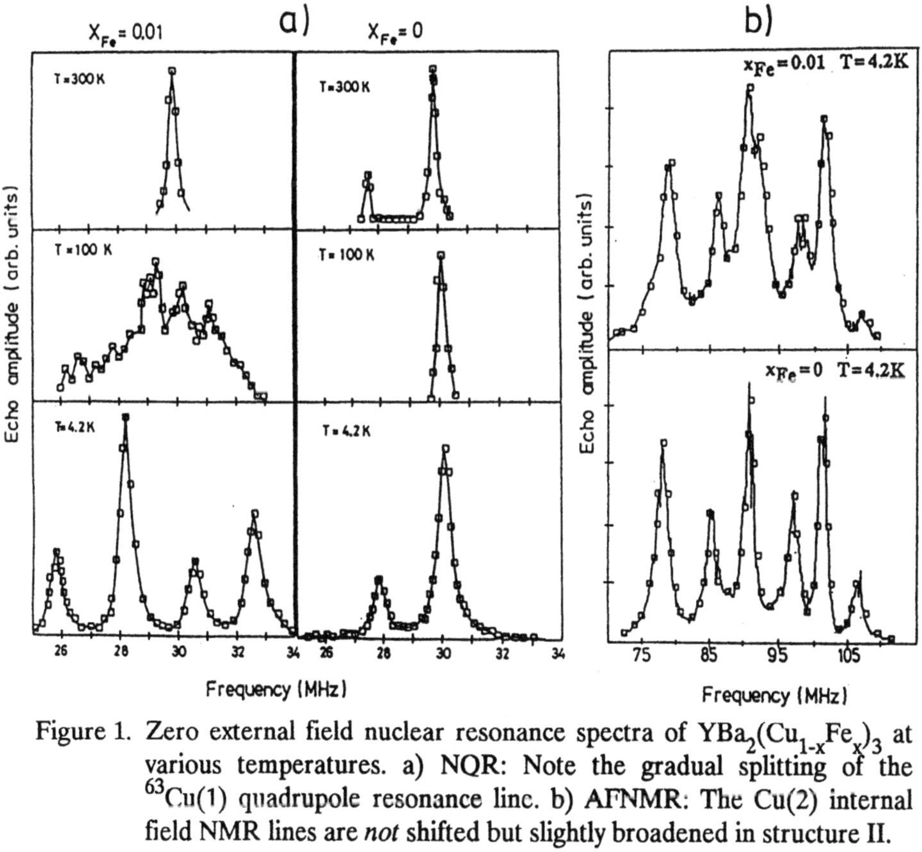

Figure 1. Zero external field nuclear resonance spectra of $YBa_2(Cu_{1-x}Fe_x)_3$ at various temperatures. a) NQR: Note the gradual splitting of the $^{63}Cu(1)$ quadrupole resonance line. b) AFNMR: The Cu(2) internal field NMR lines are *not* shifted but slightly broadened in structure II.

The Gd^{3+} EPR spectrum of magnetically oriented powder of $Y_{1-u}Gd_uBa_2Cu_3O_{6.02}$ shows a well- resolved fine structure [3]. Gd^{3+} is an $S=7/2$ state ion; a series of crystalline field parameters describe the shift of the Zeeman transition frequencies from the free ion value. The (-1/2, 1/2) central transition for magnetic fields parallel to the crystallographic *c*-axis is unperturbed by the crystalline field to high order and at 77 K and below (where the relaxation broadening is unimportant) the linewidth of this transition is small. In the ideal structure of the 1:2:3 compound, Gd sites have a mirror plane symmetry which is not broken by the doubling of the AF unit cell and no uniform internal field arises at this site. The randomness of the Fe positions may, however, give rise to a fluctuating internal magnetic field, H_i.

9.25 GHz EPR spectra of $Y_{0.99}Gd_{0.01}Ba_2(Cu_{0.99}Fe_{0.01})_3O_{6.02}$ and $Y_{0.98}Gd_{0.02}Ba_2Cu_3O_{6.02}$ at 77 K are compared in Fig. 2 to find if there is an internal field at Gd sites. The linewidth of the central (-1/2, 1/2) line is $H_{1/2}= H_Q{}^o+ H_D+ H_{Fe} + < H_i>$, where $H_Q{}^o$ is the residual linewidth with no Fe doping due to crystalline field inhomogeneities and misalignment of crystallites in the oriented powder sample, H_D is the Gd-Gd dipolar width (proportional to the Gd concentration), H_{Fe} is the linewidth due to iron induced crystalline field fluctuations as well as Fe-Gd dipolar or exchange interactions. $< H_i>$ is the contribution of internal fields arising from fluctuations in the staggered magnetization of Cu(2) planes.

For an applied magnetic field H, with H parallel to *c*, where the linewidth is only sensitive to H_i parallel to *c*, $H_Q{}^o=30$ G and $H_D=15$ Gauss/mole% Gd was found [3]. The amplitude of such fluctuations is less than 10 G since the Fe doping increased the EPR linewidth by only $H_{Fe} + < H_i> =20$ G (Fig. 2). The linewidth in the case of H tilted from *c* is sensitive to internal fields perpendicular to *c*. In this case the variations of crystalline fields in the (*a,b*) plane contribute to the linewidth since crystallites are oriented only along the *c*-direction. Since for tilted fields there is no additional linewidth in the Fe doped sample within the experimental error (Fig. 2), we conclude that internal fields are fluctuating by less than 15 G around zero at the Gd site.

Thus, despite the randomness of Fe sites, the staggered magnetizations of nearest neighbour Cu(2) planes rotate rigidly together below T_{N2}. Assuming a dipolar interaction between Cu and Gd moments, a misalignment by 1° would result in an internal field of $H_i=40$ G. Gd-Cu first neighbour exchange fields are probably larger than the dipolar ones and the misalignment would have an even larger effect. On the other hand the misalignment of Cu(2) sandwiches which broadens the Cu(1) NQR line at the phase transition little affects the Gd^{3+} linewidth because of the large distance between the Gd ions and second nearest neighbour Cu(2) AF layers.

Results of ^{57}Fe Mössbauer spectra of $Y_{0.9995}Gd_{0.0005}Ba_2(Cu_{1-x}Fe_x)_3O_{6.02}$ in Fig. 3 at room temperature and 4.2 K are similar to those reported earlier in this iron and oxygen concentration range [4]. RT spectra reveal three species, viz. A, E and C with isomer shifts of 0.1, 0.0, 0.30, quadrupole splittings of

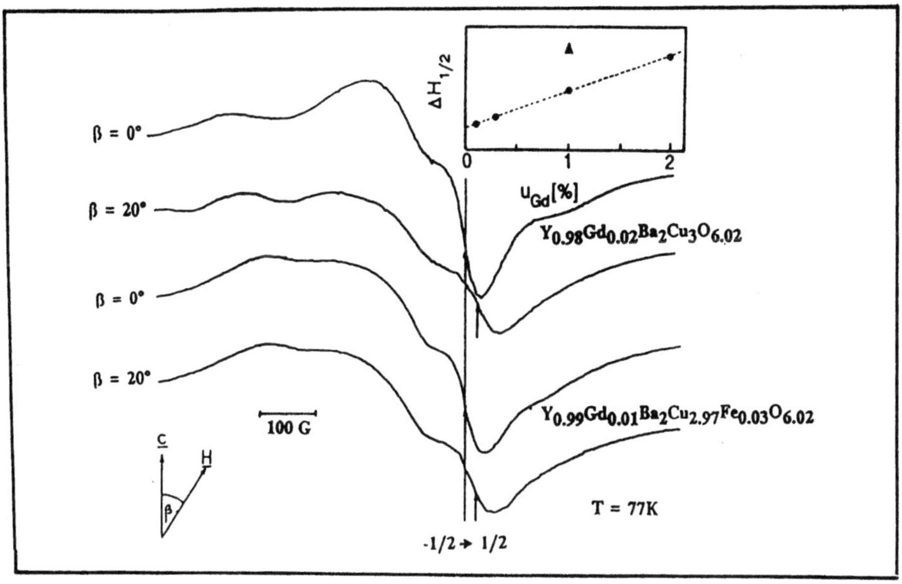

Figure 2. Gd EPR spectra of *c*-axis oriented $Y_{1-u}Gd_uBa_2(Cu_{1-x}Fe_x)_3O_{6.02}$ powder at 77 K with the external field H parallel to *c* and tilted by 20 degrees. The linewidth of the central transition vs Gd concentration for $x_{Fe}=0$ (circles) and $x_{Fe}=0.01$ (triangles) is shown in the inset.

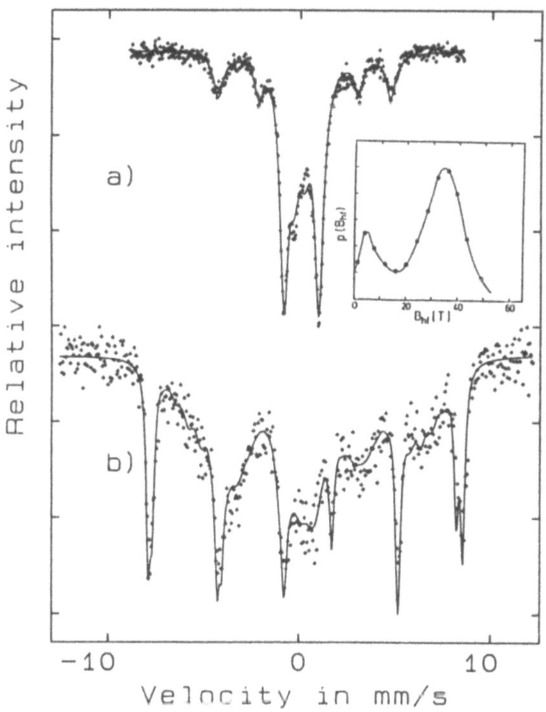

Figure 3. Room temperature (a) and 4.2 K (b) ^{57}Fe Mössbauer spectra of $Y_{0.9995}Gd_{0.0005}Ba_2(Cu_{1-x}Fe_x)_3O_{6.02}$. Inset shows the hf magnetic field distribution at 4.2 K.

$E_Q{}^A=1.84$, $E_Q{}^E=0.49$, $E_{Q,eff}{}^C=0.21$ mm/s, magnetic hyperfine splitting (hfs) of $H_{hf}{}^C=28.0$ T and relative resonance areas of $R^A=0.56$, $R^E=0.19$ and $R^C=0.25$, respectively. Component C is assigned to high spin Fe^{3+} at Cu(2) sites. (A slight splitting of this component is observed with magnetic hfs of 49.7 and 51.1 T, respectively at 4.2 K.) Species A and E are assigned to Fe substituted for Cu(1). At 4.2 K (where a well-defined transfer hfs is seen by Cu(1) NQR with narrow resonance lines in Fig.1.a), magnetic hfs at sites A and E show a relatively broad double cusp (double site) random field distribution with an average value of 37 T and width of 15 T of the higher maximum (inset in Fig.3). Such broadening can be explained *neither* by a distribution of the quadrupole frequencies *nor* by magnetic relaxation; the latter is unlikely anyway in a magnetically ordered system. Therefore, we suggest, that the magnetic hfs distribution is due to inhomogeneities in Fe-Fe coordination (Fe pairs, triplets, etc.) and *not* to an inhomogeneous distribution of the residual oxygen with preference of iron vicinity and may be due to competing AFM Cu(2)-Fe(1) and FM Fe(1)-Fe(1) coupling in those very small clusters.

In conclusion, we studied the degree of randomness in hf fields in $Y_{1-u}Gd_uBa_2(Cu_{1-x}Fe_x)_3O_6$ above and below the magnetic transition induced by Fe in a concentration as low as 0.01 using four different local probes. We found that although the staggered magnetization of the next nearest neighbour Cu(2) planes rotates within the *(a,b)* plane with decreasing temperature from antiparallel (high T) to parallel (low T) relative spin orientation, the Cu(2) magnetic moment remains unchanged. Also, the in-plane AF order and the first nearest neighbour Cu(2) plane AF coupling is maintained. On the other hand there is a distribution in the local hyperfine interactions of Fe in Cu(1) sites in the magnetically ordered structure which is due to inhomogeneities in the Fe-Fe coordination, i.e. to small Fe clusters.

3. References

[1] H. Kadowaki, M. Nishi, Y. Yamada, H. Takeya, H. Takei, S.M. Shapiro, G. Shirane, Phys. Rev. B 37, 7932 (1988).

[2] H. Lütgemeier, VIIIth International Conference on Hyperf. Interactions, Prague 1989, to be published in Hyperfine Interactions, 1990.

[3] A. Jánossy, A. Rockenbauer, S. Pekker, this conference.

[4] R. A. Brand, Ch. Sauer, H. Lütgemeier, P.M. Meuffels, W. Zinn, Proceedings of ICAME'89, Budapest, Hyp. Int. **53-58**, 1229 (1990).

Electron Paramagnetic Resonance in Gd Doped $YBa_2Cu_3O_{6.02}$

A. Jánossy[1], *A. Rockenbauer*[2], *S. Pekker*[1], *and Y. Ishihara*[1,*]

[1]Central Research Institute for Physics, P.O. Box 49, H-1525 Budapest, Hungary
[2]Central Research Institute for Chemistry, P.O. Box 17, H-1525 Budapest, Hungary
* On leave from the Kanazawa University, Department of Physics, 1-1 Marunouchi, Kanazawa, Ishikawa 920, Japan

Abstract. The Gd EPR of the 2D antiferromagnet $Y_{1-x}Gd_xBa_2Cu_3O_{6.02}$ magnetic field oriented powders with $0.001 < x < 0.05$ is studied at 9.2 GHz in the temperature range of $7\ K < T < 430\ K$. Below 77 K a well resolved temperature independent fine structure with no exchange narrowing is observed and crystalline field parameters are determined. The low temperature residual linewidth of the $-1/2 \to 1/2$ transition for $x = 0.001$ is only 28 G. Above 140 K a relaxation broadening of all lines occur. The $-7/2 \to -5/2$ resonance shifts to higher fields with a rate of 1 Gauss/K between 160 and 370 K. Near 400 K a small anomaly in the shift might be an indication of the Neél temperature.

The use of Gd as an EPR probe of local fields in high T_c superconductors and related materials has been limited by the unexpectedly broad lines found [1,2,3]. The linewidth is dipolar for highly concentrated compounds [1], otherwise it is due to a temperature dependent relaxation from an interaction with the Cu-O layers via phonons and a temperature independent broadening by defects or other disorder in the structure. We reported recently [4] that in magnetic field oriented powders of Gd doped $YBa_2Cu_3O_{6.02}$ relatively narrow lines and a well resolved fine structure has been found at low temperatures.

The energy levels of the $Y_{1-x}Gd_xBa_2Cu_3O_{6.02}$ salts ($0.001 < x < 0.05$) studied in this contribution are determined by the Zeeman interaction with the magnetic field, the crystalline field of the first neighbour oxygens and to a lesser extent copper atoms. The rare earth ions are sandwiched between two antiferromagnetic sheets to which, in zeroth order, there is no exchange coupling by symmetry. Gd^{3+} is an $S = 7/2$ state ion with little admixture of orbital moment, $g \simeq 1.99$ for most Gd salts [5]. The spectrum observed at 77 K and 9.25 GHz for external magnetic field H_0 parallel to the $\underline{c}$ crystallographic direction is shown in Fig. 1. Besides the "allowed" Zeeman transitions a number of other lines are observed; most are probably "forbidden" transitions, lines A and B arise from misoriented crystallites. An exception is a weak but narrow line at 8240 G which arises most likely from a Gd^{3+} site near a defect in the $Y_{1-x}Gd_xBa_2Cu_3O_{6.02}$ structure. This line has been observed up to 400 K, its resonance field decreases by 0.4 G/K

Electronic Properties of High-T_c Superconductors and Related Compounds
Editors: H. Kuzmany · M. Mehring · J. Fink

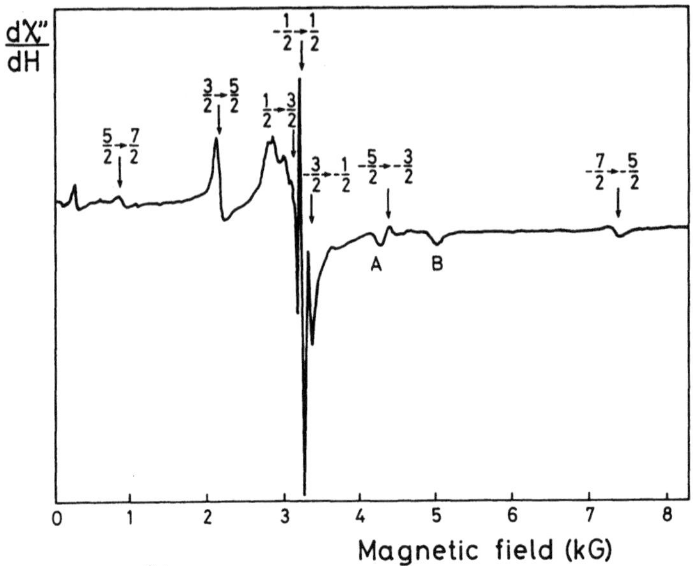

Fig. 1. Gd^{3+} 9.25 GHz EPR spectrum of $Y_{0.997}Gd_{0.003}Ba_2Cu_3O_{6.02}$ at T = 77 K. Arrows indicate the allowed transitions for $H_o // \underline{c}$.

with increasing temperature. Contributions from impurity phases (e.g. Cu^{2+} resonance from "green phase") are unimportant at 77 K.

The $1/2 \rightarrow 3/2$ and $-3/2 \rightarrow -1/2$ transitions are clearly resolved from the central $-1/2 \rightarrow 1/2$ transition showing that exchange interaction between Gd^{3+} and the antiferromagnetic layers is small. The residual linewidth of the central transition at low temperatures extrapolated to x = 0 is as narrow as $\Delta H_{1/2} = 28$ G. It is most probably due to an inhomogeneous broadening by defects and imperfect alignment of crystallites. The crystalline field D of defects perturb the $-1/2 \rightarrow 1/2$ transition in second order (i.e. proportional to D^2/H_o) only, thus the main broadening effect may be the breaking of the symmetry of the sandwiching magnetic layers. For the low concentration range investigated the Gd-Gd dipolar linewidth is linear in x and is 15 Gauss/% Gd.

Crystal field parameters from the 77 K spectrum were calculated using the Hamiltonian

$$H = \sum B_k^\ell O_k^\ell + g \mu_B \vec{S} \vec{H}_o$$

and the perturbation formulae of Rosenthal [6]. Since only the $H_o // \underline{c}$ spectrum could be resolved, the determination of all B_k^ℓ -s is not possible unambiguously. Two sets of input parameters were used: i.) the 4 experimentally observed $m \rightarrow m+1$ transition fields with $m \leq -1/2$ and fixing $g = 1.99$; ii.) 5 transition fields with $m \leq +1/2$. In case ii.) the g factor is also determined. Tetragonal symmetry and $B_6^4 = 0$ was assumed in both cases. The results of both methods (Table I.) do not differ much and are in an overall agreement with those of Shaltiel et al.[3]. The value, g = 1.998 found from the fitting procedure is in prin-

Table I. Crystal field parameters in MHz of Gd in $YBa_2Cu_3O_{6+\delta}$ at 77 K.

	$3B_2^0 = D$	B_4^0	B_4^4	B_6^0	$g_{\parallel}$
Method i.)	1266	3.1	(-)13.8*	$-2.3 \cdot 10^{-3}$	1.990+
Method ii.)	1274	3.1	(-)11.0*	$-2.8 \cdot 10^{-3}$	1.998
Shaltiel et al.[3]	1307	3.01	-11.4	-	1.989

*The sign could not be determined. +Fixed input parameter.

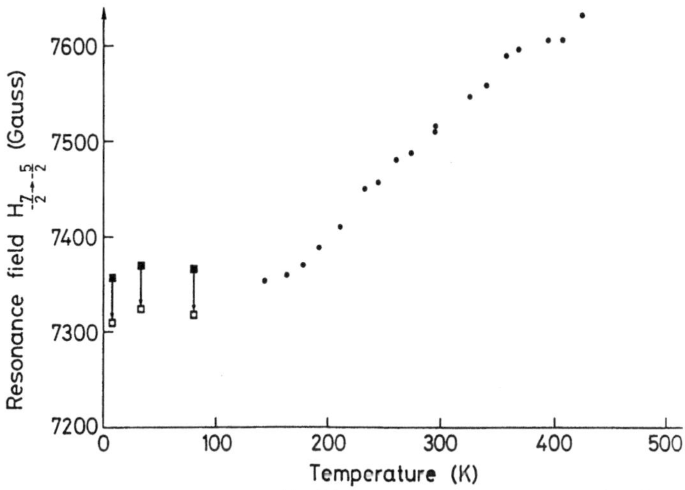

Fig. 2. Resonance field vs. temperature of the $-7/2 \to -5/2$ transition. Note the slight anomaly around 400 K. ● : x = 0.05, f = 9.16 GHz, ■ : x = 0.003, f = 9.26 GHz, □ : values corrected to f = 9.16 assuming magnetic field independent crystalline field splitting.

ciple shifted from the true g value by local polarization at the Gd sites by the antiferromagnetic layers which does not have tetragonal symmetry. The experiment shows, however, that at 77 K local polarization must be small.

Below 77 K the spectrum is almost independent of temperature. At higher temperatures all lines broaden and for some a shift is also observable. At room temperature the central line is not resolved from the first fine structure lines and the Cu^{2+} impurity phase line. The $-7/2 \to -5/2$ transition was followed up to 430 K for x = 0.05 (Fig. 2.). Between 170 and 370 K this line shifts to higher fields with a rate of 1.2 G/K. Around 400 K the field increases somewhat less, a more precise measurement is in progress to see whether this is a real effect due to the change of local susceptibility when passing the Neél temperature. Alloul et al. [7] showed this effect unambiguously at the Y site by NMR. The peak to peak linewidth of the $-7/2 \to -5/2$

transition increases from 375 G at 140 K to 500 G at 430 K for the $x = 0.05$ sample. It is assymmetric - especially at low temperatures - i.e. a large part of the width arises from misalignment of crystallites. The width at 430 K is much less than expected for dipolar coupling between Cu^{2+} and Gd^{3+} ions in the paramagnetic phase. It may be exchange narrowed, it is remarkable, however, that the fine structure is at least partially resolved at the Neél temperature, showing that exchange interactions are surprisingly small.

This work has been supported by grants of the Hungarian Academy of Sciences AKA 86-292 and OTKA 89-1-678.

References

1. H. Kikuchi, Y.Ajiro, Y.Ueda, K.Kosuge, M.Takano, Y.Takeda, M.Sato, J.Phys.Soc. Japan, 57, 1887 (1988)
2. M.T.Causa, C.Fainstein, G.Nieva, R.Sanchez, L.B.Steren, M.Tovar, R.Zysler, D.C.Vier, S.Schultz, S.B.Oseroff, Z.Fisk, J.L.Smith, Phys.Rev. B38, 257 (1988)
3. D.Shaltiel, S.E.Barnes, H.Bill, M.Francois, H.Hagemann, J.Jegoudez, D.Lovy, P.Monod, M.Peter, A.Revcolevschi, W.Sadowski, E.Walker, Physica C161, 13 (1989)
4. A.Jánossy, A.Rockenbauer, S.Pekker submitted to Physica C (Superconductivity) in press
5. A.Abragam and B.Bleaney in Electron Paramagnetic Resonance of Transition Ions, Clarendon Press-Oxford University Press, 1969.
6. J. Rosenthal, Phys. Rev. B164, 363 (1967)
7. H. Alloul, T.Ohno and P.Mendels, Proceedings of this Conference.

EPR of a Paramagnetic Probe on the Surface of $YBa_2Cu_3O_{7-\delta}$

B. Rakvin[1], *M. Požek*[1], *and A. Dulčić*[1,2]

[1]Rudjer Bošković Institute, University of Zagreb, Bijenička 54, P.O. Box 1016, YU-41001 Zagreb, Croatia, Yugoslavia
[2]Faculty of Natural Sciences and Mathematics, University of Zagreb, P.O. Box 187, YU-41001 Zagreb, Croatia, Yugoslavia

Abstract. A paramagnetic probe was placed at various positions on the surface of sintered samples of $YBa_2Cu_3O_{7-\delta}$ superconductor. Below the superconducting transition temperature T_c, the EPR line shifts due to local fields which arise on the probe location. The analysis of the EPR spectra can provide information on the induced local fields due to Meissner-Ochsenfeld demagnetization effects and due to pinning effects after exposing the sample to the various fields.

1. Introduction

Recently it was proposed that Electron Paramagnetic Resonance (EPR) can also be employed to study the Meissner-Ochsenfeld (MO) effect /1,2/, and the flux distribution /3/ in the mixed state of type-II superconductors which is developed at magnetic field H_o ($H_{C1} \ll H_o \ll H_{C2}$). In the mixed state a regular lattice of vortices is present and gives rise to a periodically modulated magnetic field. It is known that an accurate measurement of this field enables the evaluation of one of the most fundamental properties of a superconductor, the magnetic penetration depth (λ). However, several different techniques (μ^+SR /4,5/, NMR /6,7/) which have been used to deduce λ in the mixed state of a superconductor were applied under the assumption that only the effect due to vortices was present. Various other sources of magnetic fields inhomogeneities were neglected. In a granular structure of sintered superconductors one can expect the variation of local magnetic fields due to MO effect /1,2/ and internal field induced in the field sweep experiment due to pinning. These effects lead to a macroscopic field inhomogeneity which can be studied by a solid EPR probe placed at various positions on the surface of the sample. We have examined the behavior of the local magnetic field at the surface of a ceramic $YBa_2Cu_3O_7$ superconductor under various conditions: i) variation of temperature in the range from 91 to 12 K, ii) magnetic field dependence in the range from 0 to 8600 G, iii) dependence on different geometries of the sample.

2. Experimental

The experiments were carried out with a Varian E-109 EPR spectrometer operating at 9.5 GHz (X-band). We found it convenient to use a dual cavity with a reference sample in one cavity and the superconductor under study

Electronic Properties of High-T_c Superconductors and Related Compounds
Editors: H. Kuzmany · M. Mehring · J. Fink

in the other. By switching the field modulation (100 kHz) from one cavity to the other, the relative shift between resonance lines can be obtained with the accuracy of ±0.5 G. In this study organic conductor N-methylphenazinium tetracyanoquinodimethane (TCNQ) /8/ was used as EPR probe. Intensive and narrow EPR signal ($\Delta H_{pp} \approx 1$ G) can be obtained by using only few tiny crystals of TCNQ. Needle shapes of these crystals with typical dimension ($\approx$1 mm x 0.1 mm) have been placed on the surface of the sample. The samples were mounted on the cold finger of a closed cycle helium cryostat CTI-Cryogenics model 22 C refrigeration system and temperature was stabilized within ±0.2 K.

3. Results

Typical temperature and magnetic field dependence of the local magnetic field, $\Delta H = (H - H_o)$, where H_o is the resonance field observed when the sample is not in the superconducting state, while H is the resonance field for $T < T_c$ which can be detected at the edge of the superconducting disk with magnetic field perpendicular to the disk is shown in Fig 1. It is interesting to note the change of ΔH which is caused by the pinning effect and can be obtained by an exposure of the sample to low external magnetic field followed by up-field sweep so that resonance is met at H_{MIN}. Similar dependences of ΔH on pinning effect can be detected with exposure of the sample to a higher external magnetic field followed by down-sweep and resonance at H_{MAX}. However, one can assume that in the field cooled (FC) samples ($\approx H_o$) the appearance of ΔH_E is only due to the partial exclusion of the flux from the sample (MO effect) /1,2/. The temperature dependence of ΔH_E for the geometry described above is presented by open circles in Fig. 2. The local magnetic fields $\Delta H_{MAX} = (H_{MAX} - H_o)$ and $\Delta H_{MIN} = (H_{MIN} - H_o)$ which represent induced fields after exposure of the sample to magnetic field $H_{MAX} = 8600$ G or $H_{MIN} = 0$ G, respectively, are also shown in Fig. 2 for the same temperature interval. It is evident that the data for ΔH_{MAX} and ΔH_{MIN} exhibit an increase in deviations from ΔH_E when the temperature is lowered.

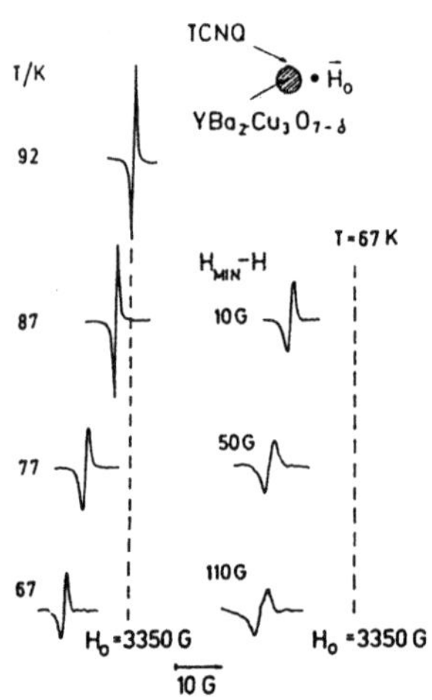

Fig. 1 Temperature and magnetic field dependences of the local magnetic field which can be detected at the edge of a superconducting disk with the applied magnetic field perpendicular to the disk.

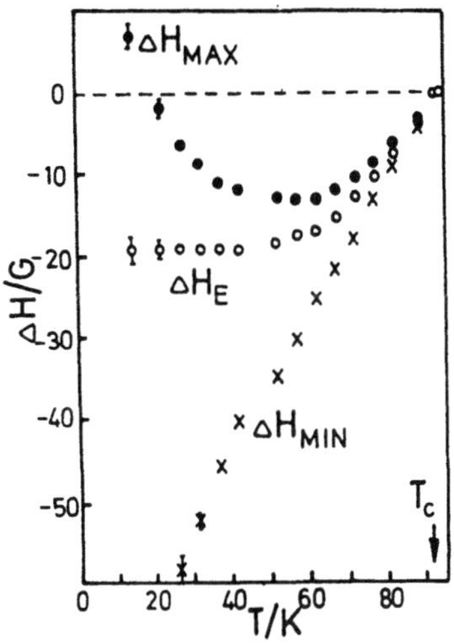

Fig. 2 Temperature dependence of the local magnetic field shift (ΔH_E) in a field cooled sample (•). ΔH_{MAX} and ΔH_{MIN} represent local field shifts after exposure of the sample to magnetic fields H_{MAX} = 8600 G, or H_{MIN} = 0 G, respectively.

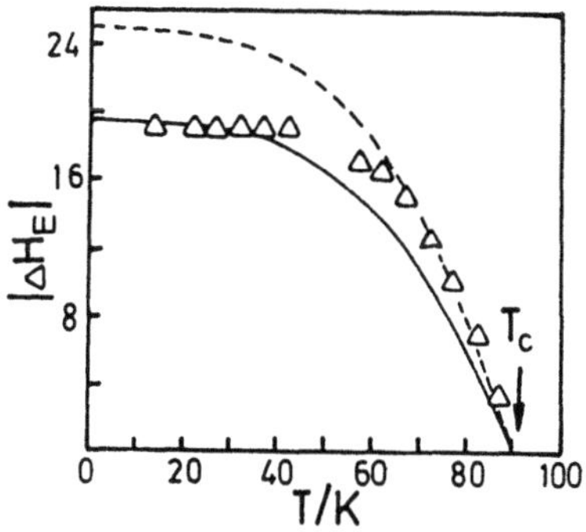

Fig. 3 The temperature dependence of ΔH_E from fig. 2 (Δ). The lines show the best fit of the temperature dependence of the equilibrium magnetization for $\lambda(0)$ = 5300 Å (dashed), and $\lambda(0)$ = 6000 Å (solid).

By subtraction of ΔH_E from ΔH_{MAX} (or ΔH_{MIN}) one can simply obtain the induced local field which can be related to the effects of pinning in the superconducting sample as a function of temperature. It should be noted that in a similar experiment /2/ with normal configuration (plane is normal to the field direction) both higher and lower internal local fields show higher strength than ΔH_E. In the case of an ideal type-II superconductor in the mixed state, the temperature dependence of MO field can be described with the equilibrium magnetization /9/

$$4\pi M = (\Phi_o/4\pi\lambda^2) \times |\ln(H_{C2}/B)| \qquad (1a)$$

$$\lambda(T) = \lambda(0)\,|1-(T/T_c)^4|^{-1/2};\quad H_{C2}(T) = H_{C2}(0)\,|1-(T/T_c)^2| \qquad (1b)$$

where Φ_o is the flux quantum and H_{C2} is the upper critical field. By using the two-fluid relation /9/ for the temperature dependence of $\lambda(T)$ and the corresponding temperature dependence of $H_{C2}(T)$ given in relation (1b), one can calculate the temperature dependences of local magnetic field due to MO effect. The lines in Fig. 3 show the temperature dependence of the equilibrium magnetization for two different parameters of $\lambda(0)$, ($\approx$ 5300 A dashed line, and $\approx$ 6000 A solid line) assuming that $H_{C2}(0) \approx 80$ T^{10}. As one can note, the local magnetic field strength due to MO effect exhibits a moderate agreement with the theoretical prediction. However, it should be pointed out that a certain contribution to ΔH_E in a sintered sample might be expected from the Josephson medium (weak links) which is not considered in this case.

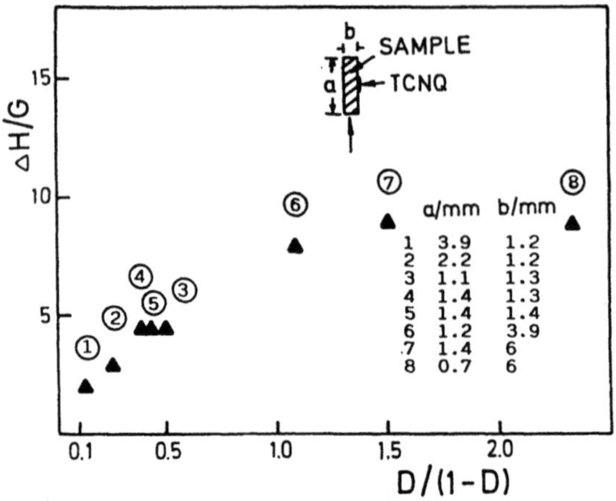

Fig. 4 Induced local magnetic fields as a function of demagnetization factors.

In order to study the local magnetic field as a function of demagnetization factors the geometry of samples was changed. An ellipsoid was used as a model to calculate the demagnetization factor (D) /11/. In this model one can expect the change in the local field for a perfect superconductor is given by /11/

$$H = H_o \times D/(1 - D)\ . \qquad (2)$$

For different shapes of the samples shown in Fig. 4 one expects a linear increase of the local field according to relation (2). Generally, the local field behaves as expected for a given geometry ($H_o \pm \Delta H$), but the strength of this field is far from the linear dependence predicted for a perfect superconductor.

4. Conclusion

EPR can be used to probe the local field at the surface of a superconductor. For a ceramic $YBa_2Cu_3O_7$ sample one can observe the induced fields due to the Meissner-Ochsenfeld and demagnetization effects, and the change due to pinning after exposure of the sample to higher or lower fields. It can be noted that the total change of a local field at the surface can exhibit a large variation, up to 60 G as shown in Fig. 2 for a typical sintered sample.

These effects should be present in the linewidth broadening in μ^+SR /4,5/, NMR /6,7/ and EPR /3/ measurements, and have to be considered carefully when the penetration depth is due to be determined.

References

1. Z. Frait, D. Fraitova, E. Pollert and L. Pust, Phys. Stat. Sol. (b) **146**, K119 (1988).
2. Z. Frait, D. Fraitova and L. Pust, J. Phys. **49**, C8-2235 (1988).
3. B. Rakvin, M. Požek and A. Dulčić, Solid State Commun. **72**, 199 (1989).
4. D.R. Harshman, E.J. Ansaldo, B. Batlogg, J.H. Brewer, J.F. Carlon, R.J. Cava, M Celio, A.C.D. Chaklader, W.N. Hardy, S.R. Kreitzman, G.M. Luke, D.R. Noakes and M. Senba, Phys. Rev. **B 36**, 2386 (1987).
5. D.R. Harshman, L.F. Schneemeyer, J.V. Waszczak, G. Appeli, R.J. Cava, B. Batlogg, L.W. Rupp, E.J. Ansaldo and D. Lj. Williams, Phys. Rev. **B 39**, 851 (1989).
6. H. Niki, T. Suzuki, S. Tomiyoshy, H. Hentona, M. Cmori, T. Kamiyama and R. Igei, Solid State Commun. **69**, 547 (1989).
7. F. Hentsch, N. Winzek, M. Mehring, Hj. Mattausch and A. Simon, Physica **C 158**, 137 (1989).
8. A.I. Vistnes and L.R. Dalton, J. Mag. Res. **54**, 78 (1983).
9. M. Tinkham, Introduction to Superconductivity (International Series in Pure and Applied Physics), McGraw-Hill, New York, 1975.
10. T.P. Orlando, K.A. Delin, S. Foner, E.J. McNiff, Jr., J.M. Tarascon, L.H. Greene, W.R. McKinnon and G.W. Hull, Phys. Rev. **B 35**, 7249 (1987).
11. T. Van Duzer and C.W. Turner, Principles of Superconductive Devices and Circuits, Elsevier, New York, 1984.

Pressure-Induced Charge Redistribution in $ErBa_2Cu_3O_x$ Determined by Neutron Crystal-Field Spectroscopy

J. Mesot[1], *U. Staub*[1], *P. Allenspach*[1], *A. Furrer*[1], *H. Mutka*[2], *and C. Vettier*[2]

[1]Laboratory for Neutron Scattering, ETH Zürich, CH-5232 Villigen, Switzerland

[2]Institut Laue-Langevin, F-38042 Grenoble, France

Abstract. The application of neutron crystal-field spectroscopy to probe the local symmetry and charge distribution around the Er^{3+} ions in $ErBa_2Cu_3O_x$ is described. As the anomalous behavior of the 2nd order crystal-field parameters versus x is correlated to the two-plateau structure of T_c, we investigated the effects of pressure on the crystalline electric field transitions. From the observed data we derived significant changes of the 2nd order crystalline electric field parameters as a function of pressure, which may again be interpreted in terms of charge redistributions in the CuO_2 planes.

1. Neutron crystal-field spectroscopy

In the high-T_c compounds $ErBa_2Cu_3O_x$ (123) the Er^{3+} ions are sandwiched between the CuO_2 planes and experience the crystalline electric field (CEF) potential :

$$V_{CEF}(\mathbf{r}) = \sum_i \rho(\mathbf{R}_i)/ \left|\mathbf{r}-\mathbf{R}_i\right| , \qquad (1)$$

where $\rho(\mathbf{R}_i)$ denotes the charge distribution at the site $\mathbf{R}_i$ of the ith ligand ion. Using the operator equivalents method we can construct the following Hamiltonian :

$$H_{CEF} = \sum_{n=1}^{3} \sum_{m=0}^{n} B_{2n}^{2m} O_{2n}^{2m} , \qquad (2)$$

where the B_{2n}^{2m} denote the CEF parameters and the O_{2n}^{2m} are the Stevens operators. In first order perturbation theory the Hamiltonian (2) splits the ground-state J-multiplet $^4I_{15/2}$ of Er^{3+} in $ErBa_2Cu_3O_x$ into eight Kramers doublets Γ_5.

By means of inelastic neutron scattering the energies of the CEF levels can be directly determined. The

Springer Series in Solid-State Sciences, Vol. 99
Electronic Properties of High-T_c Superconductors and Related Compounds
Editors: H. Kuzmany · M. Mehring · J. Fink

differential cross section for CEF transitions is given in the dipole approximation by

$$\frac{d^2\sigma}{d\Omega d\omega} \sim F^2(Q)\ \exp[-\frac{E_i}{kT}]\ |\langle\Gamma_j|J_p|\Gamma_i\rangle|^2\ \delta(E_i-E_j+h\omega) \qquad (3)$$

where F(Q) is the magnetic form factor, E_i the energy of the ith CEF state with irreducible representation Γ_i and J_p the component of the total angular momentum operator perpendicular to the scattering vector **Q**.

Using a least-squares fitting procedure we have been able to determine the CEF parameters for several oxygen concentrations x of the 123 compounds [1].

2. Internal-pressure dependence of the CEF parameters

The CEF parameters B_2^0, B_4^0, B_6^0 were found to exhibit a considerably different behavior as a function of the oxygen concentration x (fig. 1). B_4^0 is roughly constant in the whole concentration range, whereas B_2^0 and B_6^0 appear to be correlated to the two-plateau structure of T_c versus x observed in the 123 compounds.

According to the CEF potential (1), the changes in B_2^0 and B_6^0 may be due to both structural modifications and charge redistributions at the ligand sites. The structure-induced changes of the CEF parameters can be

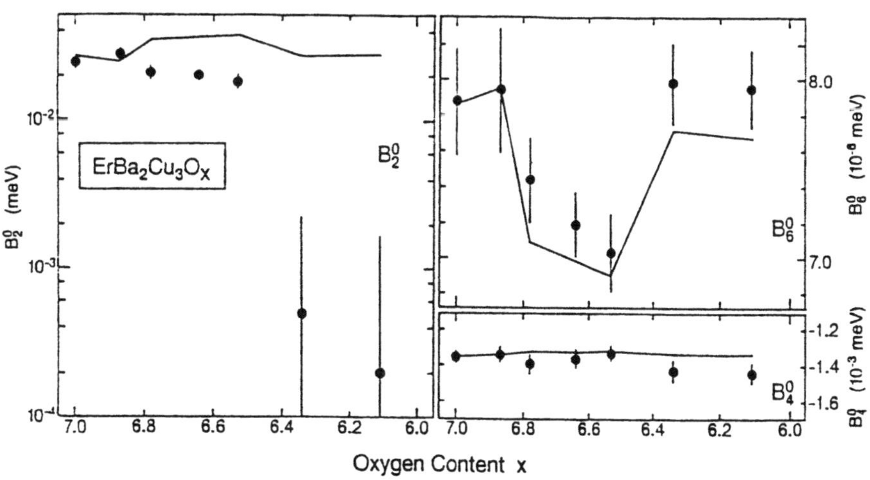

Fig. 1: Observed and calculated (extrapolated from x = 0 as explained in the text) CEF parameters B_{2n}^{2m} versus oxygen content x.

calculated, considering a simple electrostatic model, by writing them in the form $B_{2n}^{2m} = A_{2n}^{2m} \gamma_{2n}^{2m}$, where γ_{2n}^{2m} denote the geometrical coordination factors and A_{2n}^{2m} are reduced CEF parameters reflecting the charge distribution alone. Assuming constant values of A_{2n}^{2m} over the whole range of oxygen concentration x and extrapolating the CEF parameters B_{2n}^{2m} from x = 0 to the other values of x, yields the behavior displayed in fig.1 by the full curves. Any deviation of the CEF parameters from the extrapolated curves may then immediately be interpreted in terms of charge redistributions in the CuO_2 planes. This is the case for the 2nd order parameter whose observed values drastically deviate from the geometrical coordination curve.

The relation between B_2^0 and T_c versus x can be understood in terms of charge transfer from the chains to the planes. The net effect of the observed decrease of B_2^0 upon lowering the oxygen content x is to reduce the hole concentration in the CuO_2 planes and thereby reducing T_c.

3. External-pressure dependence of the CEF parameter

Another very important parameter controlling the critical temperature is the external pressure. Indeed the pressure dependence of T_c was found to be drastically different in different superconducting phases [2].

We have therefore investigated the effects of external pressure on the CEF interactions in $ErBa_2Cu_3O_x$ (x=6.98, 6.78, 6.54) and $ErBa_2Cu_4O_8$. As seen in fig.2 the application of pressure in $ErBa_2Cu_3O_{6.98}$ slightly shifts the CEF transitions to higher energies. Similar effects were observed for the other compounds. Using the above mentioned least-squares fitting procedure we found that the 4th and 6th order CEF parameters are practically independent of pressure, whereas the application of pressure significantly changes the 2nd order parameter.

Calculations based on the structural data indicate that the pressure-induced structural changes do not produce major changes of the CEF parameters, thus we may tentatively interpret the pressure-induced changes of the 2nd order parameter in terms of charge redistributions in the superconducting CuO_2 planes, consistent with our previously raised conjecture [1].

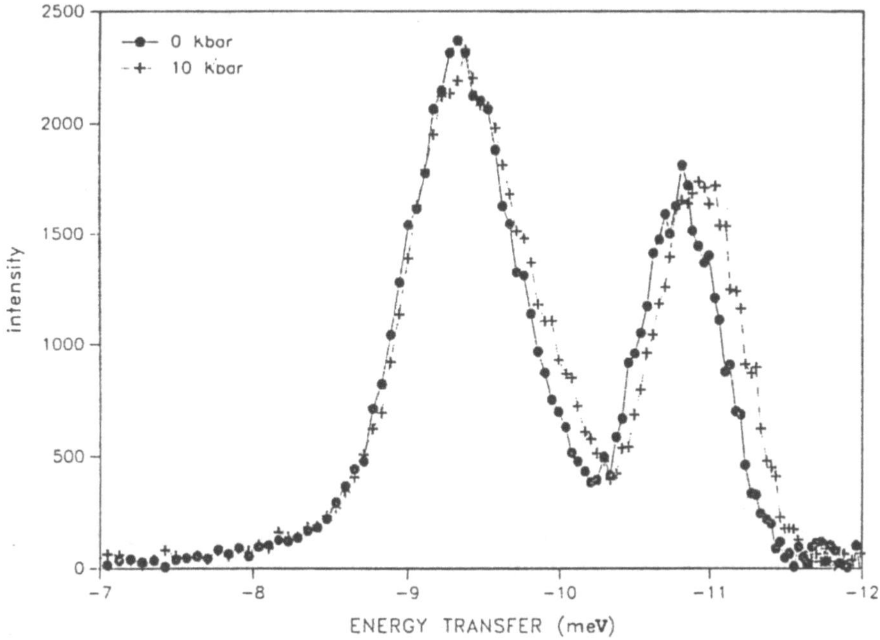

Fig. 2: Energy spectra of neutrons scattered from $ErBa_2Cu_3O_{6.98}$, T = 10 K, P = 0 and 10 kbar.

References

[1] P. Allenspach, A. Furrer, B. Rupp and H. Blank, Physica C161, 671 (1989).
[2] B. Bucher, J. Karpinski, E. Kaldis and P. Wachter, Physica C157, 478 (1989).

Magnetic Anisotropy of High-T_c Superconductors in the Normal Phase

M. Miljak, V. Zlatić, and I. Kos

Institute of Physics, University of Zagreb, Bijenička c 46, YU-41001 Zagreb, Yugoslavia

Abstract. Magnetic anisotropy measurements of high temperature superconductors reveal that the normal phase susceptibility has a temperature dependent diamagnetic component which increases with decreasing temperature. The temperature interval in which the variation of the susceptibility anisotropy is observed and total anisotropy change in the normal phase do not seem to be related to the superconductivity. Rather, the data remind one of the behaviour of some other quasi two dimensional metals like HMTSF-TCNQ or Zr_3Te_5 in which the temperature dependence of the susceptibility anisotropy is the consequence of the Landau-Peierls diamagnetism.

1. Introduction. The magnetic properties of high Tc superconductors are very anisotropic[1,2,3]. Recently, we observed[4] large anisotropy in the magnetic susceptibility of various superconducting Y-Ba-Cu-O and Bi-Ca-Sr-Cu-O single crystals. The anisotropy is defined as $\Delta\chi(T)= \Delta\chi_c - \Delta\chi_a$, where $\Delta\chi_c$ and $\Delta\chi_a$ are the principal values of the susceptibility tensor for c and a directions. The measurements are performed on a sensitive torque magnetometer and the torque, $\Gamma(T)$, which at a given temperature is a function of the applied field and the sample magnetization, allows us to obtain $\Delta\chi(T)$ from the ratio $\Gamma(T,H)/H^2$. The data indicate that the perpendicular response (the field perpendicular to the Cu-O planes) is the most paramagnetic one in the normal phase and the most diamagnetic one in the superconducting phase.

Typical results for $\Delta\chi(T)$ are shown in Fig.(1) for Y-Ba-Cu-O and Bi-Ca-Sr-Cu-O samples having the mass of 6.82 mg and 6.2 mg and T_c=87.6 K and T_c=89 K, respectively. Four different temperature regimes in the behaviour of the anisotropy can be distinguished. First, at high temperatures the anisotropy is constant: at 300 K we have $\Delta\chi(T)\approx\Delta\chi(\infty)=1.4*10^{-4}$ (emu/mole) for Y-Ba-Cu-O and $\Delta\chi(T)\approx\Delta\chi(\infty)=0.6*10^{-4}$ (emu/mole) for Bi-Ca-Sr-Cu-O. These values can be

Springer Series in Solid-State Sciences, Vol. 99
Electronic Properties of High-T_c Superconductors and Related Compounds
Editors: H. Kuzmany · M. Mehring · J. Fink

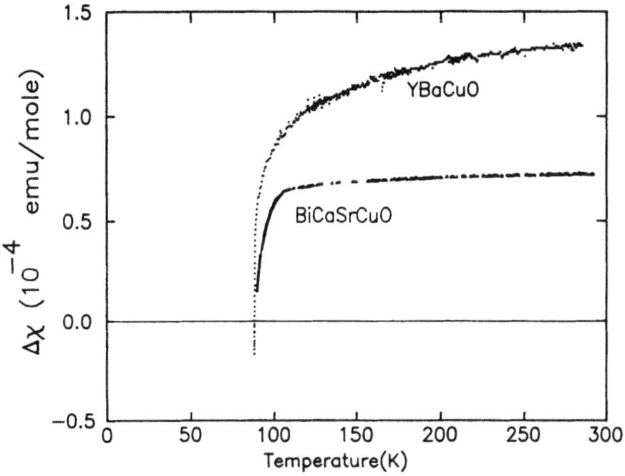

Figure 1.

Susceptibility anisotropy of Y-Ba-Cu-O sample, measured in H=208 Gauss, and of Bi-Ca-Sr-Cu-O, measured in 3 kGauss, is plotted as a function of temperature.

explained by assuming anisotropic Van Vleck susceptibility of Cu ions in Cu-O planes. Next, below 250 K the reduction of temperature results in the decrease of $\Delta\chi(T)$, with the slope which increases very fast as the superconducting transition is approached. However, the total change of $\Delta\chi(T)$ between 300 K and T_c does not exceed $\Delta\chi(\infty)$. This reduction of the transverse paramagnetism, and the associated decrease of the anisotropy, is discussed below. The third anisotropy regime is found close above T_c. For temperatures such that $T_g>T>T_c$ $\Delta\chi$ becomes field dependent. T_g is the nucleation temperature at which the superconductivity is induced in isolated grains of the sample.In the high field limit $\Delta\chi$ becomes field independent again and assumes positive values typical of the normal phase. As temperature decreases below T_g, the correlation between the phases of the superconducting wave functions of different grains increases and, in the second step, different phases lock together giving rise, at T_c, to the bulk superconductivity[5]. The bulk superconducting state is very anisotropic. The transverse response is much more diamagnetic than the parallel one and the torque values are a few orders of magnitude larger than in the normal phase. Studying the anisotropy of various quantities which characterize the superconducting state, like the Meissner effect, the flux trapping , $H_{c1}(T)$, etc, we can determine T_c to a precision which greatly exceeds the difference (T_g-T_c).

2. Discussion. First, we analyze $\Delta\chi(T)$ in more detail to see if the rapid growth of the diamagnetic component at low T could be taken as an indication of the neighbouring superconductivity. In Fig.(2) this additional diamagnetic contribution, $-(\Delta\chi(T)-\Delta\chi(\infty))/\Delta\chi(\infty)$, is plotted as a function of reduced temperature, $t=(T-T_c)/T_c$, for Y-Ba-Cu-O sample. The data are normalized by the high temperature Van Vleck part. For temperatures between t=2.5 and t=0.2, i.e. for T between 300 K and 105 K, the data could be fitted by a logarithm, f(t)=a*ln(t)+b (full line in Fig.(2)), with the coefficients a=-0.12 and b=0.13. However, in the next decade, from t=0.1 to t=0.01, the logarithmic fit (see inset in Fig.(2)) requires a and b which are one order of magnitude smaller. In the limit $t \rightarrow 0$, the normal phase torque can be obtained by correcting the data for the nonlinear contribution due to the superconductivity of isolated grains. The result is shown by open squares in the inset of Fig.(2) and we see that the total anisotropy change in the normal phase does not exceed the magnitude of the Van Vleck anisotropy. The data for single crystals of Bi and Tl cupric superconductors show similar features as well. Here, we remark that the Van Vleck anisotropy depends only weakly on the value of T_c, even when T_c is changed by one order of magnitude[5]. Similarly, the temperature interval in which the growth of the additional diamagnetic susceptibility is observed does not seem to be related to T_c. We conclude that

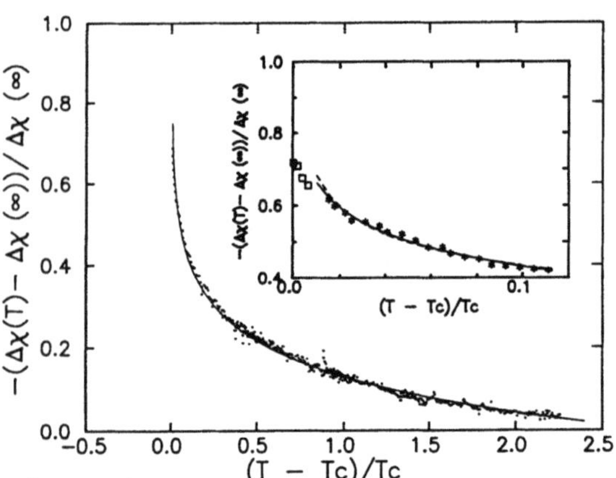

Figure 2.

Relative susceptibility anisotropy $-(\Delta\chi(T) - \Delta\chi(\infty))/\Delta\chi(\infty)$, for Y-Ba-Cu-O sample, is plotted as a function of reduced temperature $t=(T-T_c)/T_c$. The full line shows the logarithmic fit (see text). The inset shows the data in the lower t-decade. The open squares are the data corrected for the nonlinear contribution.

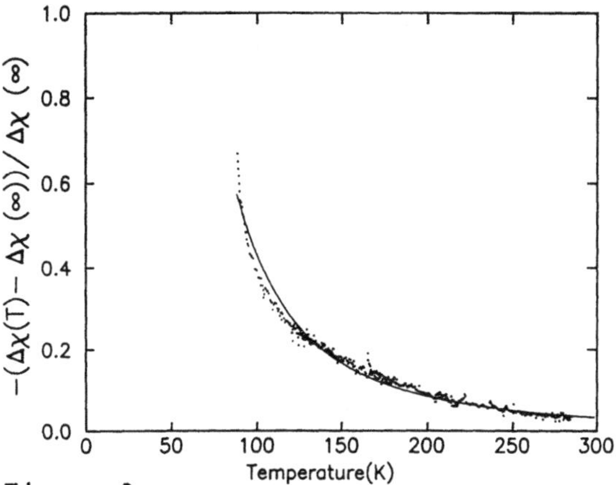

Figure 3.

Relative susceptibility anisotropy, $-(\Delta\chi(T) - \Delta\chi(\infty))/\Delta\chi(\infty)$, is plotted as a function of absolute temperature T. The full line shows the power law fit (see text).

the behaviour of $\Delta\chi(T)$ would be difficult to explain in terms of the superconducting fluctuations.

Next, we point out that the observed changes remind one on the anisotropy of typical quasi two dimensional metals like HMTSF-TCNQ[6] or Zr_3Te_5[7]. In Fig.(3) we plot again $-(\Delta\chi(T) - \Delta\chi(\infty))/\Delta\chi(\infty)$ for Y-Ba-Cu-O but now as a function of T. Above 120 K the data are described by a power law, AT^{α}, where $\alpha = -2.35$. Plotted in the same way, the HMTSF-TCNQ data would be given by a similar curve with $\alpha = -1.41$; here, the increase of the diamagnetic anisotropy is mainly due to the increase of the Landau-Peierls diamagnetism upon cooling. The deviation of the low temperature data from the power law is not surprising because the Landau-Peierls diamagnetism saturates at T=0. We can not claim that $\Delta\chi(T)$ in high T_c compounds follows simply from the band structure effects (that would require an effective degeneracy temperature of about 500 K) but we find the similarity between the high T_c data and the typical quasi two dimensional metals striking. It would be of interest to measure $\Delta\chi(T)$ in high fields, to see if $\Delta\chi(T)$ saturates as in other two dimensional metals. We remark that the data in Fig.(2) could not be fitted by a power law and the data in Fig.(3) could not be fitted by a logarithm in any reasonable temperature interval.

3. Conclusion. The magnetic measurements indicate that $\Delta\chi(T)$ of high T_c superconductors has a temperature dependent diamagnetic component which increases rapidly as temperature decreases. The temperature variation of $\Delta\chi(T)$

and its total change in the normal phase do not seem to be much affected by the presence of the superconductivity at some lower temperature and could not be accounted for by superconducting fluctuations. Rather, the data remind one on the behaviour of some quasi two dimensional metals with anisotropic Fermi surface and small degeneracy temperature. Furthermore, the measurements of the magnetic anisotropy indicate a two-step transition into the superconducting state, which is typical for granular superconductors. There exist a temperature region, $T_c<T<T_g$, in which the nonlinear effects due to the onset of superconductivity in disconnected grains are observed. These nonlinear effects have to be taken into consideration if the susceptibility data are to be compared with the theoretical results for $T>T_c$.

The financial support from Yu-USA DOE Grant No 817-31 is gratefully acknowledged.

4. References

1 K. Fukuda, S. Shamoto, M. Sato and K. Oka
Solid State Commun.**65**, 1323 (1988) (1987)
2 M. Miljak, G. Collin and A. Hamzić
Journal Mag. Mag. Mat. **76 & 77**, 609 (1988)
3 W. C. Lee, R. A. Klemm and D. C. Johnston
Phys. Rev. Lett. **63**, 1012 (1989)
4 M. Miljak, G. Collin, A. Hamzić and V. Zlatić
Europhys. Lett.**9**, 723 (1989)
5 M. Tinkham and C. J. Lobb, Solid State Phys.,Vol.42, p.91, (Turnbull and Seitz eds., Academic Press, 1989)
6 M. Miljak, J. R. Cooper and K. Bechgaard
J. Physique Coll. C38-893, **44**, (1983)
7 A. Smontara et al., Physics **143**B, 267 (1986)

Part V

Optical Spectroscopy, Raman Scattering

Optical Absorption in Non-BCS Superconductors

J. Orenstein [1,*], *S. Schmitt-Rink* [1], *and A.E. Ruckenstein* [2]

[1] AT&T Bell Laboratories, Murray Hill, NJ 07974, USA
[2] Serin Physics Laboratory, Rutgers University, Piscataway, NJ 08855, USA
* Present address: Physics Department, University of California, Berkeley, CA 94720, USA

We discuss the effects of electron-electron interactions on optical absorption of superconductors with intermediate to strong correlations. We argue that in the absence of extrinsic scattering the superconducting absorption threshold is not simply related to the order parameter or the gap in the single-particle spectrum, in contrast to BCS mean field theory.

1. Introduction

An unresolved question in the study of the new high-T_c superconductors is the spectrum of low-energy excitations, in particular the magnitude of the energy gap, Δ. In pursuit of this information, low-energy spectroscopies such as tunneling, Raman scattering, infrared absorption, and, more recently, photoemission, have been applied extensively. In many of these measurements a spectral feature suggestive of an energy gap appears below T_c, while the normal state spectrum itself is difficult to explain in the context of conventional metal theory. In this paper, we focus on the interpretation of the infrared absorption.

For several reasons, infrared spectroscopy is likely to play a central role in revealing the low-energy spectrum of the high-T_c superconductors. The photon penetration depth, δ, is much larger than the superconducting coherence length, ξ, so that infrared absorption is less surface sensitive than other techniques. In addition, experiments with polarized light are sensitive to the anisotropy of the excitation spectrum. However, despite general agreement on the main features of the optical spectra, the interpretation of the data has proved to be controversial. In this paper, we discuss several issues concerning optical absorption in superconductors with unusually strong electronic correlations, which are believed to be crucial to the understanding of the high-T_c materials. These issues are likely to be important for the ultimate interpretation of the experiments and, to our knowledge, have not been explicitly addressed in the literature.

We begin with a brief review of the current controversy. In the most thoroughly studied compound, $YBa_2Cu_3O_{7-x}$ (YBCO), experiments reveal a threshold at ~50meV in the optical absorption spectrum at low temperature [1]. Recent measurements on a detwinned crystal show that this feature is associated predominantly with the CuO_2 layers [2]. The controversy centers around its relation to the superconducting energy gap. Part of the discussion concerns the

Springer Series in Solid-State Sciences, Vol. 99
Electronic Properties of High-T_c Superconductors and Related Compounds
Editors: H. Kuzmany · M. Mehring · J. Fink

temperature dependence. However, the most important aspects, and the ones we consider in this paper, are the strength and frequency of the absorption threshold.

The notion of an absorption onset in a superconductor when the photon energy, ω, is equal to the energy gap, 2Δ, is a familiar one ($\hbar=1$). Yet this absorption is, in a sense, extrinsic to the BCS ground state. As was shown initially by MATTIS and BARDEEN [3], its strength is proportional to the rate of elastic scattering from impurities. However, several groups have argued that the absorption strength at $\omega \sim 50$meV in YBCO is too large to be consistent with the impurity scattering rate indicated by the dc resistivity [4]. Perhaps the most persuasive argument is the trend followed by the optical data as a function of sample quality. With improvements in sample preparation, the residual resistivity has decreased, indicating a smaller rate of impurity scattering. On the other hand, the normal state absorption strength near 50meV remains unchanged, and the threshold at low temperature is sharper. This behavior indicates that the absorption in this frequency range is probably not associated with impurity scattering, and that the Mattis-Bardeen calculation does not adequately describe the absorption in the superconducting state.

2. Interacting electrons on a lattice

To try to explain the absorption continuum, and the threshold which develops below T_c, we are led to focus on mechanisms for optical absorption in the absence of impurity scattering. To see how this absorption can arise intrinsically, we consider the general expression for the optical conductivity of an interacting electron system on a lattice [5,6]

$$\sigma_{xx}(\omega) = \frac{\pi}{V}[(-e^2a^2\langle T_x\rangle - \sum_n \frac{2\,|\langle 0|\,j_x\,|n\rangle|^2}{E_n-E_0})\,\delta(\omega) + \sum_n \frac{|\langle 0|\,j_x\,|n\rangle|^2}{E_n-E_0}(\delta(\omega-E_n+E_0)+\delta(\omega+E_n-E_0))]\,, \tag{1}$$

where

$$\langle T_x\rangle = -2t\sum_{\mathbf{k},\sigma}\cos(k_xa)\langle 0\,|\,c^*_{\mathbf{k}\sigma}c_{\mathbf{k}\sigma}\,|\,0\rangle \tag{2}$$

is the mean kinetic energy in the ground state and

$$j_x = -2tae\sum_{\mathbf{k},\sigma}\sin(k_xa)c^*_{\mathbf{k}\sigma}c_{\mathbf{k}\sigma} \tag{3}$$

is the current operator in the x-direction. $|0\rangle$ and $|n\rangle$ denote, respectively, the ground and excited states of the Hamiltonian, H, while E_0 and E_n are the corresponding eigenvalues. Above, we assumed for simplicity electron motion on a hypercubic lattice with lattice constant a and hopping matrix element t. By inspection of Eq. (1), the optical conductivity satisfies the f-sum rule

$$\int_{-\infty}^{\infty} d\omega\;\sigma_{xx}(\omega) = -\frac{\pi e^2a^2}{V}\langle T_x\rangle\,. \tag{4}$$

The essential feature introduced by the lattice is the possibility of optical absorption at nonzero frequencies, $\omega>0$, due to electron-electron interactions. This contrasts with translationally invariant systems in which j_x (with $\sin(k_x a)$ replaced by $k_x a$ in Eq. (3)) commutes with the Hamiltonian, H, and is diagonal in the $|n>$-basis. For these systems, all terms but the first in Eq. (1) vanish and the zero-frequency (δ-function) contribution exhausts the f-sum rule.

We are interested in the simplest model for lattice electrons for which the second term in Eq. (1) is nonzero. For noninteracting electrons, or electrons whose interactions are described by mean field theory, the Hamiltonian is of the form $H=\sum_{\mathbf{k}}\varepsilon_{\mathbf{k}} c^{*}_{\mathbf{k},\sigma} c_{\mathbf{k},\sigma}$, where $\varepsilon_{\mathbf{k}}$ is the effective single-particle energy. In spite of the presence of the lattice, the current operator Eq. (3) (with $2ta\sin(k_x a)$ replaced by $-\nabla_{\mathbf{k}}\varepsilon_{\mathbf{k}}$) commutes with this Hamiltonian and, again, the δ-function contribution to σ exhausts the sum rule, Eq. (4).

2.1 Interactions beyond mean field: normal state

Absorption at $\omega>0$ requires wave functions in which the interactions have been included beyond the mean field approximation. An illustrative example is a (zero momentum) ground state wave function of the form

$$|0> = \left[1 + \sum_{\mathbf{k},\mathbf{k}',\mathbf{q}} \phi_0(\mathbf{k},\mathbf{k}',\mathbf{q}) c^{*}_{\mathbf{k}\uparrow} c_{\mathbf{k}+\mathbf{q}\uparrow} c^{*}_{\mathbf{k}'\downarrow} c_{\mathbf{k}'-\mathbf{q}\downarrow}\right] |F> , \qquad (5)$$

where $|F>$ is the Fermi sea of the noninteracting system. Such an expression can be derived either from perturbation theory in the quasi-particle interaction or by expanding a Jastrow-type wave function to include only the two-particle-hole pair contribution. In this case, the matrix element between the ground state ($|0>$) and an excited two-particle-hole pair state ($|n>$) can be written as

$$<0|j_x|n> \propto \sum_{\mathbf{k},\mathbf{k}',\mathbf{q}} \phi_0^{*}(\mathbf{k},\mathbf{k}',\mathbf{q})\phi_n(\mathbf{k},\mathbf{k}',\mathbf{q}) <F|\ldots|F>$$

$$\times\left[\sin(k_x a)+\sin(k'_x a)-\sin([k_x+q_x]a)-\sin([k'_x-q_x]a)\right] , \qquad (6)$$

where the expectation value $<F|\ldots|F>$, the explicit form of which is not needed here, describes the phase space restrictions imposed by Fermi statistics. This nonvanishing matrix element corresponds to absorption extending from arbitrarily small frequencies, since there is no gap in the particle-hole excitation spectrum in the normal state. Notice that *both interactions and the lattice* are required for nonzero frequency absorption; as we have already implied above, in the continuum limit $\sin(k_x a)\to k_x a$ and $<0|j_x|n>\to 0$.

2.2 Interactions beyond mean field: superconducting state

We now use Eq. (1) to discuss the optical absorption in the superconducting state. It is interesting that in the BCS mean field pairing theory $\sigma(\omega)=0$ for $\omega>0$, just as in the mean field theory of the normal state, even in the presence of a lattice. To introduce notation, we briefly review this well-known result. Recall that the BCS Hamiltonian, H_{BCS}, is quadratic in the electron operators, $c_{k\sigma}$, but contains (anomalous) terms which create or destroy two electrons at a time. H_{BCS} can be diagonalized by introducing quasi-particle operators α_k and β_k through the Bogoliubov-Valatin canonical transformation $c_{k\uparrow} = u_k\alpha_k + v_k\beta^*_{-k}$ and $c_{-k\downarrow} = u_k\beta_{-k} - v_k\alpha^*_k$. The BCS ground state, $|BCS>$, is then the vacuum for the excitations α_k and β_k which in turn are described by the quasi-particle Hamiltonian

$$H_0 = \sum_k E_k(\alpha^*_k\alpha_k + \beta^*_k\beta_k) . \tag{7}$$

Here, $E_k=(\varepsilon_k^2+\Delta_{BCS}^2)^{1/2}$, $\varepsilon_k=-2t[\cos(k_xa)+\cos(k_ya)]-\mu$ (in two dimensions), μ is the chemical potential, and Δ_{BCS} is the BCS order parameter. The transformed current operator reads

$$j_x = -2tae\sum_k \sin(k_xa)[\alpha^*_k\alpha_k - \beta^*_{-k}\beta_{-k}]. \tag{8}$$

It is immediately clear that $[j_x,H_0]=0$ and, thus, the above result follows. More physically, the absence of optical absorption above zero frequency in the BCS state follows from the observations that (i) $|BCS>$ corresponds to a ground state built out of uncorrelated pairs each with zero total momentum and (ii) for *any* such pair state, the +**k** and −**k** contributions to the current cancel, as may be seen by considering the pair wave functions $|0(n)> = \sum_k \phi_{0(n)}(\mathbf{k})c^*_{k\uparrow}c^*_{-k\downarrow}|vac>$. This is, in disguise, the same argument which led to the vanishing of the finite-frequency absorption for noninteracting lattice quasi-particles.

Thus, to the extent that BCS theory is accurate, an absorption threshold at the gap frequency is expected only in the presence of extrinsic scattering. The regime of accuracy for H_{BCS} is weak coupling, where there are many electrons within the volume of a Cooper pair. In the intermediate to strong coupling regime, however, the pair radius, ξ, can be comparable to the interparticle separation, d, and correlations beyond BCS mean field theory become important. In fact, the high-T_c superconductors may be the first to manifest effects due to these correlations, since in these materials $\xi/d\sim2-3$ (which is two or three orders of magnitude smaller than in previously studied superconductors).

That correlations beyond those present in BCS lead to an absorption threshold becomes clear by generalizing the argument used for the normal state to the superconducting state. For example, we can include correlations between two pairs according to

$$|0> = \left[1 + \sum_{\mathbf{k},\mathbf{k}',\mathbf{q}} \Psi_0(\mathbf{k},\mathbf{k}',\mathbf{q})\alpha^*_{\mathbf{k}}\beta^*_{-\mathbf{k}+\mathbf{q}}\alpha^*_{\mathbf{k}'}\beta^*_{-\mathbf{k}'-\mathbf{q}}\right]|BCS> . \tag{9}$$

Such a state can be obtained by considering the conventional Jastrow projection on the BCS ground state to lowest order, as in Eq. (5), and the fact that $\alpha_k|BCS>=\beta_k|BCS>=0$. Due to the admixture of excitations with four quasi-particles (two broken Cooper pairs) into the ground state, $<0|j_x|n>$ is now nonzero if $|n\rangle$ is a four-quasi-particle excited state. The current matrix element is again proportional to $\sin(k_x a)+\sin(k'_x a)-\sin([k_x-q_x]a)-\sin([k'_x+q_x]a)$ which (as in the normal state example) vanishes in the continuum limit. A similar answer can be obtained from second-order perturbation theory in the interaction between the α_k and β_k excitations.

3. Conclusions

To summarize the main point of our discussion, many-body interactions on a lattice lead to a continuum of infrared absorption, if the interactions are treated beyond the mean field approximation. From the point of view of perturbation theory, the lowest order correction to the mean field ground state wave function contains two virtual particle-hole pairs. The matrix element of the current operator between this ground state and excited states with two real particle-hole pairs is nonzero, giving rise to an infrared continuum in the normal state. The strength of the absorption is $O(\lambda^2)$, where λ is the expansion parameter. The physics is essentially the same in the superconducting state, except that the gap leads to a minimum energy for creation of two particle-hole pairs. The onset of absorption (in the absence of impurity scattering) occurs at $4\Delta_{BCS}-O(\lambda^2)$, where the correction term reflects higher order interactions between excitations.

It is difficult to directly apply these ideas to the high-T_c superconductors, for which an adequate normal state theory does not yet exist. In particular, the small value of ξ/d suggests the presence of intermediate to strong correlations, outside the range of validity of simple perturbation theory. In addition, the infrared continuum is not weak, in the sense that the spectral weight of transitions above 50meV is of the same order as the spectral weight of the superfluid condensate (reflected in the zero-frequency δ-function). Nevertheless, independent of the strength of the interaction, the finite-frequency absorption in both normal and superconducting states always involves *at least two* (interacting) particle-hole pairs. This is most easily seen by rewriting the second term in Eq. (1) in terms of the force-force correlation function [7]

$$\sigma_{xx}(\omega>0) = \frac{1}{V\omega^3}\mathrm{Re}\int_0^\infty dt\, e^{i\omega t}\langle[\frac{dj_x}{dt}(t),\frac{dj_x}{dt}(0)]\rangle\ . \qquad (10)$$

Since the force, dj_x/dt, is a two-particle-hole pair operator, the above result follows. This simple result implies that in general the threshold energy in the superconducting state is not simply twice the order parameter, 2Δ, as in BCS theory, and hence need not be simply related to the gap seen in photoemission or tunneling. On the other hand, the optical gap is expected to be smaller than the perturbation theory result, 4Δ, since the pairs interact strongly in the excited state. A useful analogy is the biexciton state in semiconductors, which can lie well below the two-exciton continuum if exciton-exciton interactions are strong. Finally, we expect that absorption below the intrinsic onset can be seen due to single pair creation in the presence of impurity scattering [8].

Acknowledgements

We wish to thank E. Abrahams, G. Kotliar, P. B. Littlewood, A. J. Millis, and C. M. Varma for discussions. One of us (A.E.R.) would like to thank the Sloan Foundation for fellowship support.

References

1. T. Timusk, in: Physical Properties of High-Temperature Superconductors, ed. by M. Ginsburg (World Scientific, Singapore 1989), p. 339
2. Z. Schlesinger et al., unpublished
3. D. C. Mattis and J. Bardeen, Phys. Rev. 111, 412 (1958)
4. K. Kamaras et al., Phys. Rev. Lett. 64, 84 (1990); J. Orenstein et al., Phys. Rev. B, in press
5. W. Kohn, Phys. Rev. 133, A171 (1964)
6. H. J. Schulz, Phys. Rev. Lett. 64, 2831 (1990); B. S. Shastry and B. Sutherland, unpublished
7. A. J. Millis, unpublished; A. E. Ruckenstein and S. Schmitt-Rink, unpublished
8. As shown by P. B. Littlewood and C. M. Varma (unpublished), the 4Δ absorption threshold may also arise in the superconducting state of the putative "marginal Fermi liquid" recently proposed to describe the normal state of the high-T_c materials (C. M. Varma et al., Phys. Rev. Lett. 63, 1996 (1989); 64, 497 (1990)(E)). As an aside, we note that the zero-frequency δ-function contribution does *not* vanish in the normal "marginal Fermi liquid" state. Strictly speaking, as argued almost 30 years ago by Kohn [5], any metallic state at T=0 has a δ-function contribution to the optical conductivity; the coefficient of this contribution represents the sensitivity of the ground state wave function with respect to a change in boundary conditions and is analogous to ρ_s in superfluid systems. This point was reemphasized recently in an explicit Bethe Ansatz calculation of this coefficient in the one-dimensional Hubbard model [6]. Experimental verification of the vanishing of the quasi-particle peak characteristic of the "marginal Fermi liquid" requires a direct measurement of the single-particle spectral function, as done, for example, in a photoemission experiment. (This is analogous to the measurement of n_0, *not* ρ_s, in superfluids by fast neutron scattering.)

Optical Reflectivity of $Bi_2Sr_2CaCu_2O_8$ Crystals

K. Kamarás[1], *S.L. Herr*[2,*], *J.S. Kim*[2], *C.D. Porter*[2], *G.R. Stewart*[2], *D.B. Tanner*[2], *and T. Timusk*[3]

[1]Central Research Institute for Physics, P.O. Box 49, H-1525 Budapest, Hungary
[2]Department of Physics, University of Florida, Gainesville, FL 32611, USA
[3]Department of Physics and Materials Science, McMaster University, Hamilton, Ontario, L8S 4M1, Canada
*Present address: Physics Department, Virginia Commonwealth University, Richmond, VA 23284, USA

Abstract. Optical reflectivity of $Bi_2Sr_2CaCu_2O_8$ crystals has been investigated in the infrared and visible at different temperatures. The spectra in the superconducting phase show high reflectivity close to 100 per cent below 250 cm^{-1} . At 500 cm^{-1} (62 meV), a downturn in reflectivity is seen which persists at temperatures well above T_c. We analyze the spectra within a two-component model of a narrow Drude absorption and a temperature independent mid-infrared contribution. From the Drude parameters we conclude that this material, like $YBa_2Cu_3O_7$, is close to the clean limit and it is difficult to extract a gap value from the data in a straightforward way.

1. Two-component absorption and the clean limit

Infrared and optical spectra of high T_c superconductors show a complex structure already in the normal state. One of the peculiarities of these spectra is their "two-component" character[1], i.e. we cannot find a set of parameters describing the optical response as a simple metallic (Drude) process in the whole temperature and frequency range. One way of analysis is to assume a narrow free-carrier absorption and an additional bound-carrier contribution (the "mid-infrared band")[2]. Fitting procedures based on this picture have led to a conventional metallic temperature dependence of the Drude part (giving zero-frequency values in accordance with dc measurements), and little temperature variation in the mid-infrared[3]. The unusual low Drude relaxation rates found via these fits together with the high oscillator strength in the bound-carrier absorption, lead to important consequences regarding the visibility of the superconducting gap in infrared spectra[4]. Fig. 1 schematically illustrates the situation for a hypothetical case. For this simulation, we chose the relaxation rate $1/\tau$ = 150 cm^{-1} and the gap value 2Δ = 500 cm^{-1}. As Fig. 1a shows, if there were only the Drude component present, we still would see the onset at 2Δ with a small oscillator strength (moderately clean limit). However, if we add an extra component (represented here by a temperature independent Lorentzian for simplicity), the frequency where the conductivity deviates from zero reflects rather the onset of this Lorentzian, with a slight perturbation

Springer Series in Solid-State Sciences, Vol. 99
Electronic Properties of High-T_c Superconductors and Related Compounds
Editors: H. Kuzmany · M. Mehring · J. Fink

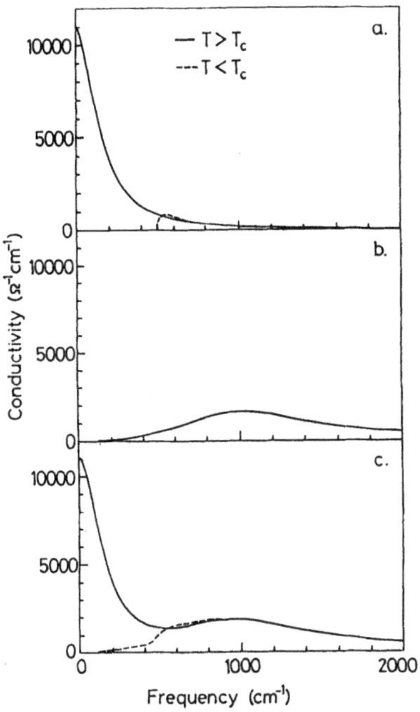

Fig. 1. Optical conductivity vs. frequency for

a) a superconductor with one-component (Drude) absorption for $T > T_c$ (ω_p = 10000 cm^{-1}, $1/\tau$ = 150 cm^{-1}, 2Δ = 500 cm^{-1})

b) a temperature-independent Lorentzian band (ω_p = 10000 cm^{-1}, ω_o = 1000 cm^{-1}, γ = 1000 cm^{-1})

c) a superconductor with two-component absorption for $T > T_c$ (same as in part a), with part b) added to both curves)

around 2Δ. Thus, the serious difficulties in determining the gap from infrared spectra in these materials, in contrast to conventional superconductors[5], are due to the combined effect of low free-carrier relaxation rate, high gap value and two--component absorption.

2. Reflectivity measurements on $Bi_2Sr_2CaCu_2O_8$

In order to further elucidate the above mentioned problems on other materials than $YBa_2Cu_3O_7$, we have investigated the reflectivity of a $Bi_2Sr_2CaCu_2O_8$ crystal grown from CuO flux, with $T_c = 84$ K. Optical reflectivity measurements were made with a Bruker IFS-113v interferometer in the far infrared and a grating monochromator in the near infrared and visible.

In Fig. 2. we plot the reflectivity and the optical conductivity (obtained by Kramers-Kronig transformation) at selected temperatures. The two-component character in the normal state is clearly evident especially from the conductivity curves in Fig. 2b. Applying a least-squares fit with a Drude--Lorentz model dielectric function to the reflectance spectra[4,6] we obtain the following Drude parameters: plasma frequency 10000 cm^{-1} (1.25 eV) at all temperatures, relaxation rate $1/\tau = 1\ cm^{-1}K^{-1}\cdot T$ (1.4 kT). The dc resistivity derived from these parameters follows closely that reported by Martin et al.[7].

In the superconducting state, there is a definite similarity to $YBa_2Cu_3O_7$[4,8]. In Fig. 3. we compare the 20 K spectra of the two materials. Both show very high (close to 100 per cent) reflectivity at low frequencies as well as a pronounced shoulder in the range 400 - 500 cm^{-1}, persisting in the normal

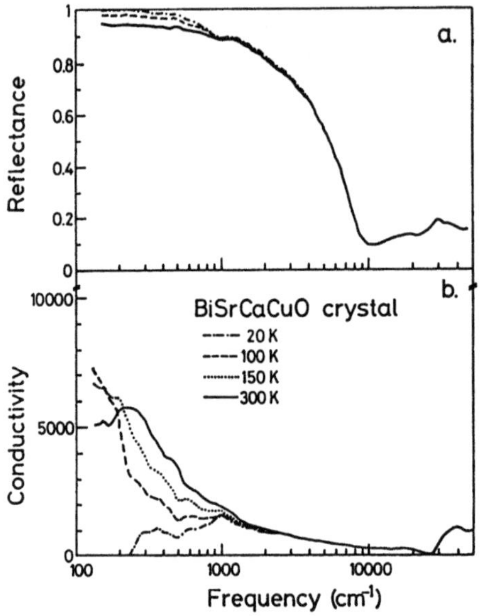

Fig. 2. Reflectivity (a) and conductivity (b) of $Bi_2Sr_2CaCu_2O_8$ at selected temperatures

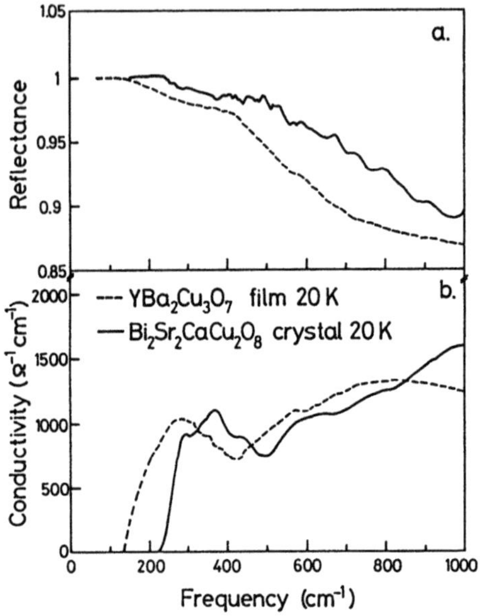

Fig. 3. Comparison of the reflectivity (a) and conductivity (b) of $Bi_2Sr_2CaCu_2O_8$ and an $YBa_2Cu_3O_7$ film (Ref. 4.)

state (best seen as a minimum in conductivity which is still visible at 150 K). This structure has been repeatedly associated with the superconducting gap[1,9]. (The exact shape of the spectrum below 300 cm^{-1} is subject to very high uncertainty[4,10] due to extremely small deviations from $R=1$ and consequently low signal-to-noise ratio; recent high-accuracy measurements indicate there is little significance in determining an "onset of absorption" in this region[11].) The striking feature of Fig. 3. is, however, that the shoulder in reflectivity (seen in conductivity as a rise after a minimum) occurs at about 25 per cent higher frequency (500 cm^{-1}=62 meV) in $Bi_2Sr_2CaCu_2O_8$ than in $YBa_2Cu_3O_7$, while T_c is 10 per cent lower in the former. These data are also in accordance with those measured previously by Reedyk et al.[12] on a different crystal. Orenstein et al.[13] have shown for oxygen-depleted crystals that the position of this feature is independent of transition temperature in $YBa_2Cu_3O_{7-x}$. A possible explanation, given the two-component character of the spectrum, would be that we are seeing the case depicted in Fig. 1c, where the structure in the superconducting state is determined by the bound-electron contribution connected to the chemical structure. The latter should certainly be different in the two substances, mainly because the "charge-reservoir" formations (Cu - O chains and Bi - O planes, respectively)[14] differ both in composition and dimensionality.

The above argument certainly does not mean there is no gap in this region, it merely implies that if it is there, it can be obscured by the intense second component of the conductivity. Recent ARPES measurements[15] estimate $\Delta = 20 - 25$ meV, so one would expect a gap in the infrared spectra at 40 - 50 meV. As seen in Fig. 3b, there is already considerable conductivity in this range which is not easy to assign unambiguously.

We also have encountered substantial difficulties in analyzing the shape of the mid-infrared band. The fit by Lorentzians used in Ref. 4. was arbitrary and required too many parameters. An alternative possibility has been described in Ref. 16, where we assumed a single electronic excitation coupled to a phonon at 433 cm^{-1}. Evidence for strong electron - -phonon coupling in this range has been observed in inelastic neutron scattering[17] and Raman[18] measurements. A fit along the same lines in $Bi_2Sr_2CaCu_2O_8$ would require a phonon of frequency 475 cm^{-1}, in order to reproduce the minimum in conductivity apparent in our spectra as well as those of Reedyk et al.[12]

Acknowledgements. The help of Drs. G.Mihály and G. Kriza in measuring the dc conductivity is greatly appreciated. Research at Florida supported by DARPA through contract MDA972-88-J-1006.

References

1. J. Orenstein, this volume
2. T. Timusk, S.L. Herr, K. Kamarás, C.D. Porter, D.A. Bonn, J.D. Garrett, C.V. Stager, J.E. Greedan, M. Reedyk, Phys. Rev. B38, 6683 (1988)
3. K. Kamarás, S.L. Herr, C.D. Porter, D.B. Tanner, S. Etemad, S.-W. Chan, in "Progress in High Temperature Superconductivity" vol. 17, p. 347 (World Scientific, 1989)
4. K. Kamarás, S.L. Herr, C.D. Porter, N. Tache, D.B. Tanner, S. Etemad, T. Venkatesan, E. Chase, A. Inam, X.D. Wu, M.S. Hegde, B. Dutta, Phys. Rev. Lett. 64, 84 (1990)
5. R.E. Glover, M. Tinkham, Phys. Rev. 104, 844 (1956)
6. K. Kamarás, S.L. Herr, C.D. Porter, D.B. Tanner, T. Venkatesan, E. Chase, A. Inam, H.D. Woo, M.S. Hegde, B. Datta, S. Etemad, in "Springer Series in Solid-State Sciences" vol. 91, p. 418 (Springer, 1989)
7. S. Martin, A.T. Fiory, R.M. Fleming, L.F. Schneemeyer, J.V. Waszczak, Phys. Rev. Lett. 60, 2194 (1988)
8. J. Schützmann, W. Ose, J. Keller, K.F. Renk, B. Roas, L. Schultz, G. Saemann-Ischenko, Europhys. Lett. 8, 679 (1989)
9. Z. Schlesinger, R.T. Collins, D.L. Kaiser, F. Holtzberg, Phys. Rev. Lett. 59, 1958 (1987)
10. G.A. Thomas, J. Orenstein, D.H. Rapkine, M. Capizzi, A.J. Millis, R.N. Bhatt, L.F. Schneemeyer, J.V. Waszczak, Phys. Rev. Lett. 61, 1313 (1988)

11. K.F. Renk, this volume
12. M. Reedyk, D.A. Bonn, J.D. Garrett, J.E. Greedan, C.V. Stager, T. Timusk, K. Kamarás, D.B. Tanner, Phys. Rev. B38, 11981 (1988)
13. J. Orenstein et al., preprint
14. B. Batlogg, this volume
15. J. Fink, this volume and private communication
16. T. Timusk and D.B. Tanner, preprint
17. C. Rietschel et al., this volume
18. C. Thomsen, this volume

Optical Excitations in Single Crystals of $YBa_2Cu_3O_{7-\delta}$ Between 5 meV and 6 eV

H.P. Geserich[1], *B. Koch*[1], *M. Dürrler*[1], *and Th. Wolf*[2]

[1]Institut für Angewandte Physik, Universität Karlsruhe, Kaiserstr. 12, D-7500 Karlsruhe, Fed. Rep. of Germany

[2]Kernforschungszentrum Karlsruhe, Institut für Technische Physik, Postfach 3640, D-7500 Karlsruhe, Fed. Rep. of Germany

Abstract. We report on the polarized reflectance spectra of single crystals of the orthorhombic superconductor $YBa_2Cu_3O_7$ and of the tetragonal semiconductor $YBa_2Cu_3O_6$ at room temperature. By fitting a Lorentz–Drude model and by Kramers–Kronig analysis the spectral distributions of the tensor components of the dielectric function parallel to the principal axes were determined. Analyzing the free carrier absorption in $YBa_2Cu_3O_7$, we could separate the contributions of the Cu–O planes and of the Cu–O chains to the electrical conductivity and could determine the anisotropy of the Fermi velocity, of the transfer integral, of the effective mass, of the Fermi surface and of other transport parameters.

1. Introduction

The optical properties of solids between the far infrared and the ultraviolet spectral region are determined by lattice vibrations, by the transport properties and by intrinsic absorption. The optical investigation of the transport properties of low–dimensional conductors, as $YBa_2Cu_3O_{7-\delta}$, offers, in particular, the opportunity to determine very directly and very precisely the anisotropy of these properties.

2. Experimental

Single crystals of $YBa_2Cu_3O_{7-\delta}$ were grown in Al_2O_3 crucibles by a technique described elsewhere [1]. The size of the crystals was about $2 \times 2 \times 1.5$ mm^3. They showed well reflecting surfaces. Annealing the crystals for several days between 450° and 650°C in an atmosphere of flowing oxygen resulted in a value of δ close to zero and a superconducting transition temperature $T_c = 90$ K. Generally, the (001) surface of these crystals was microtwinned. Some crystals, however, showed untwinned domains up to a size of $400 \times 600\,\mu$m^2. These domains were identified by a polarization microscope. The $YBa_2Cu_3O_6$ crystals were prepared by annealing for several hours at 720°C in an argon atmosphere and subsequent rapid cooling.

Reflection measurements at near–normal incidence were carried out in the spectral range between 50 meV and 6 eV with a single beam set–up, especially designed for small samples, whereas the FIR spectra were obtained by a Fourier transform spectrometer.

Electronic Properties of High-T_c Superconductors and Related Compounds
Editors: H. Kuzmany · M. Mehring · J. Fink

3. Results and Discussion

Fig.1(left side) shows in the upper part the polarized reflectance spectra of the superconducting compound $YBa_2Cu_3O_7$. It should be noticed that the plasma edge near 1 eV, reported in many papers for the electric field parallel to the (001) plane, is resolved into two plasma edges, showing a large energetic splitting of about 0.6 eV. The present result was obtained on an untwinned domain. This allows us to separate the reflectance spectra for the polarization direction parallel to the a– and b–axes [2,3] whereas in the microtwinned "single crystals", generally investigated, only an average of the two spectra can be observed [4].

It is widely accepted that the electronic transport properties of the oxide superconductors are correlated to the Cu–O network of these compounds. In $YBa_2Cu_3O_7$ two types of these Cu–O networks are present: first nearly square planes parallel to the (001) plane and second, linear chains parallel to the b-axis. Due to this crystallographic structure the spectrum $R_{\|a}$ with the direction of polarization perpendicular to the Cu–O chains reflects only the contribution of the Cu–O planes to the electronic transport properties, whereas the spectrum $R_{\|b}$ includes in addition the excitation of free carriers correlated to the Cu–O chains.

The continuous increase of the spectrum $R_{\|c}$ to lower energy shows very clearly that also for the electric field perpendicular to the Cu–O planes metallic behaviour is observed. The plasma edge of $R_{\|c}$, however, is shifted to much lower energy, indicating that $YBa_2Cu_3O_7$ is a three–dimensional but very anisotropic metallic conductor. The low metallic conductivity along the c–axis manifests itself also in the appearance of phonon structures in the spectrum $R_{\|c}$.

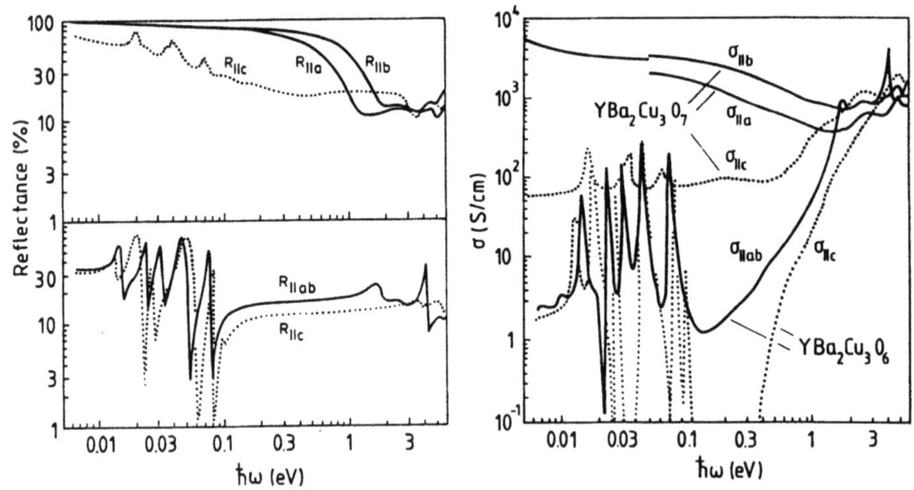

Fig.1. Polarized reflectance spectra of an $YBa_2Cu_3O_7$ single crystal (left side, upper part) and of an $YBa_2Cu_3O_6$ single crystal (left side, lower part) and the resulting conductivity spectra (right side), T = 300 K. The indices a, b. and c refer to the crystallographic axes.

That means that for this polarization direction the phonons are not completely screened, as observed for the polarization parallel to the Cu–O planes.

In the lower part of Fig.1(left side) the reflectance spectra $YBa_2Cu_3O_6$ are plotted. Both, for the polarization direction parallel and perpendicular to the Cu–O planes the metallic behaviour has completely disappeared. Instead, for the electric field parallel to the Cu–O planes in the high energy range pronounced structures at 1.7 and 4.1 eV are observed, already discussed in previous papers [5,6], and below 100 meV the optical phonons, screened in $YBa_2Cu_3O_7$, can now be seen . Comparing the spectra $R_{\|c}$ for the superconducting and the semiconducting phase, it is obvious that also for the direction of poor metallic conduction there is a large difference between the metallic and the semiconducting state.

The different absorption features, depending on the crystallographic direction and on the chemical composition, are compiled in Fig.1(right side). Here the spectral distributions of the tensor components of the conductivity $\sigma = \epsilon_0 \omega \epsilon_2$ are plotted. The figure reveals that below 50 meV the difference between the spectra $R_{\|a}$ and $R_{\|b}$ could not be resolved. It shows further that the anisotropy of the dc conductivity $\sigma_{\|ab}/\sigma_{\|c}$ of $YBa_2Cu_3O_7$ is about 70 at room temperature with a very low value of 70 S/cm for the metallic conductivity perpendicular to the Cu–O planes. The optical dc-value for $\sigma_{\|c}$, however, is close to the electrical value of high quality single crystals [7]. From the optical absorption of $YBa_2Cu_3O_6$ at higher energy an intrinsic absorption edge near 1.5 eV, as found in CuO, can be deduced.

After discussing the reflectance and the conductivity spectra qualitatively, some quantitative values of the transport parameters of $YBa_2Cu_3O_7$ will be deduced from the optical data. Supposing that the free carriers are holes in the O2p orbitals and all Cu ions are divalent, it follows from the chemical composition of $YBa_2Cu_3O_7$ that it has a p–type carrier concentration with one hole per unit cell or $5.8 \times 10^{21}\,cm^{-3}$, respectively. The Cu–O planes and the Cu–O chains are represented by two–dimensional (2–D) and one–dimensional (1–D) conduction bands. As the Cu–O distances within the planes and within the chains are nearly equal the same molecular overlap and therefore the same value of the effective mass for both arrangements is supposed. With these reasonable assumptions the components of the effective mass tensors parallel to the (001) plane and the distribution of the free charge carriers on both types of conduction bands can be determined from a Lorentz–Drude fit to $R_{\|a}$ and $R_{\|b}$. There results a value for the effective mass parallel to the (001) plane close to the free electron mass and a distribution of the free charge carriers of one third only on the Cu–O planes and two thirds on the Cu–O chains. Taking into account that there are two planes and one chain per unit cell, it results that the hole states in the 2–D conduction band are filled up to one twelfth and in the 1–D conduction band up to one third. Using a simple tight–binding model for both types of conduction bands, we can calculate further the values of the corresponding components of the Fermi wave vector k_F , of the transfer integral t, of the mean Fermi velocity $< v_F >$, and also the value of the Fermi

Table I: Electronic transport parameters of $YBa_2Cu_3O_7$

		$\parallel$ a	$\parallel$ b	$\parallel$ c
$\hbar\omega_p$	2-D	1.73	1.73	0.5
(eV)	1-D		2.40	
τ	2-D	3.5	3.5	1.9
(10^{-15} s)	1-D		1.2	
m^*/m_0	2-D	1	1	5
	1-D		1	
$< k_F >$	2-D	$0.23\pi/a$	$0.23\pi/b$	
	1-D		$0.33\pi/b$	
$< v_F >$	2-D	3.5	3.5	0.7
(10^7 m/s)	1-D		3.5	
t	2-D	320	320	65
(meV)	1-D		320	
$4(t_{\parallel a} + t_{\parallel b} + t_{\parallel c})$	2-D		2.9	
(eV)	1-D		1.6	
E_F			320	
(meV)				

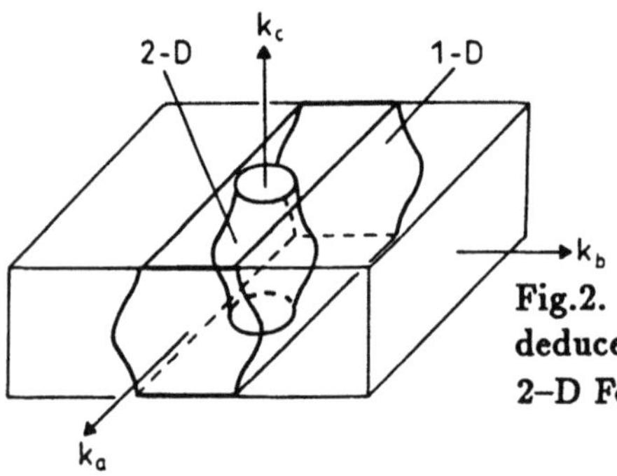

Fig.2. Fermi surfaces of $YBa_2Cu_3O_7$, deduced from the optical spectra. The 2–D Fermi surface is twofold degenerate.

energy E_F and of the total bandwidth $4(t_{\parallel a} + t_{\parallel b} + t_{\parallel c})$. The obtained data are compiled in Table I. The resulting Fermi surfaces are shown in Fig.2. The large anisotropy between the components of the plasma energy tensor for the electric field parallel and perpendicular to the Cu–O planes implies that the Fermi surfaces of $YBa_2Cu_3O_7$ are open along the c–axis and only slightly warped, compared to the strictly 2–D and 1–D cases, respectively. Along this axis, however, we cannot separate experimentally the dispersion of the 1–D and the 2–D bands.

It should be mentioned that the optical measurements do not supply directly the Fermi surfaces. They give only the mean values of the Fermi velocity along the different crystallographic axes, representing the dispersion of the conduction bands along these axes and hence the anisotropy of the Fermi surfaces. The Fermi volume, that is,the absolute values of the different Fermi wave vectors, however, must be deduced from the concentration of the free charge carriers or from band structure calculations. We prefer the low value of the free carrier concentration which follows form the O2p–hole model. The resulting small volume enclosed by the Fermi surfaces is in contradiction to the band structure calculations based on the local–density approximation (LDA). The LDA approach leads to nearly half–filled bands and hence to larger values of the Fermi wave vector [8]. But in connection with the low experimental values of the plasma energy this implies an unreasonably low dispersion of the conduction bands. These discrepancies will be discussed in detail in a forthcoming paper. It should be added that Fig.2 includes two further simplifcations. First, the coupling between adjacent Cu–O planes is neglected, taking the two–dimensional Fermi surface as twofold degenerate. Second, both Fermi surfaces were centered arbitrarily at the Γ–point of the Brillouin zone.

Acknowledgements. This work was supported by the Bundesministerium für Forschung und Technologie, Projekt-Nr FKZ:13N54864 and by the Sonderprogramm zur Grundlagenforschung auf dem Gebiet der Hochtemperatur-Supraleitung des Landes Baden-Württemberg

References

1.Th.Wolf, W.Goldacker, B.Obst, G.Roth, R.Flükiger: J.Cryst.Growth **96**, 1010(1989)
2.B.Koch, H.P.Geserich, Th.Wolf: Solid State Comm. **71**, 495(1989)
3.M.P.Petrov, A.I.Grachev, M.V.Krasinkova, A.A.Nechitailov, V.V.Poborchy,S.I.Shagin, S.V.Miridonov: JETP Letters **50**,29(1989)
4.R.T.Collins, Z.Schlesinger, F.Holtzberg, P.Chandhari, C.Feild: Phys.Rev.B **39**,6571(1989)
5.I.Kircher,M.Alouani, M.Garriga, P.Murugaraj, J.Maier, C.Thomson, M.Cardona, O.K.Andersen, O.Jepsen: Phys.Rev.B **40**,7368(1989)
6.M.K.Kelly, P.Barboux, J.M.Tarascon, D.E.Aspnes, W.A.Bonner, P.A.Morris: Phys.Rev.B **38**,870(1988)
7.G.Weigang, K.Winzer: Z.Phys. B **77**,11(1989)
8.J.Zaanen, A.T.Paxton, O.Jepsen, O.K.Anderson: Phys.Rev.Lett. **60**, 2 685(1988)

Strong-Coupling Effects in the Infrared Conductivity of Superconductors

W. Lee

Theoretical Physics III, University of Bayreuth,
D-8580 Bayreuth, Fed. Rep. of Germany

Infrared spectroscopy has been a useful method in superconductivity to determine the energy gap and to study the coupling of electrons to low lying excitations. Several groups applied this method to high T_c materials and were able to determine the dynamical conductivity of the conducting CuO_2 planes in the far-infrared range [1,2,3,4]. The far-infrared spectra, we believe, contain substantial information on the new superconducting materials. Their interpretation, however, has been very divergent and controversial [5]. One of the issues in these circumstances was, whether the measured spectrum is understandable at all in terms of the conventional theory of superconductivity.

This paper discusses strong-coupling effects in the infrared conductivity of conventional anisotropic superconductors, with the purpose to explore the typical anomalies coming from a strong coupling of the superconducting electrons to low lying excitations with energies of the order of a few $k_B T_c$.

We calculate the dynamical conductivity $\sigma(\omega)$ in the framework of Eliashberg's strong-coupling theory. Eliashberg's theory is exact to the leading order in $k_B T_c/E_F$ and $\hbar\Omega_{ph}/E_F$, where Ω_{ph} is the typical frequency of the phonons. If superconductivity is due to a *high energy* mechanism,Eliashberg's theory reduces to the BCS theory. Strong-coupling effects are in this context understood as deviations from the weak-coupling BCS results. Such deviations arise in the conventional picture of superconductivity if the pairing is mediated predominantly by low lying modes, e.g. phonons, spin fluctuations *etc.* with energies of order $k_B T_c$.

The general formula of the dynamical conductivity for a traditional superconductor can be derived using the quasiclassical linear response theory [6]. Here we will confine our discussions to a somewhat simplified situation. Namely, we model the 2-dimensional character of the oxide superconductors by taking a cylinderical Fermi surface with Fermi velocity v_F in the planes, and approximate the interactions by short range potentials. We evaluate $\sigma(\vec{Q},\omega)$ in $\vec{Q} \to 0$ limit, which is appropriate for the high temperature superconductors with very short coherence length and long penetration depth.

Springer Series in Solid-State Sciences, Vol. 99
Electronic Properties of High-T_c Superconductors and Related Compounds
Editors: H. Kuzmany · M. Mehring · J. Fink

The conductivity $\sigma(\omega)$ is then determined by four input material parameters [7]; the coupling function $\alpha^2F(\omega)$ to low lying excitations of frequency ω, the Coulomb pseudopotential μ^*, the plasma frequency ω_p, and the impurity scattering rate $1/\tau$.

The electron–electron interaction and the electron–phonon (or electron–spin-fluctuation *etc.*) interaction, which are described by the Coulomb pseudopotential μ^* and the coupling function $\alpha^2F(\omega)$, can not be evaluated reliably from first principles. So one usually takes $\alpha^2F(\omega)$ and μ^* from experiments, for instance, from electron tunneling measurements. In the absence of good tunneling data one has to rely on models for $\alpha^2F(\omega)$ and μ^*.

Once one has $\alpha^2F(\omega)$ and μ^* one can calculate all measurable quantities of the superconductor from these inputs. Strong-coupling effects in the dynamical conductivity come from the gap function $\Delta(\varepsilon)$ and the renormalization function $Z(\varepsilon)$, which are solutions of the Eliashberg's equations. If one uses constant values for $\Delta(\varepsilon)$ and $Z(\varepsilon)$, one gets the weak-coupling results. The normal state conductivity can be obtained by setting the gap $\Delta(\varepsilon)$ equal to zero, the strong-coupling effect then comes from the renormalization function $Z(\varepsilon)$ alone.

The plasma frequency ω_p enters the results only as an overall prefactor, it behaves as a scaling factor without affecting the frequency dependence. E.g., the magnitude of $\sigma(\omega)$ is scaled by ω_p^2, the slope of the temperature dependent dc-resistivity at high temperatures is given by $8\pi^2\lambda/\omega_p^2$. The impurity scattering rate $1/\tau$ is a temperature independent constant.

A characteristic feature of strong-coupling effects in the infrared conductivity is the Holstein anomaly, i.e. the anomalous absorption through the combined excitation of a particle–hole pair and a phonon mode or other low lying modes.

Fig.1 shows a typical result of the dynamical conductivity calculated for an Einstein phonon model with parameters; $\hbar\Omega = 50\ meV$, $\lambda = 1$ and $\mu^* = 0.08$ (which give $T_c = 50\ K$ and $2\Delta_o/k_BT_c = 4$). $\hbar\omega_p$ and $\hbar/\tau$ are chosen 920 and 6.5 meV, respectively. The weak-coupling Drude (dotted line) and Mattis-Bardeen (dashed line) curves are included for comparison. They are determined by λ, ω_p, $1/\tau$ and Δ_o alone, vary monotonously, and are temperature independent (beside the temperature dependence of Δ_o). The strong-coupling effect gives deviations from these weak-coupling curves and additional structures:
At low temperatures, $k_BT < \hbar\Omega/5$, and at low frequencies $\omega < \Omega$, quasiparticles cannot emit real phonons. They are fully dressed by virtual phonons. Thus the conductivity in the normal state is of the Drude form,

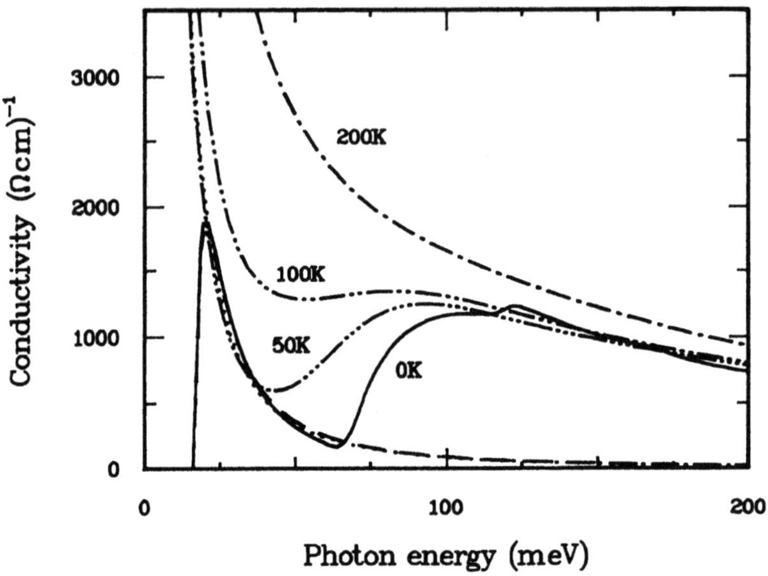

Fig.1: The real part of $\sigma(\omega)$ for various temperatures (*see* text)

with renormalized parameters, $\omega_p^* = \omega_p/(1+\lambda)^{1/2}$ and $\tau^* = \tau(1+\lambda)$. At frequencies $\omega \approx \Omega$ the Holstein mechanism sets in, and at frequencies $\omega \gtrsim \Omega$ it gives rise to additional absorption. That is, the quasiparticles can now emit real phonons and at the same time the renormalization becomes reduced. The conductivity approaches the Drude values determined by the bare plasma frequency ω_p and the effective scattering rate $1/\tau_{eff} = 1/\tau + 1/\tau_{ph}$. In the superconducting state, one needs a minimum energy of $2\Delta_o$ for the particle–hole pair excitation, and the Holstein anomaly shifts upwards by $2\Delta_o$ [8]. This shift-up, however, is hardly observable if one compares the zero temperature conductivity in the superconducting state with that of the normal state at temperatures $k_BT > \hbar\Omega/5$, above T_c: At high temperatures, the thermal excitation of phonons causes the broadening of the anomaly, and gives rise to additional scattering at low frequencies which results in a decrease of the conductivity. The decrease is proportional to $1/T$ at high temperatures at $\omega = 0$. This gives the linear temperature dependence of the dc-resistivity with a slope proportional to the coupling constant λ.

To investigate the electron–spin-fluctuation coupling, we have calculated $\sigma(\omega)$ with the assumption that the coupling in the *pairing*-channel is attractive and d-wave like [9], and the coupling in the *renormalization*-channel is of s-wave symmetry. The results [10] show essentially the same feature above the gap frequency as in the electron–phonon case. The difference from the electron–phonon case is that the typical *gap edge* .

structure in the dynamical conductivity in the superconducting state is smeared out, $\sigma(\omega)$ is nonzero down to the frequency zero, and the Holstein anomaly is somewhat rounded off, because of the distribution of gap values on the Fermi surface. If modes of different frequencies, Ω_d and Ω_s, dominate the d-wave and the s-wave channels, respectively, then the shift of the Holstein anomaly in the superconducting state is $2\Delta_0 + \Omega_d - \Omega_s$ instead of $2\Delta_0$. We note that in this simple model the Holstein-Allen shift can also be downward. The shape of $\sigma(\omega)$ within the *gap* depends on the purity of the superconductor; in the dirty limit it decreases smoothly to zero as $\omega \to 0$, and for pure superconductors it increases, showing a more Drude-like behaviour.

The detailed structure of the dynamical conductivity strongly depends on the material parameters: the coupling function $\alpha^2F(\omega)$, and the impurity scattering rate $1/\tau$.

We summarize the four characteristic features of $\sigma(\omega)$ in the strong-coupling theory:

1) The Holstein anomaly.
The anomaly is in general more clearly observable when the coupling is stronger and the impurity scattering rate is smaller. The position shifts upwards in superconducting state by the gap energy $2\Delta_0$ as compared to the $T = 0$ normal state anomaly.
2) Weakly temperature dependent high frequency conductivity ($\omega \gg \Omega$).
This part of the conductivity is Drude-like, and is governed by the bare plasma frequency ω_p and the effective life time τ_{eff}.
3) Strongly temperature dependent low frequency conductivity ($\omega \lesssim \Omega$).
This part of the conductivity reflects the temperature dependent life time and the renormalization effects of the low lying modes [11]. At $T \approx 0$ it is governed by the fully renormalized parameters, ω_p^* and τ^*. At $\omega = 0$ it gives the linear temperature dependent dc-resistivity, with the slope $8\pi^2\lambda/\omega_p^2$ for $k_BT \gtrsim \hbar\Omega/5$.
4) The enhanced gap edge $\omega_g = 2\Delta_0$.
It depends on the coupling strength and is generally larger than the BCS value $3.5k_BT_c$.

I gratefully acknowledge the valuable discussions with Prof. D. Rainer and his critical reading of the manuscript.

References

[1] G. A. Thomas, J. Orenstein, D. H. Rapkine, M. Vapizzi, A. J. Millis, L. F. Schneemeyer, and J. V. Waszczak, Phys. Rev. Lett. **61**, 1313 (1988).

[2] K. F. Renk, H. Eschrig, U. Hofmann, J. Keller, J. Schützmann, and W. Ose, Physica C **165**, 1 (1990).

[3] K. Kamarás, S. L. Herr, C. D. Porter, N. Tache, B. B. Tanner, S. Etemad, T. Venkatesan, E. Chase, A. Inam, X. D. Wu, M. S. Hegde, and B. Dutta, Phys. Rev. Lett. **64**, 84 (1990).

[4] Z. Schlesinger, R. T. Collins, F. Holtzberg, C. Felid, N. E. Bickers, and D. J. Scalapino, Nature **343**, 242 (1990).

[5] T. Timusk and D. B. Tanner, in: *Physical Properties of High Temperature Superconductors I, ed. by D. M. Ginsberg* (World Scientific, Singapore, 1989), p339.

[6] W. Lee, D. Rainer, and W. Zimmermann, to be published.

[7] W. Lee, D. Rainer, and W. Zimmermann, Physica C **159**, 535 (1989).

[8] P. B. Allen, Phys. Rev. B **3**, 305 (1971).

[9] D. J. Scalapino, E. Loh and J. E. Hirsch, Phys. Rev. B **35**, 6694 (1987).

[10] W. Lee, *Doktorarbeit*, Universität Bayreuth (1990).

[11] G. Grimvall, *The electron-phonon interaction in metals* (North-Holland, Amsterdam, 1981).

Far-Infrared Excitations in $YBa_2Cu_3O_{7-\delta}$ Thin Films

K.F. Renk[1], *J. Schützmann*[1], *A. Prückl*[1], *W. Ose*[1], *B. Roas*[2,3], *L. Schultz*[2], *and G. Saemann-Ischenko*[3]

[1]Institut für Angewandte Physik, Universität Regensburg, D-8400 Regensburg, Fed. Rep. of Germany
[2]Siemens AG, Research Laboratories, D-8520 Erlangen, Fed. Rep. of Germany
[3]Physikalisches Institut, Universität Erlangen, D-8520 Erlangen, Fed. Rep. of Germany

Abstract. Far-infrared reflectivity measurements have been performed for $YBa_2Cu_3O_{7-\delta}$ thin films. From the reflectivity data the dynamical conductivity and absorptivity are determined. The far-infrared properties can be characterized by three different absorption processes: Absorption by free carriers that condensate below T_c, a temperature independent absorption process with a threshold at 400 cm^{-1}, and low frequency rest absorption at low temperature that is possibly due to normalconducting carriers.

1. Introduction

Far-infrared spectroscopy is well suited to determine low energy excitations in high temperature superconductors; for a survey see Ref.[1]. In this paper we will concentrate on excitations in the a-b plane of $YBa_2Cu_3O_{7-\delta}$. Recent results for high quality samples, namely for single crystals [2,3] and oriented thin films [4,5], indicate that the low energy excitations are of complicated structure. In this paper we report results obtained by high accuracy far-infrared reflectivity measurements on high quality oriented thin films and we present an analysis that leads us to the conclusion that three different low energy absorption processes occur: Absorption by free charge carriers that condensate below T_c, an almost temperature independent absorption process that sets in at 400 cm^{-1}, and absorption by free charge carriers present also at low temperature that extends from microwaves into the far-infrared spectral region.

2. Electrodynamics of the normal and superconducting state

We will describe the normal state conductivity by free charge carriers by a Drude conductivity

$$\sigma = \frac{\varepsilon_0 \omega_p^2}{\Gamma - i\omega} \qquad (1)$$

Springer Series in Solid-State Sciences, Vol. 99
Electronic Properties of High-T_c Superconductors and Related Compounds
Editors: H. Kuzmany · M. Mehring · J. Fink

where $\omega_p = (Ne^2/m\varepsilon_o)^{1/2}$ is the plasma frequency, N the concentration, m the effective mass, Γ the scattering rate of the free carriers and ω the frequency of the electromagnetic radiation.

Assuming that the superconducting state corresponds for low temperature, $T \ll T_c$, to the ultra-clean limit we find from eq.(1), for the corresponding perfect conductor with $\Gamma \to 0$ (and $\omega \neq 0$) the dynamical conductivity [6]

$$\sigma = i\frac{\varepsilon_0\,\omega_p^2}{\omega} = \frac{i}{\mu_0\lambda_L^2\omega} \tag{2}$$

where $\lambda_L = c/\omega_p$ is the London penetration depth, that is the penetration depth of the magnetic field and corresponds to the range in which a supercurrent is induced by the electromagnetic field; eq.(2) can also be derived from London's equations. A superconducting energy gap at energy 2Δ manifests itself in absorption for $\omega > 2\Delta/\hbar$, then, eq.(2) has to be modified [7].

If there are, at low temperature ($T \ll T_c$), normalconducting charge carriers besides the superconducting condensate we may describe the dynamical conductivity in a two-fluid model by the sum of the dynamical conductivities of the condensate and of the normalconducting carriers

$$\sigma = \frac{i}{\mu_0\,\lambda_L^2\omega} + \frac{\varepsilon_0\,\bar{\omega}_p^2}{\bar{\Gamma} - i\omega} \tag{3}$$

where $\bar{\omega}_p = (\bar{N}e^2/\bar{m}\varepsilon_0)^{1/2}$ is the plasma frequency, $\bar{N}$ the concentration, $\bar{m}$ the effective mass and $\bar{\Gamma}$ the scattering rate of the normalconducting charge carriers. Now, absorption due to the normalconducting carriers occurs. For a thick superconductor the absorptivity A is given by

$$A = \left(\frac{\varepsilon_0\omega}{\sigma_2}\right)^{\frac{1}{2}} \frac{\sigma_1}{\sigma_2} = \frac{2}{c}\,\mu_o\,\lambda_L^3\,\bar{\sigma}_0\,\omega^2 \tag{4}$$

where σ_1 and σ_2 are real and imaginary part of σ and $\bar{\sigma}_0 = \varepsilon_0\,\bar{\omega}_p^2/\bar{\Gamma}$ the d.c. conductivity of the normalconducting carriers; we assume that $\sigma_1 \ll \sigma_2$. Eq.(4) shows that the absorptivity increases quadratically with frequency (for $T \ll T_c$). The corresponding surface resistance is $R_s = \frac{1}{4}\,A\,(\mu_0/\varepsilon_0)^{\frac{1}{2}}$.

3. Experimental results and discussion

Our $YBa_2Cu_3O_{7-\delta}$ thin films ($T_c \approx 90$ K, $j_c \approx 2\cdot 10^6$ A/cm^2 at 77 K) were prepared by laser ablation. The films, thickness about 4000 Å, had a roughness of about 40 Å and a thickness variation of about 400 Å.

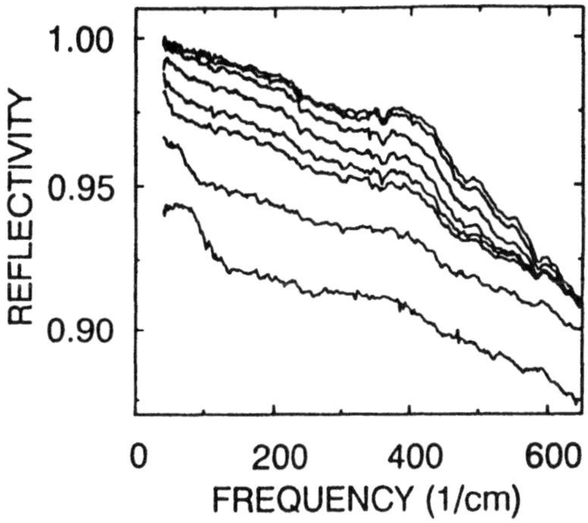

Fig. 1. Far-infrared reflectivity of a $YBa_2Cu_3O_{7-\delta}$ ($\delta \simeq 0$) film (at T = 12; 44; 69; 81; 90; 100; 200; 300 K).

The far-infrared reflectivity measurements have been performed with a Fourier infrared spectrometer. The sample was mounted together with a gold reference mirror (gold evaporated on a glass plate) in a cryostat and cooled by helium exchange gas [4].

There are three main features that characterize the far-infrared reflectivity (Fig. 1):
(1) The low frequency reflectivity (up to 400 cm^{-1}) is strongly increasing towards low temperatures, indicating an increase of conductivity. Above T_c, this conductivity is due to *free charge carriers*. Below T_c, these carriers condensate, giving rise, for $T \ll T_c$, to a reflectivity near 1.
(2) The low frequency reflectivity is, for $T \ll T_c$, however, smaller than 1, i.e., there is a *rest absorption* at low temperature. We describe the deviation of the reflectivity from 1 by an absorptivity $A = 1 - R$ and find that A increases approximately as ω^2 up to about 200 cm^{-1} .
(3) There is a *shoulder* at 400 cm^{-1}visible at all temperatures, up to room temperature. The position of the shoulder does not shift with temperature. The shoulder indicates the onset of an absorption process that extends up to very high frequencies. Our result is in agreement with various studies [2,3,5] that show that the onset of absorption at 400 cm^{-1} does not correspond to a conventional BCS superconducting energy gap; such a gap should go to zero near T_c and should not be visible above T_c. An analysis of the absorption that sets in at 400 cm^{-1} indicates that the strength may be attributed to excitation of a large amount of charge carriers.

It is known from an earlier study [4] that σ_2 is very large at small frequencies and varies as ω^{-1} according to eq.(2). From the results [4] we find the London penetration depth $\lambda_L \simeq 1800$ Å. The absorptivity (Fig. 1) can be described by eq.(4), with $\bar{\sigma}_0 \simeq 6000\ \Omega^{-1}cm^{-1}$. The experiment (Fig. 1) indicates that the absorptivity deviates from a ω^2

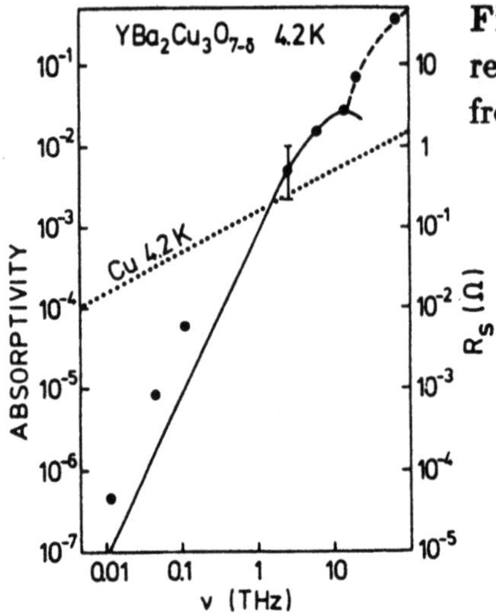

Fig. 2. Absorptivity (and surface resistivity) of $YBa_2Cu_3O_{7-\delta}$ ($\delta \simeq 0$) from microwaves to the infrared.

dependence according to a scattering rate $\bar{\Gamma} \simeq 200\ cm^{-1}$ for the normalconducting carriers at low temperature. We thus find a plasma frequency $\bar{\omega}_p \simeq 6000\ cm^{-1}$ for the normalconducting carriers as compared to $\omega_p = c/\lambda_L \simeq 9000\ cm^{-1}$. Thus, the concentration of normalconducting electrons is comparable to the concentration of condensated charge carriers.

In Fig. 2 we have plotted the absorptivity measured in the far-infrared together with the absorptivity measured at microwave frequencies [8,9]. The experimental values lie – within a factor of two – on the ω^2 curve derived from our far-infrared data. We suggest therefore that the absorption in both spectral regions can be attributed to the same absorption process, possibly caused by normalconducting carriers.

The work was supported by the Bundesministerium für Forschung und Technologie.

[1] T. Timusk and D. Tanner, in *Infrared Properties of High-T_c Superconductors*, ed. D.M. Ginsberg (World Scientific Publishing, 1989).

[2] R.T. Collins, Z. Schlesinger, F. Holtzberg, and C. Field, Phys. Rev. Lett. **63**, 422 (1989).

[3] G.A. Thomas, J. Orenstein, D.H. Rapkine, M. Capizzi, A.J. Millis, R.N. Bhatt, L.F. Schneemeyer and J.V. Waszczak, Phys. Rev. Lett. **61** (1988) 1313; G.A. Thomas, J. Orenstein, S.L. Cooper, D. Rapkine, M. Capizzi, T. Timusk, A.J. Millis, L.F. Schneemeyer, and J.V. Waszczak, Phys. Rev. B, to be published.

[4] J. Schützmann, W. Ose, J. Keller, K.F. Renk, B. Roas, L. Schultz, and G. Saemann-Ischenko, Europhys. Lett. **8**, 679 (1989).

[5] K. Kamarás, S.L. Herr, C.D. Porter, N. Tache, D.B. Tanner, S. Etemad, T. Venkatesan, E. Chase, A. Inham, X.D. Wu, M.S. Hegde, and B. Dutta, Phys. Rev. Lett. **64**, 84 (1990).
[6] M. Tinkham, *Introduction to Superconductivity*, McGraw-Hill 1975.
[7] D.C. Mattis and J. Bardeen, Phys. Rev. **111**, 412 (1958).
[8] N. Klein, G. Müller, H. Piel, B. Roas, L. Schultz, U. Klein and M. Peiniger, Appl. Phys. Lett. **54**, 757 (1989).
[9] G. Müller, *Microwave Properties of High-T_c Superconductors*, 4th Workshop on RF Superconductivity, at KEK, Tsukuba-Shi (1989).

Low Energy Optical Excitations in High-T_c Superconducting Systems from Photomodulation Spectroscopy

C. Taliani[1], *A.J. Pal*[1], *G. Ruani*[1], *R. Zamboni*[1], *X. Wei*[2], *and Z.V. Vardeny*[2]
[1]Istituto di Spettroscopia Molecolare, CNR, I-40126 Bologna, Italy
[2]Department of Physics, University of Utah, Salt Lake City, UT84112, USA

Abstract. Doping the semiconducting parent of cuprate and bismuthate perovskites by laser illumination is used to investigate the optical properties of the normal state. A new intrinsic charge photoinduced excitation having an onset at about 0.3 eV and extending over the semiconducting gap is a characteristic of all these systems. The charge excitation is linked to local lattice distortions which are typical of the normal state indicating a strong electron-phonon interaction.

1. Introduction

Several research groups have shown that in cuprate superconductors the non-Drude component of the optical conductivity is constituted by a broad mid-infrared absorption band and an additional feature at about 50 meV [1]. The nature of these optical excitations, which are present in both the normal and the superconducting state, are still unclear. A different approach to the investigation of the optical properties of high temperature superconductors is offered by the photoinduced absorption (PA) technique. It consists in the investigation of the variation of the optical properties of the semiconducting parent of the high temperature superconducting system by "doping" it with e-h pairs created by laser illumination with photons having an energy higher than the semiconducting energy gap [2-9]. It has been shown that the photoinjected charge carriers are associated with lattice distortions and with a low energy optical excitation at about 0.1eV in La_2CuO_4 [2,6], in $YBa_2Cu_3O_{7-x}$ [4,5] and in $Tl_2Ba_2Ca_{1-x}Gd_xCu_2O_8$ [6,7]. Moreover an extension of the PA study at higher energies has shown the presence of a broad optical excitation in $YBa_2Cu_3O_{7-x}$ and La_2CuO_4 which spans the whole spectral range of the semiconducting gap [9]. More recently it has been shown that $BaBiO_3$, which is the semiconducting parent of the new high temperature superconductor $Ba_{1-x}K_xBiO_3$ (T_c=30K) also displays the same PA characters [10]. In this paper we present an analysis of the lattice distortions and low energy charge excitations by means of PA and photomodulation spectroscopy (PMS) in $YBa_2Cu_3O_{7-x}$,La_2CuO_4 and $BaBiO_3$ with the aim of shedding some light on the nature of low energy charge excitations in the normal state of high temperature superconductors.

2. Experimental

Pressed pellets of the powdered ceramic materials with KBr (or CsI or polyethylene) are investigated by transmission with an FT-IR interferometer (0.03-.5eV) and a dispersive spectrometer (>0.5eV). Laser light (2-2.5eV) is allowed to illuminate the sample. The number of carriers generated by

Springer Series in Solid-State Sciences, Vol. 99
Electronic Properties of High-T_c Superconductors and Related Compounds
Editors: H. Kuzmany · M. Mehring · J. Fink

illumination is estimated in the order of 10^{19} cm^{-3}. Details of the experimental procedures are reported in Ref.4 and 8.

3. Results and Discussion

3.1 $YBa_2Cu_3O_{7-x}$

In Figure 1 the PA spectrum in the phonon spectral range and the PM spectrum of $YBa_2Cu_3O_{7-x}$ (x=0.8) at low temperature are reported. It must be noticed that while the PA (left side) reflects the steady state response of the system, the PM spectrum (right side) is taken at a photomodulation frequency (f) of 200 Hz and hence is probing the response of the subset of the photoinduced species with a lifetime of f^{-1}. The low energy is characterized by the activation of two A_g phonon modes at 435 and 500 cm^{-1}. Their activation is the result of the local symmetry suppression due to the lattice deformation caused by the photoinjected carriers. The negative band (photobleaching) of the 590 cm^{-1} marks the local suppression of a phonon mode which is characteristic of the tetragonal structure [11]. Both these effects indicate that the photoinjected carriers produce a local lattice distortion which is typical of the normal state structure. In particular the 435 cm^{-1} lattice mode is the vibration of the oxygen atoms in the Cu-O planes in which the oxygen atoms move perpendicularly to the Cu-O planes modulating the corrugation of the planes. The 500 cm^{-1} mode is also present in the normal state and has been assigned to the axial $Cu\text{-}O_4$ axial stretching vibration [12]. On the higher energy we observe the rising edge of a broad dipole allowed absorption with a rather asymmetric shape and a maximum at 0.13 eV (LE) [4]. This feature is strongly dependent on f; while its intensity in the steady state regime is of $4x10^{-3}$ it diminishes to $1.3x10^{-4}$ at f=200 Hz. A delayed PA experiment confirms this finding. In fact we observe that the 0.13 eV is still one third of the intensity of the steady state regime when the probe IR beam is delayed by as much as 350 ms from the pump laser beam. Going towards higher energies we observe a broad and flat PA with an onset at 0.25eV which extends over the whole spectral range of the semiconducting gap (HE). The crossover at 1.6 eV indicates the presence of the

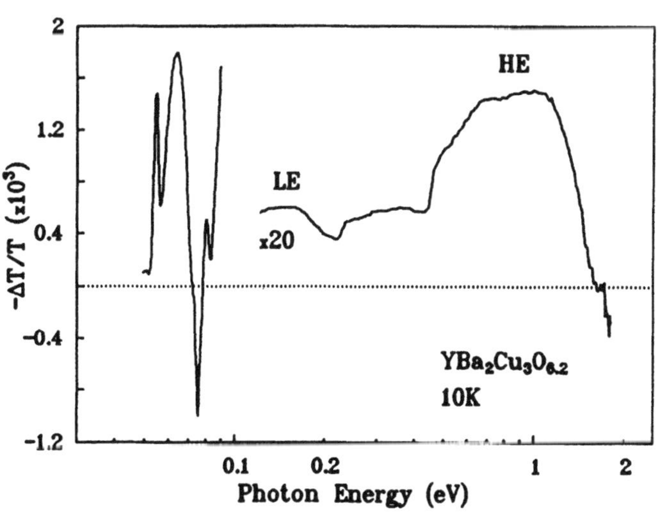

Fig.1. Photoinduced infrared absorption (left side) and photomodulation (200 Hz) absorption (right side) of $YBa_2Cu_3O_{7-x}$ (x=.8) at 10K. Positive bands indicate the activation of new bands after photoinjection of carriers;negative bands indicate band suppression.

semiconducting gap. Above this energy the PM spectrum shows the presence of a bleaching. This is in agreement with the location of the semiconducting gap at 1.6 eV [13]. The bleaching of the interband electronic absorption and the corresponding

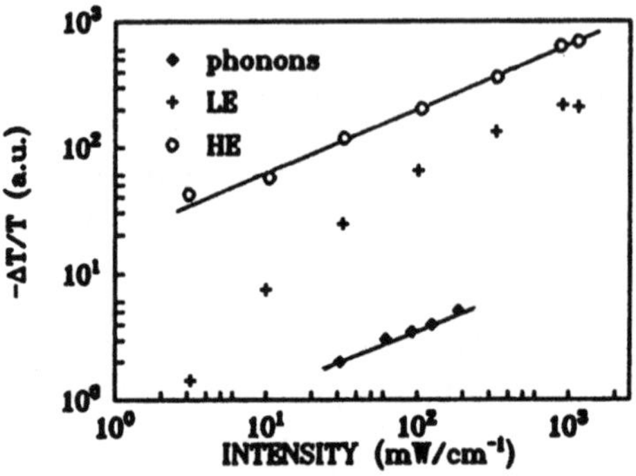

Fig. 2. Laser intensity dependence of the photoinduced phonon and electronic absorptions in $YBa_2Cu_3O_{6.2}$. Note that the LE photoinduced absorption at 0.13 eV manifests a saturation behaviour.

activation of an electronic absorption in the semiconducting gap indicates that the photoinjection of carriers induces a shift of the oscillator strength from the gap states toward new states into the gap. The PA phonon bands as well as the HE band show a square root dependence on the laser illumination intensity (see Figure 2). This suggests that contrary to ordinary semiconductors the photogenerated carriers undergo a bimolecular recombination. In contrast the LE band shows a marked tendency towards saturation above 10^3 mW/cm^2. This is characteristic of an optical excitation involving a *defect state*. Further evidence of the different nature of the LE excitation compared to the HE is given by the different temperature and *f* dependences. At this stage it is interesting to notice that these optical features have remarkable similarities with those of the normal state. This suggests that the normal state which is produced by chemical doping has the same electronic structure as the system doped by illumination.

3.2 La_2CuO_4

The PA and the PM spectra of La_2CuO_4 at 10K are shown in Fig. 3.

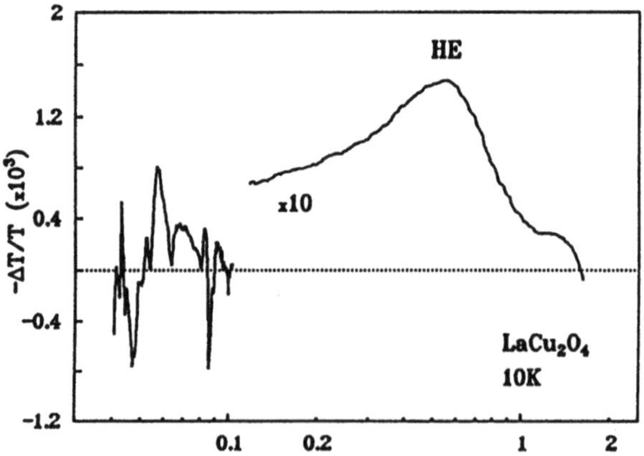

Fig. 3. Photoinduced infrared absorption (left side) and photomodulation (200 Hz) absorption (right side) of La_2CuO_4 at 10 K.

The phonon PA spectrum shows the photobleaching of two phonon modes at 380 and 690 cm^{-1} and the activation of a phonon at 460 cm^{-1}. Identical results were reported by Kim et al. [2]. Both the bleachings which are assigned to the bending mode of the oxygen atoms at the apex of the octahedra and to the planar stretching mode (Eu) respectively indicate that the system undergoes a local lattice distortion towards the normal state. In fact the 690 cm^{-1} disappears also upon Sr doping because of the screening of the free carriers in the Cu-O planes [14]. The higher energy side of the PM spectrum (f=200 Hz) is marked by a broad absorption with a maximum at about 0.5eV. The laser intensity dependence of the phonon photobleachings indicates the square root dependence as in the case of $YBa_2Cu_3O_{7-x}$.

3.3 $BaBiO_3$

Barium bismuthate is the *parent* compound of the new superconductor $Ba_{1-x}K_xBiO_3$ whose transition temperature (T_c=30K) justifies the attribute of a high temperature superconductor. It is also a perovskite but contrary to the cuprates it does not have any magnetic atom. Moreover it shows a static charge density wave (CDW) which causes a Peierls gap. The PA and PM spectra of $BaBiO_3$ at low temperature are shown in Fig. 4. Two photobleachings are observed at 130 cm^{-1} and at about 500 cm^{-1} [10].

Both modes have been shown to vanish upon Pb doping [15]. The 130 cm^{-1} mode is a translational mode of the octahedra at the zone boundary which is allowed by the doubling of the unit cell as a consequence of the CDW. Its suppression is a clear indication of the *suppression of the CDW* upon photoinjection of carriers. On the high energy side the PM spectrum (f=2000Hz) shows a broad electronic band with maximum at 0.5 eV. Once more the intensity dependence of the phonon PA features and the midinfrared absortion have the same square root dependence on the laser intensity. The crossover of the PM spectrum at about 1eV suggests that the semiconducting gap is at about 1eV rather than 2.15eV as suggested earlier [16]. A recent study does in fact confirm this conclusion [17].

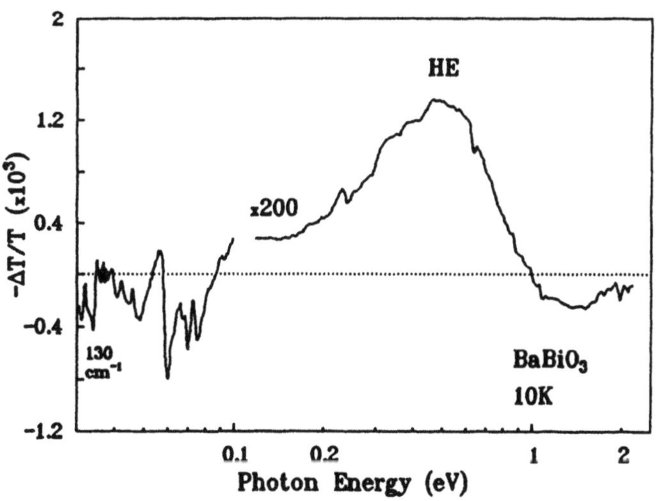

Fig. 4. Photoinduced infrared absorption (left side) and photomodulation (2000 Hz) absorption (right side) of $BaBiO_3$ at 10K.

4. Summary

The spectra of the cuprates and bismuthates in the low light "doping" regime share the following features:

1) a broad absorption band characteristic of a dipole allowed optical excitation (*charge excitation*) with an onset at about 0.3 eV (HE).

2) the charge excitations are accompanied by lattice distortions of specific lattice modes. In fact we observe the suppression of modes of the semiconducting structure and the activation of new modes (in the case of La_2CuO_4 and $YBa_2Cu_3O_{7-x}$ (x=0.8)) which indicates the formation of a local normal state distortion. In particular in $YBa_2Cu_3O_{7-x}$ (x=0.8), the 435 cm^{-1} phonon mode is activated. This mode may be responsible for the 50 meV feature in the optical conductivity spectrum [1].

3) The species responsible for the broad midinfrared absorption appear to *recombine in pairs*.

4) The common features of the layered cuprates and the 3D bismuthates indicate that the midinfrared absorption in the low doping regime is not linked to a charge excitation involving the copper atoms. Therefore we suggest that the HE charge excitation is related to the oxygen network.

Acknowledgement. We gratefully thank Dr. R. Masini and Dr. F.C. Matacotta for the sample preparation. We thank CNR Progetto Finalizzato "Tecnologie Superconduttive e Criogeniche" for financial support. One of us (A.J.P.) wishes to thank partial financial support from I.C.T.P. Trieste.

5. References

1 a) J. Orenstein et al. this issue ;b) K. Kamaras et al. this issue.
2 Y.H.Kim, A.J.Heeger, L.Acedo, G.Stucky and F.Wudl, Phys.Rev.B **36**,7252 (1987).
3 J.M.Ginder, M.G.Roe, Y.Song, R.P.McCall, J.R.Gaines, E.Ehrenfreund and A.J.Epstein, Phys.Rev.B **37**,7506 (1988).
4 C.Taliani, R.Zamboni, G.Ruani, F.C.Matacotta, K.I.Pokhodnya, Solid State Commun. **66**,487 (1988).
5 Y.H.Kim, C.M.Foster, A.J.Heeger ,S.Cox,and G.Stucky, Phys.Rev.B **38**,6478 (1988).
6 C.M.Foster, A.J.Heeger ,Y.H.Kim, G.Stucky and N.Herron, Synth. Met.**33**,171 (1989).
7 C.M. Foster, A.J.Heeger, Y.H.Kim, G.Stucky,and N.Herron, Physica C **162-164** (1989).
8 C. Taliani, R.Zamboni, G.Ruani, A.J.Pal, F.C.Matacotta, Z.V.Vardeny and X.Wei, in "High Tc superconductors: Electronic structure" ed. by A. Bianconi and A. Marcelli, Pergamon Press, Oxford (1989).
9 X. Wei, L.Chen, Z.V.Vardeny, C.Taliani, R.Zamboni, A.J.Pal and G.Ruani, Physica C **162-164**,1109 (1989).
10 C.Taliani, A.J.Pal, G.Ruani, R.Zamboni, R.Masini, to be published on Bull. Mat. Science (1990).
11 G.Ruani, C.Taliani, R.Zamboni, D.Cittone and F.C.Matacotta, J. Opt. Soc. Am.B **6**,409 (1989)

12 M.Garriga, U.Venkateswaran, K.Syassen, J.Humlicek, M.Cardona, Hj.Mattausch and E.Schoenherr, Physica C **153-155**,643 (1988).
13 H.P.Geserich et al. this issue.
14 G.Burns, F.H.Dacol, G.Kliche, W.Konig and M.W.Shafer, Phys.Rev.B **37**,3381 (1988).
15 S.Uchida, S.Tajima, A.Masaki, S.Sugai, K.Kitazawa and S.Tanaka,J.Phys.Soc.Jap. **54**,4395 (1985).
16 S.Tajima, S.Uchida, A.Masaki, H.Takagi, K.Kitazawa, S.Tanaka, and A.Katsui, Phys.Rev.B **32**,6302 (1985).
17 G.Ruani, A.J.Pal, R.Zamboni and C.Taliani, this issue.

Raman Scattering and Infrared Spectroscopy of Substituted and Oxygen-Doped $YBa_2Cu_3O_y$

M. Matus, B. Sumetsberger, and H. Kuzmany

[1]Institut für Festkörperphysik, Universität Wien, Strudlhofgasse 4, A-1090 Wien, Austria and
Ludwig Boltzmann Institut für Festkörperphysik, Kopernikusgasse 15, A-1060 Wien, Austria

Abstract. We report IR and Raman measurements on the ceramic system $YBa_2(Cu_{1-x}Ga_x)_3O_y$. The IR-reflectance was measured at temperatures between 10 and 300 K, covering 50–500 cm^{-1}. A gap-like structure was seen in the IR-spectra which scales with T_c. The BCS-ratio was found to be 1.7, independent of T_c. Samples with activated $\sigma(T)$ showed more pronounced structure (dirty limit).

1. Introduction

Among the high temperature superconductors, $YBa_2Cu_3O_y$ is well known for the strong dependence of its properties on the oxygen content. Oxygen reduction leads to a metal-insulator transition around $y = 6.5$. In the same region the material changes from orthorhombic to tetragonal.

The superconducting properties of this material can also be altered by substitution of different elements. Among the many known solid solutions, the substitution of *Cu* by *Ga* shows some interesting features. Besides the variation in the carrier concentration due to the higher valency of *Ga*, there is an influence on the crystal-structure. Gallium concentrations higher than 5% switch the X-ray pattern of the material to a tetragonal structure [1]. We found in addition that the substituted samples are very sensitive to thermal treatment.

In order to elucidate the mechanisms by which the substitutions influence the superconducting properties, we prepared *Ga*-substituted samples with various oxygen stoichiometries [2] and studied their IR and Raman properties.

2. Experimental

Ceramic superconductors were prepared by the well known solid state reaction method. Well defined oxygen concentrations were obtained using the equilibration technique described elsewhere [3]. The annealing conditions were 420°C for 40 hours followed by a linear decrease of temperature down to 300 K in 10 hours.

X-ray analysis of the different samples was performed with a Huber-Guinier camera using CuK_α radiation. Resistivity was measured between 4 and 300 K by the standard four-probe technique. Raman spectra were collected using an Ar-laser and a double grating monochromator with a Peltier-cooled pho-

Springer Series in Solid-State Sciences, Vol. 99
Electronic Properties of High-T_c Superconductors and Related Compounds
Editors: H. Kuzmany · M. Mehring · J. Fink

tomultiplier. For the IR-experiments a BRUKER IFS 66v Fourier transform spectrometer with a Si-bolometer was used. The optical system was evacuated to less than 0.02 torr. A built-in cryostat provided the temperature-range of 10 to 300 K for the sample.

3. Results

The lattice parameters of the system $YBa_2(Cu_{0.94}Ga_{0.06})_3O_y$ are summarized in [3]. The results of the conductivity measurements and the values for T_c can be found in [2].

Raman lines of the substituted samples showed characteristic changes with oxygen concentration. It was observed in particular, that the intensity ratio of the out of plane vibration at 340 cm^{-1} and the $Cu(1)$-$O(4)$ vibration at 505 cm^{-1} increase continuously with decreasing oxygen concentration. This is in agreement with previous results from unsubstituted samples [4].

An important difference was observed for the temperature dependence of the resistivity before and after annealing as can be seen in Fig.1. In the annealed samples T_c drops without loss of oxygen and $\rho(T)$ changes from a metallic to a semiconductor like behaviour.

The results of IR-measurements from two samples, one before and one after annealing, are compiled in Figs.2-4. The ratio $R(T)/R(300\,K)$ is plotted in Fig.2. for various temperatures. Besides the phonon features one can observe in the reflectivity a structure below 100 cm^{-1} with an interesting temperature dependence. Fig.3 shows the reflectivity ratios of Fig.2 at 50 cm^{-1} versus the normalized temperature. Obviously the appearance of the structure scales with T_c. The frequencies where this structure begins to develop are plotted in Fig.4, again as a function of normalized temperature. The frequency for the sample with the higher T_c is higher as compared to the one with lower T_c. In both cases the features decrease approaching T_c but extend beyond this limit.

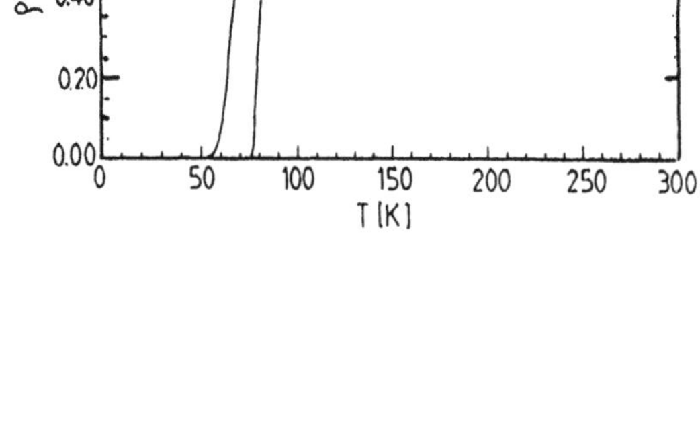

Fig.1: Resistivity curves of YBa_2-$(Cu_{0.94}Ga_{0.06})_3O_7$, before (BT) and after annealing (AT).

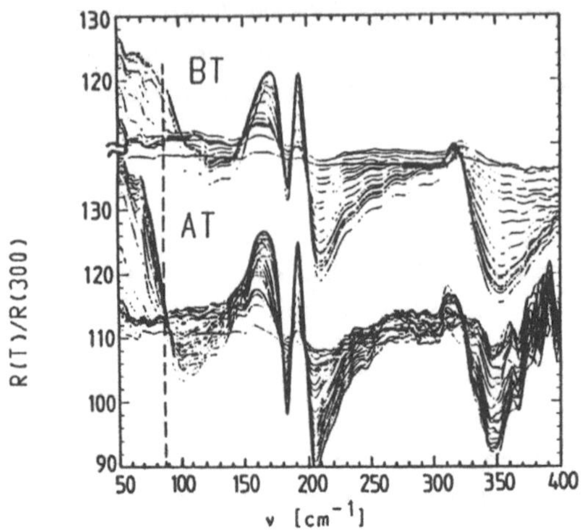

Fig.2: As measured reflectivity normalized to room temperature reflectivity, at various temperatures for the two samples of Fig.1.

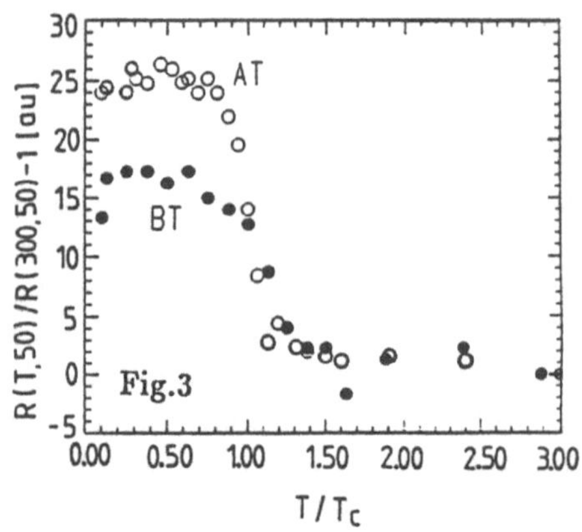

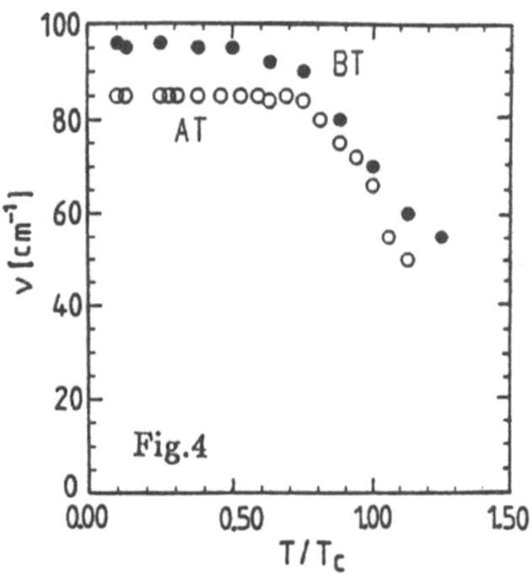

Fig.3: Normalized reflectivity at 50 cm^{-1} versus reduced temperature for samples BT and AT.

Fig.4: Frequency for the beginning of the gap-like feature versus reduced temperature for samples BT and AT.

4. Discussion

The fact that the observed low energy IR-structure scales with the critical temperature of the sample indicates a correlation to supercondutivity. Annealing the sample drives the system from metallic to activated which can be interpreted as a strain induced electron localisation [2], and thus decreases the ratio between mean free path and coherence length. This process increases the gap transition oscillator strength [5].

The two indicated results suggest that the formation of an energy gap is observed. For both samples a BCS-ratio $2\Delta(0)/k_BT_c$ of 1.7 is obtained, which is

considerably smaller than values reported previously [6,7]. The results indicate a strong gap anisotropy and suggest that only its c-axis value is observed [8]. The extension of the gap features beyond T_c is not clear at present – it may have to do with superconducting gap precursor effects.

Acknowledgement. Work supported by the the Fond zur Förderung der Wissenschaftlichen Forschung in Austria (grant P 7059). We acknowledge Dr. Zachmann from the Bruker company for assisting with reflectivity measurements.

References

1. G. Xiao, M.Z. Cieplak, A. Gavrin, F.H. Streitz, A. Bakhshai and C.L. Chien, Phys. Rev. Lett. 60, (1988) 1446.
2. H. Kuzmany, T. Pichler, M. Matus, P. Rogl and S. Pekker, in this volume.
3. M. Matus, H. Kuzmany, B. Sumetsberger and T. Pichler, Physica C 162–164, (1989) 967.
4. H. Kuzmany, M. Matus, E. Faulques, S. Pekker, Gy. Hutiray, E. Zsoldos and L. Mihaly, Phys. Scripta 38, (1988).
5. T. Timusk, and D.B. Tanner, Phys. Prop. High Temp. SC, World Scientific, Singapore, (1989) 339.
6. T. Zetterer, W. Ose, J. Schützmann, H.H. Otto, P.E. Obermayer, N. Tasler, H. Langfellner, G. Lugert, J. Keller, and K.F. Renk, J. Opt. Soc. Am. B6, (1988) 421.
7. L. Genzel, A. Wittlin, M. Bauer, M. Cardona, E. Schönherr, and A. Simon, Phys. Rev. B, 40, (1989) 2170.
8. R.T. Collins, Z. Schlesinger, F. Holtzberg, and C. Feild, Phys. Rev. Lett. 63, (1989) 422.

Optical Investigation of the Superconducting Phase Transition in $YBa_2Cu_3O_7$ Crystals

B. Koch[1], *M. Dürrler*[1], *H.P. Geserich*[1], *Th. Wolf*[2], *G. Roth*[3], *and G. Zachmann*[4]

[1]Institut für Angewandte Physik, Universität Karlsruhe, Kaiserstr. 12, D-7500 Karlsruhe, Fed. Rep. of Germany
[2]Kernforschungszentrum Karlsruhe, Institut für Technische Physik, Postfach 3640, D-7500 Karlsruhe, Fed. Rep. of Germany
[3]Kernforschungszentrum Karlsruhe, Institut für Nukleare Festkörperphysik, Postfach 3640, D-7500 Karlsruhe, Fed. Rep. of Germany
[4]Fa. Bruker, Analytische Messtechnik GmbH, Wikingerstr. 13, D-7500 Karlsruhe, Fed. Rep. of Germany

Abstract. We have performed polarized reflection measurements above and below the superconducting phase transition temperature Tc. The E-vector of the incident light was parallel to the ab–plane ($R_{\|ab}$) or perpendicular to this plane ($R_{\|c}$). Both spectra, $R_{\|ab}$ and $R_{\|c}$, show plasma edges at different energies. Below Tc a sharp increase of $R_{\|c}$ to unity reflectivity occurs in the far infrared, evidently correlated with a superconducting gap of 1.4 kTc.

1. Introduction

A special interest of optical studies of high–Tc superconductivity is the investigation of the substantial anisotropy between the in–plane (ab–plane) and the out–of–plane (c–axis) transport properties. Up to now, limitations in crystal geometry have made reflection measurements for an electric field along the direction of the c–axis extremely difficult, and only a few papers with energy ranges limited to the far infrared have dealt with preliminary c–axis data [1]. Therein, below Tc, a spectral feature at 21 meV ($170\,cm^{-1}$) has been assigned to the superconducting gap. This frequency corresponds to 3 kTc, which is twice the value given by tunneling spectroscopy [2]. Within the ab–plane, the interpretation of the spectra is controversial as well [1-4].

2. Experimental

All investigated crystals were grown in Al_2O_3 crucibles [5]. The exceptionally thick crystals, the typical size of which was $2 \times 2 \times 1.5\,mm^3$, exhibited transition temperatures up to Tc = 90 K with a width of $\Delta Tc = 1\,K$, inductively measured. The pseudo–domain sizes of the twinning were between 50 and 250 μm in diameter. In order to gain mirror–like surfaces the investigated crystals were polished. The reflection measurements were performed on the surface of only one crystal. In the energy range between 50 meV and 6 eV a single–beam setup, between 3 meV and 100 meV a Fourier spectrometer, model IFS 113 V from Bruker Analytische Messtechnik, was used.

Springer Series in Solid-State Sciences, Vol. 99
Electronic Properties of High-T_c Superconductors and Related Compounds
Editors: H. Kuzmany · M. Mehring · J. Fink

3.Results and Discussion

In Fig.1 a survey of the anisotropy of the room temperature reflection spectra $R_{\|ab}$ and $R_{\|c}$ is given between 6 meV and 6 eV, while in Fig.2, upper and lower part, the reflectivities of both crystallographic directions in the far infrared at different temperatures are shown. The accuracy of the data was 1 percent. For the direction parallel to the ab–plane our data are very close to those measured on the ab–plane [3,4]. In contrast, the spectra $R_{\|c}$ deviate considerably from previous work [1].

Beginning with Fig.1, the room temperature measurements will be discussed briefly. Both spectra, $R_{\|ab}$ and $R_{\|c}$, are characterized by strongly damped plasma edges at different energies, indicating a strong anisotropy in the transport properties for the directions parallel and perpendicular to the Cu–O planes. Due to the low screening effects in the spectrum $R_{\|c}$, several phonon modes could be observed. In the high energy range differences in reflectivity are rather small. In both spectra several features assigned to interband transitions are visible. The conductivity function for the spectra of Fig.1 will be shown elsewhere [6].

Turning our attention to the temperature dependence of the infrared spectrum $R_{\|ab}$ in Fig.2, upper part, we note that recently there has been some controversy about the assignment of two spectral features to superconducting gaps. The smaller one with a gap energy of 2.1 kTc was assigned to unity reflection onset near 17.4 meV (140 cm^{-1}) [3,4], while the second with a gap energy of 8 kTc was assigned to the midinfrared absorption shoulder near 62

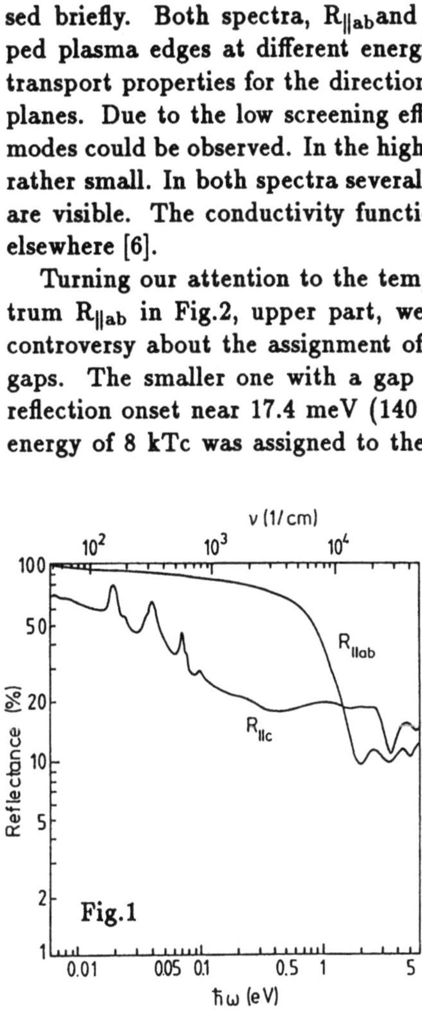

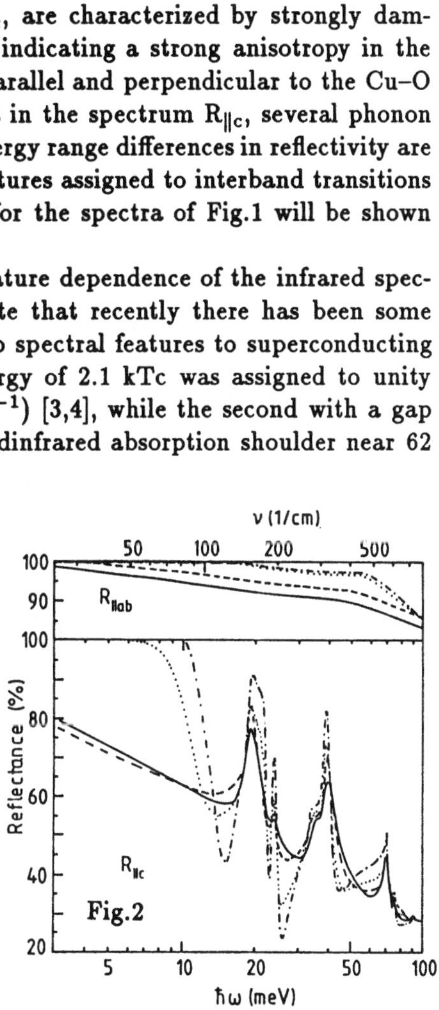

Fig.1. Reflectivity of a $YBa_2Cu_3O_7$–crystal at 300 K. The incident light vector was parallel ($R_{\|ab}$) and perpendicular ($R_{\|c}$) to the twinned ab–plane

Fig.2. Far infrared reflectivity at 300 K(——), 110 K (– – –), 60 K (·····) and 10 K (–·–·). Upper panel: Incident electric field parallel to the ab–plane direction. Lower panel: Incident electric field parallel to the c–axis direction.

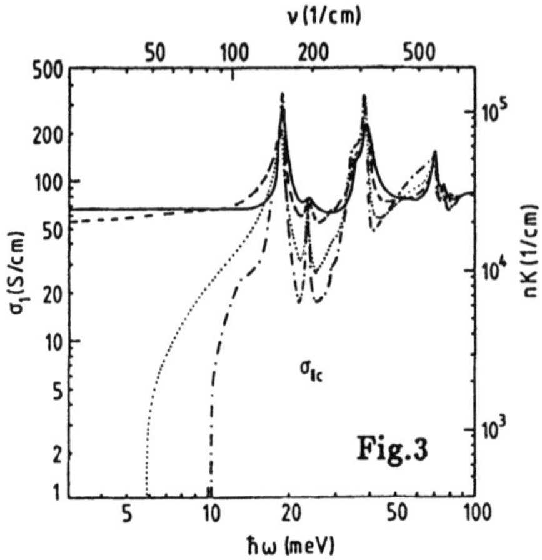

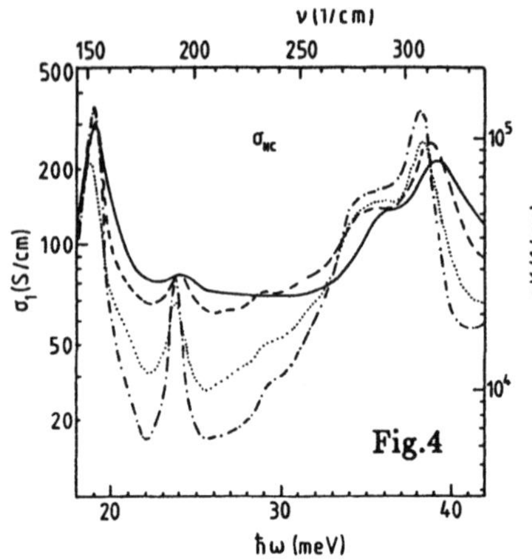

Fig.3. Real part of the conductivity perpendicular to the ab–plane at 300 K (——), 110 K (– – –), 60 K (·····) and 10 K (–·–·).

Fig.4. Enlarged view of the phonon part of Fig.3.

meV (500 cm^{-1}) [4]. Both values are in the range following from tunneling data [2]. As a weak shoulder at 62.5 meV (500 cm^{-1}) is already seen in room temperature spectra, some authors disagree in the assignment of the feature at 8 kTc [3]. From the conductivity function, obtained by a Kramers–Kronig analysis and shown elsewhere, no clear evidence is given, as the function starts oscillating when **R** reaches unity. We are convinced that without measurements performed on untwinned ab–planes a decision with respect to a gap assignment cannot be made. Untwinned reflection spectra performed at room temperature between the MIR and UV resulted in a considerable anisotropy of the Fermi surface [7]. Therefore, in a clean limit the gap anisotropy of an untwinned ab–plane should be pronounced. Following this, the direction of the crystallographic a–axis should be correlated with the 8 kTc value, while the gap energy of 2.1 kTc should be assigned to the crystallographic b–axis.

Now the temperature dependence of the spectrum $R_{\parallel c}$ will be discussed (Fig.2, lower part). The real part of the conductivity, shown in Fig.3, is obtained by a Kramers–Kronig analysis, using the high–frequency data of the 300 K spectrum. A widely spread energy scale of the conductivity function for the phonon region is given in Fig.4. At 300 K the c–axis conductivity extrapolates to a dc value of 65 S/cm. A comparison with the extrapolated dc conductivity for the ab–plane yields an anisotropy ratio of 50:1, which is comparable to the range following from transport studies [8]. As a consequence of the low conductivity, the lattice vibrations are not well screened. When the superconducting phase transition is reached, a sudden sharp increase to unity reflectivity occurs in the spectra in the vicinity of 8 meV (60 cm^{-1}) at 65 K and near 12 meV (90 cm^{-1}) at 10 K. In the related real parts of the conductivity functions, cut–off frequencies were observed with zero absorption of the free carriers below and reduced absorption of the free carriers above them, evidently correlated with the superconducting phase transition. We interpret the cut–off frequencies as gap energies with values above them, the paired electrons being cracked by photon absorption. Well above these frequencies, in the vicinity of 60 meV (480 cm^{-1}), a broad shoulder in the reflection spectra arises at the same energy as the midinfrared absorption is found within the ab-plane spectra. In correspondence, the c–axis feature increases much more rapidly below Tc than above Tc and may have a related origin. Normally, the gap energy is determined from optical measurements by the cut–off frequency in the conductivity ratio σ_{1s}/σ_{1n}. Using this approach, a gap energy of 1.4 kTc could be estimated [6]. This rather low value is in excellent agreement with tunneling data [2], but only half of the value which results from other infrared data [1].

Finally, a brief discussion of the phonon–related contribution will be made. In the 300 K spectrum 8 phonon modes were observed. The resulting frequencies (Lorentz oscillators) in meV (cm^{-1}) are: 98.6 (795), 76.3 (615), 70.2 (566) 39.2 (316), 35.6 (298), 24.0 (193), 20.6 (161), 18.9 (152). To gain better insight into the electron–phonon coupling mechanism, a part of the conductivity function is widely spread (Fig.4). Therein, the Cu(1)-O(1) streching mode at 76.3 meV (615 cm^{-1}) and the Cu(1)-O(4) bending mode at 70.2 meV (566 cm^{-1}) are not included, as they exhibit no energy shift from 10 K to 300 K. Beginning with the in–phase bending mode of the Cu(2) plane at 39.2 meV (315 cm^{-1}), the softening of 0.5 meV (4 cm^{-1}) of the eigenfrequency is apparently coupled to the superconducting phase transition, while for the out–of–phase bending mode at 35.6 meV (289 cm^{-1}) no clear evidence is seen. For the next mode at 24.0 meV (193 cm^{-1}), assigned to a vibration of the Y and Cu ions, a 0.2 meV softening is observed, while the two low–lying Ba modes probably show a little hardening. A more detailed discussion will be found in ref. [6].

References

1.R.T.Collins, Z.Schlesinger, F.Holtzberg, and C.Feild: Phys.Rev.Lett. **63**,422 (1989)
2.J.Geerk, X.X.Xi, and G.Linker: Z.Phys.B-Condensed Matter **73**, 329 (1988)
3.K.Kamaras, S.L.Herr, C.D.Porter, N.Tache, D.B.Tanner, S.Etemad, T.Venkatesan, E.Chase, A.Innan, X.D.Wu, M.S.Hedge, and B.Dutta: Phys.Rev.Lett. **64**,84 (1990)
4.J.Schützmann, W.Ose, J.Keller, K.F.Renk, B.Roas, L.Schulz, and G.Saeman–Ischenko: Europhys.Lett. **8**,679 (1989)
5.T.Wolf, W.Goldacker, B.Obst, G.Roth, and R.Flükiger: J.Cryst.Growth **96**, 1010 (1989)
6.B.Koch, M.Dürrler, T.Wolf, and G.Roth: to be published
7.B.Koch, H.P.Geserich, and T.Wolf: Solid State Comm. **71**,495 (1989)
8.G.Weigang, and K.Winzer: Z.Phys.B. - Condensed Matter **77**, 11(1989)

FIR Measurements of Zn-Substituted YBCO Ceramics

G. Strasser[1], *H. Hertle*[1], *G. Vogl*[2], *W. Seidenbusch*[2], *and E. Gornik*[1]

[1]Walter Schottky Institut, TU München,
D-8046 Garching, Fed. Rep. of Germany
[2]Institut für Experimentalphysik, Universität Innsbruck,
A-6020 Innsbruck, Austria

Abstract. Far infrared reflectivity measurements of polycrystalline $YBa_2Cu_{3-x}Zn_xO_7$ with up to 5% Cu substituted by Zn (Tc varying from 92K to 46K) are performed between 10K and 300K. We find evidence for an energy gap, moving to lower energies with decreasing Tc.

1. Introduction

Since the breakthrough in superconductivity by Bednorz and Müller/1/ and the discovery of the 90K structures /2/ worldwide activities have started to find an energy gap or at least some spectral features responsible for the superconducting pairing by infrared spectroscopy. More or less sophisticated theories are used to describe the reflection behavior of the ceramics. By using single crystals or orientated thin films, reflectivity parallel and perpendicular to the planes can be observed. Due to a high normal state reflectivity /3/ parallel – and due to phonons perpendicular – to the planes, the obtained spectroscopic results are not satisfactory yet. Transmission data would be more definite, but most of the substrates for thin films are not transparent in the FIR region.

2. Experimental Details

One approach to get different Tc's is to change the oxygen content, as done by Thomas et al. /4/ and other groups. However, in this case also the conductivity of the material due to the metal–semiconductor transition is changed. Our intention was to change the critical temperatures of the samples without changing carrier concentration, phonon positions and oscillator strengths. This is achieved in samples where Cu is substituted by Zn. We used four samples with different concentrations (x) of Zn, produced within one run. This assures identical growth conditions like temperature and quenching rate, and therefore leads to comparable

Springer Series in Solid-State Sciences, Vol. 99
Electronic Properties of High-T_c Superconductors and Related Compounds
Editors: H. Kuzmany · M. Mehring · J. Fink

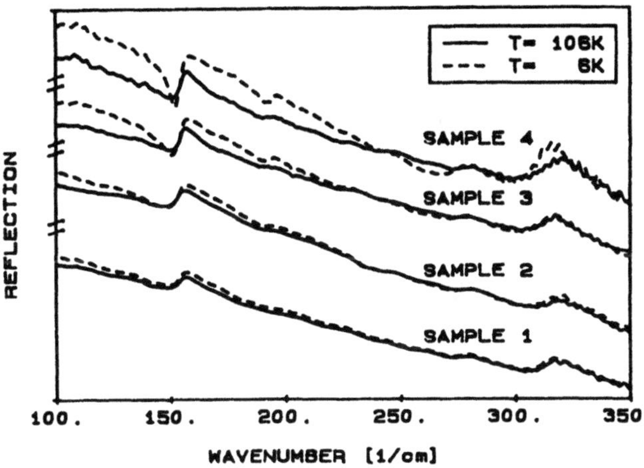

Fig 1.: Reflectivity data of $YBa_2Cu_{3-x}Zn_xO_7$. Solid lines: 106K; dashed lines: 6K. Samples 1 to 4: Tc = 92K, 78K, 61K, 46K

surface quality and unique oxygen content. Fig.1 shows the room temperature reflection behavior of these samples (solid lines). The phonon peak positions are exactly the same for the various x–values (above Tc).

Although the phonon peaks confuse the reflectance spectrum, they provide an indirect measurement of an energy gap. Phonon absorption features have finite widths caused by impurities, anharmonicity, and electron–phonon interactions /5/. The response of the low–frequency phonons changes from dispersive to absorptive in the superconducting state.

From a narrowing of the phonon lines with decreasing temperature, Bonn et al. /6/ suggested a gap between 195 and 279 wavenumbers according to the observed changes in the damping of the phonons. This gives an energy for a gap between 3 and 4.3 kTc, which fits excellently to our former results on YBaCuO /7/ and the data from various other groups /8,9/.

Dashed lines in Fig. 1 are reflectivity data at 6K. The phonon at 155 wavenumbers changes from dispersive (above Tc) to absorptive behavior (below Tc) for samples 3 and 4 (according to the Tc's 78K and 92K, respectively), it stays dispersive for samples 1 and 2. This is in perfect agreement with an expected gap, moving with Tc. For the phonon peak at 195 wavenumbers a narrowing cannot be clearly seen from the reflectivity data in Fig.1. To demonstrate the effect better difference reflectivity spectra are shown in Fig.2. In a (superconducting minus normalconducting) difference reflectivity plot, sharp dips at the phonon positions for phonon energies below the gap can be seen. Plotting the difference of two reflectivity measurements at two

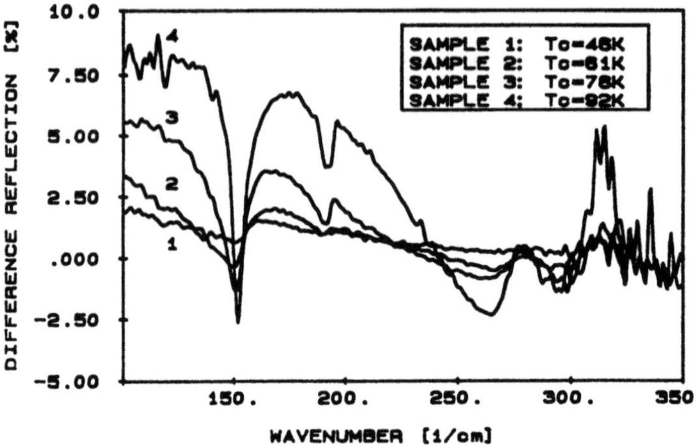

Fig 2.: Difference reflectivity data: 6 K minus 106 K for all 4 samples

temperatures well above Tc leads to featureless and flat curves. The lower the temperature of the (superconducting) sample, the narrower and deeper the dip at 155 wavenumbers becomes, due to more superconducting carriers.

The spectra in Fig.2 reflect the reflectivity changes (samples 1 to 4) when the critical temperature and thus the energy gap is changed. For sample 3 (Tc 78K) the sharp dip at 155 wavenumbers is still there. The dip at 155 wavenumbers is not so narrow and deep as for sample 4. If an energy gap were well above 195 wavenumbers, the phonon dips of samples 3 and 4 at 195 wavenumbers would look the same.

Assuming (like /6,7,8/) a gap between the highest absorptive and the lowest dispersive phonon, i.e. for sample 4 (Tc 92K) between 195 and 260 wavenumbers, for sample 3 (Tc 78K, Fig.2) a gap somewhat lower is expected.

We estimate a gap for sample 3 between 155 and 195 and close to 195 wavenumbers.

In the difference reflectivity of sample 2 (T 61K) the sharp dip of the phonon at 155 wavenumbers vanishes, giving an indication of an energy gap below 155 wavenumbers. In sample 1 (Tc 46K) the narrowing of this phonon is completely gone. If there is a gaplike behavior, it is well below 155 wavenumbers.

3. Conclusions

Summarizing all these estimated values for the gap and converting it into temperatures gives values between 3 kTc and

3.6 kTc. This is in good agreement with a BCS gap. The spectra are fitted with Lorentz oscillators for phonons, Drude conductivity above Tc and Mattis Bardeen conductivity for superconducting carriers. A very good overall agreement (better than 3%) with the measured reflectivities is achieved with an energy gap value of 3.5 kTc.

This work was sponsored by Fond zur Förderung der wissenschaftlichen Forschung.

/1/ J.G. Bednorz and K.A. Müller, Z.Phys. **B64**, 189 (1986)
/2/ M.K.Wu,J.R.Ashburn,C.J.Torng,P.H.Hor,R.L.Meng,L.Gao, Z.J.Huang,Y.Q.Wang,C.W.Chu,Phys.Rev.Lett. **58**,908 (1987)
/3/ J. Schützmann, W. Ose, J. Keller, K.F. Renk, B. Roas, L. Schultz, G. Saemann–Ischenko, Europhys. Lett. **8**, 679 (1989)
/4/ G.A. Thomas, J. Orenstein, D.H. Rapkine, M. Capizzi, A.J.Millis, R.N. Bhatt, L.F. Schneemeyer, and J.V. Waszczak, Phys. Rev. Lett. **61**, 1313 (1988)
/5/ Thomas Timusk, David B. Tanner, The Physical Properties of High–Temperature Superconductors, editor D.M. Ginsberg World Scientific Publishing Company, Singapore (1989)
/6/ D.A. Bonn, A.H. O'Reilly, J.E. Greedan, C.V. Stager, T.Timusk,K.Kamaras,D.B.Tanner,Phys.Rev.**B37**,1547(1988)
/7/ H. Krenn, G. Bauer, G.Vogl, G.Strasser, and E. Gornik, Phys. Rev. **B39**, 6716 (1989)
/8/ Z.Schlesinger, R.T. Collins, B.A. Scott, and J.A. Calise, Phys. Rev. **B38**, 9284 (1988)
/9/ L. Genzel, A. Wittlin, M. Bauer, M. Cardona, E. Schoenherr, and A. Simon, Phys. Rev. **B40**, 2170, (1989)

Anomalies of Infrared Active Phonons in Superconducting $Tl_2Ba_2Ca_2Cu_3O_{10}$

T. Zetterer, M. Franz, J. Schützmann, W. Ose, H.H. Otto, and K.F. Renk

Fakultät für Physik, Universität Regensburg, Universitätsstr. 31,
D-8400 Regensburg, Fed. Rep. of Germany

Abstract. We report on a study of the far-infrared reflectivity of a highly reflecting $Tl_2Ba_2Ca_2Cu_3O_{10}$ ceramic sample (T_c = 112 K). The reflectivity spectra show high metallic reflectivity and, additionally, strong phonon structures. We observe eight infrared active phonons. We find that the transition into the superconducting state leads to an abrupt shift in frequency by about 7% of one of the phonons (at $\sim$ 300 cm^{-1}) while the resonance frequencies of the other phonons remain almost uninfluenced by the transition. For all phonons we find very high and temperature dependent oscillator strengths. For a discussion of the anomalous behavior of the oscillator strengths we present a model which is based on phonon-induced charge transfer processes.

1. Introduction

Numerous studies of optical properties of high-temperature superconductors have been performed [1]. In several publications phonon anomalies related to superconductivity have been reported. Frequency shifts of phonons at T_c are known for $YBa_2Cu_3O_{7-\delta}$ from Raman scattering studies [2]; sample dependent softening by 1-3% of the phonon frequency ($\sim$ 330 cm^{-1}) has been found. Far-infrared reflectivity measurements also show softening ($\sim$ 1%) of a phonon (near 300 cm^{-1}) in $YBa_2Cu_3O_{7-\delta}$ at T_c, and an increase of oscillator strength [3]. Also for conventional superconductors changes of phonon frequencies are known from inelastic neutron scattering on Nb ($\sim$ 1,5%) [4] and Raman scattering on Nb_3Sn (6%) [5]. In this paper we will report on an analysis of the anomalies observed in the far-infrared reflectivity of a ceramic sample of the high-T_c superconductor $Tl_2Ba_2Ca_2Cu_3O_{10}$. We will discuss the observed phenomena and suggest that phonon induced charge transfer processes are responsible for the anomalies.

2. Experimental results

Detailed results and the analysis of our far-infrared reflectivity measurements are reported elsewhere [6]. For brevity we shall only mention the main points and perform a discussion of the observed anomalies. The far-

Springer Series in Solid-State Sciences, Vol. 99
Electronic Properties of High-T_c Superconductors and Related Compounds
Editors: H. Kuzmany · M. Mehring · J. Fink

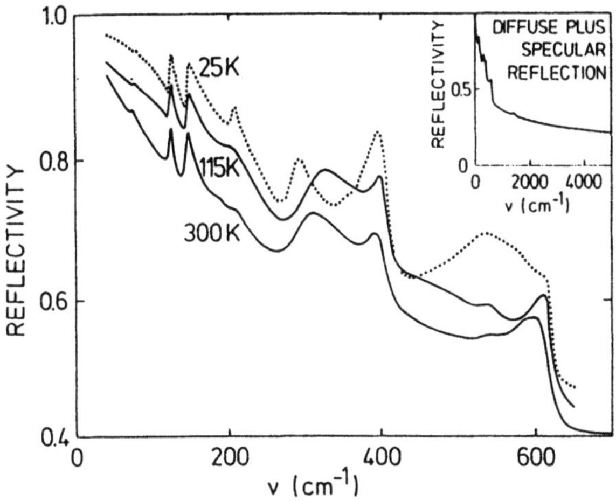

Fig. 1. Far-infrared reflectivity of a $Tl_2Ba_2Ca_2Cu_3O_{10}$ ceramic sample.

infrared reflectivity of our $Tl_2Ba_2Ca_2Cu_3O_{10}$ ceramic sample is shown in Fig. 1 for various temperatures. An anomalous phonon shift of a 300 cm^{-1} phonon mode is seen as the temperature decreases below T_c (112 K). We find a sudden decrease of frequency by about 7%, that is much larger than known from infrared and Raman studies on other high-T_c compounds.

An analysis of the reflectivity data using Kramers-Kronig transformations yields further information about the phonons. In Fig. 2 we show

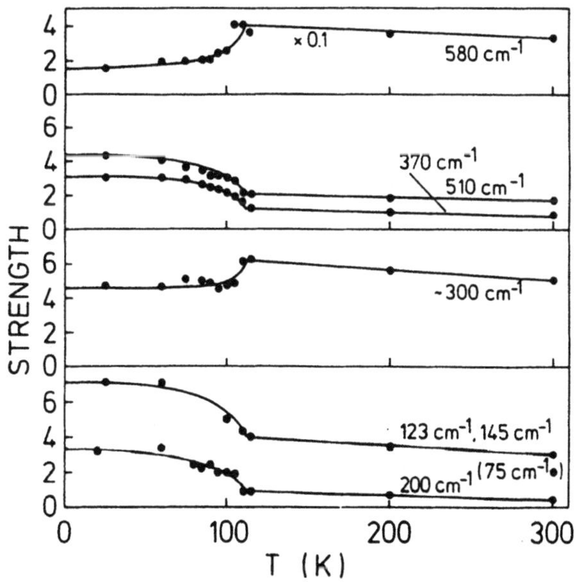

Fig. 2. Strengths of the infrared active phonons in $Tl_2Ba_2Ca_2Cu_3O_{10}$.

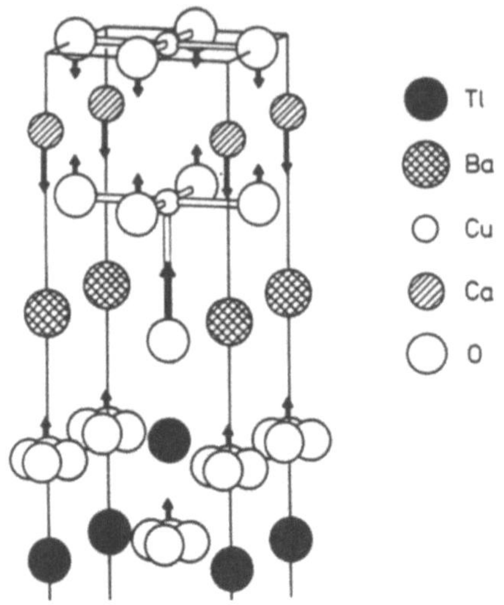

Fig. 3. Displacements for the 300 cm^{-1} phonon.

the strengths of the different phonons; each strength corresponds to the change of dielectric function caused by the corresponding phonon.

The strengths of all phonons increase between room temperature and T_c, and most phonons show a sudden further strong increase just below T_c. There are two exceptions, the 300 cm^{-1} mode that shows the frequency shift, and also the highest frequency phonon (at 580 cm^{-1}) shows a strong decrease of strength. According to lattice dynamical calculations the 300 cm^{-1} phonon mode has been associated to a mode, where mainly the bridging oxygen ions (in the BaO layer) vibrate against the calcium ions (Fig. 3).

3. Discussion

Anomalous behavior of oscillator strength of the "Barium mode" (at 155 cm^{-1}) has been explained for $YBa_2Cu_3O_7$ by assuming that the oscillator strength is strongly enhanced by charge transfer processes [3]; charge transfer was assumed to occur between neighboured bridging oxygen ions; i.e. between adjacent BaO layers. We suggest that charge transfer processes are responsible also in the case of our Tl compound and we apply the charge-transfer picture for a qualitative discussion of our results. For simplicity we treat a very simple model; we assume that only oxygen vibrates and neglect the vibrations of the other ions. The dipole moment for a vibrating bridging O^{2-} ions is given by $p(t) = (-2e) \cdot u_0 \sin \omega_0 t$ (Fig.4), where u_0 is the vibrational amplitude at zero temperature. If we assume that a transfer of charge q occurs from the CuO_2 plane to the bridging oxygen, as the oxygen ion vibrates against the CuO_2 plane, then the total dipole moment is given by $p(t) =$

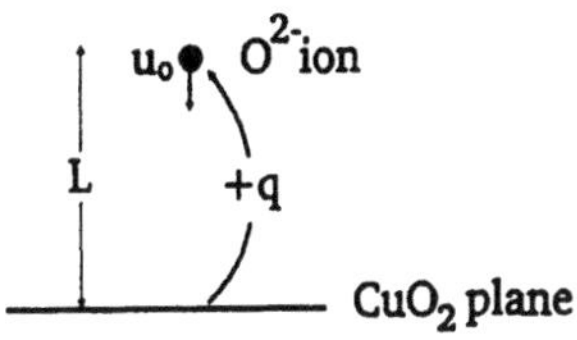

Fig. 4. A simple model for charge transfer.

$(-2e\,u_0 - q \cdot L)\sin\omega_0 t$, therefore the enhancement of the oscillator strength is $1 + S_{CT}/S \simeq (1 + qL/2eu_0)^2$. With $L \approx 2.7$ Å and assuming $u_0 \approx 0.1$ Å we see that only a small amount of charge needs to be transferred to enhance the strength of a phonon considerably.

We suggest that the increase of oscillator strength at low temperatures is due to a higher mobility of the charge carriers within the CuO_2 planes.

The decrease of strength, below T_c, for the 300 cm^{-1} and the 580 cm^{-1} modes may be due to the opening of a gap leading to a suppression of charge transfer.For all other phonon modes a rapid increase of strength is observed at T_c. A consideration of the eigenvectors of these modes [7] shows that atoms of the CuO_2 planes are involved in all these vibrations. A phonon induced transfer of Cooper pairs between two adjacent CuO_2 planes may be responsible for the strongly increased oscillator strengths.

From the charge transfer picture the frequency shift of the 300 cm^{-1} phonon mode is also understandable. Since the bridging oxygen ions are displaced about 0.7 Å from the planes of the Ba^{2+} ions, an additional repulsive force in c-direction is caused by a positive charge carrier that is present on an oxygen ion as it vibrates against a CuO_2 plane. Therefore, the vibration hardens and the frequency of the mode in the normalconducting state, where charge transfer is allowed, is higher than in the superconducting state.

The work was supported by the Bundesministerium für Forschung und Technologie.

[1] *Infrared Properties of High-T_c Superconductors*, ed. D.M.Ginsberg (World Scientific Publishing, 1989).
[2] R.M. MacFarlane, H. Rosen, and H.J. Seki, Sol.St.Comm. **63**, 831 (1987); R. Thomsen, M. Cardona, B. Gegenheimer, R. Liu, and A. Simon, Phys. Rev. B **37**, 9860 (1988).
[3] L. Genzel, A. Wittlin, M. Bauer, M. Cardona, E. Schönherr, and A. Simon, Phys. Rev. B **40**, 2170 (1989).
[4] S.M. Shapiro, G. Shirane, and J.D. Axe, Phys. Rev. B **12**, 4899 (1975).
[5] R. Hackl, thesis 1987.
[6] T. Zetterer, M. Franz, J. Schützmann, W. Ose, H.H. Otto, and K.F. Renk, Phys. Rev. B, to be published.
[7] A.D. Kulkarni, F.W. de Wette, J. Prade, U. Schröder, and W. Kress, Phys. Rev. B, to be published.

Probes of Superconductivity: Raman Scattering

C. Thomsen and B. Friedl

Max-Planck-Institut für Festkörperforschung, Heisenbergstr. 1
D-7000 Stuttgart 80, Fed. Rep. of Germany

Abstract. The anomalous phonon-frequency shift in the superconducting state of $EuBa_2Cu_3{}^{16}O_{7-\delta}$ and $YBa_2Cu_3{}^{18}O_{7-\delta}$ is found to be 2.8 and 5.0 cm^{-1} compared to 6.6 cm^{-1} in $YBa_2Cu_3{}^{16}O_{7-\delta}$.

1. Introduction

Raman scattering has proven to be a useful tool for the investigation of high-T_c superconductors. Fundamental physical properties as well as analytical information may be obtained from it. Among the latter are the detection of impurities which are Raman active, a determination of the crystal orientation and of the oxygen content if, as in $YBa_2Cu_3O_{7-\delta}$, the phonon frequencies depend sufficiently strongly on δ [1].

The method is non-destructive, requires no contacts to the sample and is fairly simple to apply once sufficiently sensitive equipment is available. On the other hand, light scattering is also a direct probe of superconductivity and has yielded information relevant to the understanding of the high-T_c materials [1]. Intuitively most obvious is perhaps the search for evidence of electron-phonon interaction in the high-T_c materials. The experimentally observed asymmetric phonon lineshape as well as frequency and linewidth anomalies near T_c are evidence for a significant interaction between the crystal lattice and the superconducting electronic system [2,3]. The lineshape asymmetry or Fano effect is the result of an interference of a discrete (phonon) state with a continuum (electronic) background, similar to what occurs in heavily doped semiconductors [4]. Evidence for the background being due to light scattering by electrons is generally taken from the reduction at low temperature in that scattering below a supposed superconducting gap energy.

The frequency and linewidth anomalies are comparable to those found in the conventional superconductor Nb for acoustic phonons at points in the Brillouin zone away from $q = 0$ [5]. In the $ReBa_2Cu_3O_{7-\delta}$ family the shifts amount to $1 - 2\%$ of the phonon frequency and are negative, i.e. correspond to a softening, for the phonon at 340 cm^{-1} (Re = Y). Applying magnetic fields larger than H_{c2} at a given temperature reverses the shift and thus unambiguously shows its relation to superconductivity [6]. Such a measurement may, in principle, be used to determine the temperature dependence of H_{c2} since it is a flux-motion free method.

Springer Series in Solid-State Sciences, Vol. 99
Electronic Properties of High-T_c Superconductors and Related Compounds
Editors: H. Kuzmany · M. Mehring · J. Fink

2. Results and Discussion

In Fig. 1 we show the frequency anomaly of one of the CuO_2-plane phonons in $EuBa_2Cu_3O_{7-\delta}$. The experimental details are as in Ref. [9]. Plotted as a function of temperature is the bare phonon frequency of the phonon which corresponds to the out-of-phase vertical vibration of oxygen in the CuO_2 planes. Corresponding atoms in different planes are out of phase as well since displacements in a Raman-active mode must be even upon inversion. This mode is of A_g symmetry in the orthorhombic point group but may be approximated by B_{1g} in a tetragonal description. The polarization selection rules of this mode are consistent with the approximate B_{1g} symmetry [1]. In the figure we see that between room temperature and approximately T_c the frequency remains nearly constant, then shows a discontinuous drop below T_c and finally saturates below ~50 K. The bare frequencies have been extracted from the Raman spectra by a fit to a Fano lineshape [4]. The total softening defined as the frequency difference between 90 K and 10 K is 2.8 cm^{-1} or 0.9%. Data similar to these have been reported by Macfarlane et al. [7]. It is remarkable that the frequency anomaly is smaller than for Re = Y by more than a factor of two. A sensitive dependence of these shifts on oxygen concentration has been reported by Krantz et al. [8] in $YBa_2Cu_3O_{7-\delta}$ and one may suspect that the oxygen concentration in the Eu samples has not been adjusted to $\delta \approx 0$. The samples, have, however, been treated in the same fashion and the oxygen content has been controlled by thermogravimetry. We thus do not believe different oxygen contents to be the predominant reason for the varying frequency shifts.

Instead, we have put forward an alternative explanation which is consistent with the smaller shift of the B_{1g} mode in both the oxygen-reduced Re = Y sample and the fully oxygenated Re = Eu material. The essential ingredient is a resonance in the phonon shifts when a phonon is near the superconducting gap energy [9]. As a test of this assumption we show here the phonon shift in oxygen-isotope replaced $YBa_2Cu_3{}^{18}O_{7-\delta}$. The oxygen was exchanged under careful control of the samples weight (thermogravimetrically) and we can as-

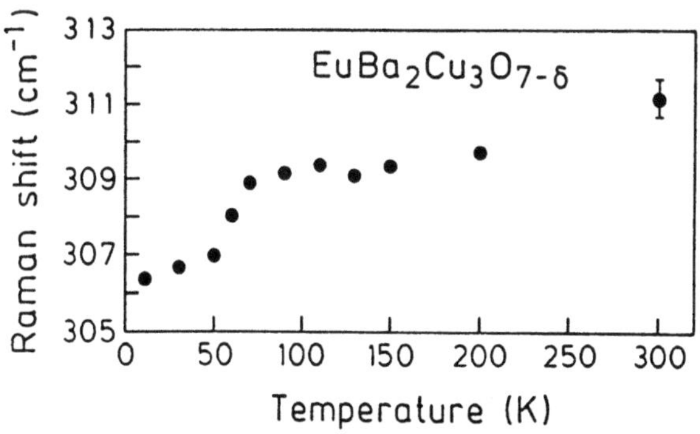

Figure 1: Phonon-frequency anomaly in $EuBa_2Cu_3O_{7-\delta}$. Note the discontinuous change below T_c in the phonon frequency.

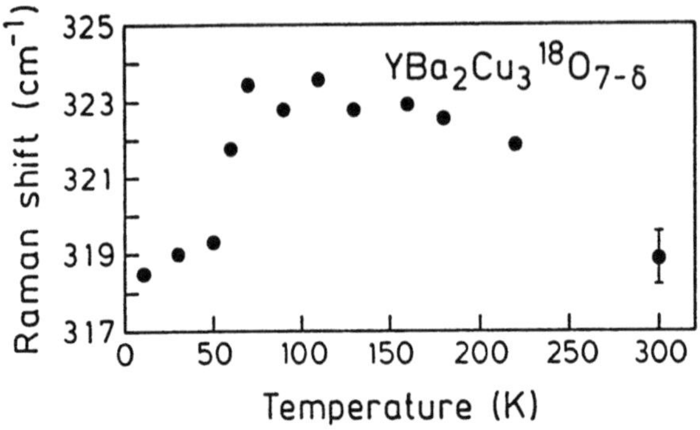

Figure 2: Phonon-frequency anomaly in isotope-replaced $YBa_2Cu_3{}^{18}O_{7-\delta}$.

certain that δ is the same in the ^{18}O sample as for the initially prepared ^{16}O material. The replacement of ^{16}O by ^{18}O in the high-T_c materials has been performed before for mainly two reasons: 1) to measure the isotope shift of T_c, which was found to be small ($\Delta T_c \leq 0.3$ K) and 2) to analyze the oxygen participation in eigenmodes of the crystal lattice [1,10]. Here we make explicit use of the fact that the eigenfrequency is lowered by 6% by the isotope replacement. If a gap lies near or above the B_{1g} phonon in $YBa_2Cu_3{}^{16}O_{7-\delta}$, the frequency *and* linewidth anomaly should be reduced in the ^{18}O sample. Figure 2 shows the experimental results. We find that the temperature anomaly $\Delta\omega/\omega$ is 1.5% compared to 1.9% in $YBa_2Cu_3{}^{16}O_{7-\delta}$ [9]. The increase in linewidth below T_c is also substantially reduced in the ^{18}O-sample, but not zero as we and also Macfarlane [7] observe for R = Eu. The anomalies are thus not related to the particular rare earth per se.

3. Conclusions

We thus show that the magnitude of the phonon softening below T_c as well as the additional linewidth broadening due to superconductivity appear to decrease with distance from a supposed "gap" energy in the high-T_c superconductors. A detailed quantitative analysis of this effect in terms of a strong-coupling theory by Zeyher and Zwicknagl [11] is presented in Ref. [9]. In conclusion, we have shown how phonon shifts and linewidths as probed by light scattering may give information about the energy gap in the high-T_c superconductors.

We benefitted in this work from discussions with M. Cardona and acknowledge support from the Bundesminister für Forschung und Technologie and the European Communities.

References

[1] C. Thomsen and M. Cardona, in *Physical Properties of High Temperature Superconductors I*, ed. by D.M. Ginsberg (World Scientific Publ., Singapore, 1989), p. 409.

[2] C. Thomsen, M. Cardona, B. Gegenheimer, R. Liu, and A. Simon, Phys. Rev. B **37**, 9860 (1988).

[3] S.L. Cooper, M.V. Klein, B.G. Pazol, J.P. Rice, and D.M. Ginsberg, Phys. Rev. B **37**, 5920 (1988).

[4] M.V. Klein, in *Light Scattering in Solids I*, ed. by M. Cardona, *Topics of Applied Physics*, Vol. 8 (Springer, New York, 1983), p. 173.

[5] S.M. Shapiro, G. Shirane, and J.D. Axe, Phys. Rev. B **12**, 4899 (1975).

[6] T. Ruf, C. Thomsen, R. Liu, and M. Cardona, Phys. Rev. B **38**, 11985 (1988).

[7] R.M. Macfarlane, M.C. Krantz, H.J. Rosen, and V.Y. Lee, Physica C **162-164**, 1091 (1989).

[8] M.C. Krantz, H.J. Rosen, R.M. Macfarlane, and V.Y. Lee, Phys. Rev. B **38**, 4992 (1988).

[9] C. Thomsen, M. Cardona, B. Friedl, C.O. Rodriguez, I.I. Mazin, and O.K. Andersen, to be published.

[10] M. Cardona, R. Liu, C. Thomsen, W. Kress, E. Schönherr, M. Bauer, L. Genzel, and W. König, Solid State Commun. **67**, 789 (1988).

[11] R. Zeyher and G. Zwicknagl, Solid State Commun. **66**, 617 (1988).

Theory of Raman Scattering in Superconductors*

A. Zawadowski

Institute of Theoretical Physics, Eötvös Loránd University,
H-1088 Budapest, Puskin u. 5–7, Hungary

Abstract. The theory of Raman scattering in isotropic superconductor is reviewed. The role of the final state interaction including the long range Coulomb force is discussed. The Raman spectra with different symmetries consist of two parts: a bound state below the threshold ($\omega<2\Delta$) and the continuum. In the Raman spectra measured on $YBa_2Cu_3O_7$ single crystals there are continua of electronic origin. It is proposed that in addition to the intraband scattering there is an interband contribution due to the charge transfer between the bridging oxygen and the plane. The entirely gapless feature of the spectra with B_{1g} symmetry is a probe of nodes in the superconducting gap. Finally, the theory of the Raman spectra of dirty metals is discussed.

1. Introduction

In 1961 Abrikosov and Fal'kovskii [1] suggested that the Raman scattering is an appropriate tool to measure the superconducting gap. The first measurement which shows the gap structure was carried out by Sooryakumar and Klein [2] in 1980. The light with energy transfer $\omega>2\Delta$ breaks a Cooper pair and a continuum is expected above 2Δ. Abrikosov and Fal'kovski [3] have also suggested that the final state interaction between the excited pair plays an important role in formation of the Raman spectra. A more elaborate theory was developed by Klein and Dierker [4] whose main conclusions have been confirmed by Monien and the present author [5].

2. Raman scattering on BCS superconductors

The light is coupled to the electronic charge and the current. The light is scattered by the charge but the current coupling contributes also in second order perturbation theory. The scattering can be described as a scattering on an effective charge

$$\rho_{eff}(\vec{q}) = \sum_{\vec{k},\sigma} \gamma(\vec{k})\, a_{\vec{k}+\vec{q},\sigma}^{+}\, a_{\vec{k},\sigma} \tag{1}$$

where $a_{\vec{k},\sigma}$ is the annihilation operator of an electron with momentum $\vec{k}$ and spin σ and $\gamma(k)$ is the scattering amplitude in which the momentum transfer q can be ignored, $\gamma(k)$ is the expec-

*The research was partially supported by the Hungarian National Research Fund No. OTKA 1787.

Springer Series in Solid-State Sciences, Vol. 99
Electronic Properties of High-T$_c$ Superconductors and Related Compounds
Editors: H. Kuzmany · M. Mehring · J. Fink © Springer-Verlag Berlin, Heidelberg 1990

tation value of a tensor taken between the polarization directions of the incident and the scattered light. The momentum dependence of $\gamma(k)$ can be expanded in terms of crystal harmonics $\varphi_L(k)$ as

$$\gamma(\vec{k}) = \sum_L \gamma_L \ \varphi_L(\vec{k}) \,. \tag{2}$$

In the case of spherical or cylindrical Fermi surfaces, the spherical functions $Y_{\ell m}(k)$ or the Fourier functions, $\exp\{im\varphi\}$ can be used, where φ is the azimuthal angle. The created pair excitation can be characterized by the quantum number L, L=0 corresponds to excitations uniform over the Fermi surface.

The two excitations forming the created pair orthogonal to the condensed Cooper pairs can interact by the residual interaction, which is not treated in the BCS theory. That final state interaction can be treated by the Bethe-Salpeter equation which is equivalent with the summation of the bubble diagrams. That theory results in collective excitations which have been first suggested by Anderson [6] and Bogoliobuv [7]. These excitations are of two kinds:

(i) the phase mode with L=0 corresponds to charge oscillation in the real space and therefore, they are coupled to the long range Coulomb forces. The collective excitation is the plasmon with energy ω_p. At low energies the Raman spectrum is proportional to $(v_F q)^2/\omega_p^2$ in the channel L=0, where v_F is Fermi velocity, therefore, in the small q limit the contribution of the channel L=0 can be ignored.

(ii) the amplitude modes appear in channel L≠0, where $\varphi_L(k)$ is orthogonal to $\varphi_{L=0}(k)$, therefore the total charge is zero in each elementary cell, and the long range Coulomb force does not play any role. The different parts of the Fermi surface contribute to charge oscillation with different phases in the real space and considering the total charge on a larger length scale a complete cancellation occurs. The only charge oscillations which may occur are between the different atoms of the elementary cell as the wave functions taken at given parts of the Fermi surface carry different weights at the positions of different atoms.

The penetration depth δ of the light into the material determines the averaged value of the momentum transfer, q. In contrary to the normal superconductors, $(v_F q) \ll 2\Delta$ holds for the A-15 compounds and the high-T_c superconductors, thus the penetration depth is large compared to the BCS coherence length. In this case the q=0 limit can be taken.

The Raman spectrum has been calculated in the entirely energy range [4], [5]. The electron-electron and the electron-hole interactions can be characterized by two couplings g_L^C (Copper channel) and g_L^Z (zero sound channel) in channel L, respectively, and their signs are chosen to be attractive for positive sign. As the ground state is the BCS one, therefore, for the BCS coupling $g^* > g_L^C, g_L^Z$ (L≠0) holds, where $g^* = g_{L=0}^C$.

Without final state interaction there is a square root singularity at 2Δ in the spectrum (see Fig. 1(a)).

The spectrum calculated with final state interaction consists of two parts:

(i) A bound state below the threshold, 2Δ is formed with binding energy $E_{B,L} \sim \Delta\ \alpha_L^2\ (0<\alpha_L \ll 1)$ where α_L is defined as

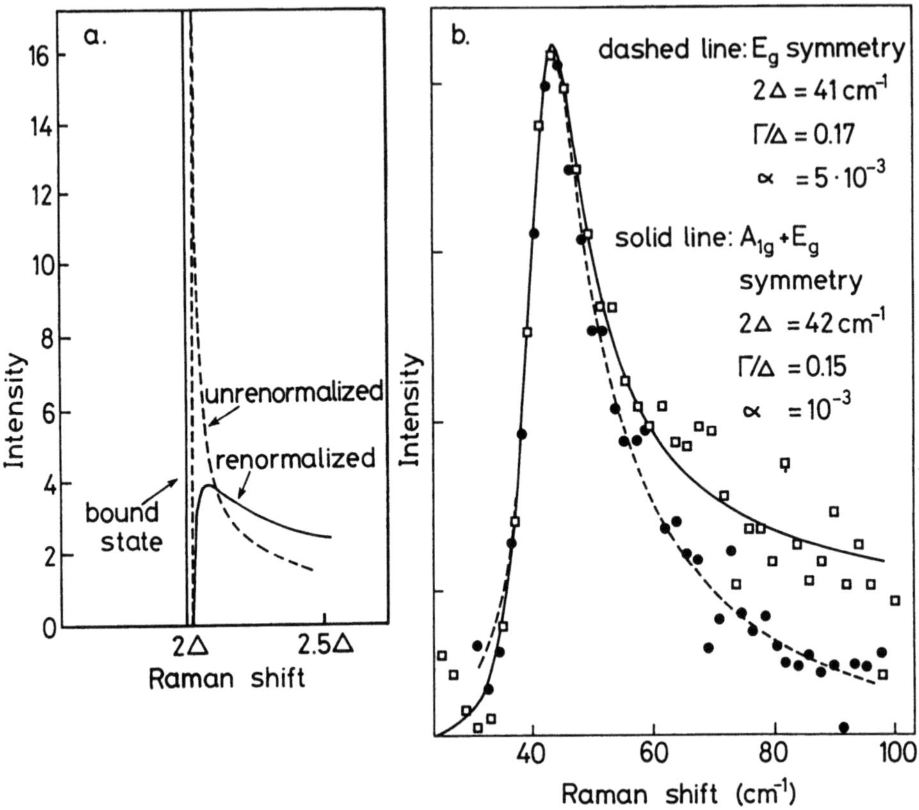

Fig. 1. Calculated Raman spectra; (a) with parameters $\alpha=0$ and $\alpha=0.05$ (b). The theoretical fit of the spectra of Ref. 4 measured in V_3Si .

$$\alpha_L = \frac{1}{2}\left[\rho_o g_L^Z + \frac{\rho_o g_L^C}{1-\dfrac{g_L^C}{g^*}}\right] , \tag{3}$$

which is attractive for all values of $g_L = g_L^C = -g_L^Z$ and ρ_o is the density of states for the electrons [4],[5],[8]. Depending on the signs of the couplings, the bound state may have rather two electron or exciton character [8],[5].

(ii) A continuum formed above 2Δ, which starts with zero amplitude at 2Δ and exhibits a maximum. The larger the coupling the stronger the bound state is. The spectral weight is transfered to the lower energy part of spectrum. The spectrum is depressed well above 2Δ by increasing α_L .

The Raman spectra measured on V_3Si with different light polarizations are shown in Fig. 1. (b). The different polarizations correspond to channels with different symmetries and different quantum numbers L and couplings α_L . The theoretical curves shown are calculated as a convolution of the theoretical Raman spectrum and a Lorentzian curve with half-width at half-height Γ. That width, Γ, may be due to experimental resolution or finite recombination time of the excited pair (likely due to the decay of the pair into two phonons). In the presence of smearing, the bound state and the continuum may not be resolved. The difference between the two curves has been previously attributed to different distributions of the gap characteristic for anistropic superconductors. The fits presented in Fig.1.(b) were obtained, however, by almost the same isotropic gap values and broadenings, but with different coupling strengths, α_L. The two curves were normalized together to have the same strength in their maxima. The large difference in their amplitudes at higher energies can not be attributed to the distribution of the gap, but it is rather a clear evidence for their difference in the coupling strengths, α_L, thus it demonstrates the importance of the final state interaction.

A very similar situation is realized in superfluid ^{4}He, where the attractive interaction between two rotons results in a two-roton resonance state formed below twice the roton energy, where the resonance is formed instead of the bound state because of the decay into two phonons (see Ref. 5).

3. Interband scattering

The suggestion that interband scattering may play a role in the Raman scattering in high-T_c materials has been first made by S.L. Cooper et al. [9]. Monien and the author [10] proposed that the Raman spectra measured in $YBa_2Cu_3O_7$ [9] may be decomposed into intraband and interband scattering mainly with A_{1g} and B_{1g} symmetry, respectively (see Fig. 2). The part of the spectrum attributed to the interband scattering shows a profound Fano line shape in the interference with the 330 cm^{-1} phonon with B_{1g} symmetry depicted in Fig. 3 (b) and shows no trace of the superconducting gap exhibited by the spectrum with A_{1g} symmetry (see Fig. 2 (a)). The interband transition is attributed to the charge transfer between the $p_y\pi$ orbital of the bridging oxygen and the oxygens in the plane. The charge transfers are indicated by arrows in Fig. 3 (a). That charge transfer is coupled to the phonon in the CuO_2 plane with B_{1g} symmetry depicted in Fig. 3 (b) as the $p_y\pi$ orbitals breaks the $x \leftrightarrow y$ symmetry. Charge transfer with very small energy is feasible only between two bands where one of them is flat and is located very near the Fermi energy (see Fig. 3 (c)), furthermore, in the neighbourhood of those lines where the bands are crossing without hybridi-

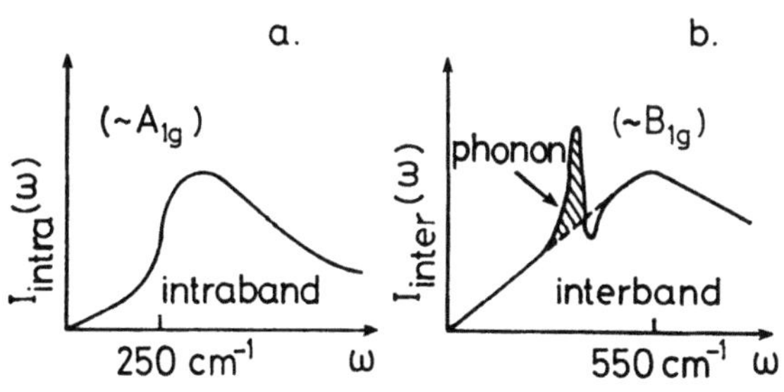

Fig. 2. The proposed line shape for the (a) intraband and (b) interband electronic Raman scattering in $YBa_2Cu_3O_7$.

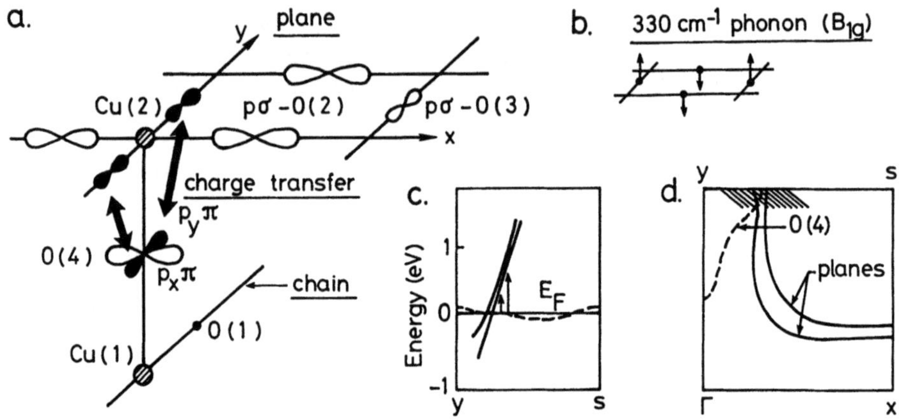

Fig. 3. (a) Part of the $YBa_2Cu_3O_7$ elementary cell with arrows representing the charge transfer. (b) The B_{1g} phonon in the CuO_2 plane. (c) and (d) Band structure of the crossing bands. The arrows indicate the interband transition and the hatched area is with small hybridization.

zation due to symmetry. Considering the bands associated with the CuO_2 plane and the bridging oxygen (see e.g. Ref. 11) the hybridization disappears at Brillouin zone boundaries perpendicular to the y-axis (see Fig. 3 (d)). The interband transition and the Fano lineshape were calculated with linearized hybridization at the zone boundary. The lineshape obtained in Ref. 10 is actually the one depicted in Fig. 2 (b).

The theory described above is developed without taking into account the existence of the superconducting gap. If a gap is developed at the band crossing lines, then a lower threshold must occur in the spectrum shown in Fig. 2 (b). In order to avoid that discrepancy, the assumption must be made that the gap also disappears along the band crossing lines. That invokes strong limitations on the possible types of superconductivity (d- or p-wave with nodes in the gap).

According to our present knowledge the only interband transition which can be realized at very small energies is the one described above. Whether the two strong assumptions on the gap nodes and the position of the flat band nearby the Fermi energy is realistic or not must be the subject of further studies. Especially, changing the oxygen concentration, applying pressure or doping could clarify that point, as the distance of the bridging oxygen from the plane must be a crucial parameter in the theory presented.

4. Theory of Raman scattering in metals with impurities

Recently, Cardona and the author [12] have devolved a theory of Raman scattering in normal metals, where the channel L=0 is treated and the effect of impurities is taken into account by perturbation theory. The spectrum has the form

$$\frac{\omega\tau_L^{*-1}}{\omega^2 + \tau_L^{*-2}} \qquad \text{where} \qquad \frac{1}{\tau^*} = \frac{1}{\tau} - \frac{1}{\tau_L} \quad . \tag{4}$$

The inverse relaxation time τ^{*-1} is the total scattering rate τ^{-1} reduced by the rate of those processes τ_L^{-1} in which the created electron-hole pair scatters inside the channel, L. The later processes can not be observed by Raman scattering. That theory differs from Fal'kovsky's work [13], where the channel L=0 is treated but without the long range Coulomb forces.

The author is especially grateful to his collaborators M. Cardona and H. Monien. He thanks R. Hackl, M.V. Klein and R.L. Martin for useful discussions and many suggestions.

References

1. A.A. Abrikosov and L.A. Fal'kovskii, Zh.Eks.Teor.Fiz. 40, 262 (1961) [Sov.Phys.-JETP 13, 179 (1961)].
2. R. Sooryakumar and M.V. Klein, Phys.Rev.Lett.44, 660 (1980).
3. A.A. Abrikosov and L.A. Falkovsky, Physica C156, 1 (1988).
4. M.V. Klein and S.B. Dierker, Phys.Rev.B29, 4976 (1984).
5. H. Monien and A. Zawadowski, Phys.Rev.B in print.
6. P.W. Anderson, Phys.Rev. 110, 827 (1958).
7. N.N. Bogoliubov, Nuovo Cimento 7, 6 (1958).
8. A. Bardasis and J.R. Schrieffer, Phys.Rev.121, 1050 (1961).
9. S.L. Cooper, M.V. Klein, B.G. Pazol, J.C. Rice and D.M.Ginsberg, Phys.Rev.B37, 5920 (1988); S.L. Cooper, F. Slakey, M.V. Klein, J.P. Rice, E.D. Bukowski and D.M. Ginsberg, Phys. Rev.B38, 11934 (1988).
10. H. Monien and A. Zawadowski, Phys.Rev.Lett.63, 911 (1989).
11. Jajeun Yu, S. Massida, A.J. Freeman and D.D. Koeling, Phys. Lett. A122, 198 (1987).
12. A. Zawadowski and M. Cardona to be published.
13. L.A. Fal'kovsky, Zh.Eks.Teor.Fiz. 95, 1146 (1989) [Sov.Phys.- - JETP 68, 661 (1989)].

A Comparative Raman Study of Classical and Oxide Superconductors

R. Hackl

Walther-Meißner-Institut für Tieftemperaturforschung,
Walther-Meißner-Str. 8, D-8046 Garching, Fed. Rep. of Germany

Abstract. Vibronic and electronic excitations of $YBa_2Cu_3O_7$ (YBCO) were investigated by Raman scattering as a function of temperature. The electronic continuum above T_c is found to be almost temperature independent. Below T_c new structures appear in the electronic spectra being related to the superconducting gap. The temperature dependence of this scattering deviates from the one observed in classical superconductors.

1. Introduction

Raman-scattering experiments revealed a lot of new insights into the physical properties of oxide superconductors and related compounds [1]. Many of the results, in particular the scattering from electronic excitations, remain puzzling so far and a quantitative analysis appears to be considerably beyond the scope of present theories. However, most of the excitations investigated in the high-T_c compounds have already been studied in metallic superconductors, especially in A15 materials like Nb_3Sn or V_3Si [2,3], and a comparison of oxide and classical superconductors may be useful.

2. Phonons

Remarkable phonon anomalies are observed in the A15 compounds above T_c [2]. A softening of the E_g phonon by some 10% is observed upon cooling from 300 K to 100 K in the "high-T_c" materials like Nb_3Sn, while the usual hardening occurs in e.g. Nb_3Sb with $T_c \approx 0$. Calculations of the electronic band structure and of the electron-phonon interaction [4] show that the trends of both the phonon anomalies and the superconductivity may consistently be explained in terms of the peculiar electronic

Springer Series in Solid-State Sciences, Vol. 99
Electronic Properties of High-T_c Superconductors and Related Compounds
Editors: H. Kuzmany · M. Mehring · J. Fink

structure near the Fermi energy E_F. The E_g phonon clearly experiences the superconducting transition. A sharp kink is observed at T_c and the peak frequencies increase by 5 and by 7% in V_3Si and Nb_3Sn, respectively, upon cooling from T_c down to $T \ll T_c$.

Compared to the A15 compounds the anomalies are less pronounced in the oxides, although the correlation with superconductivity is experimentally well established [1]. Parallel trends of T_c and anomalous frequency changes, however, were not observed so far either in compounds from various families with differing transition temperatures or as a function of doping.

3. Normal State Electronic Continuum

The continuum which the phonon lines are superimposed on in both oxide and metallic superconductors is assigned to scattering from electrons. Although the scattering intensity is expected to be reduced upon cooling at energy transfers $\hbar\omega < kT$ due to the Bose-Einstein factor, the spectra from YBCO appear to be almost temperature independent (Fig.1). Since the Raman intensity is proportional to the imaginary part of the response function $\chi''(T,\omega)$ times the thermal occupation number $\{1 + n(T,\omega)\}$ being approximately equal $kT/\hbar\omega$ at low energies, a constant scattering cross section implies $\chi''(T,\omega) \propto \hbar\omega/kT$ for $\hbar\omega < kT$ (Fig.2).

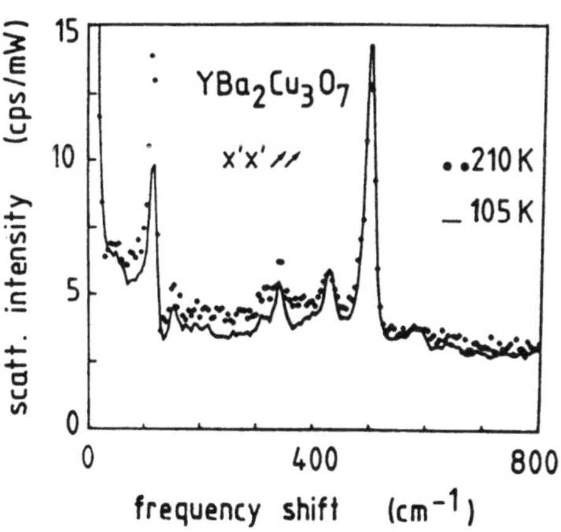

Fig.1. Raman spectra for YBCO. The a-b twinned single crystal has a T_c of 88 K [5]. Sample temperatures and polarizations of the incident and the scattered light are indicated; x = (100), x' = (110), etc.

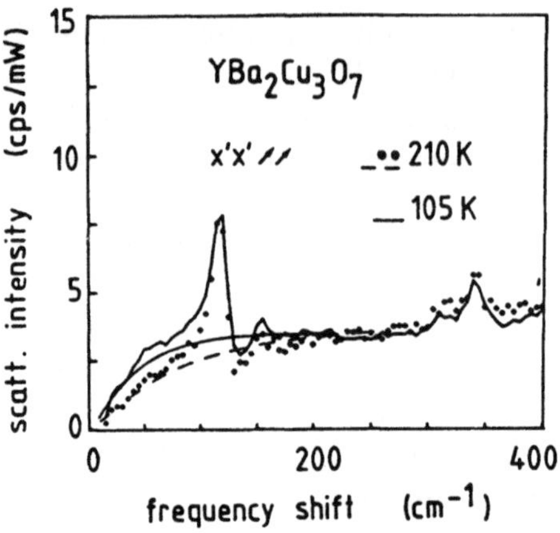

Fig.2. Spectra from Fig.1 on an expanded scale. After subtracting the elastic line the spectra are divided by the appropriate thermal factors $\{1 + n(T,\omega)\}$. The smooth lines are a guide to the eye.

In Nb_3Sn the low energy scattering closely follows the behaviour anticipated due to the Bose–Einstein thermal factor, however V_3Si comes out to be similar to the YBCO. Since there are even discrepancies in YBCO [6] further careful experiments including all families of the oxide superconductors are required to clarify the facts.

4. Electronic Scattering Below T_c

In the superconducting state a redistribution of the electronic scattering intensity to higher energies is observed. The effect obviously disappears very close to T_c (Fig.3). If the BCS assumption of fairly overlapping pairs holds and if final state interaction of excited electrons is neglected in a first approximation, the electronic Raman cross section should exhibit a square root singularity at 2Δ and cut off for $\hbar\omega < 2\Delta$ [7]. Anisotropy may broaden the structures. The smooth curves in Fig.4 represent the according fits to the $T \approx 0$ electronic spectra of YBCO single crystals at different measuring polarizations. They closely follow the experimental points [8] and may be considered as smoothed data, if one prefers to not believe in the BCS model. Intensity, maximum frequency and width strongly depend on the polarizations. The data suggest a large distribution of gap energies ranging from well below the weak coupling BCS limit of 3.53 k_BT_c to values in excess of 7 k_BT_c. The high ratio of 7 does not automati-

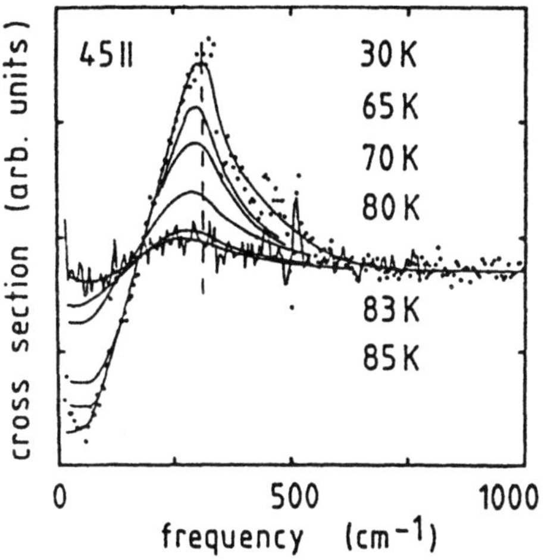

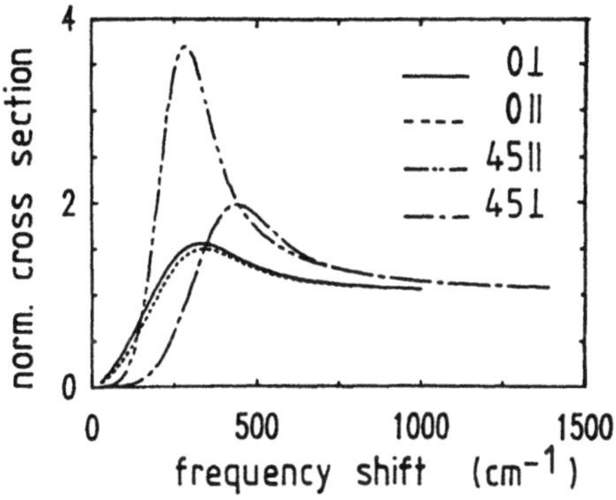

Fig.3. (left) (x'x') polarized electronic scattering (phonons subtracted) from YBCO at $T < T_c$. For clarity the experimental points are omitted at intermediate temperatures. The lines are drawn by hand.

Fig.4. Symmetry dependent electronic scattering from YBCO (normalized theoretical fits). Polarization symbols from top to bottom: (xy), (xx), (x'x'), (x'y').

cally imply strong coupling, since very small gaps are apparently also present. The finite slope of the (xx) and the (xy) spectra near zero shift could indicate line nodes of the gap, unfortunately, the data are not at all conclusive. In contrast to YBCO and other high-T_c materials the A15 compounds exhibit a sharp onset of the scattering in the vicinity of 2Δ independent of symmetry.

The maximum frequencies of the gap structures should scale with the energy gap. In Nb_3Sn (Fig.5) and in V_3Si a nice agreement with the BCS prediction is found. The deviation observed in YBCO is striking (Fig.6). There are, however, only a few experiments also indicating a similar deviation from conventional behaviour [9] and results from other compounds are ardently desired.

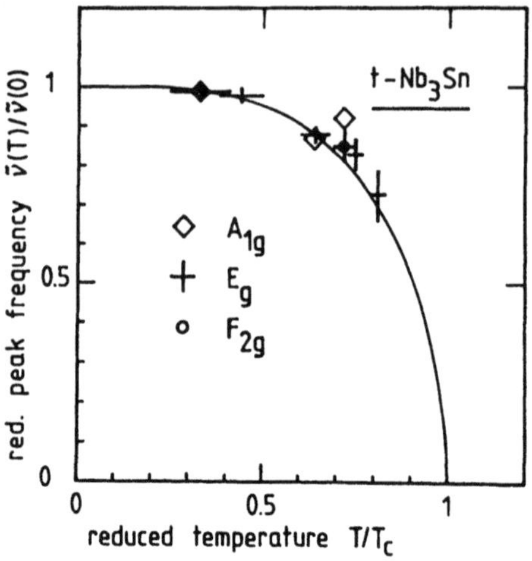

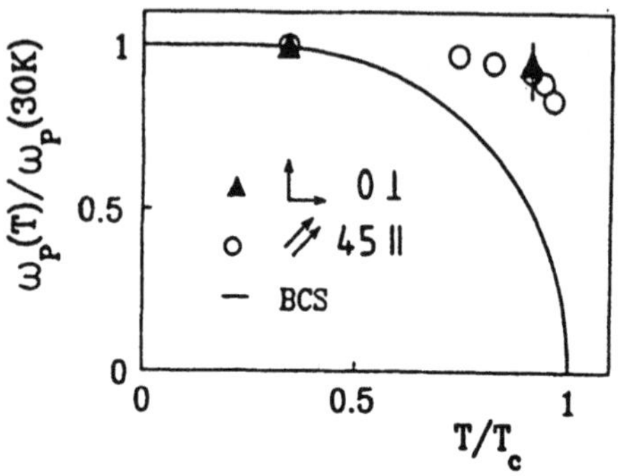

Fig.5. (left) Reduced maximum frequencies of gap structures in Nb_3Sn as a function of temperature. The line represents the BCS prediction for the energy gap.

Fig.6. Reduced peak frequencies of YBCO; (xy) and (x'x').

Acknowledgement. The work was done in collaboration with K. Andres, D. Einzel, W. Gläser, R. Kaiser, P. Müller and T. Staufer. We are greatly indebted to A. Zawadowski for communicating his theoretical work prior to publication and for elucidating discussions. Financial support by the Bundesministerium für Forschung und Technologie is gratefully acknowledged.

References

[1] C. Thomsen and M. Cardona in "Physical Properties of High Temperature Superconductors" D.M. Ginsberg ed. (World Scientific, Singapore, 1989).

[2] H. Wipf, M.V. Klein, B.S. Chandrasekhar, T.H. Geballe, J.H. Wernick, Phys. Rev. Lett. **41**, 1752 (1978); S. Schicktanz, R. Kaiser, W. Weber, C. Politis, R. Flükiger, P. Bujard, Phys. Rev. B **26**, 4909 (1982); R. Hackl, R. Kaiser, A. Müller, W. Weber, J. Phys. F **16**, 1299 (1986).

[3] S.B. Dierker, M.V. Klein, G.W. Webb, Z. Fisk, Phys. Rev. Lett. **50**, 853 (1983); R. Hackl, R. Kaiser, S. Schicktanz, J. Phys. C **16**, 1729 (1983); R. Hackl, R. Kaiser, W. Gläser, Physica C **162–164**, 431 (1989).

[4] W. Weber in "Electronic Structure of Complex Systems" NATO–A.S.I. series, P. Phariseau and W. Temmerman eds. (Plenum Press, New York, 1984) and references therein.
[5] P. Müller, K. Andres, F. Gross, H. Veith, R. Hackl, Physica C 153–155, 421 (1988).
[6] S.L. Cooper, M.V. Klein, B.G. Pazol, J.P. Rice, and D.M. Ginsberg, Phys. Rev. B 37, 5920 (1988).
[7] A. Zawadowski, this volume.
[8] R. Hackl, W. Gläser, P. Müller, D. Einzel, and K. Andres, Phys. Rev. B 38, 7133 (1988).
[9] J. Geerk, this volume.

Two-Magnon Scattering in the $REBa_2Cu_3O_6$

*P. Knoll**

Max-Planck-Institut für Festkörperforschung, Heisenbergstr. 1
D-7000 Stuttgart 80, Fed. Rep. of Germany and
Institut für Experimentalphysik der Universität Graz,
Universitätsplatz 5, A-8010 Graz, Austria

ABSTRACT: The spin excitations of the antiferromagnetic $REBa_2Cu_3O_6$ (RE=Y,Eu) have been studied by inelastic photon scattering in the temperature range 11K - 600K. The scattering mechanism is described for several polarization directions and has been calculated including an additional damping due to spin-lattice interaction. For all temperatures good agreement with experiments has been obtained.

1. Introduction

A common feature of all known HT_c cuprates is the occurrence of an antiferromagnetic phase. The strong Coulomb interaction of the 3d electrons of the Cu^{2+} cause the Mott-Hubbard insulator. The metallic configuration can be obtained by a kind of doping process. So far, it is still unclear if properties of the antiferromagnetic material (as e.g. the Hubbard U or the spin interaction) are also important in the metallic regime and, maybe, are involved in the superconducting mechanism. However, a complete understanding of the unconventional superconducting properties has to explain in the same way the occurrence of the antiferromagnetism and the continuous transition to the metallic compound.

A theoretical model of the CuO_2 planes has been given by Emery and Reiter [1]. Starting from a three-band Hubbard Hamiltonian the low-lying excitations can be described by a Heisenberg Hamiltonian:

$$H = J \cdot \left[\sum_{ij} \vec{S_i}\vec{S_j} - \Delta \cdot z \cdot S \left(\sum_i S_i^z - \sum_j S_j^z \right) \right] . \tag{1}$$

In order to obtain a more realistic description which also predicts a finite Neel temperature an anisotropy term has been added ($\Delta = 4.37 \cdot 10^{-4}$). z is the number of nearest neighbors. The exchange energy J can be expressed by the parameters of the three-band Hubbard model: t_{pd}, ϵ_{pd}, U_{pd}, U_d, and U_p:

$$J = \frac{4 \cdot t_{pd}^4}{\left[\epsilon_{pd} + U_{pd}\right]^2} \left[\frac{1}{U_d} + \frac{2}{U_p + 2\epsilon_{pd}} \right] . \tag{2}$$

* work carried out as an Alexander von Humboldt fellow.

Springer Series in Solid-State Sciences, Vol. 99
Electronic Properties of High-T_c Superconductors and Related Compounds
Editors: H. Kuzmany · M. Mehring · J. Fink

Equation (1) is solved mainly following the work of Balucani and Tognetti [2] by expanding in spin wave operators. Taking into account the higher order magnon interactions, a complex magnon energy E_k is obtained including energy renormalization by the Oguchi correction C [3] (corresponds to quantum fluctuations) and the complex temperature dependent renormalization $\zeta(T)$ [4].

$$\begin{aligned} E_k &= \left(\zeta(T) + \frac{C}{2S}\right) \cdot \Omega_k + \frac{(J(0) \cdot S)^2}{\Omega_k} \cdot \Delta(1+\Delta)\left(1 - \zeta(T) + \frac{C}{2S}\right) \\ \Omega_k &= S \cdot \sqrt{J(0)^2 \cdot (1+\Delta)^2 - J(k)^2} \qquad (3) \\ J(k) &= J \sum_r e^{ikr} \qquad r = \text{nearest neighbor vector} \end{aligned}$$

Raman scattering has been shown [5] to investigate excitations of the spin system. Fig.1 shows a typical Raman spectrum up to high energy shifts. Besides the phonon features at low energies a quite strong scattering intensity around 3000 cm^{-1} is observed and has been assigned to scattering on spin fluctuations [6]. In contrast to conventional antiferromagnets [7] as K_2NiF_4 in the HT_c cuprates the spin fluctuation scattering is much broader and has a tail at high energies. By calculating spectral moments it has been shown [8] that the occurrence of scattering intensity at quite high energies is intrinsic to the $S = 1/2$ quantum Heisenberg system and corresponds to high order magnon processes. In recent work it was demonstrated [9] that, in addition, spin-lattice interaction significantly broadens the spin excitation spectrum. Fig.1 compares the low temperature spectrum to calculated pure two-magnon scattering including additional damping which arises from magnon-phonon interaction. The high energy tail is not present in the pure two-magnon spectrum and, according to the findings by calculating spectral moments [8], we assign that tail to contributions from four-magnon processes. Usually, that high order magnon process is well separated in energy from the two-magnon scattering intensity; but due to the strong magnon damping they can merge and form the asymmetric tail.

2. The interaction of light with the spin system

It has been worked out by Fleury and Loudon [5] that in antiferromagnets the light scattering on spin fluctuations can become quite strong due to a special spin pair excitation mechanism: Incoming radiation excites the electron on site i. Because of the antiferromagnetic coupling to site j, both electrons (the excited one at site i and that on site j) flip their spins. That corresponds to the creation (or destruction) of at least two magnons, one at each sublattice. Because of the opposite spin directions of the two electrons the total spin is not changed. The excited, spin flipped electron relaxes to the ground state by emission of a photon.

In the HT_c cuprates the antiferromagnetic coupling is due to a superexchange. For that case light can excite the 3d hole on the Cu sites via a charge

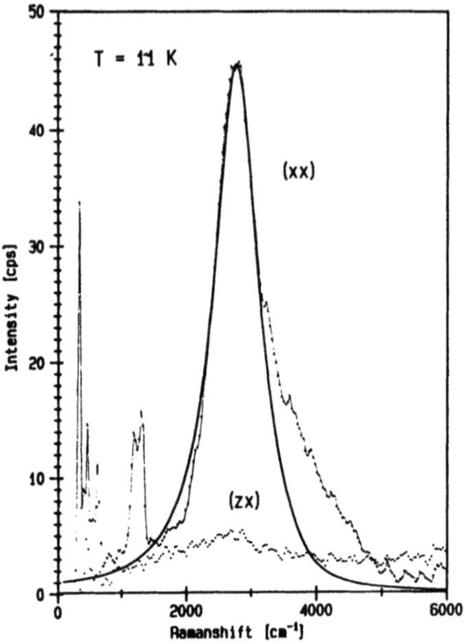

Fig.1: Pure two-magnon contribution to the spin fluctuation Raman spectrum in (xx) polarization.

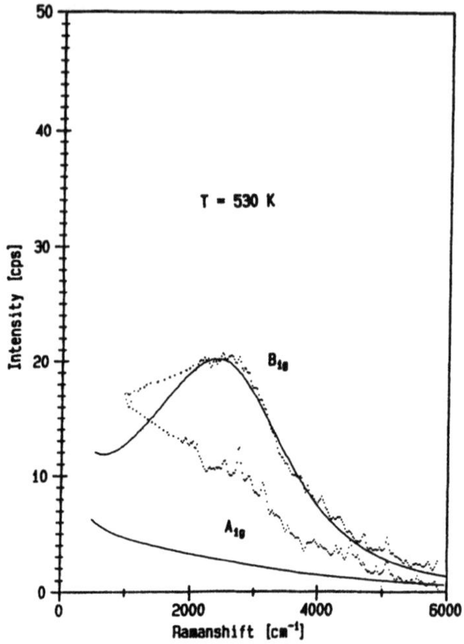

Fig.2: Calculated (solid line) and measured (dots) spin fluctuation spectra for B_{1g} and A_{1g} symmetry.

transfer excitation to the oxygen 2p state; after the flip of the spins the hole returns to the copper. In both cases the scattering intensity is expected to scale with the exchange energy, which in the HT$_c$ materials is quite high (~1000 cm^{-1}). That may explain why the spin fluctuation scattering is much stronger than scattering by phonons.

Starting with the intersite interaction Hamiltonian [5]

$$H_{ij} = -e\left(\vec{r}_i + \vec{r}_j\right)\left(\vec{E}_l + \vec{E}_s\right) + \frac{e^2}{|\vec{r}_i - \vec{r}_j|} \tag{4}$$

the intersite matrix element turns out to be proportional to a pair of spin operators. With the electric fields of incoming E_l and scattered radiation E_s the matrix element is expressed as

$$M_{ij} = \vec{E}_l T_{ij} \vec{E}_s S^- S^+ \ . \tag{5}$$

Assuming the superexchange mechanism the scattering tensor T_{ij} only contains dipole transitions along the nearest neighbor directions. Performing the sum over the nearest neighbors and transforming to the reciprocal space the symmetry factor f_k is calculated. Applying the superexchange mechanism to spins with D_{4h} site symmetry f_k only allows scattering in B_{1g} and A_{1g} symmetry (D_{4h} point symmetry). Magnons mainly contribute at the Brillouin zone edge in the nearest neighbor direction (X and Y points) and at the zone center to

B_{1g} and A_{1g} scattering symmetry, respectively. The experiments indeed exhibit no significant scattering in any polarization containing the z direction, as is shown in Fig.1 for the z and x polarizations of incident and scattered light, respectively. At low temperatures intensity only in B_{1g} symmetry is observed. But with increasing temperature also A_{1g} scattering occurs (Fig.2). The A_{1g} light scattering interaction operator commutes with the isotropic nearest neighbor Heisenberg Hamiltonian (Eq. (1) without anisotropy) and, therefore, no scattering intensity occurs. But due to additional damping mechanisms and the anisotropic crystal field the commutator rule is broken and the arising A_{1g} scattering intensity can be qualitatively calculated (Fig.2). The calculated spectra are obtained by using

$$\begin{aligned} I(\omega) &= (n(\omega)+1)\, S^2\eta(T)^2 Im\frac{-L(\omega)}{1-J\cdot L(\omega)}\,, \\ L(\omega) &= \frac{1}{n}\sum_k f_k^2(u_k^2+v_k^2)^2\frac{-2n(E_k)-1}{\hbar\omega-2E_k-2i\Gamma_k}\,. \end{aligned} \tag{6}$$

The scattering intensity scales with the square of the first order (Hartree Fock) renormalization factor $\eta(T)$. u_k and v_k are the Bogoliubov coefficients. Additional spin interactions have been considered by increasing the imaginary part of the complex one-magnon energy E_k by Γ_k.

3. Temperature dependence

On increasing the temperature well above the Neel temperature of 430 K the observed spin fluctuation features strongly broaden (Fig.3). Using equation (6) these spectra are fitted with an exchange energy of 810 cm^{-1} and an additional temperature dependent damping as shown in Fig.4. At higher temperatures the high energy tail due to four-magnon scattering vanishes because the temperature dependent one-magnon damping much more strongly influences the four-magnon feature which relaxes in a broad featureless background and does not contribute to the spin excitation lineshape.

The temperature dependence of the damping of the zone edge magnon is shown in Fig.4. The intrinsic damping due to spin-spin interaction turns out to be too weak to explain the broad spectral features. It has been shown [9] that magnon-phonon interaction becomes important in the HT_c cuprates because of the unusually high exchange energy. In first order that interaction leads to a process where a magnon is inelastically scattered by a phonon. Taking into account only phonon creation processes the influence on the one magnon damping was calculated with an interaction constant of 5000 cm^{-1}/Å. Fig.4 shows that magnon-phonon interaction can explain the observed one magnon damping within a wide temperature range. Because of the neglect of higher order processes our calculation of magnon-phonon interaction underestimates the magnon damping at very high temperatures.

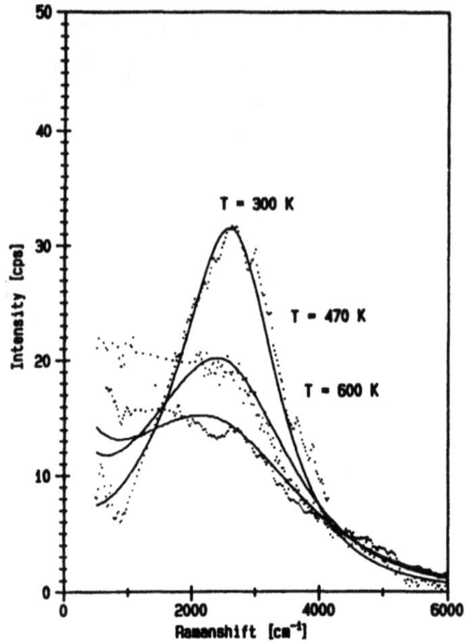

Fig.3: Measured (dots) and calculated (full line) B_{1g} two-magnon spectra.

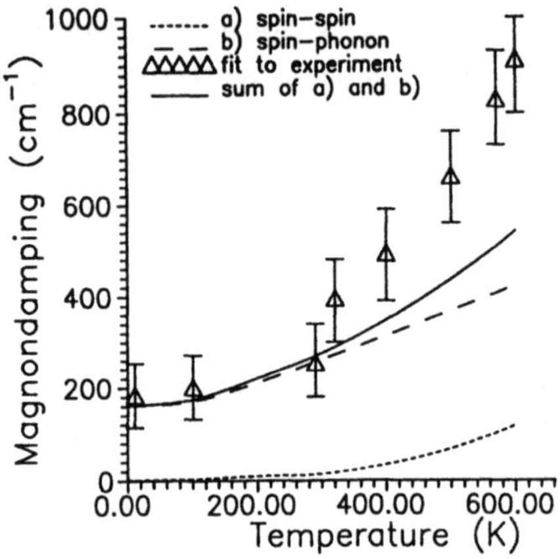

Fig.4: Temperature dependent damping of zone boundary magnons compared with contributions of spin-spin and magnon-phonon interaction.

4. Conclusions

Our experiments and calculations identify the observed Raman feature at 3000 cm^{-1} as scattering by spin fluctuations. At low temperatures contributions from four-magnon scattering cause the asymmetric line shape. The broad linewidth due to strong magnon damping together with the observed temperature dependence can be explained by magnon-phonon interaction.

It has been worked out [11] that due to a Jahn-Teller distortion the degenerate d levels of Cu^{2+} split. That effect may be important for the superconducting properties of the cuprates. Very recently [12] the influence of the d orbitals on the spin excitation spectrum has been theoretically investigated. So far, in our experiments we do not find a behavior which corresponds to d-d transitions. We assume that due to the strong exchange energy the scattering intensity is dominated by the spin character of the excitations and the orbital contributions are, maybe, too weak to be clearly identified by measurement.

References

1. V.J.Emery and G.Reiter, Phys.Rev.B38, 4547 (1988)
2. U.Balucani and V.Tognetti, Riv. Nuovo Cim.6, 39 (1976)
3. T.Oguchi, Phys.Rev.117, 117 (1960)
4. G.Wenzel and M.Wagner, J.Phys.C 11, 137 (1978)
5. P.A.Fleury and R.Loudon, Phys.Rev.166, 514 (1967)

6. K.B.Lyons, P.A. Fleury, J.P. Remeika, A.S. Cooper, and T.J. Negran, Phys.Rev.B37, 2353 (1988)
7. P.A.Fleury and H.J.Guggenheim, Phys.Rev.Lett.24, 1346 (1970)
8. R.R.P.Singh, P.A. Fleury, K.B. Lyons, and P.E. Sulewski, Phys.Rev. Lett. 62, 2736 (1989)
9. P.Knoll, C.Thomsen, M.Cardona, and P.Murugaraj, Phys. Rev. B **42** (in press)
10. M.G.Cottam, J.Phys.C7, 2901 (1974)
11. W.Weber in Advances in Solid State Physics 28, 141 (1988)
12. L.F.Feiner and J.Zaanen, this volume

Resonant Raman Scattering in $YBa_2Cu_3O_7$, $YBa_2Cu_3O_6$ and $YBa_2Cu_4O_8$

E.T. Heyen, R. Liu, C. Thomsen, and M. Cardona

Max-Planck-Institut für Festkörperforschung, Heisenbergstr. 1, D-7000 Stuttgart 80, Fed. Rep. of Germany

ABSTRACT We have investigated optical transitions in $YBa_2Cu_3O_7$, $YBa_2Cu_3O_6$, and $YBa_2Cu_4O_8$ using the resonant Raman scattering technique on single crystals. This method not only allows one to find the energies of characteristic features in the dielectric function, but also to assign the atomic orbitals that contribute most to the bands participating in electronic transitions. For $YBa_2Cu_3O_6$, we conclude that in a/b polarization almost all atoms are involved in the 1.7 eV transition, while only plane copper and oxygen orbitals lead to a 2.45 eV peak (at 10 K), suggesting that the latter is an in-plane charge transfer excitation; on the other hand the c-spectrum just shows a broad peak near 2.2 eV. The electronic structure of $YBa_2Cu_3O_7$ is very different: in a/b polarization, the 2.35 eV transition involves mostly plane oxygen and apical oxygen atoms, while a peak in the near UV seems to correspond to an apical oxygen - chain transition. In addition, we observe a Ba-only resonance below 1.9 eV. The c-polarized spectra show transitions at 2.35 and $\approx$2.8 eV involving plane-to-chain charge transfers. We will also discuss differences between a and b polarizations measured on an untwinned $YBa_2Cu_4O_8$ crystal. The results and assignments on the superconducting crystals agree well with recent LMTO LDA band structure calculations.

1. INTRODUCTION

The electronic structure of high T_c superconductors is still not well understood. Using high energy spectroscopy [1] or ellipsometry [2][3][4], the occupied and unoccupied densities of states as well as the dielectric function have been determined. It is, however, rather difficult to determine the origin of these states since there are as many as 36 bands [5] (of which only three cross the Fermi surface [6]) formed by the hybridized Cu 3d and O 2p orbits. All these bands lie within a 6 eV range below the Fermi level and

Springer Series in Solid-State Sciences, Vol. 99
Electronic Properties of High-T_c Superconductors and Related Compounds
Editors: H. Kuzmany · M. Mehring · J. Fink

lead to rather broad structures observed in spectroscopy. Also the applicability of one-electron band structure calculations like the LMTO local density method to the high T_c materials is still disputed since correlations of the Cu 3d electrons are important. It has recently been demonstrated [4] that the measured dielectric function in $YBa_2Cu_3O_{7-\delta}$ agrees remarkably well with LMTO calculations [7]. It seems to be clear that these simple theories fail for the corresponding insulating antiferromagnets.

A technique that yields further information on the electronic structure is resonant Raman scattering which has been applied with great success to semiconductors [8]. The Raman scattering intensity for a certain phonon is given by

$$I(\omega) \sim \omega^4 \left| \vec{e}_L \cdot \frac{d\tilde{\epsilon}}{du} \cdot \vec{e}_S \right|^2 ,$$

where $\vec{e}_L$ and $\vec{e}_S$ are the polarization unit vectors of incident and scattered light, ω is the laser photon frequency, $\tilde{\epsilon}$ is the complex dielectric tensor and u the normal coordinate of the phonon under consideration. If we plot the Raman peak intensities of a certain phonon over the photon energy of the exciting laser, we should get a spectrum showing not only the features of the dielectric function, but also the coupling of the bands participating in the electronic transition to this phonon. We can expect this coupling to be large only if the bands involve a large admixture of the atomic orbitals of those atoms that move most if a particular phonon is generated. A strong long-range Coulomb interaction might still modify this picture. In order to separate the ω^4 factor and the detector and spectrometer response from changes in $d\tilde{\epsilon}/du$ we divided the obtained intensities by those obtained from the Raman phonon in BaF_2 which are known to be nearly proportional to ω^4 [9]. Concerning high T_c superconductors this technique was first applied to $Pb_2Sr_2Y_{1-x}Ca_xCu_3O_{8+\delta}$ [10] and then extended to $Nd_{2-x}Ce_xCuO_{4+\delta}$ [11].

2. RESONANCES IN $YBa_2Cu_3O_7$

The Raman-active phonons with A_g symmetry in $YBa_2Cu_3O_7$ are well known [12]: they have energies of 115, 150, 335, 435, and 500 cm^{-1} and correspond primarily to motions in the c-direction of the Ba, Cu2 (plane), O2/O3 (plane) out-of-phase, O2/O3 in-phase, and

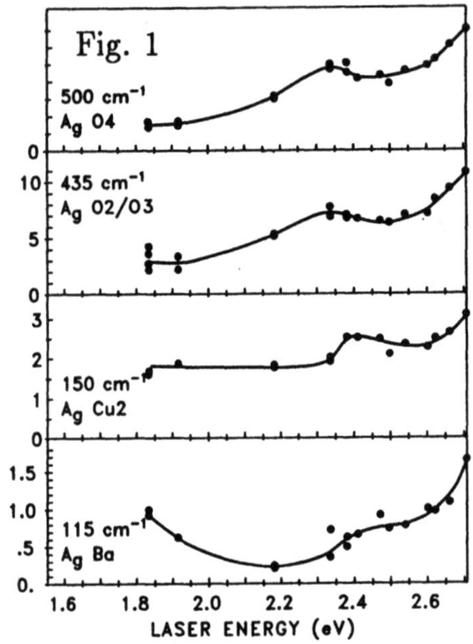

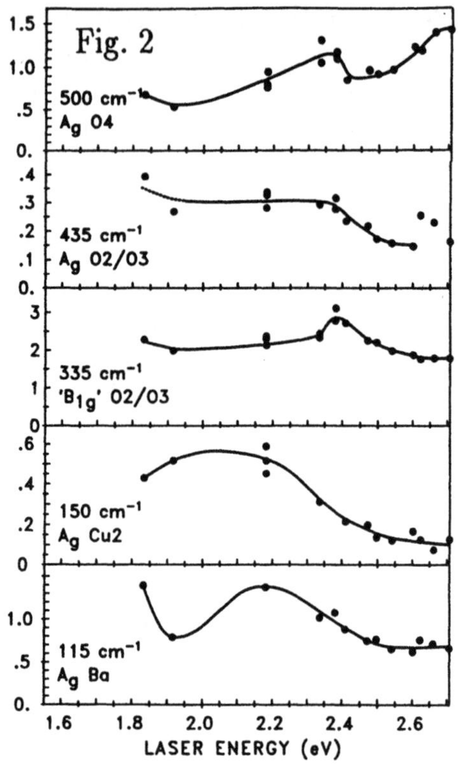

Fig. 1: Resonance profiles in $YBa_2Cu_3O_7$ taken at 10 K with *c*-polarized light
Fig. 2: Resonance profiles in $YBa_2Cu_3O_7$ taken at 10 K with *a*/*b*-polarized light.

O4 (apical oxygen) atoms, respectively. The 335 cm^{-1} phonon is only weakly influenced by the orthorhombic distortion of the unit cell; for a tetragonal point group it would have B_{1g} symmetry. The resonance profiles of these phonons obtained as described above with the light polarized in the *c* direction as well as in *a*/*b* direction are shown in figures 1 and 2. (The crystal is twinned, we can therefore not distinguish between *a* and *b* polarizations.) For each phonon, its energy, symmetry and assignment are indicated in the figures. We used as excitation 13 discrete lines of Ar^+ and Kr^+ lasers. From ellipsometry measurements [2][4] it is known that this material shows a strong *c*-polarized resonance at 2.8 eV while the *a*/*b* spectrum is almost featureless. Our measurements also show the rise towards the 2.8 eV resonance, but in addition to that also *a*/*b* polarized transitions below 1.9, near 2.35 , and above 2.7 eV and another c-polarized transition at 2.35 eV. Since resonant Raman scattering is a form of modulation spectroscopy [13], it is not surprising to see more features than in ellipsometry. The resonance energies observed are in excellent agreement with recent LMTO LDA calculations of the dielectric function [7]. We will therefore use results from such calculations [14] also for the interpretation of our data.

All phonons probed with c-polarization resonate at 2.35 and above 2.7 eV. This indicates that both transitions have a similar origin and that they both involve highly hybridized states with admixtures from the Cu2, O2/O3, O4, and Ba atoms. The calculations show that these transitions have such hybridized bands (with additional Cu1 (chain) admixture) as the initial state and the quasi one-dimensional Cu1 - O1 band as the final state. They occur between S and Y points of the Brillouin zone [14].

The resonances in a/b directions, however, are of a different origin. Only the B_{1g}-like O2/O3 phonon and the O4 phonon resonate near 2.35 eV while Cu2 and Ba phonons do not show any features at this energy. This indicates that this transition corresponds to a charge transfer from O2/O3 to O4 or also to the chain. Similarly, only the O4 phonon resonates above 2.7 eV, hinting at a O4-chain transition. Finally, we observe a Ba-only resonance below 1.9 eV. The two data points showing this resonance have been reproducibly measured. According to preliminary LMTO calculations, all these transitions belong to different "spaghetti"-like bands with varying orbital admixtures as initial states and the Cu $3d_{x^2-y^2}$ O $2p_{x,y}$ plane bands as final states. The transitions seem to occur directly at the Fermi surface crossing of the latter bands half-way between S and X or S and Y points.

Measurements on untwinned $YBa_2Cu_4O_8$ crystals allowed us not only to determine the resonances in this material, but also to separate a and b polarized transitions. The Raman-active phonons in this material have recently been investigated [15]. We found [16] that the O2-O4 transition is now shifted to 2.45 eV while also the other a/b resonance above 2.7 eV has moved up, out of the energy range measured. All these transitions are a- as well as b-polarized. In b-polarization, we observe no further sharp feature but only additional broad resonances, mainly concerning the Cu1 phonon. All phonon peaks, with the remarkable exception of the B_{1g}-like O2/O3 phonon, have much larger intensities in b-

polarization. A detailed discussion of the electronic structure in $YBa_2Cu_3O_7$, $YBa_2Cu_4O_8$, and $YBa_2Cu_3O_6$ will be published elsewhere [16]. We remark that the plane oxygen phonon resonances in $YBa_2Cu_3O_7$ and $YBa_2Cu_4O_8$ are similar to those observed in the $Pb_2Sr_2Y_{1-x}Ca_xCu_3O_{8+\delta}$ superconductor, while in the latter we see additional resonances from Pb and Sr orbitals influencing the other phonons [10].

3. RESONANCES IN $YBa_2Cu_3O_6$

The insulating and antiferromagnetic compound $YBa_2Cu_3O_6$ is tetragonal and has also 5 Raman-active phonons corresponding to *c*-direction motion, of which 4 are of A_{1g} and one of B_{1g} symmetry. Their energies are 110, 145, 340, 455, and 475 cm^{-1}. Using ellipsometry, it has been shown [3][4] that there exist two *a*/*b*-polarized resonances at 1.7 and 2.5 eV, while the *c*-spectrum is featureless. The 1.7 eV resonance has often been attributed to the Cu2 $3d_{x^2-y^2}$–O2/O3 $2p_{x,y}$ charge transfer transition [1]. The corresponding resonant Raman profiles are shown in figures 3 and 4.

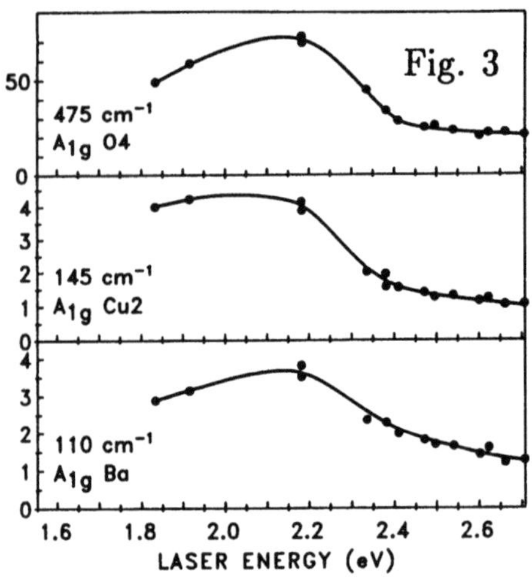

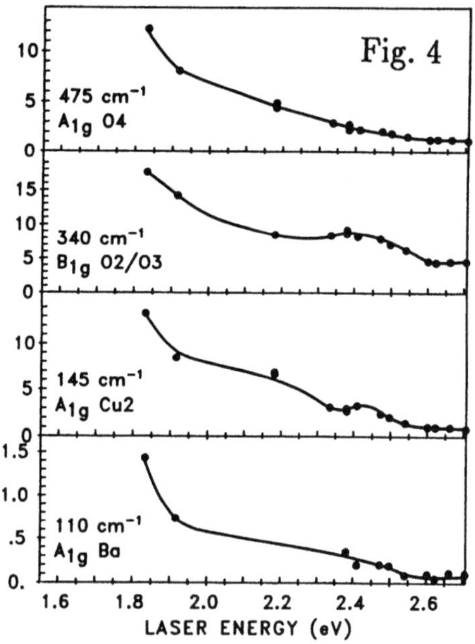

Fig. 3: Resonance profiles in $YBa_2Cu_3O_6$ taken at 10 K with *c*-polarized light.
Fig. 4: Resonance profiles in $YBa_2Cu_3O_6$ taken at 10 K with *a*/*b*-polarized light.

As expected, the c-polarized spectrum does not display any sharp resonances, but just a broad band around 2.2 eV for all observed phonons. In comparison to $YBa_2Cu_3O_7$, it appears that the transitions with the one-dimensional chain band as the final state have disappeared. The remaining structure seems to belong to electronic states with admixtures from all atoms.

The a/b polarized resonance profiles also clearly show the two resonances observed in ellipsometry as a rise towards the infrared region and a peak at 2.45 eV, respectively. These two features, however, exhibit a very different character. All phonons, particularly the Ba-mode, resonate towards 1.7 eV indicating that this transition involves states with strong admixtures from the CuO_2 plane as well as from O4 and especially Ba. Thus it seems to be unlikely that this is the predicted in-plane charge transfer transition. The 2.45 eV peak appears to be significant only for the Cu2 and O2/O3 phonons while it is at least much weaker for the other phonons. We therefore suggest that the transition at 2.45 eV is an in-plane charge transfer excitation.

We thank H. Hirt, M. Siemers, and P. Wurster for expert technical help. We acknowledge financial support from the Bundesministerium für Forschung und Technologie, Bonn, and the European Community.

4. REFERENCES

[1] See for a review: J. Fink et al., to be published in IBM J. of Research and Development
[2] M.K. Kelly, P. Barboux, J.M. Tarascon, P.E. Aspnes, Phys.Rev.B 40, 6797 (1989)
[3] J. Kircher, M. Alouani, M. Garriga, P. Murugaraj, J. Maier, C. Thomsen, M. Cardona, O.K. Andersen, and O. Jepsen, Phys. Rev. B 40, 7368 (1989)
[4] J. Kircher, M. Kelly, M. Garriga, P. Murugaraj, M. Cardona, E. Schönherr, to be publ.
[5] See for a review: W.E. Pickett, Rev. Mod. Phys. 61, 433 (1989)
[6] C.O. Rodriguez, A.I. Liechtenstein, I.I. Mazin, O. Jepsen, O.K. Andersen, to be published
[7] E.G. Maksimov, S.N. Rashkeev, S.Yu. Savrasov, and Yu.A. Uspenski, Phys. Rev. Lett. 63, 1880 (1989)
[8] M. Cardona, in "Light Scattering in Solids II", p.128ff, Springer-Verlag

[9] J.M. Calleja, H. Vogt, and M. Cardona, Philos. Mag. A 45, 239 (1982)
[10] E.T. Heyen, R. Liu, M. Garriga, B. Gegenheimer, C. Thomsen, and M. Cardona, Phys. Rev. B 41, 830 (1990)
[11] E.T. Heyen, R. Liu, B. Gegenheimer, C. Thomsen, M. Cardona, S. Piñol, and D.McK. Paul, p.346ff in *PHONONS'89*, ed. S. Hunklinger, W. Ludwig, and G. Weiss, World Scientific (1990)
[12] R. Liu, C. Thomsen, W. Kress, M. Cardona, F.W. de Wette, J. Prade, A.D. Kulkarni, and U. Schröder, Phys. Rev. B 37, 7971 (1988)
[13] M. Cardona, Surface Science 37, 100 (1973)
[14] S. Rashkeev, private communication
[15] E.T. Heyen, R. Liu, C. Thomsen, R. Kremer, M. Cardona, J. Karpinski, E. Kaldis, and S. Rusiecki, Phys. Rev. B 41, 11058 (1990)
[16] For $YBa_2Cu_3O_7$: E.T. Heyen, S.N. Rashkeev, I.I. Mazin, O.K. Andersen, R. Liu, M. Cardona, and O. Jepsen, submitted to Phys. Rev. Lett.,
For $YBa_2Cu_3O_6$and $YBa_2Cu_4O_8$: E.T. Heyen and M. Cardona, to be published

FT Raman Scattering and Luminescence in $BaBiO_3$

G. Ruani, A.J. Pal, R. Zamboni, and C. Taliani

Istituto di Spettroscopia Molecolare, CNR, I-40126 Bologna, Italy

Abstract - Unconventional Raman scattering, exciting at 1.16 eV and photoluminescence measurements with different laser sources in $BaBiO_3$ were carried out. The semiconducting gap opened up by the Peierls distortion in the undoped compound is identified as an indirect gap with E_g = 1.1 eV. The 360 cm^{-1} band observed in both these experiments, which in a recent paper is shown to interact strongly with the electronic continuum, is attributed to the out of phase LO bending mode of the oxygens of the tetrahedron.

1. Introduction

$BaBiO_3$ doped with potassium is a new superconductor which shows a T_c = 30 K and may be considered a new high T_c superconductor (HT_cS) [1,2]. This material presents several peculiarities with respect to the other HT_cS compounds. It is a three dimensional system, it has no magnetic ions and no Cu-O planes. Finally, in the undoped phase, a static charge density wave (CDW) coupled with the lattice opens a gap at the Fermi level making this compound a Peierls insulator; furthermore the superconductivity appears when the CDW is destroyed because of doping. In view of this we have performed photoluminescence measurements on the undoped material with different laser sources (from 1.2 to 2.5 eV) and FT-IR Raman measurements in order to investigate the nature of both the optical gap and the interaction between phonons and electronic states.

2. Experimental

Stoichiometric proportions of powdered metal nitrates dissolved in distilled water were employed to synthesize $BaBiO_3$ by the citrate-polymer route [3]. The resulting solution was stirred up and heated to 90°C, and allowed to react. Upon complete reaction the resulting gel was subsequently decomposed to a solid by heating. The powders were calcined at 450°C and then ground, pressed into pellets, sintered and annealed at 950°C. X-ray diffraction measurements showed the product to be X-ray pure [4].

An FT interferometer (Bruker IFS 88) was used for luminescence measurements. Three different lasers (Nd-YAG at 1.16 eV, HeNe at 1.96 eV and Ar^+ at 2.41 eV) were used to excite the sample and a liquid N_2 cooled Ge detector (active range 0.75 - 1.5 eV) was used to detect the luminescence. The same interferometer was used for IR (1.16 eV) excited Raman experiments.

3. Results and Discussion

3.1 FT-Raman

The Raman spectrum of $BaBiO_3$, exciting at 1.16 eV, shows two Raman bands at 360 and 570 cm^{-1} (Fig. 1). We used a total power of 300 mW in order not to heat the sample. To estimate the actual temperature of the sample we calculated it from the Stokes and Antistokes intensity ratio of the 570 cm^{-1} mode. The temperature obtained was 300 K demonstrating that the sample was not overheated by the Nd-YAG laser.

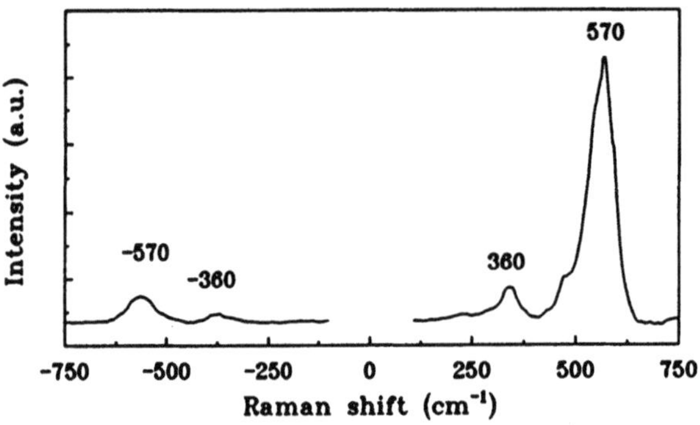

Figure 1. The Fourier-transform Raman spectrum of $BaBiO_3$ excited with an IR Nd-YAG laser at 1.16eV showing Stokes and anti-Stokes sides.

In the Raman spectra with a visible laser excitation (2.41 eV) Sugai [5] observed a very strong 570 cm^{-1} band (27 times the intensity of the 360 cm^{-1} band), assigned as the breathing mode of the oxygen, with three overtones of this mode. In the IR excited Raman spectrum, the resonant enhancement of the 570 cm^{-1} band is strongly renormalized (see Fig. 1): the ratio of the 570 and 360 cm^{-1} modes is only 8; also the overtones are no more detectable. The assumption that only the 570 cm^{-1} band depends on resonance conditions is justified because the 360 cm^{-1} mode intensity does not depend upon doping [5,6].

In our previous paper on FT-Raman of $YBa_2Cu_3O_{7-y}$ [7], the semiconducting sample showed a multiphonon Raman scattering of the 505 cm^{-1} band. This was due to preresonance conditions of the exciting IR laser with the semiconducting gap. In the case of $BaBiO_3$ we also excited (by Nd-YAG laser) at the tail of the first excited electronic band [8]. The absence of resonance may suggest that the onset of the intergap transition is due to phonon assisted transitions through an indirect gap. In fact in absorption, a phonon assisted transition from the top of the VB to the bottom of the CB can occur in an indirect gap semiconductor but, because of conservation of momentum selection rules in Raman backscattering ($k_{ph} = 2k_v$, where k_{ph} and k_v are the momentum of phonon and photon respectively) the transition is approximately vertical, and the event is far from resonance. A gap around 1.1 eV is in agreement with the finding of an optical gap observed by photomodulation absorption technique [9].

Mattheiss and Hamman [10] proposed that the stability of the monoclinic structure at low temperature ($\approx$400 K) of $BaBiO_3$ is enhanced by the splitting of the electronic band around E_F decreasing the average one electron energy for these states. In this picture the splitting would give rise to an indirect gap semiconductor with a static CDW. From this calculation the splitting is approximately 1 eV and the indirect gap is just opened up. The same is obtained by Blaha et al. [11]. The shape of the density of state of $BaBiO_3$ obtained with XPS [12] is in agreement with that obtained from these calculations except that the centroid of the VB is shifted by about 1 eV to a higher binding energy and that the distance from the top of the VB to the E_F is more than 0.2 eV. The discrepancy in energy values between theory and experiments is not critical, in fact it is normal to have a factor of 3 in energy calculations around E_F.

3.2 Luminescence

In the photoluminescence spectrum the main feature is a doublet at 1.04 eV separated by 46 cm^{-1} and is followed by two equally spaced (360 cm^{-1}) bands decreasing in intensity at lower energy. The observation of luminescence in $BaBiO_3$ at about 1 eV when the sample is excited with photons of 1.16 eV gives a direct evidence that the semiconducting gap is located between these two values. The same spectrum is obtained using different laser excitations (Nd-YAG, He-Ne and different lines of Ar^+), this proves the luminescence nature of the spectra itself (Fig. 2). Only the efficiency of the process depends on the energy of the pump, in fact the luminescence with the same photon density is 50 times more intense when the sample is excited at 1.9 eV than at 1.16 eV. This is further evidence that the gap is indirect i.e. transitions at 1.16 eV are phonon assisted and those at 1.9 eV are not.

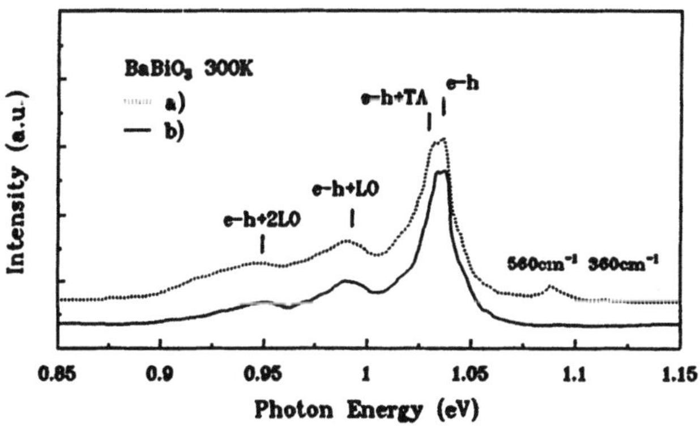

Figure 2. Luminescence spectrum of $BaBiO_3$ at room temperature a) exciting at 1.16 eV, b) exciting at 1.96 eV.

The 46 cm^{-1} feature fits well with the TA mode observed with inelastic neutron scattering and is strongly emphasized in tunneling measurements [13] in $Ba(Pb_{1-x}Bi_x)O_3$. The energy separation between the strongest features of the photoluminescence spectrum (360 cm^{-1}) coincides with the energy of the phonon which appears in the Raman spectrum (Fig. 1) [5,6]. In the undoped sample this mode appears as a peak in the phonon density of state determined by neutron scattering. When the system is doped with Pb this mode remains as a shoulder at Γ point; but it reappears in a longitudinal geometry for phonon wave vectors away from Γ point [14]. In this way the luminescence spectrum can be interpreted as due to electron-hole recombination with multiphonon emission. This is typical [15] of a recombination process between a free electron (hole) and a trapped hole (electron). The traps should be shallow levels which, estimating the difference between the excitation and the emission energy, will be less than 0.12 eV. In this way the recombination takes place in the field of the imperfection relaxing the momentum conservation selection rule and a strong multiple emission of longitudinal optical phonons becomes possible. The anomalous width of the structures, even at 10 K, is probably due to relaxation processes of the lattice around the trapping centers or lifetime broadening.

The presence of the 360 cm^{-1} phonon in luminescence is an indication that the mode is a LO. Moreover, the 360 cm^{-1} band in $Ba_{0.6}K_{0.4}BiO_3$ shows a Fano resonance, i.e. is strongly coupled with the electronic continuum. The same Fano behaviour is observed also in $YBa_2Cu_3O_7$ for a phonon in the same spectral region (337 cm^{-1})[16]. This phonon is the B_{1g} out of phase bending mode of the O_2 O_3 oxygens. The similarity of the oxygen squares in $YBa_2Cu_3O_7$ and $BaBiO_3$, the same Fano shape, lead us to associate the 360 cm^{-1} band of the latter compound to the same bending mode; unfortunately in $BaBiO_3$ at Γ point this mode is silent. The fact that it appears as a strong feature in photoluminescence would confirm that the band observed in Raman is the LO component of the out of phase silent mode activated because of the local breaking of symmetry due to defects. Moreover, Rice and Wang [17] suggested that in a phonon assisted mechanism of superconductivity in such a material, this LO mode should be the most relevant mediator in the electron-electron pairing.

Acknowledgments. One of us (AJP) has carried out this work with the support of the ICTP Programme for Training and Research in Italian Laboratories, Trieste, Italy. We thank CNR Progetto Finalizzato "Tecn. Superconduttive e Criogeniche" for financial support.

References

1. R.J.Cava, B. Batlogg, J.J Krajewski, R. Farrow, L.W. Rupp Jr., A.E. White, K. Short, W.F. Peck and T. Kometani, Nature 3322, 814 (1988).
2. L.R. Mattheiss, E.M. Gyorgy and D.W.Johnson Jr., Phys Rev. B 37, 3745 (1988).
3. H.H. Wang, K.D. Carlson, U. Geiser, R.J. Thorn, H.I. Kao, M.A. Beno, M.R. Monaghan, T.J. Allen, R.B. Proksch, D.L.Stupka, J.M. Williams, B.K. Flandermeyer and R.B. Poepple, Inorg. Chem. 26, 1476 (1987).
4. G. Thornton and A.J. Jacobson, Acta Cryst. B 34, 351 (1978).
5. S. Sugai, S. Uchida, K. Kitazawa, S. Tanaka and A. Katsui, Phys. Rev. Lett. 55, 426 (1985).
6. K.F. McCarty, H.B.Radousky, D.G. Hinks, T.J. Folkerts and R.N. Shelton, Phys. Rev. B 40, 2662 (2662).
7. R. Zamboni, G. Ruani, A.J. Pal and C. Taliani, Solid State Commun. 70, 813 (1989).
8. S. Tajima, S. Uchida, A. Masaki, H.Takagi, K.Kitazawa, S. Tanaka and A.Katsui, Phys. Rev. B. 32, 6302 (1985).
9. C. Taliani, A.J. Pal, G. Ruani, R. Zamboni, X. Wei and Z. V. Vardeny, this issue.

10. L.F. Mattheiss and Hamman, Phys. Rev. B 28, 6629 (1983).
11. P. Blaha, K. Schwarz, G. Vielsack and W. Weber, this issue.
12. Z.-X. Shen, P.A.P. Lindberg, B.O. Wells, D.S. Dessau, A. Borg, I. Lindau, W.E. Spicer, W.P. Ellis, G.H. Kwei, K.C. Ott, J.-S Kang and J.W. Allen, Phys. Rev. B 40, 6912 (1989).
13. W.Reichardt, B.Batlogg and J.P.Remeika, Physica 135B, 501 (1985).
14. W.Reichardt and W.Weber, Jpn. J. Appl. Phys, 26, suppl. 3, 1121 (1987).
15. J.J. Hopfield, J. Phys. Chem. Solids 10, 110 (1959).
16. C. Thomsen, M. Cardona, B.Gegenheimer, R.Liu and A.Simon, Phys. Rev. B 37, 9860 (1988).
17. M.J. Rice and Y.R. Wang, Physica C157,192 (1989).

Part VI

Theory

Charge Distribution and Electric Field Gradient Calculations for $YBa_2Cu_3O_{7-x}$

C. Ambrosch-Draxl, P. Blaha, and K. Schwarz

Institut für Technische Elektrochemie, TU Wien,
Getreidemarkt 9/158, A-1060 Wien, Austria

Abstract. The electric field gradients (EFGs) of $YBa_2Cu_3O_7$, $YBa_2Cu_3O_{6.5}$ and $YBa_2Cu_3O_6$ are calculated on a first principles basis using the full potential linearized augmented plane wave (LAPW) method in which exchange and correlation effects are treated by the local density approximation (LDA). Good agreement between calculated and experimental EFGs is obtained. The EFG is shown to originate from the anisotropy of the electronic charge and thus can be related to difference densities and symmetry decomposed partial charges. The LDA calculations yield reliable charge distributions to which the EFG is very sensitive.

1. Introduction

In the high temperature superconductor $YBa_2Cu_3O_7$ (YBCO7) there is strong evidence that superconductivity occurs mainly in the CuO_2 layers and the carriers are holes, while the Cu-O chains with varying oxygen content control the charge transfer into this plane. It is important to know where the holes are and how the charge is distributed between Cu and O.

The electronic structure of these systems has been obtained by various band structure calculations (reviewed by Pickett [1]) which are based on the density functional theory (DFT) in combination with the local density approximation (LDA). Since the results of such calculations fail to reproduce the insulating and antiferromagnetic ground state in $YBa_2Cu_3O_6$ (YBCO6), and show discrepancies with experimental spectra, it has been argued that LDA calculations are not able to describe these systems. Therefore it is important to investigate to which extent one can rely on LDA band structures.

The electric field gradient (EFG) is a ground state property of a solid and depends sensitively on the asymmetry of the electronic charge in a crystal. Experimentally YBCO7 and YBCO6 have been investigated by nuclear magnetic resonance (NMR) and nuclear quadrupole resonance (NQR) spectroscopy.

Blaha et al. [2] developed a first principles method to compute EFGs from an all electron band structure calculation. They have used the full potential linearized augmented plane wave (LAPW) method and have calculated the EFG directly from the self-consistent charge density without further approximations such as the use of Sternheimer antishielding factors. This method was successfully applied to various compounds, so that the study of the new class of ceramic superconductors is well founded.

We have performed such EFG calculations for YBCO7 and YBCO6 and have published first results in two short papers [3,4]. A detailed analysis of all three compounds has been submitted for publication [5].

YBCO7 (YBCO6) is orthorhombic (tetragonal) with space group Pmmm (P4/mmm). For the additional calculation for YBCO6.5 we assumed an ordered structure in which Cu(1) has three oxygen neighbours, a coordination which Lütgemeier proposed to interpret his NMR data [6].

We utilize the well known full potential LAPW method (WIEN code [7]) and treat the Y-4s, Y-4p, Ba-5s and Cu-3p states as semi-core states in an

Springer Series in Solid-State Sciences, Vol. 99
Electronic Properties of High-T_c Superconductors and Related Compounds
Editors: H. Kuzmany · M. Mehring · J. Fink

additional band calculation. All three compounds, YBCO7, YBCO6.5, and YBCO6, are studied by non-spin-polarized calculations.

2. The Electric Field Gradient

All nuclei with a nuclear-spin quantum number I≥1 have a nonspherical nuclear charge distribution and an electric quadrupole moment Q. The nuclear quadrupole interaction between this Q and the electric field gradient (EFG) determines the NQ coupling constant eQV_{zz}/h, where e is the electric charge, h is Planck's constant and V_{zz} represents the principal component of the EFG, which is defined as the second derivative of the electrostatic potential written as a traceless tensor.

In the LAPW calculation the Coulomb potential is obtained numerically for a given charge density by solving Poisson's equation in form of a boundary value problem. When this potential is expanded in an LM representation, only L=2 terms contribute to the EFG [2,5,8]:

$$V_{2M} = - C_{2M} \int_0^R \rho_{2M}(r)\, r^2 / r^3\, dr \quad + \textit{lattice} \text{ contributions.} \qquad (1)$$

From these components the diagonal elements V_{aa}, V_{bb}, and V_{cc} of the traceless EFG tensor are calculated. Ordering these terms according to their magnitudes, such that $|V_{zz}| \geq |V_{yy}| \geq |V_{xx}|$, the EFG tensor is characterized by the largest component V_{zz} (in short EFG) and the anisotropy parameter η defined as $\eta = (V_{xx} - V_{yy}) / V_{zz}$.

In Table I our calculated EFGs and their asymmetry parameters η are compared with experimental data for the Cu positions [6,9,10] and for oxygen [11]. Recent measurements on Ba by Lütgemeier [12] are included; the

Table I: Theoretical and experimental (see text) electric field gradients (in $10^{21}Vm^{-2}$) and asymmetry parameters for YBCO7, YBCO6.5 and YBCO6. The sign of the EFG is not known experimentally.

		YBCO7		YBCO6.5		YBCO6	
		Theory	Exp.	Theory	Exp.	Theory	Exp.
Y	EFG	3.4		3.8		3.6	
	η	0.9		0.0		0.0	
Ba	EFG	7.3	9.4	8.0		11.0	11.2
	η	0.8	1.0	0.3		0.0	0.0
Cu(1)	EFG	7.4	7.5	-9.5	8.4	-10.9	11.8
	η	0.8	1.0	0.9	1.0	0.0	0.0
Cu(2)	EFG	-5.6	12.3	-2.9		-3.6	9.0
	η	0.1	0.0	0.1		0.0	0.0
O(1)	EFG	18.3	17.3	23.3		-	-
	η	0.3	0.4	0.2		-	-
O(2)	EFG	11.8	10.5	11.3		10.7	
	η	0.2	0.2	0.2		0.2	
O(3)	EFG	11.9	10.2	11.4		10.7	
	η	0.2	0.2	0.2		0.2	
O(3')	EFG	11.9	10.2	11.3		10.7	
	η	0.2	0.2	0.2		0.2	
O(4)	EFG	11.7	11.6	7.8		6.7	
	η	0.2	0.3	0.2		0.0	

NMR frequencies are converted to the EFGs using the nuclear quadrupole moment Q=0.2 barn, an average over several (strongly deviating) values quoted in literature. Since there is no isotope with a nuclear quadrupole moment for Y, the values for this position cannot be measured. Good agreement is found between our first principles calculations and the experimental values for all positions except for Cu(2) where theory predicts only half the experimental value. An analysis of this situation [3,5] has shown that it is sufficient to transfer only 0.07 electrons from the $d(x^2-y^2)$ to the $d(z^2)$ orbital to reproduce the experimental value. The orientation of the EFG tensor agrees for all positions [4,5].

For Cu(1) we properly describe the increase of the EFG from YBCO7 to YBCO6. From the components of the EFG tensor (Table II) it becomes apparent that the value for YBCO6.5 cannot be obtained by linear interpolation, since the main component changes sign from YBCO7 to YBCO6.5 and the orientation of the principal axis changes from b to c between YBCO6.5 and YBCO6. These results confirm the threefold coordinated Cu(1) proposed by Lütgemeier [6].

Table II: EFG components in $10^{21}Vm^{-2}$ for Cu(1) in YBCO7, YBCO6.5, and YBCO6:

	V_{aa}	V_{bb}	V_{cc}
YBCO7	- 6.7	**+ 7.4**	- 0.7
YBCO6.5	+ 0.6	**- 9.5**	+ 8.9
YBCO6	+ 5.4	+ 5.4	**-10.9**

3. Relation between EFG and Charge Distribution

Cu has an almost filled d shell and interaction with oxygen leads to bonding and antibonding orbitals most of which are essentially filled; i.e. about 1.86e (1.25e) inside the copper (oxygen) sphere. Whenever this interaction is strong, part of these antibonding states will not be occupied (favouring short distances), whereas weak interaction yields higher occupation numbers (Table III). The former case leads to holes which show up in a difference density (Fig.1), e.g. the hole state of σ symmetry between the Cu(2)-$d(x^2-y^2)$ and O(3)-$p(y)$ in the CuO_2 plane. The anisotropy in form of a large negative difference density around the oxygens causes the EFG (both in magnitude and direction), since for oxygen only p-like densities are involved. At the copper sites the 4p and 3d wave functions contribute to the radial density coefficient ρ_{20} which via Eq.(1) leads to $V_{zz}=V_{20}$, the main component of the Cu(2) position. Fig.2 shows the

Table III: Partial charges (in electrons) of the valence states inside the respective atomic spheres and their symmetry decomposition for YBCO7.

	s	p	d	p_x	p_y	p_z	d_{z^2}	$d_{x^2-y^2}$	d_{xy}	d_{xz}	d_{yz}
Y	0.18	0.26	0.87	0.09	0.08	0.09	0.09	0.23	0.08	0.25	0.22
Ba	0.06	5.57	0.29	1.86	1.86	1.85	0.06	0.04	0.09	0.06	0.04
Cu(1)	0.23	0.20	8.60	0.03	0.07	0.10	1.41	1.65	1.84	1.84	1.86
Cu(2)	0.20	0.17	8.69	0.07	0.07	0.03	1.76	1.44	1.85	1.82	1.82
O(1)	1.55	3.34	0.01	1.18	0.91	1.25					
O(2)	1.54	3.40	0.01	1.01	1.21	1.18					
O(3)	1.54	3.39	0.01	1.21	1.00	1.18					
O(4)	1.54	3.36	0.01	1.18	1.19	0.99					

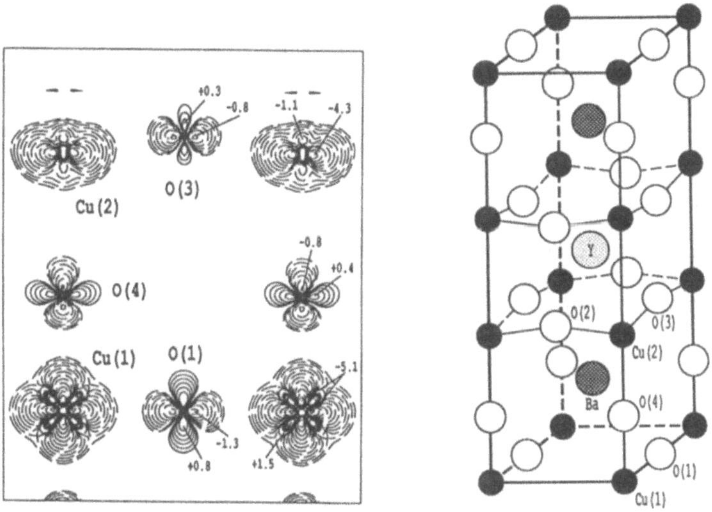

Fig.1: Difference electron density taken between the crystalline (LAPW) and the superposed ionic densities assuming the ions Y^{+3}, Ba^{+2}, Cu^{+1} and $O^{-10/7}$. A contour plot is shown in a plane parallel to the b- and c-axis through the Cu(1)-O(1) chains cutting the atoms O(4) Cu(2) and O(3) as seen from the crystal structure; maxima and minima are labeled in units of $e / Å^3$, contour lines start with ±0.1 and differ by a factor √2 for positive (solid) and negative (dashed) values.

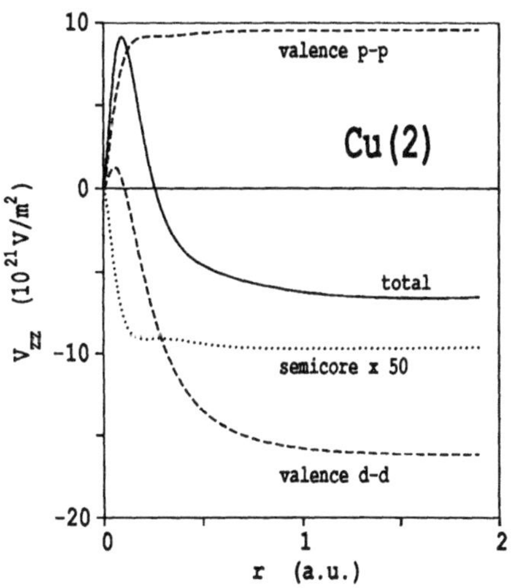

Fig.2: The EFG contribution Vzz of Cu(2) is shown as a function of radius: a decomposition of the total component into contributions from the Cu-3p semi-core states (50 times enlarged) and the valence p-p and d-d parts is displayed.

integral according to Eq.(1) as a function of the upper integration limit. The p- and d-terms have opposite sign and thus partly cancel each other. Core polarization effects are small, since already the Cu-3p semicore states contribute about 50 times less than the valence states. The EFG is determined by the anisotropy of the charge distribution near the nucleus up to a radius of about 1 a.u., so that the *lattice* part (from the neigbouring atoms) brings a few percent for oxygen, but less than 1 percent for copper.

4. Conclusion

It was questioned whether or not the present LDA version can properly describe these highly correlated systems. The present analysis has shown that within the LDA the charge distribution and the electric field gradients are obtained in close agreement with experimental data. The small change in occupation numbers of 0.07e between two symmetries of the Cu(2)-d states, which would be needed to make the agreement perfect, indicates that the LDA is rather good, but some intraatomic angular correlations might be underestimated presumably by the implicitly assumed spherical averaging over the exchange correlation hole. In conclusion, we confirm the picture of the chemical bond in this class of compounds with strong covalent interactions rather than fully localized Cu-d states.

This project was supported by the 'Fonds zur Förderung der wissenschaftlichen Forschung', project number P7063P.

References

1. W.E.Pickett: Rev.Mod.Phys. 61, 433 (1989)
2. P.Blaha, K.Schwarz and P.H.Dederichs: Phys.Rev.B 37, 2792 (1988)
3. C.Ambrosch-Draxl, P.Blaha and K.Schwarz: J.Phys.: Condens. Matter 1, 4491 (1989)
4. C.Ambrosch-Draxl, P.Blaha, and K.Schwarz: Physica C 162-164, 1353 (1989)
5. K.Schwarz, C.Ambrosch-Draxl, and P.Blaha: submitted to Phys.Rev.B
6. H.Lütgemeier: Physica C 153-155, 95 (1988)
7. P.Blaha, K.Schwarz, P.Sorantin and S.B.Trickey: Comp. Phys. Commun. (in print) (1990)
8. P.Blaha, P.Sorantin, C.Ambrosch and K.Schwarz: Hyperfine Interactions 51, 917 (1989)
9. C.H.Pennington, D.J.Durand, D.B.Zax, C.P.Slichter, J.P.Rice and D.M. Ginsberg: Phys.Rev.B 37, 7944 (1988)
10. H.Yasuoka, T.Shimizu, Y.Ueda and K.Kosuge: J.Phys.Soc.Japan 57, 2659 (1988)
11. M.Takigawa, P.C.Hammel, R.H.Heffner, Z.Fisk, K.C.Ott and J.D.Thompson: Phys.Rev.Lett.63, 1865 (1989)
12. H.Lütgemeier: this volume

Total Energy Calculations of the Breathing Mode in $BaBiO_3$

P. Blaha [1], *K. Schwarz* [1], *G. Vielsack* [2], *and W. Weber* [2]

[1] Institut für Technische Elektrochemie, TU Wien, A-1060 Wien, Austria
[2] Institut für Physik, Universität Dortmund,
D-4600 Dortmund, Fed. Rep. of Germany

Abstract. Undoped $BaBiO_3$, the parent compound of the new high Tc $Ba_{(1-x)}K_xBiO_3$ superconductor, deviates from the ideal perovskite structure and shows a charge disproportionation of the Bi ions where the oxygen atoms move towards one Bi but away from the other. We have investigated this breathing distortion of the oxygen octahedra by an all electron self-consistent band structure calculation using the full-potential linearized augmented plane wave method in combination with the local density approximation. We find this structure to be more stable and obtain a gap at the Fermi energy and semiconducting behavior in agreement with experiment.

1. Introduction

The discovery of high Tc superconductivity in the Cu free $Ba_{(1-x)}K_xBiO_3$ system [1] has raised considerable interest in the electronic properties of $BaBiO_3$, the parent compound of this class of superconductors. This system shows some remarkable similarities to the La_2CuO_4 based class of materials. In both cases the undoped compound is semiconducting, but on doping a phase transition and superconductivity occurs. La_2CuO_4 exhibits an antiferromagnetic ground state, but in $BaBiO_3$ a charge density wave (CDW) gives rise to the semiconducting state.

Mattheiss and Hamann [2] performed band structure calculations for $BaBiO_3$ assuming the ideal perovskite structure and found it to be metallic with an almost perfectly nested Fermi surface. When they used the experimentally observed distorted monoclinic structure a band splitting near the Fermi energy was obtained, which can be interpreted as a Peierls instability. Tight-binding calculations of the electron-phonon coupling show an unstable 'breathing' phonon [3], but so far total energy pseudopotential calculations based on the local density approximation (LDA) have not found the corresponding minimum for the breathing distortion [4], but encountered an unstable rotational mode.

We are investigating this problem by an all electron full-potential linearized augmented plane wave (LAPW) calculation to see whether or not LDA calculations can reproduce the observed breathing distortion. Frozen phonon calculations in the fcc unit cell ($Ba_2Bi_2O_6$) are performed utilizing the WIEN-code [5], a scalar relativistic full-potential version of the LAPW

Springer Series in Solid-State Sciences, Vol. 99
Electronic Properties of High-T_c Superconductors and Related Compounds
Editors: H. Kuzmany · M. Mehring · J. Fink

method (neglecting spin-orbit coupling). The treatment of semi-core (Bi-5d and Ba-5s) states required special attention. It was found to be essential to expand these states in a separate energy window, so that for the valence states diffuse Bi-d and Ba-s functions could be provided in the basis. Since nesting features of the Fermi surface could be essential in these calculations, the Brillouin zone integration was monitored carefully and finally 89 **k** points in the irreducible wedge together with a tetrahedron scheme were used in all calculations.

2. Results

The energy band structures near the Fermi energy for the distorted and undistorted structures are displayed in Fig.1, where they are aligned in energy such that the flat O-p complexes (at the bottom of the figure) coincide. Fig.1 shows that the antibonding Bi-s O-p bands along X - W - L and K - X split upon distortion leading to a semiconductor and also to a decrease of the Fermi energy with respect to the undistorted structure. This lowering in energy could explain the observed stabilization of the distorted structure, a hypothesis which we proved by additional total energy calculations.

Fig.2 shows the total energy as a function of the displacement u of the oxygen atoms. We find a minimum at a distortion corresponding to u=0.007 (in units of the lattice parameter a_0=16.3 a.u.) which is close to the experimentally observed value of 0.009 [6]. Note that in our calculation the small rotation of the oxygen octahedra which gives rise to a monoclinic structure is not included. Nevertheless we find a stabilization of about 7 mRy of the distorted structure with respect to the ideal perovskite. From a parabolic fit of the total energy curve at the equilibrium we obtain a frequency of the breathing phonon of 590 cm^{-1}, which is close to the observed bond-stretch modes [7] and a value of 575 cm^{-1}

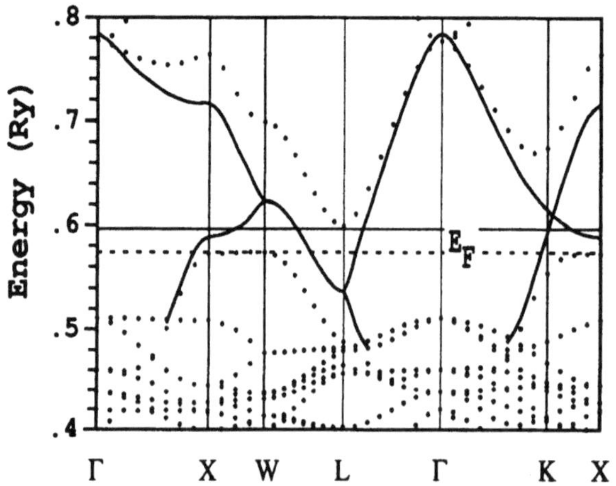

Figure 1. Energy band structures of distorted (u=0.0125, dotted lines) and undistorted (solid lines) $BaBiO_3$ around the Fermi energy. The flat O-p bands coincide for both structures.

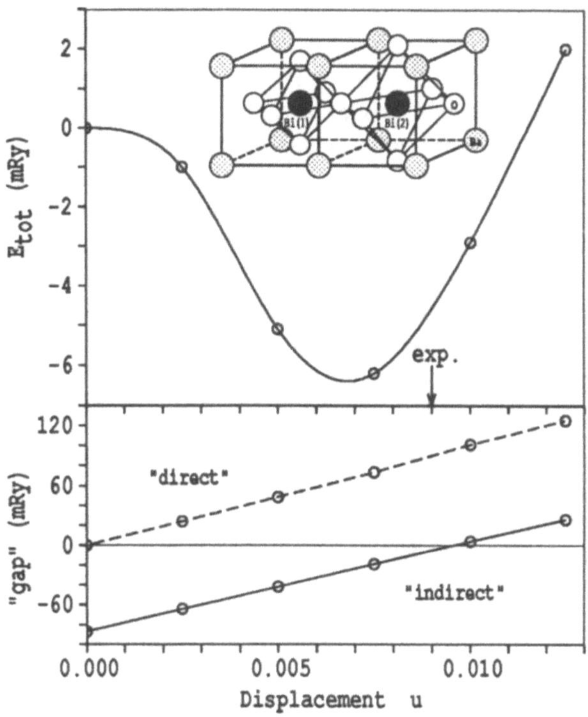

Figure 2. Top: Total energy difference with respect to undistorted $BaBiO_3$ (in mRy) as a function of distortion u (in units of a_o=16.3 a.u.).
Bottom: "Indirect gap" (overlap up to u=0.0095) at W - L (solid) and direct gap at W (dashed line) as a function of distortion u.

observed with reflectivity mesurements [8]. However, the thus obtained first phonon excitation energy lies above the energy gain of the distortion, so that one has to include large anharmonic effects.

In the bottom panel of Fig.2 the "indirect gap" (or overlap, see Fig.1) L - W as well as the direct gap at W is plotted versus the displacement u. An almost perfect linear relation is found, which can be interpreted in terms of tight-binding calculations, in which the Bi-s O-p hopping matrix elements scale with the inverse square of the distance. Close to the experimental displacement an indirect gap opens up and a semiconductor is obtained. The direct gap at this point is about 1.2 eV, while from optical data one finds less than 1.9 eV for the direct and 1 eV for the indirect gap. [9]

For the interpretation it is useful to investigate the amount of charge transfer between the two inequivalent Bi sites. It is commonly assumed in the literature [6], that a disproportionation into Bi^{3+} and Bi^{5+} ions can explain the CDW, but photoemission has shown that only small charge fluctuations exist [10]. Within the LAPW method the charge transfer is given in terms of integrated charges inside the atomic spheres with equal radii around each Bi atom. Our results reveal a fairly small charge difference of 0.1 electrons between the two sites (for u=0.01) in agreement with photoemission data. The Bi site surrounded by the smaller O octahedra, nominally the Bi^{5+}, has less charge than the other site as is to be expected.

3. Summary

We have investigated the breathing distortion of the oxygen atoms in undoped $BaBiO_3$ and found this phonon mode unstable in agreement with experiment. Upon displacing the oxygen atoms a gap at the Fermi energy is formed in accordance with the semiconducting behavior of this material. Thus, in contrast to La_2CuO_4 where the antiferromagnetic phase transition could not be obtained with LDA calculations, accurate band structure calculations for $BaBiO_3$ can describe the observed CDW very well. In this context it is interesting to note that simple tight-binding calculations predict in both cases an unstable breathing phonon, while full LDA calculations find this mode to be stable in La_2CuO_4 [11].

Acknowledgements: This project was supported by the 'Fonds zur Förderung der wissenschaftlichen Forschung', project number P7063P and by the State of Northrhine Westphalia IV A5 Projekt no. 106 019 89 and by the European Community ESPRIT-project no. 3041-MESH.

References

1. L.F.Mattheiss, E.M.Gyorgy, D.W.Johnson, Jr.: Phys.Rev.B37, 3745 (1988)
2. L.F.Mattheiss, D.R.Hamann: Phys.Rev. B28, 4227 (1983)
3. Masafumi Shirai, Nasoshi Suzuki and Kazuko Motizuki: J.Phys.: Condens.Matter 2, 3553 (1990)
4. R.Zeyher, K.Kunc: Solid State Commun. 74, 805 (1990)
5. P.Blaha, K.Schwarz, P.Sorantin, S.B.Trickey: Comp.Phys.Commun.59, 399 (1990)
6. D.E.Cox, A.W.Sleight: Acta Cryst. B35, 1 (1979)
7. W.Reichert and W.Weber: Jap. J. Appl. Phys. 26,1121 (1987)
8. S.Sugai, S.Uchida, K.Kitazawa, S.Tanaka, A.Katsui: Phys.Rev.Lett. 55, 426 (1985)
9. G.Ruani, A.J.Pal, R.Zamboni, C.Taliani: this volume
10. Z.X.Shen, P.A.P.Lindberg, B.O.Wells, D.S.Dessau, A.Borg, I.Lindau, W.E. Spicer, W.E.Ellis, G.H.Kwei, K.C.Ott, J.S.Kang, J.W.Allen: Phys.Rev. B40, 6912 (1989)
11. R.E.Cohen, W.E.Pickett, H.Krakauer: Phys.Rev.Lett. 52, 831 (1989)

Antiferromagnetism – Beyond Band Theory

T.C. Leung [1], *B.N. Harmon* [1], *A.J. Fedro* [2], *and S.K. Sinha* [3]

[1]Ames Laboratory – USDOE and Department of Physics, Iowa State University, Ames, IA, USA
[2]Northern Illinois University and Argonne National Laboratory – USDOE, Argonne, IL 60439, USA
[3]Brookhaven National Laboratory – USDOE, Upton Long Island, NY 11973, USA

Abstract. A review is given of the antiferromagnetic ground state encountered for the undoped CuO_2 planes in the high temperature superconductors. The need to go beyond band theory is explained, and a "mean field" correlation theory is presented and shown to reproduce many results for the magnetic properties obtained from quantum Monte Carlo simulations on finite lattices.

1. Introduction

Neutron scattering experiments have shown that the 3D long range antiferromagnetic (AF) order in La_2CuO_4 and $YBa_2Cu_3O_6$ is quickly suppressed by doping [1-3]. The AF correlations within the 2D CuO_2 planes persist over a larger doping range, but the interactions among the local moments become weaker [3-5]. The degree to which these AF correlations extend into the same doping region of the crystal where superconductivity develops is still unclear. Since such correlations may enhance or suppress superconductivity (depending on the mechanism), they are of interest. NMR experiments on Cu and O sites indicate the pairing of spins occurs on both sites below T_c [6], so the electrons on Cu do participate in the superconducting ground state and should not be considered as independent localized orbitals with spin=1/2.

Analysis of magnetic neutron scattering data suggest that the net ordered moment in the undoped materials is on the order of 0.6 Bohr magnetons [1-3]. It is believed that 2D quantum fluctuations are partially responsible for reducing this from the 1.06 value expected for a pure S=1/2 3D ordered antiferromagnet. In fact, the spin wave spectrum for $YBa_2Cu_3O_6$ has been analyzed remarkably well by Rossat-Mignod *et al.* by using a 2D Heisenberg model with small XY-anisotropy [3]. These authors conclude that covalency effects of the Cu orbitals with the neighboring oxygen orbitals could help reduce the ordered moment by 20% at the most. The rather small amount of oxygen 2p admixture into the Cu 3d "conduction bands" is also found with electron energy-loss studies reported by Romberg *et al.* at this conference [7]. The same conclusion

Editors: H. Kuzmany · M. Mehring · J. Fink

was obtained by analysis of induced neutron magnetic form factor measurements on La_2CuO_4 by Stassis *et al.*[8]. All of these measurements are consistent with an oxygen admixture on the order of 15% +/- 10%; a value in stark contrast to the one obtained by band structure calculations.

The first spin polarized band structure calculations for the 214 structure were expected to find an antiferromagnetic insulator similar to the case of NiO; however, the results revealed a rather small magnetic susceptibility which was far from an AF ground state [9], a result which has since been confirmed by many other groups. Furthermore, by applying a large staggered magnetic field to the Cu sites in the band theory Hamiltonian, the induced magnetic moment on the Cu sites saturated at about 0.3 Bohr magnetons[9]. The reason this is considerably less than the value of 1.0 one might expect, is because of the strong admixture of oxygen character into the conduction band. Near the Fermi level the conduction band is composed of antibonding Cu-O orbitals with about 60% from the in-plane oxygen and only 40% from the Cu. The magnetic contribution from the oxygen (between two oppositely polarized Cu sites) is zero by symmetry. The failure of the band theory approach to find a magnetic ground state and its prediction of too much oxygen admixture is believed to arise from an inadequate treatment of the on site Coulomb correlations among the tightly bound Cu 3d-state electrons.

A recent calculation which treated the 3d states as core-like and constrained their occupation to 9.0 electrons gave a Cu density of states in better agreement with XPS data than band theory where the d occupation is found to be closer to 9.5 [10]. In fact the calculated XPS peak was at a slightly larger binding energy than the experimental one, and a calculation with a constrained $d^{9.1}$ configuration would probably have yielded better agreement. When one then dopes these materials with holes, one certainly expects the 3d electron occupation to either remain the same or to slightly decrease [11]. On the other hand standard band theory for the doped metallic phase seems to yield Fermi surface [12] and other properties (e.g. phonon frequencies [13]) in reasonably good agreement with experiment in spite of having a $3d^{9.5}$ configuration on Cu. We believe an explanation of these apparently contradictory viewpoints will require understanding of the strong correlations, and we have therefore retreated temporarily from the standard band structure approach and have been investigating models that allow a more precise treatment of the correlations.

2. The Model and Method

Since the treatment of strong intra-atomic correlations in solids is difficult, it is appropriate at this stage to study model systems which hopefully reveal the correct qualitative physics. Quantum Monte Carlo methods have recently been used to study finite 2D Hubbard models for various band fillings [14], and have even been extended to 3 band models to better study the CuO_2 interactions [15]. The quantum Monte Carlo methods are presently restricted to finite lattices, finite temperatures, and very large computers. We have developed a Green's function equation of motion method (outlined previously [16]) which is able to treat the correlations for such models in a much faster but approximate fashion. The method also has the desirable feature that it can be generalized all the way to many atoms per unit cell with s,p, and d orbitals included. As in the Hubbard I approach, we use separate operators to keep track of singly and doubly occupied states:

$F_{ls} = n_{l-s}c_{ls}$, destroys a spin **s** electron in a doubly occupied state,

$f_{ls} = (1-n_{l-s})c_{ls}$ destroys a spin **s** electron in a singly occupied state,

where n_{ls} is the usual number operator at site **l** for spin **s**, and c_{ls} is the usual annihilation operator. In the implementation of the method we assume **k** and **ω** independent self energies (in Hubbard I these are zero) which are determined self-consistently using a mesh of about two hundred **k** points in the irreducible Brillouin zone. We evaluate expectation values for

the number of electrons	$n_s = f_s{}^+f_s + f_s{}^+F_s + F_s{}^+f_s + F_s{}^+F_s$
the moment	$S_z = n_s - n_{-s}$
the double occupancy	$n_s n_{-s} = F_s{}^+F_s$
and the moment squared	$S_z{}^2 = f_s{}^+f_s + F_s{}^+f_s + f_{-s}{}^+f_{-s} + F_{-s}{}^+f_{-s}$.

Unfortunately the approximation of **k** and **ω** independent self-energies results in the expectation value of $f_s{}^+F_s$ being nonzero. This expectation value nevertheless is small (or zero) for most regions of the band occupancy of interest, and we have applied the method to study the AF ground state in the 1 and 3 band Hubbard models.

3. Results

In the first application of our method [16] we were not sure of its accuracy or how severe would be the problems associated with the inconsistency mentioned above. When Monte Carlo results became avialable [14,15,17] we were able to test the method with well defined models. For the 2D Hubbard model on a square lattice at half filling [17] we have calculated as a function of U the $S_z{}^2$ and the double occupancy (they are not independent). The results are shown in figure 1.

At half filling the inconsistency with the expectation value of f^+F cancels within the unit cell so the good comparison with the Monte Carlo

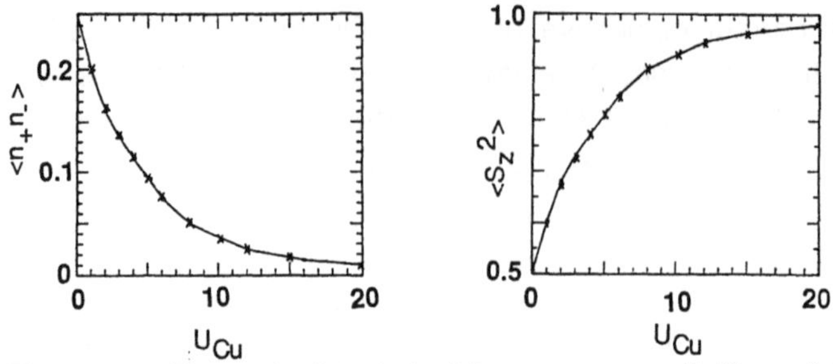

Figure 1. The calculated double occupancy (left) and the average square of the local magnetic moment (right) as a function of the Hubbard U parameter. The crosses are for the present mean field results while the curves are straight line segments between results obtained by the Monte Carlo method [17] for a 4x4 lattice, 1/(kT) = 16, and hopping parameter t=1.

Table 1. The total energy for the 2D Hubbard model (t=1.0) as a function of band filling and the Hubbard parameter, U.

	<n>				
	0.25	0.5	0.80	1.0	1.5
U					
0	-0.082	-0.133	-0.157	-0.164	-0.133
1	-0.081	-0.128	-0.141	-0.141	-0.072
2	-0.080	-0.124	-0.127	-0.120	-0.012
4	-0.078	-0.118	-0.107	-0.088	+0.104

results might not hold for <n> not equal to 1.0. However when comparing total energies for different values of doping and different values of U, our mean field results tracked the Monte Carlo results [14] to within a few percent. The squared moment and double occupancy also agreed well as expected when the total energy is reasonable. Table 1 lists our total energy results for various band fillings and U values. These should be compared with the Monte Carlo results of figure 3, reference [14]. A more detailed presentation will be made in a larger paper.

While the results presented in figure 1 and table 1 are encouraging, there is one aspect of the method which has failed so far. Luttinger's theorem is violated away from half filling, and the Fermi surface is not in agreement with the Monte Carlo [14] or simple band theory results.

We should also mention that in modeling the CuO_2 planes and comparing with the Monte Carlo results of Dopf *et al*.[15], we obtained good agreement for $\langle S_z^2 \rangle$ vs U_{Cu} only if we assumed the paramagnetic state. The $\langle S_z^2 \rangle$ values for the AF state were always significantly higher. The authors do not discuss if their results are for the paramagnetic or AF state.

Acknowledgments

This work was supported by the U.S. Department of Energy, BES Materials Sciences, under contracts #W-31-109-ENG-38, #W-7405-ENG-82, and DE-AC02-76CH00016.

References

1. D. Vaknin, S.K. Sinha, D.E. Moncton, D.C. Johnston, J.M. Newsam, C.R. Safina, and H.E. King Jr., Phys. Rev. Lett. **58**, 2802 (1987).
2. J.M. Tranquada, A.H. Moudden, A.I. Goldman, P. Zolliker, D.E. Cox, G. Shirane, S.K. Sinha, D. Vaknin, D.C. Johnston, M.S. Alvarez, A.J. Jacobson, J.M. Newsam, and J.T. Lewandowski, Phys. Rev. B**38**, 2477 (1988).
3. J. Rossat-Mignod, J.X. Boucherle, P. Burlet, J.Y. Henry, J.M. Jurgens, G. Lapertot, L.P. Regnault, J. Schweizer, F. Tasset, and C.Vettier, Proceedings of the International Seminar on High Temperature Superconducting, Dubna, July 1989, (World Scientific, in press).
4. D.C. Johnston, Phys. Rev. Lett. **62**, 957 (1989).
5. A. Rigamonti, M. Corti, T. Rega, J. Ziolo, and F. Borsa, (this conference).
6. Panel Discussion, Proc. Int. Conf. High Temp. Supercond., Interlaken, Ed. by J. Mueller qnd J.L. Olsen, Physica C **153-5**, 75 (1988).
7. H. Romberg, M. Alexander, N. Nucker, P. Adelmann, and J. Fink, (this conference).
8. C. Stassis, B.N. Harmon, T. Freltoft, G. Schirane, S.K. Sinha, K. Yamada, Y. Endoh, Y. Hidaka, and T. Murakami, Phys. Rev. B**38**, 9291 (1988).
9. T.C. Leung, X.W. Wang, and B.N. Harmon, Phys. Rev. B**37**, 384 (1988).
10. T.C. Leung and B.N. Harmon, Solid State Commun. **74**, 79 (1990).
11. H. Tolentino, A. Fontaine, A.M. Flank, P.Lagarde, J.-Y. Henry and J. Rossat-Mignod, (this conference).
12. Y. Petroff (this conference).
13. O.K. Andersen (this conference).
14. A. Moreo, D.J. Scalapino, R.L. Sugar, S.R. White, and N.E. Bickers, Phys. Rev. B**41**, 2313 (1990).
15. G. Dopf, A. Muramatsu, and W. Hanke, Invited lecture at the Workshop on Quantum Simulation of Condensed Matter Phenomena, Los Alamos National Laboratory, August 8-11, 1989 (preprint).
16. A.J. Fedro, S.K. Sinha, B.N. Harmon, and T.C. Leung, Int. J. Mag. and Mag. Mat. **76-77**, 582 (1988).
17. S.R. White, D.J.Scalapino, R.L. Sugar, E.Y. Loh, J.E. Gubernatis, and R.T. Scaletar, Phys. Rev. B**40**, 506 (1989).

Orbital Excitations in the Copper Oxides

W. Weber

Institut für Physik, Universität Dortmund,
D-4600 Dortmund 50, Fed. Rep. of Germany

Abstract. In the limit of strong electronic correlations, and in local octahedral environment, Cu^{++} is well known to be a strong Jahn-Teller ion. As a consequence, Cu d-d orbital excitations may become very important for the physical properties of the superconducting and insulating copper oxides. In this paper, we review some of the work on the Cu d-d excitation model, mainly for the insulating limit, and compare the results with various experimental findings.

1. Introduction

In octahedral symmetry, an isolated Cu^{++} ion with a d^9 configuration does not only exhibit spin 1/2 degeneracy, but also degeneracy of the two $d(e_g)$ type orbitals. This orbital degeneracy leads to the occurrence of the Jahn-Teller effect. In a regular lattice with Cu^{++} ions, the resulting states do not only show spin excitations, but may also exhibit orbital excitations. The latter may play an important role for the superconductivity in the copper oxides. This idea is the basis of the Cu d-d excitation model for an excitonic pairing mechanism in the copper oxide superconductors [1].

The Cu d-d excitation model can be seen as an extended Hubbard model, and all problems which are not solved in the simple Hubbard model, are not solved in the Cu d-d model either. Superconductivity due to the pairing of oxygen p holes mediated by excitations of the Cu^{++} Jahn-Teller levels appears possible, yet none of the studies [2-4] has included the spin flip terms of the model in any satisfactory manner. This problem appears to be unsurmountable without further detailed knowledge of the properties of the simpler Hubbard-type models. Therefore, recent studies of the model have focussed on the insulating limit. In this paper, we will survey these studies and also will discuss the experimental situation.

2. The σ-τ Hamiltonian for the insulating limit

For a Cu-O square lattice, the Hamiltonian of the Cu d-d excitation model can be transformed for the insulating case to a generalized Heisenberg Hamiltonian $H_{\sigma-\tau}$ with interactions of spins $\vec{\sigma}_l$ and pseudo-spins $\vec{\tau}_l$ which represent the orbital degrees of freedom. The pseudo-spin components are defined by

$$\tau^z|(x^2-y^2)\rangle = (+1)\ |(x^2-y^2)\rangle; \quad \tau^z|(3z^2)\rangle = (-1)\ |(3z^2)\rangle;$$

Springer Series in Solid-State Sciences, Vol. 99
Electronic Properties of High-T_c Superconductors and Related Compounds
Editors: H. Kuzmany · M. Mehring · J. Fink

$\tau^x|(x^2-y^2)\rangle = |(3z^2)\rangle; \quad \tau^x|(3z^2)\rangle = |(x^2-y^2)\rangle$. Then,

$$H_{\sigma-\tau} = -\Delta/2 \sum_l \tau_l^z + J_{sx}/2 \sum_{l,m} (\vec{\sigma}_l\vec{\sigma}_m+1)(1+\tilde{\tau}_l^z)(1+\tilde{\tau}_m^z). \quad (1)$$

Here, l and m denote Cu sites and $\tilde{\tau}^z = \cos\phi\,\tau^z \pm \sin\phi\,\tau^x$. The + (-) signs apply for bonds (l,m) along the x(y) directions. The angle $\phi = \pi/3$, and

$$\Delta = \Delta_0 + 8t^2/(E_p-E_d) + \mathcal{O}(t^4) \quad (2)$$

$$J_{sx} = 2t^4 [(E_p-E_d)^{-3} + U^{-1}(E_p-E_d)^{-2}] \quad (3)$$

$$t = t_{(j,3z^2)} = (1/\sqrt{3})\, t_{(j,x^2-y^2)}. \quad (4)$$

In eq. (2), Δ_0 is the "genuine" splitting of the two d(e_g) levels,;e. g., due to an electric field gradient of the lattice, E_p-E_d is the separation of the O p and Cu d levels, and t represents the (pdσ) hopping between O p and Cu d orbitals. Any Hopping $t_{jj'}$ between the O p orbitals has been neglected. The Hubbard U in eq.(3) excludes both double occupancy of up and down spin orbitals of one kind as well as that of the two d(e_g) orbitals (the Hund's rule coupling J_H has been ignored).

The σ-τ Hamiltonian is very similar to one derived by Kugel and Khomskii [5] from a Hubbard model with orbital degeneracy, using direct d-d hopping of σ-type between the Jahn-Teller orbitals. Note that there are no terms of the form $\tau^y\tau^y$. If we neglect the orbital degrees of freedom, we obtain the usual spin superexchange Heisenberg Hamiltonian.

It is possible to consider a finite width $t_{jj'}$ of the oxygen p bands for the derivation of $H_{\sigma-\tau}$. When terms up to $t_{jj'}^2/(E_p-E_d)^2$ are taken into account, a somewhat more general form of $H_{\sigma-\tau}$ is obtained. For first neighbors, the exchange coupling now contains not only effective (ddσ) but also (ddδ) type hopping. This leads to small $\tau_l^y\tau_m^y$ terms, to deviations of ϕ from $\pi/3$, and to second neighbor exchange terms.

$H_{\sigma-\tau}$ contains terms of the form $(\vec{\sigma}_l\vec{\sigma}_m)$, $(\tau_l^\alpha\tau_m^\beta)$, and coupling terms $(\vec{\sigma}_l\vec{\sigma}_m)(\tau_l^\alpha\tau_m^\beta)$. It thus represents a model of two coupled spin systems. Its two parameters are the effective d level splitting Δ and the 1NN superexchange J_{sx}. As any orbital excitation changes the charge density of the Cu^{++} ions, it is obvious that that pseudo-spins $\vec{\tau}_l$ have to couple to the octahedral distortions; i. e. to the lattice. This coupling is described by the Hamiltonian [6]

$$H_{ph} = g\,(Q_3\,\tau^z + Q_2\,\tau^x), \quad (5)$$

where Q_2 and Q_3 are the two quadrupolar deformations of the octahedron. When this interaction is incorporated in a lattice, all phonons with a Q_2 or Q_3 component in their displacement pattern will mediate a coupling between neighbor pseudo-spins, resulting in a spin-independent superexchange J_{ph} for 1NN and also higher neighbors.

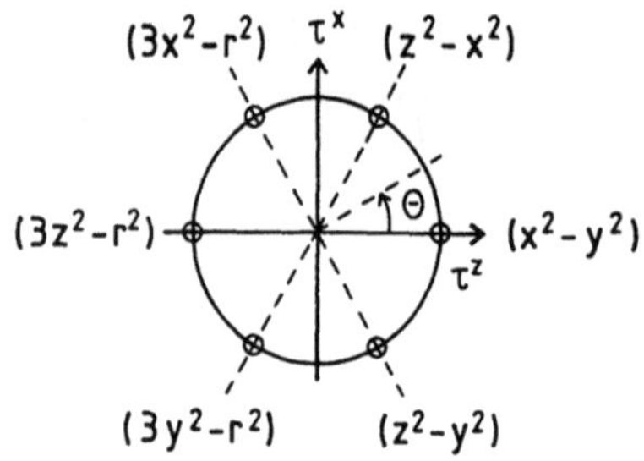

Fig. 1: τ^z-τ^x plane. Various $d(e_g)$ orbital combinations as a function of the angle θ are indicated.

The σ-τ model (without phonons) has so far been studied on a rather limited basis only. It is obvious that large values of Δ/J lead to "ferromagnetic" pseudo-spin or orbital ordering. In the classical limit, there is a boundary $(\Delta/J)_{max}$, beyond which no $d(3z^2-r^2)$ admixture is found. This corresponds to an angle θ=0 in the (τ^z,τ^x) plane (see Fig. 1). Because of the $(\tau^x_l\tau^x_m)$ terms in $H_{\sigma-\tau}$, there is a finite rotation angle θ of the pseudo-spins in the (τ^z,τ^x) plane below $(\Delta/J)_{max}$. For small (Δ/J) θ can be as big as $\pm\pi/3$ in materials such as in K_2CuF_4, where a static orbital order of alternating $d(z^2-x^2)$ and $d(z^2-y^2)$ orbitals has been observed, forming orbital "antiferromagnetism" in a $\pm\theta$ sequence (see Fig. 1). This order, a static cooperative Jahn-Teller effect with an ABAB.. or $\vec{q}$ = (π,π) superstructure, manifests itself in the corresponding quadrupolar distortions of the fluorine octahedra. Complementary to the orbital order, the spin order in K_2CuF_4 is ferromagnetic and thus $< \vec{\sigma}_l\vec{\sigma}_m > > -1$. Note that, because of the finite values for $\langle\tau^x\rangle$, the site symmetry of the Cu^{++} ions has to be *less than tetragonal*.

From numerical simulations on small one-dimensional systems [7] it seems that for $\Delta \gg J$ the spin correlations are identical to those of a Heisenberg spin chain, while in the limit $J \gg \Delta$, the orbital degrees of freedom distort as to allow the formation of spin singlets (a form of "discrete" spin-Peierls distortion).

For realistic estimates of J ($\approx$ 0.15 eV) and Δ ($\approx$ 0.5 - 1.5 eV) for the copper oxides, $\Delta/J \approx (\Delta/J)_{max}$, so that $\theta \approx 0$. In the classical limit, $\langle \tau^z \rangle \approx 1$ and thus $\langle \tau^x \rangle \approx 0$; i. e., almost no $(3z^2-r^2)$ admixture is expected: the $(3z^2-r^2)$ orbital would be "dead". Quantum fluctuations may bring it "alive" again, yet the results from the 1-dim numerical simulations do not predict large values of $\langle\tau^x\rangle$ near $(\Delta/J)_{max}$ either.

A further, very interesting study on the σ-τ model has been presented by Zaanen, Oleś and Feiner [8,9]. The σ-τ model was first treated in the classical limit, where a phase diagram was derived as a function of Δ/J and of the Hund's rule coupling J_H. Further, the coupled spin and orbital excitations were investigated and the role of quantum fluctuations was estimated. It was found that for significant admixture of the $(3z^2-r^2)$ orbital in the ground state, the quantum spin fluctuations should strongly increase.

3. Comparison to experiment

A variety of experimental results indicate the presence and/or importance of the $(3z^2-r^2)$ orbitals in the cuprates. Most prominent are probably the data from Cu 3p core level spectroscopy, either by EELS [10] or by XAS [11]. These data suggest a ≈ 10% admixture of the $(3z^2-r^2)$ orbital, even in insulating, tetragonal $YBa_2Cu_3O_6$; corresponding to a value of $\langle \tau^z \rangle \approx 0.8$ or an angle $\theta \approx 20°$. In the superconducting Bi-cuprates, the admixture may be even larger, although the results of the two groups differ to some extent. Furthermore, the XAS data indicate a strong increase of the $(3z^2-r^2)$ admixture in the superconducting Bi-cuprates, in parallel with the increase of T_c there. The latter data have yet to be confirmed by other researchers.

There are also reports on structural distortions in the Cu-O planes, resulting in variations of the Cu-O bond lengths. A static superstructure with a wavevector $q = (\pi/2,\pi/2)$ has been observed in $YBa_2Cu_3O_6$ by electron diffraction [12]. The centerpiece of the rather complicated superstructure are rhombic distortions of the Cu-O squares, reminiscent of a co-operative Jahn-Teller effect. EXAFS data taken on superconducting Bi-cuprates also indicate, among other features, two different in-plane Cu-O bond lengths [13].

Finally, inelastic neutron scattering data exhibit an extra band of longitudinal optic phonons in La_2CuO_4 [14,15] through most of the Brillouin zone. This extra branch near 50 meV seems to have oxygen in-plane breathing character (basically like the Q_3 mode). Similar features also appear in $YBa_2Cu_3O_{6+\delta}$.

It is very tempting to take all these experimental findings as a strong support for the importance of the Jahn-Teller effect and of the resulting orbital excitations in the cuprates. However, there are quite a few question marks.

Firstly, when we use realistic estimates for the superexchange coupling J_{sx} and for the level splitting Δ of the σ-τ Hamiltonian, the $(3z^2-r^2)$ orbital should really be "dead". An angle $\theta \approx 20°$ appears to be much too large, even when quantum fluctuations were included. One way out may be that the phonon mediated superexchange J_{ph} (via H_{ph}, eq. 5) is comparable to J_{sx}, as is indeed found in first estimates [9]. The combined coupling $J_{sx} + J_{ph}$ may then be sufficient to yield values of $\theta \approx 20°$.

Secondly, the $(3z^2-r^2)$ admixture is observed in crystals with a tetragonal site symmetry of Cu^{++}. $H_{\sigma-\tau}$ provides only couplings of τ^z and τ^x pseudo-spins, and the same holds for the phonon mediated exchange. As a consequence, the Cu^{++} site symmetries should be less than tetragonal in all cuprates. This may indeed be the case, when we consider the results of refs. [12,13]. Then however, the orbital superstructures should exhibit wavevectors $q = (\pi,\pi)$ (concomitant with ferromagnetic spin order [12]) but *not* a wavevector $q = (\pi/2,\pi/2)$.

An orbital superstructure as proposed by [12] may indeed provide a relatively simple explanation of the "extra" phonon band in La_2CuO_4. Of all the phonon bands folded back in the new Brillouin zone, only the modes most strongly coupling to the static distortions should exhibit a large gap at the new zone boundaries, while all the other phonon branches may remain rather unaffected by the superstructure. Yet again, according to [8,9] those strongly coupling phonon modes should rather be the in-plane quadrupolar type (Q_2) and not Q_3, in contradiction to the neutron data.

Finally, we should point to an experimental result still missing. The d-d excitations in the insulating cuprates should be accessible by electronic Raman scattering. However, in the range up to ≈ 0.6 eV, no signal has been observed (one would have to re-interpret the two-magnon line near 0.3 eV as the d-d excitation - which would fit the selection rules, but a further, detailed experimental and theoretical study seems to rule that out [16]). Thus it is very interesting to scan the energy range beyond ≈ 0.6 eV for the d-d excitations.

Acknowledgments. This work was supported by the Minister für Wissenschaft und Forschung, Land Nordrhein-Westfalen, FRG, and by the European Community ESPRIT project No 3041-MESH

References

1. W. Weber: Z. Phys. B70, 323 (1988)
2. W. Weber, A. L. Shelankov, and X. Zotos: Physica C 162-164, 307 (1989)
3. M. Jarrell et al.: Phys. Rev B38, 4584 (1988)
4. D. L. Cox et al.: Phys. Rev Lett. 62, 2189 (1989)
5. K. I. Kugel and D. I. Khomskii: Usp. Fiz. Nauk. 136, 621 (1982), [Sov. Phys. Usp. 25, 231 (1982)
6. A. L. Shelankov, X. Zotos, and W. Weber: to be published
7. J. Kanamori: J. Appl. Phys. (Japan) 31 Suppl., 145 (1960)
8. J. Zaanen, A. M. Oleš, and L. F. Feiner: preprint
9. L. F. Feiner and J. Zaanen: this volume
10. J. Fink et al.: this volume
11. P. Lagarde et al.: this volume
12. T. Krekels et al.: preprint
13. J. Röhler and A. Larisch: this volume
14. L. Pintschovius et al.: to be published in Festkörperprobleme Vol. 30 (1990)
15. H. Rietschel et al.: this volume
16. P. Knoll: this volume

Orbital Dynamics and Jahn-Teller Coupling in the CuO_2 Planes of the High-T_c Cuprates

L.F. Feiner[1] *and J. Zaanen*[2]

[1]Philips Research Laboratories, P.O. Box 80000, 5600 JA Eindhoven, The Netherlands
[2]Max-Planck-Institut für Festkörperforschung, Heisenbergstr. 1 D-7000 Stuttgart 80, Fed. Rep. of Germany

Abstract. We have investigated the Jahn-Teller coupling in the CuO_2 plane between the Cu $d_{3z^2-r^2}$ orbital degrees of freedom and the lattice modes. We show that it can lead to $d_{3z^2-r^2}$ occupancy, accompanied by a static local quadrupolar distortion (relevant for the cuprates if the *d-d* excitation energy is $\lesssim$0.8 eV, with the associated excitonic polaron at $\approx$0.2 eV), and to a phonon anomaly caused by harmonic mode coupling of the $d_{x^2-y^2} \rightarrow d_{3z^2-r^2}$ orbital excitation (dressed by breathing mode phonons) with the quadrupolar mode (rather than with the breathing mode as apparently observed).

1. Introduction: Model

Recent experimental evidence [1] indicates that the Cu orbital momentum is not completely quenched in the high T_c cuprates, and it has been proposed [2] that low-lying *d-d* excitations might then mediate pairing of oxygen *p* holes, thus providing a mechanism for superconductivity. If so, the Cu $d_{3z^2-r^2}$ orbital degrees of freedom should manifest themselves already in the undoped systems, in particular through their coupling to the lattice.

We therefore consider a single CuO_2 plane in the insulating limit. The *electronic structure* is represented by the Kugel-Khomskii [3] model Hamiltonian $\mathscr{H}_{KK}(J, \varepsilon_z)$, which describes the coupling between Cu orbital and spin degrees of freedom with the help of spin operators $\mathbf{s}_i$ and pseudo-spin operators τ_i^z, τ_i^x (Pauli matrices acting on the orbital states). Further, J is the (super)exchange coupling and ε_z is the (bare) splitting between the $d_{3z^2-r^2}$ and $d_{x^2-y^2}$ orbital. In the classical (mean field) picture the groundstate is antiferromagnetic with only the $d_{x^2-y^2}$ orbital occupied, but quantum fluctuations can produce a finite occupancy of the $d_{3z^2-r^2}$ orbital [4] .

Concerning the *lattice distortions* we consider only the in-plane motion of Cu and O ions, which gives us three strains and six phonon branches. Explicit expressions can be obtained by using a simple model with two force constants (Cu-O stretch and O-O stretch). The forces exerted by the rest of the crystal are modelled by 'clamping' the CuO_2 plane between stiff elastic layers (like the La_2O_2 layers in La_2CuO_4). At each Cu^{2+} ion there is a *local Jahn-Teller (JT) coupling* $V_a Q_{ai} \tau_i^z + V_b Q_{bi} \tau_i^x$ between the orbitals and the distortions of the square of oxygen neighbours. Here Q_{ai} is the breathing coordinate and Q_{bi} is

Springer Series in Solid-State Sciences, Vol. 99
Electronic Properties of High-T_c Superconductors and Related Compounds
Editors: H. Kuzmany · M. Mehring · J. Fink

the quadrupolar cooordinate (two oxygens moving outward, two inward), and $V_b \approx -\sqrt{3}\, V_a$ because of the square planar coordination. By summing over sites and expressing the local coordinates in terms of strains and phonons one obtains the electron-lattice coupling $\mathcal{H}_{\rm JT}(V_a, V_b)$.

2. Electronic Ordering: Ground State in Mean Field

Transformation to displaced strains and phonons [5] generates an effective interaction $\mathcal{H}^{\rm eff}_{\rm JT} = -(1/2)\Sigma_{\mathbf{q}}[J^a(\mathbf{q})\tau^z(\mathbf{q})\tau^z(-\mathbf{q}) + J^b(\mathbf{q})\tau^x(\mathbf{q})\tau^x(-\mathbf{q})]$ between orbitals, where the coupling strength $J^a(\mathbf{q})$ ($J^b(\mathbf{q})$) obtains contributions from the local breathing (quadrupolar) character in the various phonon branches and strains and is proportional to V_a^2 (V_b^2). This Hamiltonian together with $\mathcal{H}_{\rm KK}$ now determines the electronic ground state. In mean field approximation and on the assumption of a two sublattice solution, there are four relevant coupling constants, $J^a(\mathbf{0})$, $J^b(\mathbf{0})$, $J^a(\mathbf{q}_M)$, and $J^b(\mathbf{q}_M)$, associated with isotropic strain, orthorhombic strain, M-point ($\equiv (\pi/a, \pi/a)$ in the unreduced Brillouin zone) breathing mode, and M-point quadrupolar mode, respectively. Because of the clamping $J^\alpha(\mathbf{0}) < J^\alpha(\mathbf{q}_M)$, and furthermore $J^b(\mathbf{q}_M) > J^a(\mathbf{q}_M)$ because $V_b > V_a$.

The lattice induced coupling is thus dominated by the zone boundary quadrupolar mode and favours antiferro orbital order (AFO), with mixed $d_{x^2-y^2}$ and $d_{3z^2-r^2}$ orbitals. This competes with the ferro orbital order (FO) with only the $d_{x^2-y^2}$ (or $d_{3z^2-r^2}$) orbital occupied favoured by the electronic interactions of $\mathcal{H}_{\rm KK}$. The resulting phase diagram is shown in Fig. 1. All phases are antiferromagnetic. In the (AFM,AFOm) phase the orbitals in the two sublattices are given by $\cos\theta\,|d_{x^2-y^2}\rangle \pm \sin\theta\,|d_{3z^2-r^2}\rangle$. with $\cos 2\theta = (\varepsilon_z + 2E_0)/2\bar{J}^b$, where $E_0 = 8J/9$ and $\bar{J}^b = J^b(\mathbf{q}_M) - J^a(\mathbf{0})$. We see that sufficiently strong JT coupling will actually lead to a finite $d_{3z^2-r^2}$ occupancy accompanied by a static local quadrupolar distortion.

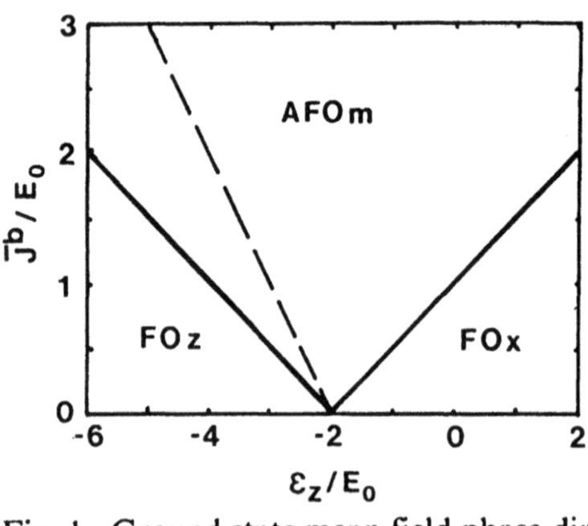

Fig. 1. Ground state mean field phase diagram of Kugel-Khomskii model with Jahn-Teller coupling. Along the dashed line the (AFM,AFOm) phase is degenerate with the (FM,AFOm) phase.

Although the cuprates do not appear to exhibit a macroscopic M-point quadrupolar distortion, superstructures showing such distortion locally have recently been observed [6] . This suggests a location in the (AFM,FOx) phase region, but close to the (AFM,FOx)-(AFM,AFOm) phase boundary. *i.e.* $2\bar{J}^b \approx \varepsilon_z + 2E_0$. Is this consistent with estimated values for the parameters? In the (AFM,FOx) phase the energy of the *d-d* orbital excitation (in the static lattice) is $E_z = \varepsilon_z + 2E_0 + 2J^a(\mathbf{0})$, and the lattice model relates $J^b(\mathbf{q}_M)$ to the local JT stabilization energy: $J^b(\mathbf{q}_M) \approx 5E_{JT}$. Since JT energies for Cu^{2+} in oxides generally do not exceed 500 cm^{-1} [7] , one expects $J^b(\mathbf{q}_M) \lesssim 0.4$ eV. We therefore conclude that, unless the JT coupling strength is unusually large. the *d-d* excitation energy must be $\lesssim 0.8$ eV , *i.e.* well inside the insulating gap. or else the JT coupling is not the origin of the observed distortion and $d_{3z^2-r^2}$ occupancy.

3. Dynamics: Excitations in RPA

The JT interaction $\mathscr{H}_{JT}$ leads to mode coupling between the phonons and the orbital excitations, which turns the latter into dispersing vibrons. In the random phase approximation *only the quadrupolar component* of the phonons enters the equations of motion in the (AFM,AFOx) phase, and the vibron dispersion relation is $\hbar\omega_{\mathbf{q}}^{\mathrm{vibr}} = [E_z(E_z - 2J^b(\mathbf{q}))]^{1/2}$. Whenever the vibron branch intersects a quadrupolar active phonon branch there is anticrossing and mode mixing, which could produce an observable phonon anomaly. However, this is only symmetry allowed in branches of quadrupolar character, not in those of breathing character, contrary to what is apparently observed in neutron scattering along Γ-M. [8] . Moreover, for the parameters estimated above the vibron would be extremely dispersive with energy at Γ of $\gtrsim 0.7$ eV, so that such an anomaly could only occur very close to the M-point.

The energetics could be considerably different for the *d-d* excitonic polaron, *i.e.* the orbital excitation dressed with local breathing mode phonons. As illustrated in Fig.2, the polaron corresponds to the zero phonon mode in the

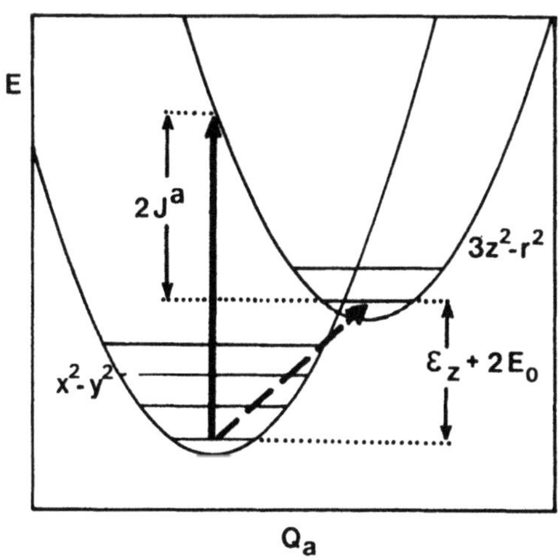

Fig. 2. Configuration coordinate diagram of the *d-d* excitonic polaron.

phonon progression of which the vibron is the vertical transition. Its dispersion is therefore reduced compared with the vibron by (at low T) a factor $\exp(-S)$, where S is the number of phonons comprised in the dressing. Also, its static energy is $\varepsilon_z + 2E_0$ instead of E_z, but since $J^a(0) \ll J^a(q_M) \lesssim J^b(q_M)/4 \lesssim 0.1$ eV this polaron would still be at high energy. However, if ε_z were completely due to tetragonal distortion, as could be the case in La_2CuO_4, the energy could be further reduced to $2E_0$ if the polaron were dressed as well by phonons that de-elongate the oxygen octahedron. One would be left with an essentially dispersionless *d-d* excitonic polaron, at an energy of about 200 meV, determined by the superexchange $J \approx 0.1$eV, *i.e.* due to the less effective hybridization of the $d_{3z^2-r^2}$ orbital compared with the $d_{x^2-y^2}$ orbital. This would therefore be the minimum energy for the zero phonon line of the *d-d* orbital excitation.

References

1. N. Nücker, H. Romberg, X.X. Xi, J. Fink, B. Gegenheimer, and Z.X. Zhao, Phys. Rev. B 39, 6619 (1989); and this volume.
2. W. Weber, Physica C 162-164, 307 (1989); and this volume.
3. K.I. Kugel and D.I. Khomskii, Sovi. Phys. Usp. 25, 231 (1982).
4. J. Zaanen, A.M. Olesz, and L.F. Feiner, Proceedings NATO ARW 'Magnetic Fluctuations in High T_c Superconductors', (in press).
5. G.A. Gehring and K.A. Gehring, Rep. Progr. Phys. 38, 1 (1975).
6. T. Krekels, T.S. Shi, J. Reyes-Gasga, G. van Tendeloo, J. van Landuyt, and S. Amelinckx, Physica C 167, 677 (1990).
7. R. Englman, *The Jahn-Teller Effect in Molecules and Crystals*, Wiley, London, 1972.
8. H. Rietschel, L. Pintschovius and W. Reichardt, Physica C 162-164, 1705 (1989), and references therein; and this volume.

A Polaronic and Bipolaronic Theory of Superconductivity

D.K. Ray

Laboratoire des Propriétés Mécaniques et Thermodynamiques des Matériaux, Université Paris-Nord, F-93430 Villetaneuse, France

Abstract. Against the background of the recent experimental results on high Tc copper oxide superconductors, we propose a polaron and bipolaron model adapted to the characteristics of these materials and general hamiltonians for polaron pairing and boson condensations are presented. Solutions of the polaronic hamiltonian for different regimes of values of the Hubbard U are discussed. It is shown that in $2 + \epsilon$ dimensions the copper oxide systems might be in a region of crossover from the fermion-pairing mechanism of s-wave type to boson condensation thus indicating very unusual fermi liquid properties.

There is now renewed interest in the importance of electron-phonon interaction for the mechanisms of superconductivity in the copper-oxide materials.

Recent studies of infra-red photoinduced absorption spectram of the semiconducting phases of La_2CuO_4, $YBa_2Cu_3O_{7-\delta}$ and $Tl_2Ba_2Ca_{1-x}Gd_xCu_2O_8$ (with x = 0.02) showed indirect evidence of the presence of lattice deformations as well as low-lying electronic transitions associated with the photoinduced holes [1]. Similar experiments have been done by Taliani et al. [2] on the La- and Y-compounds. These experiments and analysis of their results gave indications of the existence of polarons, their effective masses and energies. Neutron measurements of the orthorhombic distortions as a function of temperature for $La_{1.85}Ba_{0.15}CuO_4$ by Paul et al. [3] showed that the charge carriers are related to the orthorhombic distortion in this compound. Recent analysis of the experimental results on the temperature dependence of Raman lines in 123-compounds by Zwicknagl [4] indicates that λ, the electron-phonon coupling parameter in Eliashberg-McMillan theory, needs a value of 2.9 to fit the shift of the Raman lines with temperature and the value of T_c = 90K in this compound. This large value of λ signifies that electron-phonon coupling is strong enough to go beyond the Migdal approximation, which neglects corrections to the vertex part of the electron-phonon interaction [5]. Also, in view of recent CuL_3XAS experiments on the copper oxide materials, the

Springer Series in Solid-State Sciences, Vol. 99
Electronic Properties of High-T_c Superconductors and Related Compounds
Editors: H. Kuzmany · M. Mehring · J. Fink

significance of the orbital degeneracy and consequent pseudo Jahn-Teller effect in these compounds is apparent [6].

Against the background of these experimental facts we shall present an effective polaron and bipolaron model appropriate to these compounds and study superconductivity on the basis of pairing of polarons and condensation of bosons out of bipolarons. The electronic structure of the copper oxide compounds is fairly well-established now [7]. Due to large U_d, the d-density of states (DOS) is split. The p DOS lies inside this Hubbard gap and due to its large hybridization with d-holes (the Cu^{2+} ion has one d-hole) a charge transfer gap is created between the upper d-hole DOS and the occupied p DOS. In the semiconducting phase the Fermi energy E_F lies within this gap leaving the upper d DOS to be occupied by a d-hole which gives rise to an antiferromagnetic phase. When these compounds are doped, holes are created either in the p DOS below the charge transfer gap or inside the charge transfer gap. This gives rise to a hole band which is predominantly of p character. As a first approximation we consider a single band for the p hole polarons. A more general consideration of the two-band character of hole polarons will be developed later on. The total hamiltonian for such a single band can be written as

$$\mathcal{H} = \mathcal{H}^{o}_{H} + \mathcal{H}^{on-site}_{e-p} + \mathcal{H}^{inter-site}_{e-p} + \mathcal{H}_{phonon} \tag{1}$$

where

$$\mathcal{H}^{o}_{H} = \sum_{\langle i,j\rangle,\sigma} t^{o}_{ij}\, C^{+}_{i\sigma} C_{j\sigma} + \sum_{i} \epsilon^{o}_{i} \hat{n}_{i\sigma} + \sum_{i} U^{o} \hat{n}_{i\uparrow} \hat{n}_{i\downarrow} + \sum_{\substack{\langle i,j\rangle,\\ \sigma,\sigma'}} U^{o}_{ij} \hat{n}_{i\sigma} \hat{n}_{j\sigma'}$$

is the extended Hubbard hamiltonian for the holes. The meaning of the different terms is now well-known. The last term, the inter-site coulomb interaction, neglected by Hubbard [8], is kept here to take account of the pairing interaction between two polarons.

$$\mathcal{H}^{on-site}_{e-p} = \sum_{iq\sigma} V_{iq} \left(b_{q} + b^{+}_{-q}\right) \hat{n}_{i\sigma} \tag{2}$$

is the local or diagonal electron-phonon (e-p) interaction so that

$$V_{iq} = \sum_{t} \sqrt{\frac{\hbar}{2M_{t}\omega_{t}}} \left\langle i \left| \frac{\partial V}{\partial R_{t}} \right| i \right\rangle_{o} e^{-iq.R^{o}_{t}} \tag{3}$$

and the inter-site or non-diagonal e-p interaction is given by

$$\mathcal{H}^{inter-site}_{e-p} = \sum_{\langle i,j\rangle,q,\sigma} V_{ijq} \left(b_{q} + b^{+}_{-q}\right) C^{+}_{i\sigma} C_{j\sigma} \,. \tag{4}$$

This is the phonon assisted hopping of holes, so that following Barisic et al. [9] we have

$$V_{ijq} = \sum_t \sqrt{\frac{\hbar}{2M_t\omega_t}} \left\langle i \left|\frac{\partial V}{\partial R_t}\right| j \right\rangle_o e^{-iq.R^o_t} , \qquad (5)$$

where t is either i or j. The on-site interaction gives rise to BCS-type superconductivity in the weak-coupling case and to small polarons in the strong coupling limit [9].
The second e-p interaction term has been shown to be proportional to the band width [9] and can explain the magnitude of T_c for transitional metals and their alloys. To derive a more general polaronic hamiltonian we have considered both types of e-p interactions, and after polaronic transformation [10] we get

$$\mathcal{H}_{Pol} = \sum_{\langle ij\rangle,\sigma} t^p_{ij} C^+_{i\sigma} C_{j\sigma} + \sum_{i\sigma} \epsilon^p_i \vec{n}_{i\sigma} + \sum_{i\sigma} U^{eff} \hat{n}_{i\uparrow} \hat{n}_{i\downarrow} + \sum_{\substack{\langle ij\rangle \\ \sigma,\sigma'}} U^{eff}_{ij} \hat{n}_{i\sigma} \hat{n}_{j\sigma'} + \sum_q \hbar\omega q \left(b^+_q b_q + 1/2\right) , \qquad (6)$$

where $t^p_{ij} = t^o_{ij} \ \hat{L} \exp \left[\sum_q \frac{V_{iq} - V_{jq}}{\hbar\omega q} \left(b^+_q - b_{-q}\right)\right]$

$$\epsilon^p_i = \epsilon_o - \sum_q \frac{V_{iq} V_{i-q}}{\hbar\omega q}$$

$$U^{eff} = U^o - \sum_q \frac{2V_{iq} V_{i-q}}{\hbar\omega q} \qquad (7)$$

$$U^{eff}_{ij} = U^o_{ij} - \sum_q \frac{V_{iq} V_{j-q}}{\hbar\omega q}$$

with $\hat{L} = 1 + \sum_q \left(\frac{V_{ijq}}{t^o_{ij}}\right) \left(b_q + b^+_{-q}\right) - 2 \sum_{l,q,\sigma'} \frac{V_{lq}}{\hbar\omega q} \frac{V_{ijq}}{t^o_{ij}} \hat{n}_{L\sigma'}$.

Using a model for the interaction of the holes with an optical phonon we have worked out the expressions for the coupling parameters V_{iq} and V_{ijq} and estimated the effective mass of the polarons, which is reduced by a factor of 3-4 from the usual small polaron value of 30, in agreement with experimental values [11].

The polaronic hamiltonian is quite general and the values of the parameters depend upon the system under consideration. Several types of systems are being studied : (1) A-15 and Chevrel phase compounds where

U^{eff} is small and positive and the d band is nearly full; (2) superconductivity in semiconductors where U^{eff} can be negative and small (e.g. (Ba, K) BiO_3 system) and (3) doped copper oxide systems where U^{eff} is repulsive and large. We would like to mention here the most pertinent result concerning the dimensionality effect on the polaron pairing. We solved the gap equations concerning s-wave and d-wave pairings in $2 + \epsilon$ dimensions where we related ϵ to the hopping along the c-axis in the multilayered compounds. The numerical results show that s-wave pairing dominates over the d-wave pairing for hole concentrations ~ 0.3 which is in agreement with experimental results [12]. As ϵ increases, T_c decreases both for the s-wave and the d-waves but it is more drastic for the former. Next, we considered the boson condensation temperature from a bipolaronic hamiltonian [13]. This temperature is zero for 2D but increases fast with the increase of ϵ. So, for the compounds of bismuth and thallium with layers of CuO_2, a boson band may lie close to a polaron band and this would have important perturbing effects on the polaron pairing mechanism with significant experimental consequences [14].

A detailed account of this work will be published elsewhere. The author would like to thank Espirit project 3041 for their financial support.

References

1 C.M. Foster, A.J. Heeger, G. Stucky and N. Herron : Solid State Commun. 71, 945 (1989).

2 C. Taliani, R. Zamboni, G. Ruani, F.C. Matacotta and K.I. Pokhodnya Solid State Commun. 66, 487 (1988).

3 Mc K. Paul, G. Balakrishnan, N. Bernhoeft, W.I.F. David and W.T.A. Harrison : Phys. Rev. Lett. 58, 1976 (1987).

4 G. Zwicknagl : this volume.

5 A.B. Migdal : Sov. Phys. J.E.T.P. 1, 996 (1958).
A.S. Alexandrov and E.A. Mazur : Sov. Phys. J.E.T.P. 69, 1001 (1989).

6 A. Bianconi, P. Castrucci, A. Fabrizi, M. Pompa, A.M. Flank, P. Lagarde, H. Katayama-Yoshida and G. Calestani : In "Earlier and Recent Aspects of Superconductivity", edited by J.G. Bednorz and K.A. Müller (Springer-Verlag, Berlin, Heidelberg 1989).

7 O.K. Anderson : this volume; G.A. Sawatzky : this volume.

8 J. Hubbard : Proc. Roy. Soc., A 276, 238 (1963).

9 S. Barisic, J. Labbé and J. Friedel : Phys. Rev. Lett. 25, 919 (1970).

10 A. Alexandrov and J. Ranninger : Phys. Rev. B 23, 1796 (1981); ibid B 24, 7164 (1981); T. Holstein, Ann. Phys. 8, 325 (1959).

11 T. Ray : to be published.
12 J.-M. Imer, F. Patthey, B. Dardel, W.-D. Schneider, Y. Baer, Y. Petroff and A. Zettl : Phys. Rev. Lett. 62, 336 (1989).
13 A.S. Alexandrov, J. Ranninger and S. Robaszkiewicz : Phys. Rev. B 33, 4526 (1986).
14 D.K. Ray, J. Konior, A.M. Oles and A.N. Das : communicated to Phys. Rev. B.

Magnetic Order and High-T_c Superconductivity

E. Sigmund [1] *and V. Hizhnyakov* [2]

[1] Institut für Theoretische Physik, Universität Stuttgart, Pfaffenwaldring 57, D-7000 Stuttgart 80, Fed. Rep. of Germany
[2] Institute of Physics, Estonian Academy of Sciences, Riia 142, 202400 Tartu, Estonian SSR, USSR

Abstract. Doping the antiferromagnetically (AF) ordered perovskites with holes leads to the formation of small immobile ferromagnetic (F) clusters. At higher concentrations the clusters start to percolate. Based on some general properties of fractal percolation networks the critical doping concentrations for destruction of long-range AF order and for the insulator - conductor transition are calculated. A pairing mechanism via deformational coupling of holes to the lattice is briefly discussed.

1. Introduction

For our considerations of the electronic and magnetic properties of the new perovskite-type superconductors we use a Hubbard hamiltonian containing the Cu (ε_d) and O (ε_p) on-site energies ($\varepsilon = \varepsilon_p - \varepsilon_d$), the Cu-O (T) and O-O (t) transfer terms and the Coulomb repulsion U for two electrons on the same Cu site. For the calculations the relations $U \gg T, \varepsilon, t$; $4|T| \geq |\varepsilon|$; $T \geq 2t$ are assumed and the explicit values $U \sim 4$ eV, $T \sim 1.5$ eV and $\varepsilon \sim T$ are taken [1].

2. Cluster Formation

Neglecting the O-O transfer a hole with spin σ in the AF-ordered lattice is practically immobile and confined to the totally symmetric state of a CuO_4 unit with Cu spin $-\sigma$. Destruction of AF order near the hole leads to its delocalization via CuO bonds and non-AF aligned Cu spins. For example, if one of the AF aligned Cu spins is turned the hole will move in a cluster consisting of 5 neighboring Cu ions with parallel spins and 16 O ions. For $|\varepsilon| \leq T$ the corresponding gain of delocalization energy is 0.45 T. Additional gain of energy is given by the interaction between the hole spin and the Cu spins, which is of the order of $0.2\,T^2/U$. On the other

Springer Series in Solid-State Sciences, Vol. 99
Electronic Properties of High-T_c Superconductors and Related Compounds
Editors: H. Kuzmany · M. Mehring · J. Fink

hand one loses energy by turning one Cu spin, which causes an increase of magnetic energy of 7.5 T^4/U^3. For $U \sim 2.5\,T$ the loss of energy is smaller than the gain and therefore the creation of a hole-cluster is energetically favorable. For $U > 2.5\,T$ larger clusters with more turned Cu spins are formed [1].

Creation of clusters leads to a reduction of the hole transport via the O-O bonds. For larger O-O transfer ($t \sim T$) binding energy and size of the clusters decrease. This behavior can be taken into account exactly by the Lifshitz formalism, where the energy gain $E=E_n$ due to hole delocalization after turning n Cu spins is given by subtracting the solution of the equation

$$\det\left| E-\varepsilon-(T^2/2)\, G(E)_{m,m'} \right| = 0$$

from the lowest band energy. G is the Green function of the unperturbed lattice projected according to the Cu-O interaction (at positions m,m′) induced by turning n Cu spins. The total gain of energy due to the formation of the F cluster is ($T_1=1.5\, T^4/U^3$)

$$\delta E_n = E_n + \xi T^2/U - 2\left[4n-(1/2)\left(1-\varepsilon/(\varepsilon^2-16T^2)^{1/2}\right)-2\zeta\right]T_1 .$$

ζ and ξ are the probabilities of finding the hole on the n Cu ions with turned spins and the surrounding O ions, respectively (for $|\varepsilon| \leq T$, $t \ll T$ $\zeta \approx \xi \approx 0.25$). For $U \approx 5T$, $|\varepsilon| \leq T \geq 2t$ hole-clusters with around 13 parallel (4 turned) Cu spins have lowest energy.

In the considerations presented here the quantum fluctuations of the Cu spins are neglected. Recently the role of these fluctuations has been studied numerically by Sherman [2] using the one-band Hubbard model in a two-dimensional lattice. He showed that ferromagnetically ordered hole-clusters exist in this model also.

3. Optical Properties

Evidence of the existence of hole-clusters can be found in optical experiments on electronic states in high-T_c cuprates. In the range of very small hole concentrations an optical band at $h\omega \approx 0.5$ eV was found in La_2CuO_4 and $YBa_2Cu_3O_6$ [3,4]. The position of the observed band is consistent with the energy $h\omega$ of the first excitation level of the average clusters with size $L \sim a\,(U/T)^{3/4}$:

$$h\omega \sim 3\,T(\pi a/L)^2 / 4\sqrt{2} \sim 0.3\,T \text{ (for } U \approx 4T);$$

a is the Cu-O distance. It is important to note that the intensity of the band increases and its position shifts to lower energies when with the increase of the hole concentration more and more clusters come into contact (see following section), as a result of which a Drude-type absorption of free carriers is built up.

Another experimental hint towards hole-cluster formation emanates from optical absorption spectra of La_2CuO_4, where a weak sharp feature at $h\omega \sim 1.5$ eV is observed on the low-energy side of the charge-transfer band. This feature can be associated with the charge-transfer optical transition accompanied by a spin-flip process: the hole spin (σ) and the nearest Cu spin ($-\sigma$) are turned, leading to the creation of a minimal hole-cluster. The corresponding energy is $h\omega = h\omega_0 - \delta E$, where $h\omega_0 \approx 1.7$ eV is the low-energy edge of the charge-transfer band. For $U/T \approx 4$, $|\varepsilon| \leq T$ and $t \leq 0.5\,T$ one finds $\delta E \sim 0.2$ eV and $h\omega_1 \approx 1.5$ eV in agreement with experiment [5].

4. Percolation Thresholds

The hole-clusters are practically immobile due to the small probability of spin-flip processes which have to accompany the motion of the clusters. On increasing the concentration of the holes, and with it of the immobile clusters, at a critical concentration c_c the clusters build up a percolation network along the lines of which the holes can move freely, and the system goes from the insulating to the conducting state [6]. Within the percolation model the magnetic properties of the perovskite-type materials can be understood by taking the three-dimensional magnetic ordering of the systems into account characterized by strong in-plane and weak inter-plane interactions. Three different percolation limits are obtained [7]:

Firstly, percolation lines in the planes are built up from clusters of both spin directions. This limit characterizes the destruction of long-range AF order caused by strong in-plane magnetic interactions and corresponds to a hole concentration

$$c_M^{(1)} \approx 2.2 \,/\, (8N).$$

Secondly, percolation lines in the planes are built up from clusters with the same spin orientation. This limit is important for the motion of charge carriers and characterizes the transition to electrical conductivity corresponding to a hole concentration

$$c_c \approx 4.4 / (7N).$$

Thirdly, at higher concentrations the three-dimensional inter-plane percolation of AF-ordered areas will be destroyed. At this second magnetic threshold the system finally loses also the weak long-range AF order

$$c_M^{(2)} \approx 15.8 / (18N).$$

These considerations lead to the following relation between the three critical concentrations:

$$c_M^{(2)} \geq 1.4\, c_c \approx 3.2\, c_M^{(1)}.$$

For $La_{2-x}(Sr,Ba)_xCuO_4$, where c=2x, this leads to $x_M^{(1)} \approx 0.02$, $x_c \approx 0.05$, $x_M^{(2)} > 0.08$ in agreement with experiment. In the case of $YBa_2Cu_3O_{6+x}$ a phase transition from tetragonal to orthorombic structure appears at $x \approx 0.4$ connected with the ordering of oxygens in the chains. This leads to the fact that for $x<0.4$ most holes are kept in the chains, whereas at $x \approx 0.4$ a large number of holes are introduced into the planes, bringing the system immediately above the threshold for destruction of strong AF order and the beginning of electrical conductivity (see [6] and [7]).

5. Superconductivity

Due to the low dimensionality of the carrier motion the deformational-type hole-phonon interaction leads to the formation of 'large size polarons'. Concerning the superconducting state the main contribution to the correlation energy between two quasi-particles moving in the percolation net is caused via this deformational-type coupling, which is attractive for all particles. A rough estimate for the gap leads to [1]

$$\Delta \sim 2E_F \exp(-1/2\rho W),$$

where ρ is the density of states near the Fermi level, W the hole-phonon interaction and $E_F \sim T(T/U)^{3/2}$ the Fermi energy.

Acknowledgement. The authors would like to thank G. Zavt for fruitful discussions.

References

1. V. Hizhnyakov, E. Sigmund, Physica **C156**, 655 (1988)
2. A. Sherman, private communication
3. S. Tajima, S. Tanaka, T. Ido, S. Uchida, Proc. of 2nd Int. Symp. on Superconductivity, Tsukuba, Japan, Springer-Verlag, 1989
4. X. Wei, L. Chen, Z.V. Vardeny, C. Taliani, R. Zamboni, A.J. Pal, G. Ruani, Physica **C162-164**, 1109 (1989)
5. S.Tajima, H. Ishii, T. Nakahashi, T. Takagi, S. Uchida, M. Seki, S. Suga, Y. Hidaka, M. Suzuki, T. Murakami, K. Oka, H. Unoki, J. Opt. Soc. Am. **B6,** 475 (1989)
6. V. Hizhnyakov, N. Kristoffel, E. Sigmund, Physica **C160**, 119 (1989) and Physica **C161**, 435 (1989)
7. V. Hizhnyakov, E. Sigmund, in: Phonons, 252, eds. S.Hunklinger et al. World Scientific Publ. (1990) and Materials Lett. **9**, 425 (1990)

Variational Monte Carlo Study for Generalized Flux Phases in the t-J Model

M. Ogata and T.M. Rice

Theoretische Physik, ETH Hönggerberg, CH-8093 Zürich, Switerland

We study new wavefunctions for generalized flux phases by the variational Monte Carlo method, in which the motion of holes produces additional phase factors besides those due to the motion of electrons. The phase factors cancel each other at some special electron densities and, as a result, there are no orbital currents in these states. We investigate especially the 1/3-filled case and find a wavefunction with lower kinetic energy than the original commensurate flux phase. We also calculate the chirality to confirm that time reversal symmetry is broken also in these states.

1. Introduction

Various generalizations of the Affleck-Marston flux state [1] have been proposed as variational wavefunctions for the two-dimensional t-J model away from 1/2-filling [2-4]. They are composed of the Gutzwiller projection operator acting on a Slater determinant of non-interacting wavefunction describing fermions on a square lattice under a fictitious or statistical magnetic flux (Hofstadter problem [5]). A renormalized mean-field theory [3] and variational Monte Carlo calculation [6] have shown that the exchange energy is minimized in the commensurate flux phase (CFP) where the fictitious magnetic flux per plaquette is equal to the electron density.

On the other hand, the kinetic energy is frustrated and the expectation value $\langle c_{i\sigma}^{\dagger} c_{j\sigma}\rangle$ in this wavefunction is complex,

$$\langle c_{i\sigma}^{\dagger} c_{j\sigma}\rangle \propto e^{-i\phi_{ij}}, \tag{1}$$

with $\phi_{ij} = 2\pi \int_i^j \vec{A} \cdot d\vec{\ell}$, $\vec{A}$ being the vector potential. Due to this frustration, the total kinetic energy is raised above the zero-flux state [6,7].

The complex value of $\langle c_{i\sigma}^{\dagger} c_{j\sigma}\rangle$ leads to another important result. Since the current operator for a bond has the form,

$$j_{ij} = ieta \sum_{\sigma} (c_{i\sigma}^{\dagger} c_{j\sigma} - c_{j\sigma}^{\dagger} c_{i\sigma}), \tag{2}$$

it has a finite expectation value in the CFP [8], which has not been observed in the μSR experiment [9].

The purpose of this paper is to study new variational wavefunctions of generalized flux phases which do not have orbital currents. In these wavefunctions, an additional phase is introduced which depends on the positions of the holes. As a result, the motion of the hole produces an additional phase factor in $\langle c_{i\sigma}^\dagger c_{j\sigma}\rangle$ and it can be arranged that the phase factors cancel each other at some special electron densities and $\langle j_{ij}\rangle$ becomes zero. The frustration of the kinetic energy is also reduced and the state has a slightly lower kinetic energy than the original CFP.

The similar idea of attaching an additional phase to the holes has been developed in the slave-boson mean- field theory [10,11]. In this paper, we use the variational Monte Carlo (VMC) method in order to take account of the local constraint unambiguously.

2. Generalized Flux Phases

It is convenient to rewrite the flux phase in a different way. The original CFP has the form [2,3],

$$\Psi = P_d \prod_{\ell}^{\text{occ.}} c_{\ell\uparrow}^\dagger c_{\ell\downarrow}^\dagger |0\rangle, \tag{3}$$

where the Gutzwiller projection operator $P_d = \prod_i (1 - n_{i\uparrow} n_{i\downarrow})$ excludes the doubly occupied sites from the wavefunction and $c_{\ell\sigma}^\dagger$ is the creation operator for the eigenstate of the Hofstadter spectrum [5]. When the fictitious magnetic flux per plaquette, ϕ, is equal to the electron density ν, the lowest Hofstadter band which corresponds to the lowest Landau level in the continuum model is completely filled . Therefore a natural simplification approximates the flux phase (3) as

$$\Phi_H = P_d \Phi_0,$$

$$\Phi_0 = \prod_{i<j} (z_i - z_j) e^{-\frac{1}{4\ell^2}\sum |z_i|^2} \prod_{k<n} (\omega_k - \omega_n) e^{-\frac{1}{4\ell^2}\sum |\omega_k|^2}, \tag{4}$$

in the first quantized form [12]. Here the complex numbers $z_j = x_j + iy_j$ and $\omega_k = x_k + iy_k$ are the locations of the j-th electron with up spin and the k-th electron with down spin, respectively. The lattice constant is set to unity and ℓ is the magnetic length given by $1/\sqrt{2\pi\phi}$. We have checked that the variational energy for the state Φ_H is similar to that for the original flux state Ψ.

Introducing an additional phase factor depending on the locations of the holes leads us study the following variational wavefunctions, [13]

$$\Phi_A = P_d \prod_{r<s} \frac{(h_r - h_s)^2}{|h_r - h_s|^2} \Phi_0,$$
$$\Phi_B = P_d \prod_{r<s} \frac{(h_r - h_s)^2}{|h_r - h_s|} e^{-\frac{1}{4\ell'^2}\sum |h_r|^2} \Phi_0, \tag{5}$$
$$\Phi_C = P_d \prod_{r<s} (h_r - h_s)^2 e^{-\frac{2}{4\ell'^2}\sum |h_r|^2} \Phi_0,$$

where $h_r = x_r + iy_r$ represents the location of the r-th hole. These wavefunctions are symmetric with respect to the exchange of holes. In the wavefunction Φ_A, the configuration of holes gives only a phase factor. In other words, a flux tube of two flux quanta is attached to each hole. In the wavefunction Φ_B and Φ_C, hard-core nature of the holes is taken into account [10].

In order to have exact cancellation of the phase factors in $\langle c^\dagger_{i\sigma} c_{j\sigma} \rangle$, we must choose $2\pi\phi - 2\pi\phi_h = 2\pi m$, with m being an integer. Since the flux tube for one hole can be any odd (even) integer flux quanta in Φ_B (in Φ_A), ϕ_h is equal to $\pm 2n\nu_h$ in general, where ν_h is the hole density, $\nu_h = 1 - 2\nu$. Combining these conditions, we find

$$\nu = \frac{2n+m}{4n+1} \quad \text{or} \quad \frac{2n-m}{4n-1}. \tag{6}$$

The same equation has been obtained in the mean-field theory by optimizing the mean-field energy [10]. In the following section, we especially focus on the 1/3-filled case where $n = 1$ and $m = 1$.

3. Variational Monte Carlo Study

The VMC calculation is performed on finite lattices with free boundary conditions, because the symmetric gauge is used in (5). (Note however these wavefunctions are invariant under simultaneous gauge transformations on electrons and holes coordinates.) The kinetic energies are shown in Fig. 1 for various lattices. Although there are large size-dependences, the energy for the state Φ_B is apparently lower than that for the CFP Φ_H.

The expectation value of the current operator (2) for each bond is found to be zero in the states Φ_A, Φ_B and Φ_C, as expected. We found that the kinetic energy is higher than that for the zero-flux state because of the large fluctuation of the phase factors in $\langle c^\dagger_{i\sigma} c_{j\sigma} \rangle$.

The chirality defined by $E_{123} = \langle \vec{S}_1 \cdot (\vec{S}_2 \times \vec{S}_3) \rangle$ [14] is also calculated. Here 1,2,3 are the nearest three sites forming a triangle. In practice, we average the expectation values near the center of the lattice to avoid the boundary effect. We obtain $E_{123} = 0.029$ (0.028) for the state Φ_H

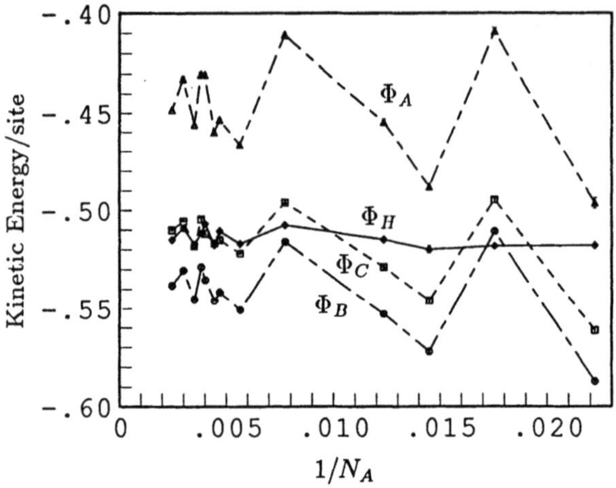

Figure 1

Kinetic energy per site in units of t for the states Φ_H, Φ_A, Φ_B and Φ_C on various lattices with N_A sites. The choice of the gauge in Φ_H has been optimized.

(Φ_B). This clearly indicates that time reversal symmetry is broken in these states.

A wavefunction

$$\Psi_s = e^{iq(\sum_j \mathrm{Re} z_j + \sum_k \mathrm{Re} \omega_k)} \Phi_B, \tag{7}$$

has also been studied, which corresponds to a uniform current carrying state in the x-direction [3]. We find that the increase of the kinetic energy is proportional to q^2 in small q region, which is consistent with the prediction in the generalized mean-field theory [3].

4. Summary and Conclusion

We have studied new variational wavefunctions for the generalized flux phases by the VMC method. Due to the additional phase factors depending on the configuration of the holes, the wavefunctions have no orbital currents and the state Φ_B has a lower kinetic energy than the original CFP. Although there is no current, time reversal symmetry is still broken. This is confirmed by calculating the chirality. The similarity of the wavefunction to the quantum Hall effect is very interesting and this will be studied in a future publication.

We wish to thank B. Doucot for stimulating the examination of the generalized flux phases. We are also grateful to E. Abrahams, P. Lederer, H. Shiba and H. Yokoyama for useful discussions and comments.

References

[1] I. Affleck and J. B. Marston, Phys. Rev. B **37** (1988) 3774; G. Kotliar, Phys. Rev. B **37** (1988) 3664.

[2] P. W. Anderson, B. S. Shastry and D. Hristopulos Phys. Rev. B **40** (1989) 8939.

[3] P. Lederer, D. Poilblanc and T. M. Rice, Phys. Rev. Lett. **62** (1989) 1519.

[4] A. B. Harris, T. C. Lubensky and E. J. Mele, Phys. Rev. B **40** (1989) 2631.

[5] D. R. Hofstadter, Phys. Rev. B **14** (1976) 2239.

[6] S. Liang and N. Trivedi, Phys. Rev. Lett. **64** (1990) 232.

[7] The zero-flux state corresponds to the Gutzwiller-projected Fermi sea, H. Yokoyama and H. Shiba, J. Phys. Soc. Jpn. **56** (1987) 3570; C. Gros, Ann. Phys. **189** (1989) 53.

[8] P. Lederer, D. Poilblanc and T. M. Rice, preprint.

[9] R. F. Kiefl *et al.*, preprint.

[10] F. Nori, E. Abrahams and G. T. Zimanyi, preprint.

[11] J. P. Rodriguez and B. Doucot, preprint.

[12] Φ_H has been studied in the half-filled case by Z. Zou, B. Doucot and B. S. Shastry, Phys. Rev. B **39** (1989) 11424.

[13] B. Doucot, private communication.

[14] X. G. Wen, F. Wilczek and A. Zee, Phys. Rev. B **39** (1989) 11413.

Combined Phonon–Acoustic Plasmon Theory of High-T_c Copper Oxides

H. Morawitz [1] *and V.Z. Kresin* [2]

[1] IBM Almaden Research Center, 650 Harry Road, San Jose, CA 95120, USA
[2] Lawrence Berkeley Laboratory, 1 Cyclotron Road, Berkeley, CA 94720, USA

Abstract. We present a model for the high T_c copper-oxides involving both phonon and acoustic plasmon contributions to the pairing. The carriers are assumed to be described as a Fermi liquid and to couple strongly to optical phonons. In addition – because of the layered structure of the cuprates – the collective excitations (plasmons) of the carriers have acoustic character and contribute to the pairing.

We discuss some normal state properties such as the optical conductivity and electron loss features, which can be related to the inverse dielectric function for the carriers in the layered electron gas model. In addition, we give a simple explanation for the experimentally observed maximum of T_c as a function of carrier concentration x in doped $La_{2-x}Sr_xCuO_4$ in terms of phase-space arguments for the electron-optical phonon coupling mechanism.

1. Introduction

In this paper we discuss several topics related to our view of the superconducting cuprates [1,2], which considers a strong coupling carrier-phonon contribution together with a weak coupling carrier-acoustic plasmon channel to cause the observed high transition temperatures.

We propose to describe the cuprates by a very anisotropic model: The layered electron gas [3,4]. The central quantity to be addressed is the quasi-particle formed in the strongly correlated state. We choose to discuss the correlations in terms of the properties of the collective modes of the carrier charge fluctuations. Because of the layering these modes are drastically different since they include metallic screening within the conducting sheets and dielectric screening between them.

The topics we will discuss here are: (1) The dramatically different structure of the collective charge density oscillation spectrum (plasmons) of layered conductors; (2) The effect of this low frequency response due to charge carrier density on the dielectric loss function and a comparison for three different high temperature superconductors; (3) The presence of a temperature dependent loss feature in an electron loss experiment (EELS); and (4) A qualitative model for the maximum in the superconducting transition temperature T_c based on strong coupling carrier-optical phonon interaction.

Springer Series in Solid-State Sciences, Vol. 99
Electronic Properties of High-T_c Superconductors and Related Compounds
Editors: H. Kuzmany · M. Mehring · J. Fink

2. Layer Plasmons Dispersion Relations

The most general structural aspect of the new high temperature cuprate superconductors is their layered structure. Most simply they consist of metallic CuO_2 layers separated by insulating spacer layers (e.g., (LaO) in La_2CuO_4). As a consequence of this layering the Coulomb interaction between carriers on the same and/or different sheets is screened in a fundamentally different way from conventional, isotropic metals. While the standard RPA screening treatment of the three-dimensional electron gas converts the long-range Coulomb interaction 1/r into a short range exp(-ar)/r interaction, the presence of insulating layers and absence of mobile screening charge in them leads to electric fields very different from the spherically symmetric, longitudinal electric field in three-dimensional metals for the collective charge density fluctuations (plasmons).

We note that the justification for using a local limit for the electron-electron interaction is in general given by the known short-range screening arguments in isotropic metal physics. Our argument is to consider the consequences of the layering on the collective plasmon modes, which is very dramatic, and to pursue unexpected consequences of the appearance of low frequency plasmon branches for normal and superconducting properties.

The price we pay for dealing with the screening in a formally exact way is that we simplify the electronic structure to the barest bone: The layered electron gas, in which the carriers have plane wave character in the CuO_2 planes and no overlap perpendicular to the planes, which are separated by a distance D. The Hamiltonian for this system then is

$$H = \sum_{\vec{k},l} \varepsilon_{\vec{k}} A^{+}_{\vec{k}l} A_{\vec{k}l} + \sum_{\vec{k},\vec{k}',\vec{q}} \sum_{l,l'} V^{0}_{l,l'}(k,k') A^{+}_{\vec{k}+\vec{q}l} A^{+}_{\vec{k}'-\vec{q}l'} A_{\vec{k}l} A_{\vec{k}'l'} , \tag{1}$$

where $A^{+}_{\vec{k}l}$, $A_{\vec{k},l}$ are creation and annihilation operators for the carriers, $\varepsilon_k = k^2/2m^*$ their kinetic energy, l labels the different layers and the second term describes the Coulomb interaction between them. We are interested in obtaining the screened Coulomb interaction between carriers, which is (within the random phase approximation) a solution of

$$V_{ll'}(\vec{k},\vec{k}') = V^{(0)}_{ll'}(k,k') + V^{(0)}_{ll'}(k,k'')\pi^0(\vec{q},\omega) V_{ll'}(k'',k') ; \tag{2}$$

the function $\pi^0(\vec{q},\omega)$ is the polarizability of the two-dimensional carriers. The real and imaginary parts of the polarizability can be analytically calculated and are given by

$$\pi^0(\vec{q},\omega) = m^*/\pi \left[\left[1 - \left(\frac{v_F q}{\omega} \right)^2 \right]^{-1/2} - 1 \right] . \tag{3}$$

The dielectric function $\varepsilon(\kappa, q_z, \omega)$ can be related to the ratio of the screened to unscreened Coulomb interaction and we obtain the layer plasmon dispersion relation by solving for the the zeroes of ε.

The dependence on the layer index of the Coulomb interaction for an infinite system depends only on (l-l′) and we can perform the sum to find $\Phi^0(\kappa;q_z) = 2\pi e^2/\varepsilon_M \kappa F(\kappa, q_z)$. Here the layer form factor is $F(\kappa, q_z) = \sinh(\kappa D)/[\cosh(\kappa D) - \cos(q_z D)]$.

Solving for the layer plasmon dispersion relations using both the analytic form of the single layer polarizability π^0 and the RPA equations for the layer screening, we find

$$\Omega_P(\kappa, q_z) = v_F \kappa\left(1 + (2F/\kappa r_B)^2 (1 + 2F/\kappa r_B)^{-1}\right)^{1/2}; \tag{4}$$

$r_B = h^2 \varepsilon_M e^2 m^*$ is the effective Bohr radius for the carriers. Note, this reduces to the single layer result for large κ, $F \rightarrow 1$ and the hydrodynamic solution first obtained by Fetter [3].

3. Layer Plasmon Density of States and Temperature Dependence of Electron Loss

In this section we pursue the consequences of the layer plasmon spectrum derived in section 2 for two common normal state properties of metals, namely their optical properties and the studies of electron loss.

The knowledge of the plasmon excitation spectrum allows the calculation of the real and imaginary parts of their contribution to the dielectric function. Actually we have so far neglected all aspects of layer plasmon damping (e.g., Landau damping into electron-hole pairs or the intrinsic carrier loss processes modifying the polarizability), which arise in a more complete treatment of the single particle excitation spectrum in a layered electron gas. Steps in this direction can be found in Ref. 4.

We calculate the layer plasmon density of states $\rho(\omega)$, which is proportional to $\mathrm{Im}(-1/\varepsilon(\omega))$:

$$\rho(\omega) = (2\pi)^{-3} \sum_{\vec{k}, q_z} \delta(\omega - \Omega_P(\vec{k}, q_z)). \tag{5}$$

Note, that we have summed over the wave-vector dependence to obtain the density of states of layer plasmons at frequency ω. To equate the obtained expression to the dielectric loss function $\mathrm{Im}(-1/\varepsilon(\omega)) = \varepsilon_2/|\varepsilon|^2$ (where $\varepsilon_{1,2}$ are the real and imaginary parts of the dielectric function), we argue that processes breaking the translational invariance of the solid are involved in coupling the long wavelength optical response to the plasmon spectrum. Reducing the three-dimensional integration in Eq. (5) to a one-dimensional integral, we obtain (in the frequency region $\omega < \omega_U$) $\rho(\omega) = \alpha\omega^2$.

Our interest in the layer plasmon contribution arose from the experiments of I. Bozovic [5] on films and single crystals of $YBa_2Cu_3O_7$, $Bi_2Sr_2CaCu_2O_8$ and $Tl_2Ba_2Ca_2Cu_3O_{10}$. This experimental work employed a combination of reflectance measurements over a large frequency range ($0.01 < \omega < 6\,eV$) and ellipsometric studies combined with Kramers-Kronig transformations. These allowed the construction of $Im(-1/\varepsilon(\omega) = \varepsilon_2/\varepsilon_1^2 + \varepsilon_2^2)$ directly from the experimental data. For all measured materials a direct proportionality to ω^2 was found.

To calculate the electron loss from a layered system, we utilize a general expression in terms of the inverse dielectric function

$$-\frac{\partial E}{\partial t} = (4\pi)^{-3}\int \omega\,|V_{\vec{k}q_z}|^2(1-e^{-\beta\omega})^{-1}\,Im[\varepsilon(\vec{k}, q_z; \omega)]^{-1}d^3\vec{k}' , \qquad (6)$$

where V_{κ, q_z} describes the Coulomb interaction between the high energy (EELS) electron and the carriers in the layered system. The standard single scattering (Born) approximation has been made, necessitating experiments to be made on films less than or about 1000Å in thickness.

In order to simplify the calculation we approximate the layer plasmon spectrum by the topmost and bottom branches, characterized by wavevectors $q_z = 0$ and $q_z = \pi/D$, respectively. The loss function γ can then be expressed as a sum of two terms, the first of which comes from the topmost branch, which is analogous to the standard 3-D plasmon term. The second term arises from the low frequency part ($\omega = v_F\kappa$) of the spectrum and is temperature dependent. Specifically we find $\gamma = \gamma_U + \gamma_L$.

The temperature dependence of the lower branch can be written $\gamma_L = \gamma_U(T/\omega)^3$, with $\omega = v_F\pi/D$ (0.5-1.0 eV), and we estimate it should contribute an amount of order 1-5% to the total loss at room temperature. Such a contribution may be experimentally accessible .

4. Carrier Concentration Dependence of T_c

A very intriguing aspect of the new high temperature superconductors is the dependence of the superconducting transition temperature on the carrier concentration observed by several groups. In this section we will demonstrate how such a dependence arises naturally in a strong coupling model for the carrier-lowest optical phonon branch. It arises from the cross-over of the maximum momentum transfer of two quasi-particle carriers on the Fermi circle being set by the diameter $2\kappa_F$ of the Fermi circle or the maximum phonon wavevector q_c.

Our starting point is the Eliashberg equation

$$\begin{aligned}\Delta(\vec{k}, \omega_n, T)Z = T\sum_{\omega_{n'}}\int d^3k'\xi_F\Omega^2/(\Omega^2 + (\omega_n - \omega_{n'})^2) \\ \times\,\Delta(\vec{k}', \omega_{n'}, T)/(\omega_{n'}^2 + \zeta^2 + \Delta^2(k', \omega_{n'}; T')) ,\end{aligned} \qquad (7)$$

where Ω is optical phonon frequency, Δ the anisotropic gap function and ξ_F the Froehlich carrier-phonon coupling. We assume that the carriers have a cylindrical Fermi surface of radius κ_F and height $2\pi/D$. By making the Ansatz $\Delta(\omega_n, T, k) = \tilde{\Delta}(\omega_n, T) f(k_z)$, which requires the order parameter Δ and hence the gap to vanish at T_c in all directions, we reduce Eq. (7) to a simpler one, directly involving the carrier-optical phonon parameter λ, which contains the key features of our explanation:

$$\lambda = \frac{2\pi m^*}{k_F D} \int_0^{k_1} \xi_F(q_\parallel) dq_\parallel / (1 - (q_\parallel 2\kappa_F)^2)^{1/2} . \tag{8}$$

The upper limit k_1 in the $q_\parallel$ integration is the smaller of $2\kappa_F$ and q_c. The existence of a maximum phonon wave-vector plays an important part in determining the peak in λ. We propose to use results of the detailed, inelastic neutron scattering experiments on $La_{2-x}Sr_xCuO_4$ to assign q_c [7]. Our reason for keeping only the single, lowest optical branch is its flat dispersion relation extending over about a third of the Brillouin zone before hybridizing strongly with the transverse acoustic branch and acquiring a quasi-linear dispersion.

For a two-dimensional Fermi cylinder, the relationship between κ_F and the three-dimensional carrier density n is given by $\kappa_F = (2\pi n D)^{1/2}$.

The functional dependence of the carrier-optical phonon parameter $\lambda(n)$ can be explicitly calculated and is given by [8]

$$\lambda(n) = \lambda_0 \begin{cases} x^2 \sim n & x \le 1 \quad x = 2\kappa_F/q_c , \\ \frac{2}{\pi}\,\frac{1}{x} \sim n^{-1/2} & x > 1 . \end{cases}$$

Using the experimental parameters for $La_{2-x}Sr_xCuO_4$, we find the critical carrier concentration n_c at the peak of T_c to be given by

$$n_c = q_c^2/8\pi D \simeq 2 \times 10^{21}\ \mathrm{cm}^{-3}$$

in good qualitative agreement with the carrier density obtained by Hall measurements.

5. Summary and Conclusion

We have addressed several questions concerning the layered electron gas aspects of the high T_c cuprate superconductors and their collective charge fluctuation spectra (layer plasmons). In particular, normal state properties such as the loss function $(-\mathrm{Im}(1/\varepsilon(\omega)))$ are shown to be in agreement with recent experimental data on $YBa_2Cu_3O_7$, $Bi_2Sr_2CaCu_2O_8$ and $Tl_2Ba_2Ca_2Cu_3O_{10}$, which were obtained by I. Bozovic from both reflectance and ellipsometric measurements combined with Kramers-Kronig transformations. We have also proposed a novel diagnostic test for the presence of the acoustic plasmon branches by demonstrating that they contribute a temperature dependent part proportional to T^3 in a transmission electron loss experiment.

Finally, by explicitly calculating the carrier concentration dependence of the carrier-optical phonon coupling parameter λ within a cylindrical (two-dimensional) Fermi surface we have shown that there is a maximum value in λ at a value in qualitative agreement with the concentration obtained from Hall data for $La_{2-x}Sr_xCuO_4$.

6. References

1. V. Z. Kresin and H. Morawitz, *Phys. Rev.* **B37**, 7854 (1988).

2. V. Z. Kresin and H. Morawitz, *Superconductivity* **1**, 79 (1988), 1, 79 (1988) Ed. D. Gubser and S. Wolf, Plenum Press, NY 1988.

3. A. L. Fetter, *Ann. Phys. (N.Y.)* **88**, 1 (1974).

4. P. Hawrylak, Gunnar Eliasson and J. J. Quinn, *Phys. Rev.* **B37**, 10187 (1988).

5. I. Bozovic, submitted to *Phys. Rev. B*, Feb. 90).

6. J. B. Torrance, Y. Tokura, A. I. Nazzal, A. Bezinge, T. C. Chuang and S. S. P. Parkin, *Phys. Rev. Lett.* **61**, 1127 (1988); H. Takagi, et al., *Phys. Rev.* **B40**, 2254 (1989).

7. B. Böni, et al., *Phys. Rev.* **B38**, 185 (1988).

8. V. Z. Kresin and H.Morawitz, submitted to *Solid State Comm.* (Feb. 90).

Optical Phonons in High-Temperature Superconductors: Evidence for Strong Electron–Phonon Coupling

G. Zwicknagl

Max-Planck-Institut für Festkörperforschung, Heisenbergstr. 1
D-7000 Stuttgart 80, Fed. Rep. of Germany

Abstract. Optical phonons in High-Temperature superconductors exhibit characteristic modifications in the superconducting state. We calculate the phonon changes within the conventional theory of strong-coupling superconductivity. Comparing our results with experimental data allows us to deduce information on the electron-phonon interaction.

1. Introduction

The optical phonons in high-temperature superconductors (HTSC) exhibit a characteristic variation with temperature in the superconducting state [1–4]. A prominent example is the anomalous softening of the Raman mode at $\simeq 340cm^{-1}$ in $RBa_2Cu_3O_{7-\delta}$ with $R = Y, Eu, Sm, Tm$. In addition, the width of this line is found to vary non-monotonically with temperature [5]. The modifications can be suppressed by applying a magnetic field $H > H_{c_2}(T)$ [6]. This fact clearly demonstrates that the anomalous behavior is induced by the formation of the superconducting state and that it provides quantitative evidence for the electron-phonon coupling. In the present paper we study the changes in the electronic excitation spectrum and its influence on optical phonons within the conventional theory of strong-coupling superconductors. In section 2 we briefly summarize the theoretical background and discuss qualitatively the change of the phonon self-energy in the superconducting state. For a detailed derivation we refer to [7]. In section 3 we specify the strong-coupling model from which the properties of the superconductor are calculated. The influence of sample imperfections is studied in section 4. In section 5 we finally compare our results with experiments.

2. Theoretical Background

The frequency shifts and the line widths are calculated from the phonon self-energy which is evaluated to lowest order conserving approximation displayed in Fig. 1. The modifications in the superconducting state result from changes in the excitation spectrum of the quasiparticles and in their effective coupling to the phonons. The bare coupling has to be corrected for the interaction among the quasiparticles which includes the attraction leading to superconductivity. HTSC represent charged superfluids and we have to explicitly include the influence of the long-range

Springer Series in Solid-State Sciences, Vol. 99
Electronic Properties of High-T_c Superconductors and Related Compounds
Editors: H. Kuzmany · M. Mehring · J. Fink

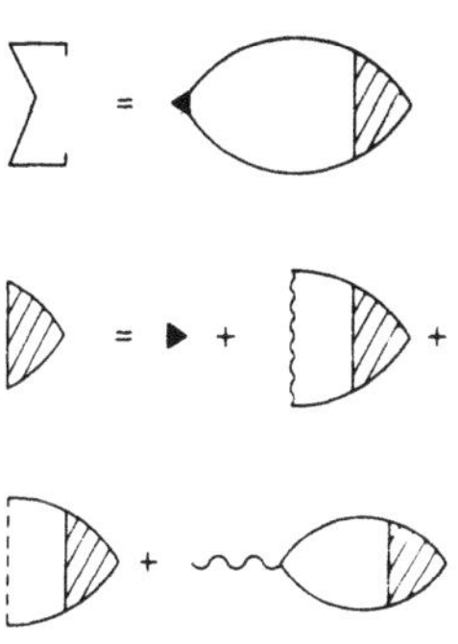

Fig. 1: Phonon self-energy in lowest order conserving approximation

Coulomb forces which in the normal state are accounted for by screened quasiparticle-phonon coupling constants. Due to the long-range nature of the interaction, $V_c(q) \sim q^{-2}$ the vertex Γ_q and concommittantly, the modifications due to superconductivity, vanish in the long-wave length limit ($\Gamma_q \sim q^2$) *unless* the polarization corrections do not contribute e.g. for symmetry reasons.

We focus on conventional superconductors where the symmetry of the order parameter is that of the underlying lattice. In that case the polarization correction vanishes identically for phonons which are not fully symmetric. As a consequence these modes should be influenced by the onset of superconductivity. Examples are the IR active modes which are odd under inversion and the Raman mode at $340cm^{-1}$ which - in the approximate tetragonal classification - transforms according to the B_{1g} representation [8]. As mentioned above, this mode shows dramatic changes in the superconducting state.

For fully symmetric modes, however, the self-energy effects are reduced due to the long-range nature of the Coulomb interaction. The polarization correction renormalizes the Fermi surface average of the electron-phonon coupling $\langle g_\nu \rangle_{FS} =< g_\nu / v_F > / < v_F >$ which therefore is screened out. Here v_F denotes the Fermi velocity and the index ν refers to the phonon branch. Only the anisotropic part of the electron-phonon coupling, $\delta g_\nu = g_\nu - \langle g_\nu \rangle_{FS}$, can lead to changes in the superconducting state. In particular, the modifications should vanish in an isotropic metal.

The superconductivity-induced changes in the phonon self-energy are given by $\langle (\delta g_\nu)^2 \rangle_{FS} \Delta\Pi(\omega)$ with the polarization change $\Delta\Pi$ displayed in Fig. 2. The general features can be understood by considering a weak-coupling (BCS) superconductor at $T = 0$ where an energy gap Δ separates the ground state from the quasiparticle excitations. As a result, a minimum energy of 2Δ is required to create a particle-hole pair. This cannot be achieved by phonons with frequencies $\omega < 2\Delta$ the line width of which therefore vanishes. The most dramatic effects occur for phonons with $\omega \sim 2\Delta$ the broadening of which reflects the large density of single-particle excitations across the gap. As the temperature T is increased the gap $\Delta(T)$ decreases and particle-hole pairs are thermally

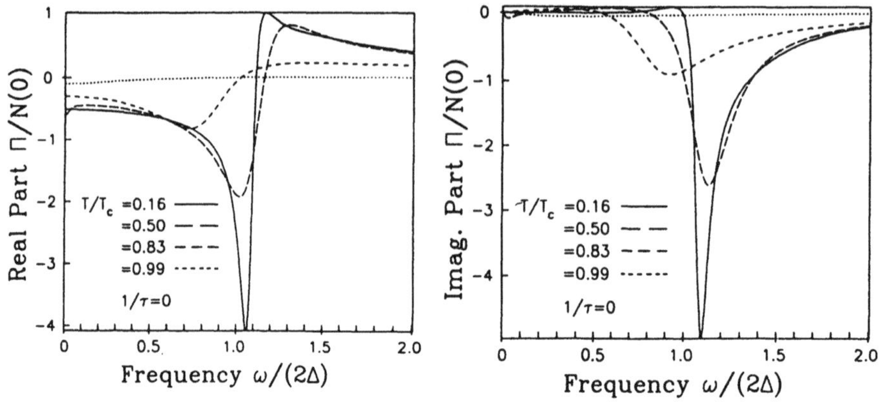

Fig. 2: Polarization of $RBa_2Cu_3O_{7-\delta}$ for various temperatures in the pure case after [7]

excited. The maximum in the phonon line width is reduced in height and shifted towards lower frequencies. The non-monotonic temperature dependence of the line width is a characteristic feature of phonons with $\omega < 2\Delta$. Strong electron-phonon coupling introduces only quantitative changes. First, for a given transition temperature, the gap Δ is enhanced over the weak-coupling value which is expressed in terms of a gap ratio exceeding the universal BCS value. Second, the low-temperature line width reaches its maximum for $\frac{\omega}{2\Delta} \simeq 1.1$. In addition, the structures in the self-energy are less pronounced.

The frequency-dependent line width is connected with a frequency shift. The most prominent feature is a sign change in the vicinity of 2Δ which can be inferred from rather simple considerations. If we approximate the sharp feature at 2Δ by a single line we see that level repulsion predicts a hardening for phonons with $\omega > 2\Delta$ whereas frequencies $\omega < 2\Delta$ are shifted towards lower frequencies. The shift is rather pronounced in the vicinity of 2Δ where the weak-coupling BCS theory predicts a singular behavior for $T = 0$.

3. Strong-coupling model for $RBa_2Cu_3O_{7-\delta}$

We assume that the quasiparticle attraction leading to superconductivity is induced by the phonons and that the standard theory of strong-coupling superconductors can be applied [9]. The order parameter is known to have no nodes, and we assume it to be spherically symmetric. For the electron-phonon spectral function $\alpha^2F(\omega)$ we adopt the model

$$\alpha^2F(\omega) = g(\omega)F(\omega) \tag{1}$$

where $F(\omega)$ is the measured phonon density of states [10]. The coupling constant is approximated by a step function

$$g(\omega) = \begin{cases} 0 & \text{for } \omega \leq \omega_0 \\ g & \text{for } \omega \geq \omega_0 \end{cases} . \qquad (2)$$

The cut-off frequency $\omega_0 \simeq 24meV$ is obtained from the assumption that the low-lying Ba- and Y-modes do not strongly couple to quasiparticles. Our assumption is motivated by the fact that no frequency shifts are observed in the superconducting state for low frequencies. Finally, the constant g is fixed by the requirement that the superconducting transition occurs at the observed temperature $T \simeq 90K$. The actual value of g depends on the specific choice of the Coulomb pseudopotential for which we adopt the value $\mu^* = 0.25$. The model yields $\lambda = 2\int d\omega \frac{\alpha^2 F(\omega)}{\omega} = 2.9$ and an excitation gap $\Delta = 19.7meV$ which corresponds to a the ratio $2\Delta/k_B T_c = 4.96$.

4. Influence of sample imperfections

The observed frequency shifts and line widths below T_c depend on sample properties. We study here the influence of non-magnetic impurities and of inhomogeneities associated with spatial variation of oxygen content. The results are summarized in Fig. 3.

Non-magnetic impurities do not alter the thermodynamic properties of a superconductor as long as the mean-free path is large compared to the Fermi-wave length. This fact which is known as Anderson's theorem implies that the energy gap remains unaffected. The phonon self-energy, on the other hand, describes the response of the quasiparticle system. It is a nonequilibrium property which does explicitly depend on the mean-free path. The frequency shifts are reduced and the structures become less pronounced. This behavior reflects the fact that the volume contributing to the $\underline{q} = 0$ response is now limited by the mean-free path l_F rather

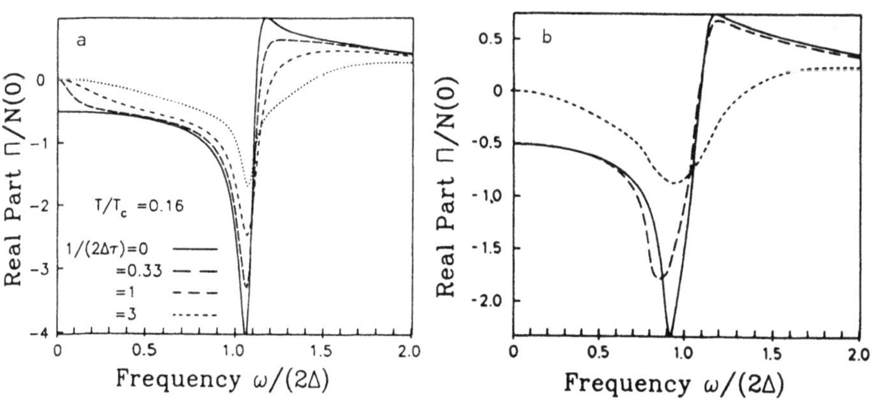

Fig. 3: Real part of the polarization for $T/T_c = 0.16$ *(a) for various values of the impurity concentration,(b) for square distribution of* λ *defined by* $[\lambda_{min}, \lambda_{max}]$. *Full line: [2.5,2.98],* $1/\tau = 0$; *broken line: [2.3,3.0],* $1/\tau = 0$; *dotted line: [2.3,3.0],* $1/(2\tau\Delta) = 3$

than the coherence length ξ which is in close analogy to the cross-over from non-local to local electrodynamics. Typical values for the inverse lifetime , $\frac{1}{\tau}$ can be determined from an analysis of the Drude behavior and are found to be comparable to 2Δ. In the presence of nonmagnetic impurities the phonons below 2Δ are expected to soften more rapidly as a function of temperature when the sample is cooled below T_c.

Macroscopic inhomogeneities are a different type of sample imperfection . Of particular importance are variations in the oxygen content which should occur on length scales $l > \xi$. As a result, we expect the electronic properties to vary slowly in space. We account for the inhomogeneities by introducing a distribution of "local transition temperatures" at which a superconducting order parameter can locally form. The width of the distribution is reflected in a broadening of the specific heat jump. The material is usually characterized by the transition measured in the resistivity. We expect the resistive transition to occur whenever the superconducting volume fraction equals the critical volume fraction for classical percolation in a three-dimensional random structure [11]. The distribution of local transition temperatures translates into a distribution of excitation gaps over which the response function has to be averaged. As a result, the pronounced features in the phonon self-energy are averaged out leading to a net reduction of the renormalization.

5. Comparison with experiment

To calculate the frequency shifts and line widths of the phonons we need the appropriate average coupling constants $\lambda_\nu = 2N(0)\langle(\delta g_\nu)^2\rangle_{FS}/\omega_\nu$. For a preliminary estimate we use an average value $\bar{\lambda} \simeq 0.1$ obtained from the mass enhancement λ divided by the number of phonon branches contributing to $\alpha^2F(\omega)$. This crude approximation overestimates the

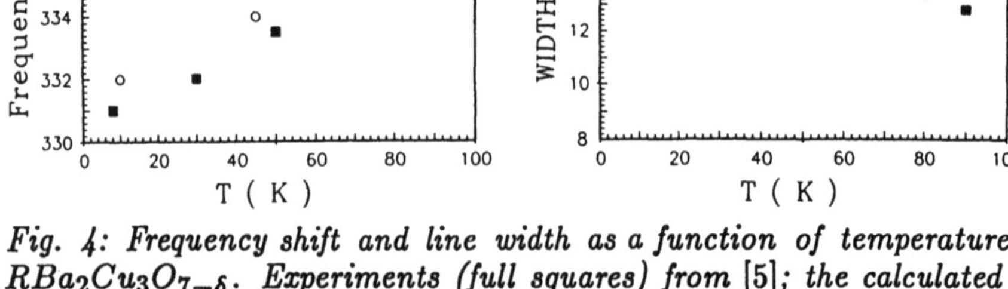

Fig. 4: Frequency shift and line width as a function of temperature for $RBa_2Cu_3O_{7-\delta}$. Experiments (full squares) from [5]; the calculated values (open circles) were obtained for $1/(2\tau\Delta) = 3$ by averaging over a square distribution of λ given by [2.3,3.0]

frequency shift of the $340cm^{-1}$-Raman mode by a factor of 5 when sample inhomogeneities are neglected. Including a finite mean-free path and inhomogeneitieties reproduces the observed temperature dependence of the frequency shift and the line width as can be seen from Fig. 4.

6. Conclusions

The anomalous behavior of the optical phonons in the superconducting state show that HTSC can be described by conventional strong coupling theory starting from an interaction with a characteristic frequency in the same energy range as the phonons. The central question now arises whether the quasiparticle-phonon coupling deduced from experiment is strong enough to explain the high transition temperature of these compounds. There are some indications that this is actually the case. We showed that the observed phonon anomalies can be obtained from the interaction leading to superconductivity provided that sample imperfections are taken into account. Additional support is coming from rather recent investigations. Comparing the observed frequency shifts $\Delta\omega/\omega$ with our polarization changes $\Delta\Pi$ for homogeneous samples with finite mean-free path $\frac{1}{\tau} \simeq 2\Delta$ yields estimates for the appropriate coupling strenths which are in excellent agreement with the predictions of ab initio electronic structure calculations [12]. Recent realistic calculations of the the quasiparticle-phonon interaction [13] show that the rather small values λ_ν deduced by the above-mentioned procedure for the $\underline{q} = 0$-phonons can indeed be reconciled with the rather large values values deduced from the transition temperature T_c.

REFERENCES

[1] R.M. MacFarlane, H.J. Rosen, and H. Seki, *Solid State Commun.* **63**, 831 (1987)
[2] C. Thomsen, M. Cardona, B. Gegenheimer, R. Liu, and A. Simon, *Phys. Rev.* **B 37**, 9860 (1988)
[3] L. Genzel, A. Wittlin, M. Bauer, M. Cardona, E. Schönherr, and A. Simon, *Phys. Rev.* **B 40**, 2170 (1989)
[4] R. Feile, *Physica* **C 159**, 1 (1989)
[5] M. Krantz, H.J. Rosen, R.M. Macfarlane, and V.Y. Lee, *Phys. Rev.* **B 38**, 4992 (1988)
[6] T. Ruf, C. Thomsen, R. Liu, and M. Cardona, *Phys. Rev.* **B 38**, 11985 (1988)
[7] R. Zeyher and G. Zwicknagl, *Z. Phys.* **b 78**, 175 (1990)
[8] R. Liu, C. Thomsen, W. Kress, M. Cardona, B. Gegenheimer, F.W. deWette, J. Prade, A.D. Kulkarni, and U. Schröder, *Phys. Rev.* **B 37**, 7971 (1988)
[9] For a review, see e.g. D. J. Scalapino, in "Superconductivity", ed. R.D. Parks, Marcel Dekker, New York (1969), Chapter 10, Vol.1

[10] B. Renker, F. Gompf, E. Gering, D. Ewert, H. Rietschel, and A. Dianoux, *Z. Phys.* b **73**, 309 (1988)
[11] R. Zallen, "The Physics of Amorphous Solids", John Wiley and Sons, New York (1983), Chapter 4.7
[12] C. Thomsen, M. Cardona, B. Friedl, I.I. Mazin, C.O. Rodriguez, and O.K. Andersen, to be published
[13] R. Zeyher, *Z. Phys.* b (1990) (to appear)

Anharmonic Lattice Dynamics and Electron–Phonon Coupling in High-T_c Oxide Superconductors

H. Rietschel

Kernforschungszentrum Karlsruhe, Institut für Nukleare Festkörperphysik, Postfach 3640, D-7500 Karlsruhe, Fed. Rep. of Germany

Abstract. We discuss experimental indications for strong electron-phonon coupling in high-T_c cuprates. Particular emphasis is laid on anharmonic features in the lattice dynamics as observed in recent neutron-scattering experiments and on their implications with regard to the applicability of standard Eliashberg theory.

There is no doubt that electron (or hole) pairs are relevant for superconductivity also in the newly discovered oxide superconductors. However, the origin of the attractive interaction required for pairing is still vague, and the numerous theories so far proposed to explain this interaction are highly controversial. In a rough way, one may classify these theories into two groups:

i. Theories postulating the attractive pairing interaction to be mediated solely by electronic degrees-of-freedom;
ii. Theories where this interaction is at least partly mediated by the dynamic polarizability of the ionic lattice.

The advocates of i. frequently forward the following argument: Microscopic calculations of the electron-phonon coupling function $\alpha^2F(\omega)$ and subsequent solutions of Eliashberg's equations can explain T_c's of at most 30-40 K [1] thus proving that electron-phonon coupling by itself cannot account for, e.g., $T_c = 92$ K in $YBa_2Cu_3O_7$. And they find this argument strongly supported by the virtually missing isotope effect in this compound [2].

Since not only such T_c calculations but also the usual evaluation of the isotope effect is based on Eliashberg theory, the validity of the above argument depends on the validity of Eliashberg theory itself. In its standard form, this theory implies two severe assumptions, namely, that the perturbation series in the ionic displacements δR_i may be cut off in leading order (harmonic approximation) and that Migdal's theorem holds [3]. For the high-T_c compounds, none of these assumptions is justified. Rather, as will be outlined below, there are strong experimental indications against these assumptions.

Strong anharmonicities in the lattice dynamics of compounds of the YBaCuO and LaCuO families were revealed in inelastic neutron-scattering experiments, and they were also found in the related

Springer Series in Solid-State Sciences, Vol. 99
Electronic Properties of High-T_c Superconductors and Related Compounds
Editors: H. Kuzmany · M. Mehring · J. Fink

compound La_2NiO_4. For detailed presentations of these results as well as for extensive discussions, the reader is referred to Refs. [4-6]. These anharmonicities manifest themselves in anomalous dispersions, splittings and line widths of selected phonon branches which are all related to Cu-O bond-bending or bond-stretching displacements and which cannot be described in any harmonic model however sophisticated it is [6].

Among these anomalies, the most striking is the appearance of extra modes in addition to the 3N allowed phonon branches (N: number of atoms per u.c.). Since the lattice degrees-of-freedom are already exhausted by 3N branches, the only interpretation of these extra modes is hybridization of lattice modes with low-lying electronic excitations thus leading to branches of mixed character. Incidentally, this phenomenon is by no means new and was observed already a long time ago, as, e.g., in $CeAl_2$ where crystal-field excitations of 4f electrons hybridize strongly with optical phonons [7].

Now, since Migdal's theorem is based on a perturbation series in the ratio (sound velocity)/(Fermi velocity), i.e., on the mismatch between phononic and electronic dispersions, strong hybridization of phononic and electronic excitations represents a manifest violation of this theorem. Thus, neither the assumption of harmonic displacements nor that of the validity of Migdal's theorem can be maintained for high-T_c oxide superconductors, and standard Eliashberg theory cannot be applied to these systems a priori. In other words, any statement on high-T_c cuprates inferred from this theory is questionable, and this should be kept in mind when discussing isotope effects or standard microscopic T_c-calculations.

So, the vanishing isotope effect in YBaCuO [2] and the low T_c's calculated in the standard way [1] only show that the usual electron-phonon scheme developed for conventional superconductors does not work in high-T_c cuprates, but by no means do they rule out the possibility that the pairing interaction is mediated by lattice displacements. In fact, there are numerous experimental findings which can be traced back unambiguously to strong interactions between the lattice dynamics and the free charge carriers in high-T_c oxide superconductors:

- In $(La,Sr)_2CuO_4$, a sizeable isotope effect *has* been found [8], and it is well conceivable that for the 92 K superconductors, this isotope effect is reduced by anharmonicity. This phenomenon, too, is not new but well known from conventional superconductivity, PdD being a very prominent example even for a reversed isotope effect.
- There is a clear-cut kink at T_c in the thermal conductivity κ of YBaCuO [9]. In this compound, κ is obviously phonon dominated, and the kink results from freezing out phonon-electron scattering below T_c by opening a gap. So, this result even shows that the *superconducting* free charge carriers strongly couple to the lattice displacements.
- In some phonon branches strong frequency renormalization is

observed on creating holes by doping. For example in YBaCuO, the highest optical branch softens by about 20% when going from the insulating O_6 to the superconducting O_7 compound, and similar renormalization effects show up when doping insulating La_2CuO_4 with Sr [6].

- Finally, Raman scattering experiments showed significant changes in frequency and linewidth of selected Γ-point phonons when decreasing the temperature from above T_c to below T_c [10]. These latter results could even be interpreted in terms of strong-coupling Eliashberg theory [11].

In conclusion we state that for the high-T_c oxide superconductors, there is clear evidence for strong interrelations between lattice dynamics and electrons (holes). In view of this, any theory involving only electronic degrees-of-freedom to explain pairing is very likely to fail. Rather, dynamic polarization of the ionic lattice must be included but in a way which allows for large (anharmonic) ionic displacements and for non-adiabatic electron-ion coupling.

The author gratefully acknowledges extensive discussions with L. Pintschovius and W. Reichardt.

References

1. W. Weber, Adv. in Sol. St. Phys. **28** (1988) 141
2. B. Batlogg et al., Phys. Rev. Lett. **58** (1987) 2333
3. J.R. Schrieffer, Theory of Superconductivity
4. H. Rietschel, L. Pintschovius and W. Reichardt, Physica C **162-164** (1989) 1705
5. W. Reichardt et al., Physica C **162-164** (1989)
6. L. Pintschovius, Festkörperprobleme **30** (1990) in print
7. M. Loewenhaupt et al., Phys. Rev. Lett. **42** (1979) 1709
8. T.A. Faltens et al., Phys. Rev. Lett. **59** (1987) 915
9. U. Gottwick et al., Europhys. Lett. **4** (1987) 1183
10. C. Thomsen et al., Phys. Rev. **B37** (1988) 9860
11. R. Zeyher and G. Zwicknagl, Z. Phys. **B78** (1990) 175

Energy Gap of Electronic Pair Formation in the One-Dimensional Attractive Hubbard Model

J. Carmelo

Max-Planck-Institut für Festkörperforschung, Heisenbergstr. 1
D-7000 Stuttgart 80, Fed. Rep. of Germany
Permanent address: CFMC–INIC, Av. P.G. Pinto 2,
P-1699 Lisboa Codex, Portugal
and Departamento de Fisica, Universidad Evora, Ap. 94,
P-7001 Evora Codex, Portugal

Abstract. The gap in the magnetic excitation spectrum of the one-dimensional attractive Hubbard model is calculated as a function of the band filling in the weak-coupling limit $|U| \ll t$. Within this limit, we use a variational function to describe the ground state of the model, which is a good starting state for the Gutzwiller wave function. Our results are consistent with the analytical expression derived by Larkin and Sak [1], which was based on a renormalization-group calculation. The present variational scheme has the advantage of giving a new insight into the origin of the gap.

1. Introduction

The attractive Hubbard model can be considered as generally resulting from a system of narrow band electrons strongly coupled to a bosonic field (phonons, excitons, acoustic plasmons, etc.), upon elimination of bosonic degrees of freedom [2].

Recently, we have extended the Gutzwiller wave function to describe the ground state properties of the model [3]. A continuation of that work, including the study of superconductivity and CDW instabilities for the square lattice, is in progress. Such study requires the introduction of broken symmetries into the starting state of the Gutzwiller wave function.

This paper is devoted to a related problem, which is the introduction of broken symmetry into the starting ground state of the one-dimensional model, to study the gap in the spin excitations spectrum.

Krivnov and Ovchinnikov [4] have shown that the spin excitations spectrum has such a gap for arbitrary n, contrary to the case of $U>0$, where the gap exists only for $n=1$. The attractive interaction favors the formation of electronic pairs, resulting in a ground state with a gap for magnetic excitations, because such excitations require the breaking of at least one of the bound pairs. The size of the gap is a measure of the strength of the effective interaction between the electrons.

Our results are restricted to the weak coupling limit $|U| \ll t$.

Springer Series in Solid-State Sciences, Vol. 99
Electronic Properties of High-T_c Superconductors and Related Compounds
Editors: H. Kuzmany · M. Mehring · J. Fink

2. The Variational Wave Function

The 1D attractive Hubbard model is given by

$$\hat{H}=\hat{T}+\hat{V},\ \hat{T}=-\sum_{j,\sigma} t(c^{+}_{j\sigma}c_{j+1\sigma}+h.c.),\ \hat{V}=-|U|\hat{D}, \tag{1}$$

$$\hat{D}=\sum_{j} n_{j\uparrow}n_{j\downarrow},\quad n_{j\sigma}^{+}=c^{+}_{j\sigma}c_{j\sigma}\ . \tag{2}$$

The model contains three parameters, the transfer integral t, U and n=N/Na, where N is the number of electrons and Na the number of sites ($K_F=\frac{\pi}{2}n$).

We consider the ground state wave function of an effective single particle Hamiltonian Heff. This has the same form as $\hat{T}$ in (3), except that the transfer integral t is replaced by a j-dependent one of the form

$$t_j=t\left\{1+\delta_o[\cos(2K_F(j+1))-\cos(2K_Fj)]\right\}, \tag{3}$$

where

$$\delta o=\frac{\Delta_o}{4t\sin K_F} \tag{4}$$

is a variational parameter and Δ_o is the Fermi surface gap which arises in the spectrum of $\hat{H}$eff after diagonalization. We omit the details of the calculations which will be presented elsewhere. Our results are not valid for densities very close to n=1, because the canonical transformation, which diagonalizes $\hat{H}$eff, neglects Umklapp scattering. We also neglect a small term which is unimportant for small and intermediate values of n. Within the weak coupling limit $|U|<<t$, $\delta o<<1$, we obtain

$$\hat{H}eff\ |\Phi_{\delta o}>=(E_o+E_{\delta o})|\Phi_{\delta o}>,\ <\Phi_{\delta o}|\Phi_{\delta o}>=1\ , \tag{5}$$

where

$$E_{\delta o}=T_{\delta o}-|U|D_{\delta o}\ , \tag{6.a}$$

$$T_{\delta o}=-\frac{\delta o^2}{2}T_o(\frac{1}{\lambda}-\frac{1}{2})+h.o.,\ D_{\delta o}=\frac{\delta o^2}{2\pi^2}Na\ \frac{1}{\lambda}2+h.o., \tag{6.b}$$

$$\frac{1}{\lambda}=\ln\ (\frac{2\sin K_F}{\delta o}) \tag{7}$$

and

$$E_o=T_o-|U|D_o\ ,\ T_o=-\frac{4tNa}{\pi}\sin K_F,\ D_o=Na(\frac{K_F}{\pi})^2 \tag{8}$$

is the ground state energy $<SL|\hat{H}|SL>$ of the Hamiltonian (1) for U=0 (|SL> - Slater Determinant).

We introduce the following wave function

$$|1>=|\Phi_{\delta o}>-|SL>. \tag{9}$$

As both $|\Phi_{\delta o}>$ and |SL> are normalized, |1> has a well defined norm <1|1>. Moreover, we derived

$$<1|\hat{T}|1>=T_o<1|1>+T_{\delta o}+h.o.\ , \tag{10.a}$$

$$<1|\hat{D}|1>=D_o<1|1>+D_{\delta o}-\theta+h.o.\ , \tag{10.b}$$

with θ given by

$$\theta = \langle \Phi_{\delta o} | (\hat{D}-D_o) | SL \rangle + \langle SL | (\hat{D}-D_o) | \Phi_{\delta o} \rangle$$
$$= \frac{\delta o^2}{\pi^2} \, C \, Na \, \frac{1}{\lambda^{\alpha}} + \text{h.o.} \qquad (11)$$

where $C \approx 1$ and $\alpha < 2$ are positive constants which we could not evaluate precisely.

Within the weak-coupling limit $|U| << t$, we introduce the following variational ansatz for the ground state of the Hamiltonian (1)

$$|\Phi_{\beta,\delta o}\rangle = |SL\rangle + \beta |1\rangle \quad , \qquad (12)$$

where β is a second variational parameter. For $\beta=1$ the state (12) reduces to $|\Phi_{\delta o}\rangle$, which corresponds to a variational approximation of Hartree Fock (H.F.) type.

3. The Gap Equation

The analytical expression derived by Larkin and Sak [1], which for small band fillings is supported by the exact results of references [4], [5], reads

$$\Delta = 8t \sqrt{\frac{2}{\pi}} \sin^2 K_F \; \sqrt{\lambda} \; e^{-\frac{1}{\lambda}}, \; \lambda = |U|/[2t\pi \sin K_F]. \qquad (13)$$

We start by evaluating the H.F. gap equation by minimizing the energy (6) with respect to δo. We obtain

$$E = E_o + E_{cond}, \; E_{cond} = - \frac{\sin K_F}{\pi} \, \delta o^2 \, Na \, t \quad , \qquad (14)$$

$$\Delta_o = 8t \sin^2 K_F \, e^{-\frac{1}{\lambda}} \; , \; \lambda = |U|/[4t\pi \sin K_F]. \qquad (15)$$

Although eqs. (13) and (15) show some similarity, the agreement on the quantitative level is poor.

The wave function (12) leads to the following energy:

$$E_{\beta,\delta o} = E_o + |U| \frac{\theta}{\langle 1|1\rangle} + \frac{\beta^2 E_{\delta o} - |U| \frac{\theta}{\langle 1|1\rangle}}{1-\beta(1-\beta)\langle 1|1\rangle} \quad , \qquad (16)$$

which for $\beta=1$ recovers the H.F. excitation energy (6.a.). The double minimization with respect to β and δo is difficult to perform, because we don't know the precise value of constants C and α of the r.h.s. of Eq. (14). Instead we fix δo by assuming for the constant λ (7) the correct value (13) which is twice the H.F. one, Eq. (15). Minimization with respect to β yields then $\beta_{min} = \lambda^{2-\alpha} C$. For $\alpha=3/2$, it leads to a condensation energy of the form (14), but with δo replaced by $\delta = \sqrt{\lambda}\, C \delta o$. The corresponding renormalized gap then reads

$$\Delta = \sqrt{\lambda} C \Delta_o = 8tC \sin^2 K_F \sqrt{\lambda} \; e^{-\frac{1}{\lambda}} \quad , \qquad (17)$$

with λ as in Eq. (13). Equation (17) agrees with the correct result provided that $C = \sqrt{\frac{2}{\pi}} \approx 0.80$.

4. The Generalized Gutzwiller Wave Function

For small values of $|U|$, $|U|<t$, we describe the ground state of the model (1) by means of the generalized Gutzwiller wave function [3]

$$|\Psi\rangle = g^{\hat{D}}\ |\Psi_0\rangle \quad , \qquad g>1 \quad , \qquad (18)$$

where g is a variational parameter (for U>0 , 0<g<1), and $|\Psi_0\rangle$ is the wave function (12) for $\beta=C\lambda^{1/2}$. The corresponding linked cluster expansion of the energy contains, besides the usual linear and quadratic terms in $|U|$ [3], [7]-[9], small exponential terms. These terms, which are present in the exact energy for $|U|<1$, but are very difficult to evaluate, correspond to the condensation energy of the bound pairs. In analogy with the energy of reference [3], which in the case of the half filled band accounts for 98% of the exact Bethe ansatz result [6] for U=-4t, our linked cluster energy is reasonably good for the range of electronic densities considered in the present study.

Introducing

$$\hat{O}_j = c_{j\uparrow}c_{j\downarrow} \quad , \qquad (19)$$

we analyzed the on-site Cooper pair-Cooper pair correlation function

$$S_{ij} = \frac{1}{2}\ \langle\left\{\hat{O}_i^{+},\hat{O}_j\right\}\rangle \qquad (20)$$

for small values of $|U|$. As expected, with increasing $|U|$ the fluctuations for on-site pairing are enhanced for all densities. The same result is obtained for the charge fluctuations. In contrast, the spin fluctuations are reduced with increasing $|U|$.

The wave function (18) leads to a gap in the spin excitation spectrum, which, for $|U|<<t$, is given by (17). A detailed version of these results will be presented elsewhere.

5. Conclusions

Exact results, obtained by Bethe ansatz techniques, are often difficult to interpret, particularly in the weak coupling limit $|U|<<t$ [4]-[6], [10],[11].

Our variational scheme, besides allowing the calculation of correlation functions, has the advantage of giving a new insight into the origin of the gap in the spin excitations of the model (1). In fact, unlike the paramagnetic state ($\beta=0$) and H.F. state ($\beta=1$), the wave function (12) ($\beta=C\lambda^{1/2}$) introduces quantum fluctuations through the matrix element (11), which lead to a lowering of the energy with respect to the $\beta=0,1$ cases. This is achieved by a renormalization of the coupling constant λ, which, due to the quantum fluctuations, is twice the one obtained for the H.F. symmetric point $\beta=1$. The H.F. approximation leads to a bad gap equation because, the corresponding energy, Eq.(16), does not depend on the matrix element θ, which introduces the quantum fluctuations.

Acknowledgement. This work was supported by the Alexander von Humboldt-Stiftung.

References

1. A.I. Larkin, J. Sak: Phys. Rev. Lett. 39, 1025 (1977)
2. R. Micnas, J. Ranninger, S. Robaszkiewcz: Rev. Mod. Phys. 62, 113 (1990)
3. J. Carmelo, R. Micnas: J. Phys.: Condens. Matter 2, 6981 (1990)
4. V.Y. Krivnov, A.A. Ovchinnikov: Zh. Eksp. Teor. Fiz. 67, 1568 (1974) (Sov. Phys. -JETP 40, 781 (1975))
5. M.A. Gusmão: Phys. Rev. B 35, 1682 (1987)
6. F. Woynarowich: J. Phys. C 16, 6593 (1983)
7. D. Baeriswyl, J. Carmelo, A. Luther: Phys. Rev. B 33, 7247 (1986); 34, 8976 (1986)
8. D. Baeriswyl, J. Carmelo, K. Maki: Synth. Metals 21, 271 (1987)
9. J. Carmelo, M. Dzierzawa, X. Zotos, D. Baeriswyl: submitted to Phys. Rev. B (1990); J. Carmelo, D. Baeriswyl: Physica C 153, 1281 (1988)
10. J. Carmelo, D. Baeriswyl: Phys. Rev. B 37 7541 (1988); J. Mod. Phys. B 1, 1013 (1988)
11. J. Carmelo, A.A. Ovchinnikov: submitted to Phys. Rev. Lett. (1990); in the proceedings of the NATO ASI "Theory of High T_c Superconductivity", P.W. Anderson, T.M. Rice, B. Doucot, eds., (1990)

The 2-D Hubbard-Peierls Model in the Self-Interaction Corrected Density Functional Formalism

J.A. Majewski [1], *P. Vogl* [2], *and A. Svane* [3]

[1] Institut für Theoretische Physik, Universität Graz, Universitätsplatz 5, A-8010 Graz, Austria
[2] Physik Department, TU München, D-8046 Garching, Fed. Rep. of Germany
[3] Physics Department, University of Aarhus, Denmark

Abstract. We present results for correlation functions, the energy gap, and the phonon-induced electronic energy gain in the half-filled two-dimensional Hubbard-Peierls model. We find that the self-interaction-corrected local spin-density functional theory (SIC-LSD) gives excellent agreement with exact results on small lattices. Noticeably, we find that lattice distortions destabilize the antiferromagnetic ground state in 2-D, in contrast to the situation in one dimension.

1. Introduction

Virtually all quantitative electronic structure calculations in solids are based on the local density (LDA) or local spin density (LSD) approximation. In spite of its impressive history of successes, this technique fails in a number of situations. It grossly underestimates the energy gaps of semiconductors and the stability of antiferromagnetic phases; this failure became particularly apparent in the insulating phase of high-temperature superconductors such as La_2CuO_4 and $YBa_2Cu_3O_6$ (for a review, see [1]). Presently there is an intense search for systematic improvements of electronic structure calculations [2].

2. Self-interaction corrected density functional formalism

One possibility to go beyond LDA or LSD was pointed out already several years ago [3] and is based on the following observation.

In local density theory, the total energy functional is a sum of the kinetic energy, the external interaction, the electronic Hartree U_C and the approximate exchange-correlation interaction E_{xc}^{LSD}. The Hartree potential U_C includes an unphysical self-Coulomb energy for each electron in the system which is exactly cancelled only by the exact – albeit unknown – self-exchange-correlation energy. The crucial point is that this cancellation is incomplete in the LSD approximation. Consequently, every electron moves in a potential of all the other electrons plus a residual and artificial self-repulsion. This effect raises the filled valence bands with respect to the empty conduction bands, since there is no self-repulsion for the empty states.

By appending to the total energy functional another term, one can remove a large part of this residual self-interaction. One subtracts the self-Coulomb and self-exchange-correlation energy of each occupied orbital. Thus one forms an orbital-dependent self-interaction corrected total energy functional E_{tot}^{SIC}:

Springer Series in Solid-State Sciences, Vol. 99
Electronic Properties of High-T_c Superconductors and Related Compounds
Editors: H. Kuzmany · M. Mehring · J. Fink

$$E_{tot}^{SIC} = E_{tot}^{LSD} - \sum_{\nu\sigma}^{occ} (U_C[n_{\nu\sigma}] + E_{xc}^{LSD}[n_{\nu\sigma}, 0]) \quad .$$

In the limit of infinitely separated atoms, the conventional LSD formalism is known to fail [4]. In contrast, the self-interaction corrected local spin density theory (SIC-LSD) correctly captures the localized nature of the ground state of isolated atoms. In the framework of the Hubbard model, this limit corresponds to $U \to \infty$. It is for this reason, that SIC-LSD can be considered as treating correlations more accurately than conventional LSD.

The presence of the individual orbital densities $n_{\nu\sigma}(r)$ in the SIC functional means that – unlike the LSD Hamiltonian – it is not invariant under a unitary transformation among the occupied orbitals. Whereas the SIC correction is zero for plane waves or delocalized orbitals, for example, it is non-zero for Wannier functions.

How can one find the "correct" orbitals? The SIC total energy has been shown to be a functional of the density and to take on a minimum for the exact density n(r) [5]. Therefore, one attempts to find a set of orthogonal localized orbitals $\psi_{\nu\sigma}$ (ν=1,..N_{el}) which minimize the SIC total energy. This leads to the following set of equations [2,6]:

$$\left(H_{0\sigma}^{LSD} - \int dr' \, n_{\nu\sigma}(r') \, |r - r'|^{-1} + \frac{\delta E_{xc}}{\delta n_\sigma}(n_{\nu\sigma},0)\right)\psi_{\nu\sigma}(r) = \sum_{\nu'} \lambda_{\nu'\nu}^{\sigma} \, \psi_{\nu'\sigma}(r). \quad (1)$$

$$\lambda_{\nu'\nu}^{\sigma} = \langle \psi_{\nu\sigma} | H_{\nu'\sigma} | \psi_{\nu'\sigma} \rangle = (\lambda_{\nu\nu'}^{\sigma})^* \quad . \quad (2)$$

The Lagrangian multipliers λ cannot be diagonalized straightforwardly, since the l.h.s of Eq. (1) depends on the individual state ν. Equations (1)-(2) can be solved by a suitable minimization technique such as the steepest descent method.

3. Tests and applications of SIC-LSD: The 2-D Hubbard-Peierls model

The discovery of high temperature superconductors has revived the theoretical interest in the Hubbard model – particularly in two dimensions – as a simple yet non-trivial model for highly correlated systems (see, e. g., [7]). Consequently, this model may provide an interesting and stringent test for SIC-LSD.

In this paper we have performed SIC-LSD calculations for the two-dimensional Hubbard-Peierls model, which incorporates both electron-electron and electron-phonon effects. We compare the results with exact numerical solutions for small finite lattices. This work generalizes earlier SIC-LSD studies on the Hubbard model with no phonon effects [5]. The Hubbard-Peierls model is characterized by the following Hamiltonian:

$$H_{el} = -\sum_{ij\sigma} [t_0 + \frac{\alpha}{t_0}(\mathbf{u}_i - \mathbf{u}_j)\cdot\mathbf{r}_{ij}] \, c^{\dagger}_{i\sigma} c_{j\sigma} + \frac{1}{2} U \sum_{i\sigma} \hat{n}_{i\sigma} \hat{n}_{i-\sigma} \, . \quad (3)$$

The electron operators $\hat{n} = c^{\dagger}c$ are labeled by a site-index i and spin σ; we take into account one s-state per site. U is the intra-atomic Coulomb repulsion parameter, t_0 the nearest-neighbor hopping matrix element with all atoms arranged in an ideal two-dimensional square lattice. A distortion or phonon displacement $\{\mathbf{u}_i\}$ changes the one-electron energy by an amount linearly proportional to the electron-phonon coupling constant α and to the change in distance between site i and j. The vector $\mathbf{r}_{ij}$ is a unit vector pointing from site i to site j.

For an infinite lattice, no solutions to this model are known. For finite systems, only a few effects of phonon distortions on the electronic energy have been studied so far [8].

We have solved the model given by Eq. (3) for half filled bands both numerically exactly for small lattices, and by employing a discrete version of LSD and SIC-LSD in analogy to [5]. Within SIC-LSD, much larger lattices can be considered than in Monte-Carlo simulations or exact diagonalizations.

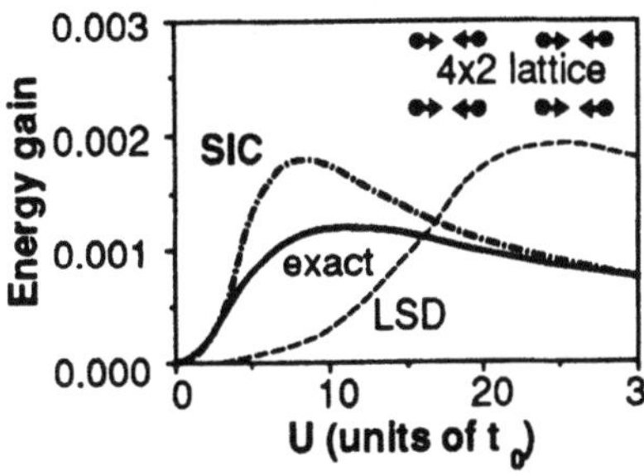

Fig. 1. Electronic energy gain as a function of the on-site Coulomb repulsion U for the phonon displacement shown in the insert.

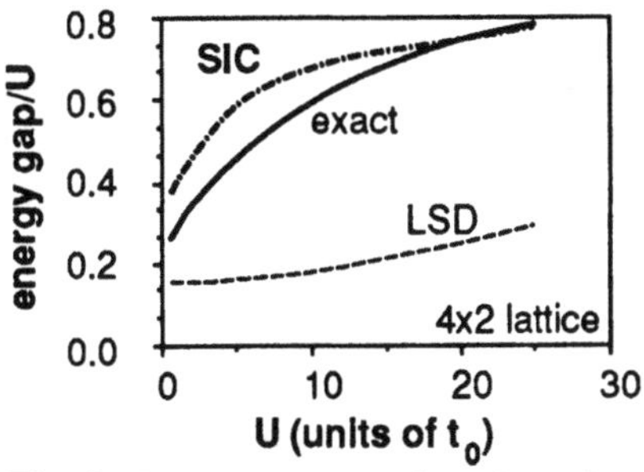

Fig. 2. Energy gap as a function of U.

In Fig. 1, we compare the electronic energy gain $E_{tot}(U,0) - E_{tot}(U,\delta)$ due to a $(\pi,0)$-phonon as a function of U for a 4×2 lattice. The change in the hopping matrix element is chosen to be $\delta t/t_0 = \alpha\,(u_{xi} - u_{xj}) = \pm 0.1$. As can be seen, LSD grossly underestimates the distortion energy gained by the correlated electrons even for small U, in contrast to SIC-LSD. We find similar good agreement between SIC-LSD and exact results for all phonon branches. The energy gap, defined by $[E_{tot}(N+1) - E_{tot}(N)] - [E_{tot}(N) - E_{tot}(N-1)]$, with $E_{tot}(N)$ denoting the ground state energy for N electrons, also agrees very well with the exact results (Fig. 2). Fig. 3 depicts a typical charge correlation function, namely the average on-site double occupancy. Even though SIC gives much improved results compared to LSD, this figure does exhibit a failure of both SIC and LSD-methods: they yield a transition to a symmetry broken Néel ground state at some finite U and therefore do not completely reproduce the strong quantum fluctuations characteristic of low dimensional systems.

The physically most interesting result of the present study is shown in Fig. 4. We have calculated the energy gap of the system as a function of a (π,π)-phonon distortion for given $U = 2\,t_0$. In a 1-dimensional Hubbard-Peierls chain, a zone-

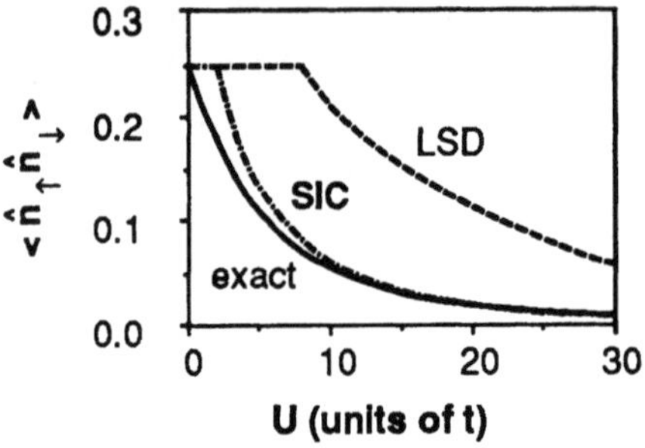

Fig. 3. Mean double occupancy of an atomic site as a function of U.

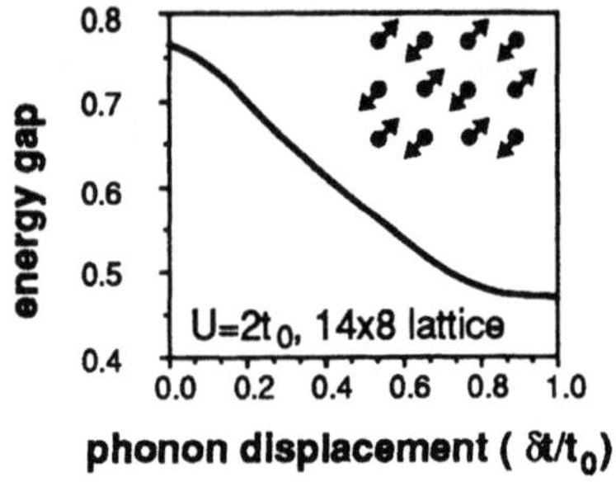

Fig. 4. Predicted energy gap for the 2-D Hubbard model by SIC-LSD, as a function of the phonon displacement.

boundary phonon induces an energy gap which is enhanced by the electron-electron interaction U. In other words, in 1-D, both the electron-phonon and the electron-electron interaction tend to stabilize an insulating ground state by enhancing the energy gap. In two dimensions, we find the contrary to be the case: **the 2-D Hubbard-Peierls model exhibits a negative U behavior.** Whereas we find any finite U produces an energy gap, any phonon distortion in the present system reduces it. Physically, a phonon distortion leads to a higher double occupancy in the short bonds which in turn corresponds to a smaller value of U.

Thus, in two dimensions, phonon distortions destabilize an antiferromagnetic ground state, in contrast to the situation in one dimension.

This work was supported by the Fonds zur Förderung der wissenschaftlichen Forschung in Österreich.

References

[1] K. C. Hass, Solid State Physics **42**, 213 (1989).
[2] R. O. Jones and O. Gunnarsson, Jones, Rev. Mod. Phys. **61**, 689 (1989).
[3] J. P. Perdew and A. Zunger, Phys. Rev. B **23**, 5048 (1981).
[4] J. P. Perdew, R. G. Parr, M. Levy, and J. L. Balduz, Phys. Rev. Lett. **49**, 1691 (1982).
[5] A. Svane and O. Gunnarson, Europhys. Lett. **7**, 171 (1988); Phys. Rev. B **37**, 9919 (1988); and private communication.
[6] R. A. Heaton, J. G. Harrison, and C. C. Lin, Phys. Rev. B **28**, 5992 (1983).
[7] S. R. White, D. J. Scalapino, R.L. Sugar, E.Y. Loh, J.E. Gubernatis, R.T. Scalettar, Phys. Rev. B **40**, 506 (1989).
[8] S. Tang and J. E. Hirsch, Phys. Rev. B **37**, 9546 (1988); ibid. B 39, 12327 (1989); S. Mazumdar, Phys. Rev. B **39**, 12324 (1989).

Observable Anomalies in Cuprate Superconductors Related to a Cross-Over from BCS Pairing to Bose-Einstein Condensation

D. van der Marel

Max-Planck-Institut für Festkörperforschung, Heisenbergstr. 1
D-7000 Stuttgart 80, Fed. Rep. of Germany
On leave from Delft University of Technology, 2628 CJ Delft, The Netherlands (from January 1, 1990 to December 31, 1990)

Abstract. The excitation spectrum of high T_c superconductors is discussed. It is pointed out that the cross-over from Bose-Einstein condensation to BCS superconductivity is characterized by a change of character of the lowest excitations of the system: Bosonic collective excitations in the former case and fermionic pair-breaking excitations in the latter case. Depending on which type of excitation dominates, the phase transition to the normal state can be regarded as Bose-Einstein condensation or as BCS superconductivity. An intermediate situation where both types of excitations contribute to the superconductor to normal transition is also possible. In both extreme cases there is a discontinuity at T_c of the chemical potential, although the detailed behaviour is quite different. It is shown, from numerical calculations, that the size of the anomaly is in a range where it can actually be observed experimentally. In the cuprate superconductors two factors are responsible for the presence of low lying plasmon branches: The low density of charge carriers, and the layered structure, resulting in an accoustic branch of charge density oscillations. In principle this branch can be responsible for a transition to the normal state, in which a finite fraction of the electrons is in a (non-condensed) paired state above T_c.

Discussion. The question whether there exists a smooth cross-over between BCS superconductivity and Bose-Einstein condensation has a long history. Already in the early days of BCS theory it was understood that no qualitative difference exists between the wavefunctions of condensed pre-existing Cooper pairs and the BCS ground-state wavefunction [1], *i.e.* the BCS ground-state wavefunction for a given phase ϕ is

$$|\psi_{BCS}(\phi)>=\prod_k(u_k+v_k e^{i\phi}c^\dagger_{k\uparrow}c^\dagger_{-k\downarrow})|\psi_{vac}> \ . \qquad (1)$$

Due to the degeneracy with respect to the choice of phase one is free to take a linear combination of all $|\psi_{BCS}(\phi)>$, resulting in a wavefunction[1] with fixed particle number $2N$:

$$\frac{1}{2\pi}\int e^{-iN\phi}|\psi_{BCS}(\phi)>d\phi=(C\sum_k\frac{v_k}{u_k}c^\dagger_{k\uparrow}c^\dagger_{-k\downarrow})^N|\psi_{vac}>\equiv(a_0^\dagger)^N|\psi_{vac}> \ . \qquad (2)$$

In the extreme dilute limit $a_0^\dagger$ obeys Bose commutation rules and Eq.(2) can be interpreted as a boson condensate wavefunction of Cooper pairs. In a number of papers it has been pointed out that for extremely dilute superconductors

Electronic Properties of High-T_c Superconductors and Related Compounds
Editors: H. Kuzmany · M. Mehring · J. Fink

there exists a special point, where the chemical potential of the ground-state wavefunction at $T = 0$ drops below the bottom of the conduction band, resulting in an enhanced binding of the Cooper pairs[2,3]. It was suggested by some of these authors[2] that this point may approximately mark the border between the BCS limit and the Bose-Einstein limit, keeping in mind that for the ground-state wavefunction at $T = 0$ there exists no sharp transition between the two limits. A better criterion can be obtained by considering the nature of the excitations which are responsible for the superconducting to normal transition, *i.e.* are they bosons (like in He^4), fermions (like in classical BCS superconductors), both, or neither of these? The answer to this question may turn out to be essential for the understanding of the superconducting transition in the cuprate oxides.

The obvious candidate for electronic degrees of freedom with boson character is the Goldstone phase fluctuation mode treated by Bogoliubov and Anderson[4]. As has been shown by Anderson, only for neutral superconductors does this mode occur below the pair-breaking gap. Because of phase/number-density conjugation, phase fluctuations result directly in density fluctuations. As a result the Bogoliubov-Anderson mode of a neutral superconductor has an acoustic dispersion for long wavelengths. In a 3D superconductor with long range Coulomb forces the Bogoliubov-Anderson mode shifts to the plasma frequency. Only in the extreme dilute limit may thermal excitation of plasmons therefore be relevant. Such conditions do however not occur in the cuprate superconductors, where the zone center bulk plasmon is situated between 1 and 2 eV. A recent claim[4d] that the charge degrees of freedom can be decoupled from the phase fluctuations is in conflict with this notion, perhaps due to an implicit assumption of screening of the longe range interactions in Ref. 4d.

At first glance one would therefore conclude that in the cuprate superconductors there is no low lying boson branch competing with the pair-breaking quasi-particle excitations across the superconducting gap. However, we must also consider the layered structure of the cuprates, with a tubular Fermi surface along the z-axis perpendicular to the planes. This shape of the Fermi surface is characteristic of an insulator-metal superlattice, of which it is known[5] that the plasmon dispersion deviates dramatically from the dispersion of an isotropic 3D free electron system: The dispersion starts at the bulk plasma frequency at the center of the Brillouin zone, curves to higher frequencies for $\boldsymbol{k}$ parallel to the planes, but curves downward for $\boldsymbol{k}$ perpendicular to the planes. For $\boldsymbol{k}_z \neq 0$ the energy approaches zero linearly as a function of $\boldsymbol{k}_{//}$, which is reminiscent of an acoustic plasmon. In order to make a theoretical estimate of the relevance of these modes for breaking up the superconducting state at finite T a full solution is required of the three coupled equations describing collective excitations in the electron-electron, electron-hole and hole-hole channels[4], using the plasmon dispersion of a quasi-2D layered structure. To my knowledge this has not been done so far.

At this stage it is only possible to sketch some experimental consequences for the case where the superconducting-to-normal transition is dominated by thermal excitations of the boson branch: In the first place the character of the transition would be non-pair breaking, *i.e.* there is still a finite fraction of paired electrons above T_c. Such pairs are similar to the bipolarons investigated by Robaszkiewitz *et al.*[6]. At higher temperatures these bipolarons break apart. Hence a mixture of bipolarons and unpaired electrons would exist at

$T > T_c$. As a result a precurser gap should exist in this temperature range, characterized by the binding energy of a pair.

Below T_c the gap would be almost temperature independent, but it fills up with thermally activated bosonic excitations. As the lowest of these excitations occur at the edge of the Brillouin zone, they have no influence on *e.g.* infrared spectra due to k-conservation in the optical selection rules. In practice collective modes inside the gap must show up in the infrared spectra of samples which don't have 100% translational symmetry due to disorder. Direct probes of the single particle spectral function such as photo-electron spectroscopy display the pair-breaking gap and are insensitive to the density of states connected to the collective modes, independent of the degree of disorder.
Another fingerprint of the nature of the superconductor-to-normal transition is the temperature dependence of the chemical potential[3], which can be determined experimentally by measuring the changes in workfunction near T_c. In the simplest of approaches to Bose-Einstein condensation one calculates the number of bosons in excited states by multiplying each state with the Bose-Einstein factor $\{\exp[(E_k - \mu)/k_B T] - 1\}^{-1}$. We see that in order to avoid diverging occupancies μ has to be below the bottom of the boson band, which is at zero energy. In fact μ is pinned to the band bottom, resulting in the macroscopic occupation of the Cooper pair state. On increasing the temperature μ stays at this value, until the point is reached where all bosons are in an excited state. Above this temperature the chemical potential is smaller than zero and is determined by the condition of number conservation:

$$N = \sum_k \frac{1}{\exp \frac{E_k - \mu}{k_B T} - 1} . \tag{3}$$

It is straightforward to calculate the behaviour of the chemical potential near T_c from this formula. For a density of states proportional to ϵ^α at low energies, the chemical potential has approximately the following temperature dependence:

$$\mu \propto -(\tau^{\alpha+1} - 1)^{1/\alpha} , \tag{4}$$

where $\tau = T/T_c$ is the reduced temperature above T_c. The exponent α depends on the dispersion of the boson band and on the dimensionality. In 3D a quadratic dispersion corresponds to $\alpha = 0.5$, and a linear dispersion corresponds to $\alpha = 2$. One can also consider the intermediate situation, where the dispersion is linear along one or two axes and quadratic along the remaining axes, corresponding to $\alpha = 1$ and $\alpha = 1.5$. In all cases a discontinuity in μ occurs at T_c, the strength of which depends on α. For higher values of τ the curves bend down with a $(\tau - 1)^{(1+1/\alpha)}$ power law. From Eq. (3) one can calculate universal plots of μ/T_c versus τ for each value of α, which are displayed in Fig. 1. We see that for $\alpha < 1$ there is only a discontinuity in the second derivative of μ. For higher values of α a kink-like feature devolops, which is most pronounced for $\alpha = 2$, *i.e.* for an acoustic dispersion.

In the opposite limit of a mean field BCS superconducting transition no critical behaviour in μ occurs near T_c except if the density of the electron gas is very low, *i.e.* if the Fermi energy is smaller than the 'Debye' cutoff energy of the electron-electron interaction. In Ref. 3 it was shown that a kink in the chemical potential appears at T_c, with the approximate relation

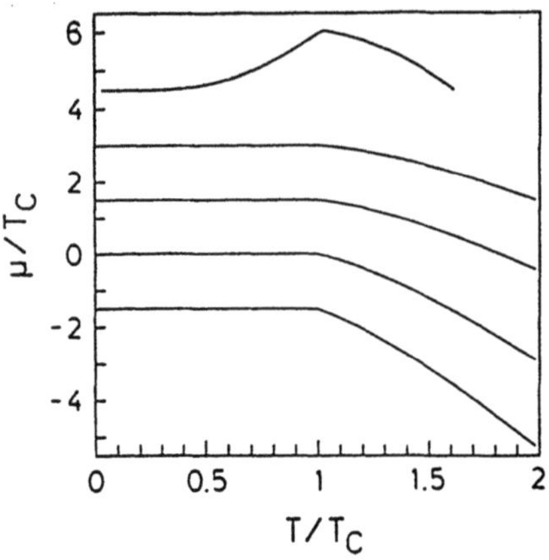

Fig. 1. Chemical potential versus reduced temperature. Top curve: A low density BCS superconductor with $E_F = 0.1eV$ and $T_c = 123.7K$, see Ref. 3 for details. The vertical scale of this curve is enlarged 20 times. Lowest four curves: universal plots for Bose-Einstein behaviour with $\alpha = 0.5$, $\alpha = 1$, $\alpha = 1.5$ and $\alpha = 2$ from top to bottom. All curves have been given arbitrary vertical offsets for clarity.

$$\mu(T) = \mu_n(T) - \frac{\Delta_0^2(1 - T/T_c)}{4\mu_n(T)} , \tag{5}$$

where μ_n is the normal state Fermi energy and Δ is the gap at $T = 0$. In Fig. 1 we compare this behaviour to the behaviour obtained from the above examples for Bose-Einstein 'evaporation'. At this stage one can only speculate on the behaviour of μ in the intermediate region where both pair breaking quasiparticle excitations and collective modes contribute equally to the superconductor-to-normal transition. As a discontinuity occurs in both limiting cases, one might expect critical behaviour of μ in the intermediate regime as well. It is interesting to compare these plots with the curves given in a recent paper by Schmitt-Rink *et al.*[7], where they used the free energy expression of Nozières and Schmitt-Rink[2c] to interpolate between the Bose-Einstein limit and the BCS limit. The results of Schmitt-Rink *et al.* give a maximum for μ near T_c without any discontinuities. The authors pointed out that their results are a good approximation for T well above T_c only, which may explain the discrepancy with Fig. 1. Clearly an experimental investigation of the behaviour of μ would shed further light on these questions. Preliminary experimental results[8] seem to favour a Bose-Einstein type of behaviour with $\alpha = 2$, but we have to await further confirmation before definite conclusions can be drawn.

Acknowledgements. The author gratefully acknowledges Prof. J. Ashkenazi for his encouragement and patience in responding to numerous lengthy e-mail discussions on this subject.

References

1. G.Rickayzen, "Theory of Superconductivity", edited by R.E.Marshak, (Interscience Publishers, New York, 1965).

2. D.M.Eagles, Phys. Rev. 186, 456 (1969); A.J.Legget,in "Modern Trends in the Theory of Condensed Matter", edited by A.Pekalski and J.Przystawa (Springer-Verlag, Berlin, 1980); P.Nozieres and S.Schmitt-Rink, J. Low Temp. Phys. 59, 195 (1985); M.Randeria, J.M.Duan and L.Y.Shieh, Phys. Rev. Lett. 62, 981 (1989).

3. D.van der Marel, Physica C165, 35 (1990).

4. N.N.Bogoliubov, Soviet Phys. JETP 34, 41 (1958); P.W. Anderson, Phys. Rev. 112, 1900(1958); T. Kostyrko, Phys. Stat. Sol. b143, 149(1987); S.H.Liu, proceedings of the third annual conference on superconductivity and applications, Buffalo, New York, September 19-21, (1989).

5. H.Morawitz, these proceedings; V.Z.Kresin and H.Morawitz, Phys. Rev. B37, 7854 (1988).

6. R.Micnas, S.Robaszkiewitz and J. Ranninger, to appear in Rev. Mod. Phys. (January 1990).

7. S.Schmitt-Rink, C.M.Varma and A.E. Ruckenstein, Phys. Rev. Lett. 63, 445 (1989).

8. G. Rietveld, M. Biglari, M. van Veenendaal and D. van der Marel, unpublished results (1989).

Part VII

Related Topics

Structural Aspects of Organic Superconductors

D. Schweitzer

3. Physikalisches Institut der Universität Stuttgart, Pfaffenwaldring 57, D-7000 Stuttgart 80, Fed. Rep. of Germany

Abstract. Some physical properties of the organic superconductors with the highest transition temperatures are presented. In addition the influence of small structural changes on the superconducting transition temperature is discussed. Finally bulk superconductivity in polycrystalline pressed pellets of organic metals is reported.

1. Introduction

In 1980 superconductivity in an organic metal was observed for the first time [1]. Today about 35 different organic metals are known which become superconducting under pressure or ambient pressure. The organic superconductors with the highest transition temperatures are all radical salts of the donor bis(ethylenedithiolo)tetrathiofulvalene (BEDT—TTF), which are at ambient pressure $(BEDT-TTF)_2Cu(NCS)_2$ (T_c= 10.4 K) [2] and α_t-$(BEDT-TTF)_2I_3$ (T_c= 8 K) [3] and under isotropic pressure β_H-$(BEDT-TTF)_2I_3$ (0.5 kbar, T_c= 7.5 K) [4,5].

In the following some of the structural and physical properties of single crystals of the above-mentioned organic metals are discussed briefly. In the second part a few new results and observations such as bulk superconductivity in polycrystalline pressed samples of the same materials are presented. The latter observation is remarkable since it shows that organic superconductors can be used in principle also for applications such as the preparation of electronic devices (SQUIDS) and might even be suitable for the preparation of superconducting cables. In addition it indicates that the observation of superconductivity in organic polymers should be possible as well.

2. Properties of Single Crystals

$(BEDT-TTF)_2Cu(NCS)_2$ is a typical quasi two-dimensional organic metal. The two-dimensional behaviour is due to conducting sheets of dimerised BEDT-TTF molecules which are separated

Springer Series in Solid-State Sciences, Vol. 99
Electronic Properties of High-T_c Superconductors and Related Compounds
Editors: H. Kuzmany · M. Mehring · J. Fink

by non-conducting anion layers. In the BEDT-TTF sheets the dimers are nearly orthogonally arranged with respect to each other [2] and the dimers do not form the usual face to face piling along the stacking axis. The latter property increases the two-dimensional character of the molecular network. An important property of $(BEDT-TTF)_2Cu(NCS)_2$ is that the crystals show - compared to other organic superconductors - a relatively sharp transition into the superconducting state. This is observed not only in resistivity [2,6] but also in the ac-susceptibility measurement [6] and in the Meissner effect [2,7]. The two latter methods have shown that superconductivity in $(BEDT-TTF)_2Cu(NCS)_2$ is a nearly 100 % bulk effect. Measurements of the upper critical fields have demonstrated the typical quasi two-dimensional character of the crystals [8,9]. It could be shown that at temperatures below 4 K the superconducting properties can be described best in the picture of layered two-dimensional superconductors as first reported by Klemm et al. [10]. This picture of layered two- dimensional superconductors is valid for the high T_c copper oxide superconductors as well and demonstrates the similarities of the structural principles of both classes of materials. These similarities will be discussed in the following in more detail for the example of the organic superconductor α_t-$(BEDT-TTF)_2I_3$.

Crystals of α_t-$(BEDT-TTF)_2I_3$ are quasi two-dimensional organic metals as well which have a stable superconducting state at 8 K and ambient pressure [3]. The crystals can be prepared by annealing crystals of α-$(BEDT-TTF)_2I_3$ above 70° C for several days, resulting in a structural phase transition. The structure of α_t-$(BEDT-TTF)_2I_3$ is very similar to that of crystals of β-$(BEDT-TTF)_2I_3$ as was shown by several spectroscopic methods [3,11,12]. Nevertheless, the exact structure is not known yet due to the fact that after the structural transformation the mechanically still very stable crystals have a mosaic type of architecture so that a detailed crystal structure investigation by X-ray has not been possible up to now. The mosaic-type arrangement is a result of strong stress during the transformation because the structures of α-$(BEDT-TTF)_2I_3$ and β-$(BEDT-TTF)_2I_3$ are quite different from each other. Nevertheless, the unit cell data of β-$(BEDT-TTF)_2I_3$ and α_t-$(BEDT-TTF)_2I_3$ and the arrangement of the molecules are the same. On the other hand the structure of α_t-$(BEDT-TTF)_2I_3$ cannot be exactly the same as in β-$(BEDT-TTF)_2I_3$ (or more correctly as in β_H-$(BEDT-TTF)_2I_3$), since β_H-$(BEDT-TTF)_2I_3$ has only a metastable superconducting state at 8 K while crystals of α_t-$(BEDT-TTF)_2I_3$ have a stable superconducting state at this temperature. The difference in the two structures exists probably only in the ordering of the terminal ethylene groups of the BEDT-TTF molecules. While in β_H-(BEDT-TTF)$_2$I$_3$ all the terminal ethylene groups have the so-called staggered form [13], in the crystals of α_t-(BEDT-TTF)$_2$I$_3$ these groups might have the so-called eclipsed form or even more plausibly they might be ordered as in the α-phase, where both forms of ordering exist in neighbouring stacks [14]. That means that very small changes in the structure such as a different ordering of the terminal ethylene groups might be responsible for the fact that the material has a stable or meta-stable superconducting state.

From the measurements of the upper critical fields of crystals of α_t-(BEDT-TTF)$_2$I$_3$ it can clearly be seen that for the region $T/T_c > 0.65$ ($T_c = 8$ K) the crystals behave as isotropic three-dimensional superconductors, since for the magnetic field perpendicular and parallel to the c*-axes the upper critical fields are identical. In the region $0.45 < T/T_c < 0.65$ an anisotropic three-dimensional behaviour is found while for $T/T_c < 0.45$ the material can be described in the picture of typical two-dimensional layered superconductors [10]. By applying the anisotropic effective mass model of the Ginzburg-Landau (GL) theory the coherence lengths $\xi_{\parallel}(0) \simeq 120$ Å and $\xi_{\perp}(0) \simeq 20$ Å and the London penetration depths $\lambda_{\parallel}(0) = 4300$ Å and $\lambda_{\perp}(0) = 750$ Å were obtained. However, since the anisotropic GL theory is only valid sufficiently close to T_c, these values can give only the right order of magnitude. Nevertheless, the value of $\xi_{\perp} \simeq 20$ Å is comparable to the unit cell dimension in the c-direction. In addition ^{13}C-NMR investigations [12,15] indicate that $\xi_{\perp}$ might even be only between 6 and 10 Å since it was shown that the spin density (and therefore to some degree the charge density) is mainly located on the middle part of the BEDT-TTF molecules. This means that the thickness of the superconducting layer is of the order of the length of the BEDT-TTF molecules, supporting the picture of the Josephson coupled two-dimensional layered superconductors as described by Klemm et al. [10].

As pointed out before, a close similarity exists between the organic superconductors and the high-T_c copper oxides. In both materials short coherence lengths of the order of the size of the unit cell and large London penetration depths exist. These properties are due to the typical quasi two-dimensional electronic conditions and structural arrangement. Certainly, in the case of the organic superconductors minor changes and differences in the structure, for instance a different ordering of the terminal ethylene groups of the BEDT–TTF molecules or a different symmetry of the anions, might strongly influence the superconducting transition temperature. On the other hand such small variations in the structure of the organic materials open up the chance to obtain a larger number of materials with different transition temperatures into a superconducting state and variation of these structural parameters might be a way of obtaining higher transition temperatures in the case of the organic materials as well.

Nevertheless, a major difference exists between samples of the high T_c copper oxide materials and the organic superconductors. The former are usually polycrystalline or granular and single crystals are difficult to obtain; in the latter case all physical investigations were done on single crystals. The electrochemical preparation of these organic metals results in single crystals suitable for physical investigations. But since the organic superconductors have very short coherence lengths as well, it should be possible to prepare pressed samples from polycrystalline organic metals - as in the case of the "ceramics" - which should exhibit bulk superconductivity.

3. Properties of Polycrystalline Pressed Pellets of Organic Metals

Polycrystalline pressed samples of size 4x1x0.5 mm^3 were prepared from grained single crystals of the organic metals α_t- and β-(BEDT-TTF)$_2I_3$ by applying a pressure of $(3\text{-}10)\cdot 10^3$ kg/cm^2 to the powder. The samples thus prepared are mechanically very stable and the temperature dependence of the resistivity can be measured easily by the usual four point method.

In the case of pressed samples prepared directly from α_t-(BEDT-TTF)$_2I_3$ no bulk superconductivity could be found. On the other hand in these samples a metal-like behaviour was observed over the whole temperature range between 300 and 1.3 K. This fact is at least worth noticing since usually in polycrystalline pressed pellets of organic conductors only a semiconducting behaviour is found (this is specially valid for quasi one-dimensional materials).

Since from the single crystal investigations it was known that small changes in the structure might suppress the superconductivity, it was assumed that during the preparation of the polycrystalline pellets of α_t-(BEDT-TTF)$_2I_3$ under pressure phase transitions might occur. Therefore, newly prepared pellets of α_t-(BEDT-TTF)$_2I_3$ were annealed for 3 days at 75^0 C before the samples were provided with contacts and measured. After this procedure a superconducting transition could be observed, but compared with the crystal of α_t-(BEDT-TTF)$_2I_3$ the superconducting transition in the polycrystalline sample is much broader [16]. Zero resistivity is found at 2.2 K and the middle of the transition appears at about 5.5 K, while in the case of the crystals of α_t-(BEDT-TTF)$_2$I$_3$, zero resistivity appears already at 6 K and the middle of the transition occurs at 8 K [3].

The most surprising observation was made by measuring the temperature dependence of the resistivity of polycrystalline pressed samples of β-(BEDT-TTF)$_2$I$_3$ (in the following called β_p-(BEDT-TTF)$_2$I$_3$) [17]. Without annealing the samples showed an onset to superconductivity at 9K, zero resistivity at 3.2 K and the middle of the resistive transition at 7.5K [17]. This observation is surprising because single crystals of β-(BEDT-TTF)$_2$I$_3$ show a rather sharp stable superconducting transition but only at 1.4 K [18] and a metastable superconducting state at 8 K [19, 20]. Here in the polycrystalline pressed pellets of β_p-(BEDT-TTF)$_2$I$_3$ the superconducting state at 7.5 K is stable and a bulk effect of the pellets [17].

The resistivity characteristics of the polycrystalline pressed pellets of β_p-(BEDT-TTF)$_2$I$_3$ [17] indicate that the detailed structure of β_p-(BEDT-TTF)$_2$I$_3$ is different to the one of the polycrystalline pressed α_t-(BEDT-TTF)$_2$I$_3$ and different to the single crystals of β-(BEDT-TTF)$_2$I$_3$, even if the unit cell data are identical and the molecular arrangement in the materials might be very similar. A possible difference might lay again in the ordering of the terminal ethylene groups of the BEDT—TTF molecules (we assume at the moment that all ethylene groups are ordered in the eclipsed form).

Acknowledgement

I would like to thank all my colleages and co-workers (see references) for very effective cooperation during all the investigations in recent years. My special thanks go to Professor H. J. Keller for many discussions and efficient teamwork.

References

1. D. Jérome, A. Mazaud, M. Ribault and K. Bechgaard, J. Phys. Lett. 4, L95 (1980).
2. H. Urayama, H. Yomochi, G. Saito, K. Nozava, T. Sugano, M. Kinoshito, S. Saito, K. Oshima, A. Kawamoto and J. Tanaki, Chem. Lett 1988, 5 (1988).

3. D. Schweitzer, P. Bele, H. Brunner, E. Gogu, U. Haeberlen, I. Hennig, T. Klutz, R. Swietlik and H. J. Keller, Z. Phys. B Condensed Matter 67, 489 (1987).
4. K. Murata, M. Tokumoto, H. Anzai, H. Bando, G. Saito, K. Jajimura and T. Ishiguro, J. Phys. Soc. Japan 54, 1236 (1985).
5. V.N. Laukhin, E.E. Kostynchenko, Yu.Y. Sushko, I.F. Shegolev and E. B. Yagubskii, Soviet Physics JETP Lett. 41, 81 (1985).
6. S. Gärtner, E. Gogu, I. Heinen, H. J. Keller, T. Klutz and D. Schweitzer, Solid State Comm. 65, 1531 (1988).
7. S. Klotz, J. S. Schilling, S. Gärtner and D. Schweitzer, Solid State Comm. 67, 1981 (1988).
8. K. Oshima, H. Urayama, H. Yanochi and G. Saito, J. Phys. Soc. Japan 57, 730 (1988).
9. E. Gogu, Ph.D. Thesis, Universität Heidelberg 1989.
10. R. A. Klemm, A. Luther and R. M. Beasley, Phys. Rev. B 12, 877 (1975).
11. G.O. Baram, L.I. Buravov, L.C. Degtariev, M.E. Kozlov, V.N. Laukhin, E. E. Laukhina, V. B. Drischenko, K. L. Pokhodnia, M. K. Scheinkmann, R. P. Shibaeva and E. B. Yagubskii, Sov. Phys. JETP Lett. 44, 2913 (1986).
12. I. Hennig, U. Haeberlen, I. Heinen, D. Schweitzer and H.J. Keller; Physica C 153-155, 493 (1988).
13. A. J. Schultz, H. H. Wang, J. M. Williams and A. Filhol, J. Am. Chem. Soc. 108, 7853 (1986).
14. K. Bender, I. Hennig, D. Schweitzer, K. Dietz, H. Endres and H.J. Keller, Mol. Cryst. Liqu. Cryst. 108, 359 (1984).
15. D. Schweitzer, K. Polychroniadis, T. Klutz, H. J. Keller, I. Hennig, I. Heinen, U. Haeberlen, E. Gogu and S. Gärtner, Synth. Metals 27, A 465 (1988).
16. D. Schweitzer, S. Gärtner, H. Grimm, E. Gogu and H. J. Keller, Solid State Comm. 69, 843 (1989).
17. D. Schweitzer, E. Gogu, H. Grimm, S. Kahlich and H. J. Keller, Angew. Chem. Adv. Mater. 101, 977 (1989).
18. E. B. Yagubskii, I. F. Shegolev, V.N. Laukhin, P.A. Kononovich, M. V. Kartsovnik, A. V. Zwarykina, L. I. Biwavov, Sov. Phys. JETP Lett. 39, 12 (1984).
19. F. Creuzet, G. Creuzet, D. Jèrome, D. Schweitzer and H. J. Keller, J. Physique Lett. 46, L -1079 (1985).
20 F. Creuzet, D. Jèrome, D. Schweitzer and H. J. Keller, Europhys. Lett. 1, 461 (1986).

Index of Contributors

Springer Series in Solid-State Sciences

Editors: M. Cardona P. Fulde K. von Klitzing H.-J. Queisser

1 **Principles of Magnetic Resonance** 3rd Edition By C. P. Slichter

2 **Introduction to Solid-State Theory** 2nd Printing By O. Madelung

3 **Dynamical Scattering of X-Rays in Crystals** By Z. G. Pinsker

4 **Inelastic Electron Tunneling Spectroscopy** Editor: T. Wolfram

5 **Fundamentals of Crystal Growth I** Macroscopic Equilibrium and Transport Concepts. 2nd Printing By F. Rosenberger

6 **Magnetic Flux Structures in Superconductors** By R. P. Huebener

7 **Green's Functions in Quantum Physics** 2nd Edition By E. N. Economou

8 **Solitons and Condensed Matter Physics** 2nd Printing Editors: A. R. Bishop and T. Schneider

9 **Photoferroelectrics** By V. M. Fridkin

10 **Phonon Dispersion Relations in Insulators** By H. Bilz and W. Kress

11 **Electron Transport in Compound Semiconductors** By B. R. Nag

12 **The Physics of Elementary Excitations** By S. Nakajima, Y. Toyozawa, and R. Abe

13 **The Physics of Selenium and Tellurium** Editors: E. Gerlach and P. Grosse

14 **Magnetic Bubble Technology** 2nd Edition By A. H. Eschenfelder

15 **Modern Crystallography I** Symmetry of Crystals. Methods of Structural Crystallography By B. K. Vainshtein

16 **Organic Molecular Crystals** Their Electronic States. By E. A. Silinsh

17 **The Theory of Magnetism I** Statics and Dynamics. 2nd Printing By D. C. Mattis

18 **Relaxation of Elementary Excitations** Editors: R. Kubo and E. Hanamura

19 **Solitons.** Mathematical Methods for Physicists. 2nd Printing By G. Eilenberger

20 **Theory of Nonlinear Lattices** 2nd Edition By M. Toda

21 **Modern Crystallography II** Structure of Crystals By B. K. Vainshtein, V. M. Fridkin, and V. L. Indenbom

22 **Point Defects in Semiconductors I** Theoretical Aspects By M. Lannoo and J. Bourgoin

23 **Physics in One Dimension** Editors: J. Bernasconi, T. Schneider

24 **Physics in High Magnetic Fields** Editors: S. Chikazumi and N. Miura

25 **Fundamental Physics of Amorphous Semiconductors** Editor: F. Yonezawa

26 **Elastic Media with Microstructure I** One-Dimensional Models. By I. A. Kunin

27 **Superconductivity of Transition Metals** Their Alloys and Compounds By S. V. Vonsovsky, Yu. A. Izyumov, and E. Z. Kurmaev

28 **The Structure and Properties of Matter** Editor: T. Matsubara

29 **Electron Correlation and Magnetism in Narrow-Band Systems** Editor: T. Moriya

30 **Statistical Physics I** Equilibrium Statistical Mechanics By M. Toda, R. Kubo, N. Saito

31 **Statistical Physics II** Nonequilibrium Statistical Mechanics By R. Kubo, M. Toda, N. Hashitsume

32 **Quantum Theory of Magnetism** 2nd Edition By R. M. White

33 **Mixed Crystals** By A. I. Kitaigorodsky

34 **Phonons: Theory and Experiments I** Lattice Dynamics and Models of Interatomic Forces. By P. Brüesch

35 **Point Defects in Semiconductors II** Experimental Aspects By J. Bourgoin and M. Lannoo

36 **Modern Crystallography III** Crystal Growth 2nd Edition By A. A. Chernov

37 **Modern Crystallography IV** Physical Properties of Crystals Editor: L. A. Shuvalov

38 **Physics of Intercalation Compounds** Editors: L. Pietronero and E. Tosatti

39 **Anderson Localization** Editors: Y. Nagaoka and H. Fukuyama

40 **Semiconductor Physics** An Introduction 4th Edition By K. Seeger

41 **The LMTO Method** Muffin-Tin Orbitals and Electronic Structure By H. L. Skriver

42 **Crystal Optics with Spatial Dispersion, and Excitons** 2nd Edition By V. M. Agranovich and V. L. Ginzburg

43 **Structure Analysis of Point Defects in Solids by Multiple Magnetic Resonance Spectroscopy** By J.-M. Spaeth, J. R. Niklas, and R. H. Bartram

44 **Elastic Media with Microstructure II** Three-Dimensional Models By I. A. Kunin

45 **Electronic Properties of Doped Semiconductors** By B. I. Shklovskii and A. L. Efros

46 **Topological Disorder in Condensed Matter** Editors: F. Yonezawa and T. Ninomiya

47 **Statics and Dynamics of Nonlinear Systems** Editors: G. Benedek, H. Bilz, and R. Zeyher

48 **Magnetic Phase Transitions** Editors: M. Ausloos and R. J. Elliott

49 **Organic Molecular Aggregates** Electronic Excitation and Interaction Processes Editors: P. Reineker, H. Haken, and H. C. Wolf

Springer Series in Solid-State Sciences

Editors: M. Cardona P. Fulde K. von Klitzing H.-J. Queisser

MIX
Papier aus verantwortungsvollen Quellen
Paper from responsible sources
FSC® C105338

If you have any concerns about our products,
you can contact us on
ProductSafety@springernature.com

In case Publisher is established outside the EU,
the EU authorized representative is:
Springer Nature Customer Service Center GmbH
Europaplatz 3, 69115 Heidelberg, Germany

Printed by Libri Plureos GmbH
in Hamburg, Germany